Antimicrobial Materials for Biomedical Applications

Biomaterials Science Series

Editor-in-chief:
Julian Jones, *Imperial College London, UK*

Series editors:
Cole DeForest, *University of Washington, USA*
Changyou Gao, *Zhejiang University, China*

Titles in the Series:
1: Stimuli-responsive Drug Delivery Systems
2: Biodegradable Thermogels
3: Biofabrication and 3D Tissue Modeling
4: Biomaterial Control of Therapeutic Stem Cells
5: Antimicrobial Materials for Biomedical Applications

How to obtain future titles on publication:
A standing order plan is available for this series. A standing order will bring delivery of each new volume immediately on publication.

For further information please contact:
Book Sales Department, Royal Society of Chemistry, Thomas Graham House, Science Park, Milton Road, Cambridge, CB4 0WF, UK
Telephone: +44 (0)1223 420066, Fax: +44 (0)1223 420247,
Email: booksales@rsc.org
Visit our website at www.rsc.org/books

Antimicrobial Materials for Biomedical Applications

Edited by

Abraham J. Domb
Hebrew University of Jerusalem, Israel
Email: avid@ekmd.huji.ac.il

Konda Reddy Kunduru
University of Hyderabad, India
Email: kkondareddy@gmail.com

and

Shady Farah
Technion-Israel Institute of Technology, Israel
Email: sfarah@technion.ac.il

ROYAL SOCIETY OF CHEMISTRY

Biomaterials Science Series No. 5

Print ISBN: 978-1-78801-188-4
PDF ISBN: 978-1-78801-263-8
EPUB ISBN: 978-1-78801-854-8
Print ISSN: 2397-1401
Electronic ISSN: 2397-141X

A catalogue record for this book is available from the British Library

The Royal Society of Chemistry is a charity, registered in England and Wales, Number 207890, and a company incorporated in England by Royal Charter (Registered No. RC000524), registered office: Burlington House, Piccadilly, London W1J 0BA, UK, Telephone: +44 (0) 20 7437 8656.

For further information see our web site at www.rsc.org

Printed in the United Kingdom by CPI Group (UK) Ltd, Croydon, CR0 4YY, UK

Preface

Microbial infections and antimicrobial resistance remain global challenges and pose serious threats to human health. Several recent studies indicate that infections are spreading rapidly and that microbial resistance to traditional antibiotics and antiviral drugs is increasing. Furthermore, at advanced stages, microorganisms including bacteria, viruses and other parasites stop being receptive to antimicrobial agents, resulting in ineffective treatments for infectious diseases. To overcome microbial infections and antimicrobial resistance, various antimicrobial materials including small molecules and macromolecules, inorganic and organic agents have been developed and evaluated. Efforts worldwide have resulted in new materials that offer promising actions towards pathogens, as they can kill or inhibit microbial growth on their surface or within the surrounding environment with superior efficacy, low toxicity and minimal environmental effects. In this book, leading antimicrobial materials and formulations reported to exhibit strong antimicrobial activity and which have potential against antimicrobial resistance have been covered. This volume contains 17 chapters covering the dominant antimicrobial materials with biomedical applications to combat infectious diseases.

Chapter 1 discusses various classes of antimicrobial materials, surface modification and design requirements, and their mode of action, antimicrobial evaluation tests and clinical status.

Chapter 2 presents an overview of microbial infections, ultimately emphasizing that due to increasing emergence of infection events as well as drug resistance, improved adoption of older or newly developed healthcare practices are required to help mitigate risk.

Chapter 3 provides an overview on the controlled release of antimicrobial small molecules. Various types of antimicrobial small-molecule platforms

Biomaterials Science Series No. 5
Antimicrobial Materials for Biomedical Applications
Edited by Abraham J. Domb, Konda Reddy Kunduru and Shady Farah

Published by the Royal Society of Chemistry, www.rsc.org

have been developed in the treatment of infections. This chapter describes the main types of these systems based on nanoparticles, fibers, dendrimers, liposomes, nanotubes, and films including novel concepts in antibiotic-loaded bioresorbable films.

Chapter 4 discusses the mechanistic, design and computational details of biomimetic antimicrobial polymers and their interaction with model membranes, particularly highlighting the effect of such polymers on structural integrity of membranes.

Chapter 5 provides an overview of synthetic cationic water-soluble antimicrobial polymers and describes the rationale and characteristics of each class of antimicrobial polymers. These polymers have distinct advantages over traditional antibiotics, including a lower probability of developing microbial resistance.

Chapter 6 discusses the etiology of common oral diseases, the characterization of infection–host interactions in oral disease, and classic dental treatment modalities. The organization of oral microbes in the form of biofilms and the intrinsic susceptibility characteristics of oral tissues, as well as the advantages of focal controlled drug delivery are discussed.

Chapter 7 focuses on photodynamic antimicrobial polymers. A detailed discussion on current research and the development of these unique materials for the production of light-activated antimicrobial biomedical devices or for anti-infective surfaces in clinical settings are discussed.

Chapter 8 discusses antimicrobial biomaterials in opthalmology. Various strategies safeguarding the vision from infectious diseases are discussed.

Chapter 9 describes the antimicrobial uses of metals and metal-based compounds. It follows the historical use of metal-based antimicrobials (MBAs), their decline with the emergence of antibiotics and subsequent rediscovery with the advent of antibiotic resistance. The potential of MBA nanoparticles, their use and the mechanisms of toxicity are briefly discussed. Current applications and formulations of a wide range of MBAs are provided.

Chapter 10 is an overview of the different antimicrobial polymers based on quaternary ammonium moieties. The chemical structure, chemical modification, bioactivity and biomedical application are summarized and discussed.

Chapter 11 discusses the polymer conjugates of antifungal drugs, their merits and demerits. The advantages of polymer–drug conjugates compared to conventional dosage forms for antifungal therapy are also discussed.

Chapter 12 discusses biopolymers sterilization methods for both the natural and synthetic biodegradable polymers. Various sterilization methods that have been applied on biopolymers include steam-autoclaving, dry heat sterilization, irradiation (gamma, X-rays, ultraviolet and electron beam), chemical treatment (ethylene oxide), gas plasma and supercritical fluid sterilization, are comprehensively reviewed.

Chapter 13 presents recent advancements in antimicrobial hydrogels. Various antimicrobial hydrogel categories possessing inherent antimicrobial activities and hydrogels loaded with antimicrobial materials are discussed.

Chapter 14 discusses antimicrobial catheters. Catheters are widely in use for addressing several medical needs and purposes. This chapter presents a detailed discussion of the recent approaches and progress in developing antimicrobial coatings and combination therapies for addressing catheter associated infections.

Chapter 15 provides an overview of antimicrobial dendrimers and hyperbranched polymers. The potential of dendrimers and hyperbranched polymers as antimicrobials are reported.

Chapter 16 discusses antimicrobial fatty acids and their derivatives. Fatty acids are naturally available compounds with low toxicity and which possess broad-spectrum antimicrobial properties. In this chapter, use of the various fatty acids and their derivatives as antimicrobial materials, their target organisms and the proposed mode of actions are all discussed.

In **Chapter 17** recent and improved drug carriers for treating intracellular pathogens, including antibiotics loaded into hydrogels, liposomes, micelles, polymeric carriers and metal nanoparticles are discussed. This chapter focuses on the role of a drug delivery system as a potential tool against intracellular bacterial pathogens.

Abraham J. Domb
Jerusalem, Israel

Konda Reddy Kunduru
Hyderabad, India

Shady Farah
Haifa, Israel

Contents

Biomaterials Science Series No. 5
Antimicrobial Materials for Biomedical Applications
Edited by Abraham J. Domb, Konda Reddy Kunduru and Shady Farah

Published by the Royal Society of Chemistry, www.rsc.org

CHAPTER 1

Antimicrobial Materials—An Overview

SHAHEEN MAHIRA,[a] ANJALI JAIN,[a] WAHID KHAN*[a] AND ABRAHAM J. DOMB[b]

[a] Department of Pharmaceutics, National Institute of Pharmaceutical Education and Research (NIPER), Hyderabad 500037, India; [b] School of Pharmacy-Faculty of Medicine, The Hebrew University of Jerusalem, Jerusalem 91120, Israel
*Email: mail4wahid@gmail.com

1.1 Introduction

The presence of harmful microorganisms in the area of human health has become a great concern, due to the variety of infections and diseases. Rapid antibiotic resistance further worsens the situation.[1] Microbial contamination, adhesion, persistence and colonization of surfaces have become detrimental to health and society. Biofilms are microbial aggregates that adhere to a substrate and these account for 80% of infections that lead to increased patient morbidity and medical expenses.[2] According to Neely and Maley, vancomycin-resistant enterococci and methicillin-resistant *Staphylococcus aureus* (MRSA) can survive for a day on materials used in healthcare systems and some microbes can survive for more than 90 days.[3] To overcome these issues, materials that can provide antimicrobial activity are being explored for biomedical use to reduce hospital-acquired infections.[4] Disinfectants such as hydrogen peroxide, hypochlorite, *etc.* have a short duration of action and environmental toxicity issues.[5] Antimicrobial materials are capable

Biomaterials Science Series No. 5
Antimicrobial Materials for Biomedical Applications
Edited by Abraham J. Domb, Konda Reddy Kunduru and Shady Farah

Published by the Royal Society of Chemistry, www.rsc.org

of inhibiting or killing the microbes on their surface or within their surroundings,[6] but clinically they have significant shortfalls such as poor antimicrobial activity, issues with microbial resistance and difficulty of functioning in a dynamic environment. Thus, there is a need for effective and long-term antibacterial and biofilm-preventing materials to meet the demands in biomedicine.[2] With that said, new macromolecules with antimicrobial activity as well as structural modification of polymers to achieve desirable physicochemical and biological properties are being developed.[7]

Biofilms are a bacterial defense mechanism that protect bacteria from being washed away and make bacteria less susceptible or ineffective towards toxins. Biomedical devices are commonly used in hospitals as part of medical practice. They can be a source of microbial infections *via* contact with body fluids and tissues and due to openings in protective barriers, such as the skin, leading to nosocomial or hospital-acquired infections. Out of 150 million intravascular devices used annually in the USA, 200 000–400 000 result in nosocomial bloodstream infections. So prevention of these infections becomes necessary to reduce patient suffering and huge associated medical costs.[8] Related to this, there is an increased need to explore long-lasting, broad-spectrum and more efficient antimicrobial agents, due to unceasing global emergence of new infectious agents.[9]

Antimicrobial polymers have wide applications in the biomedical field, especially when they are in direct contact with the human body. They should possess certain requirements and meet regulations for safe use within the body. Firstly, they should be biocompatible, unreactive to the body with good stability and resistance to bodily fluids. Moreover, as previously mentioned, higher content of microbes in biofilms can result in serious infections and health issues. Therefore, selecting appropriate polymers against microbes is essential for biomedical applications.[10] Microbes can acquire resistance easily upon use of conventional antimicrobials, and can lead to environmental contamination and toxicity to humans due to biocidal diffusion.[11,12] Antimicrobial polymeric materials can address these problems by promoting antimicrobial efficacy and reducing residual toxicity.[13,14] Some described polymeric systems can belong to more than one section in this chapter. However, the purpose of the chapter is to provide a handy overall vision of the field of antimicrobial materials.

1.2 Antimicrobial Materials

1.2.1 Antimicrobial Polymers

Antimicrobial polymers have emerged as promising candidates against microbial contamination owing to their properties. Their versatile macromolecular chemistry facilitates the tailoring of polymer physicochemical properties to be used for various applications in the biomedical field.[15]

1.2.1.1 Polymers with Intrinsic Antimicrobial Activity

In nature, most materials possess antimicrobial ability. Materials that exhibit antimicrobial action without any modification are known as intrinsic antimicrobial materials.[16]

1.2.1.1.1 Natural Polymers

1.2.1.1.1.1 Chitosan. Chitosan was discovered by Rouget in 1859 and is the most widely used polymer in biomedicine, with its broad-spectrum antibacterial activity, first proposed by Allan and Hadwinger.[17,18] It is a linear, polycationic heteropolysaccharide composed of (1–4)-2-acetamido-2-deoxy-βD-glucan (*N*-acetyl D-glucosamine) and (1–4)-2-amino-2-deoxy-β D-glucan (D-glucosamine) units obtained by partial alkaline *N*-deacetylation of chitin.[19] The physicochemical and biological properties of this biopolymer depend on its number of amine groups, thus favoring site-specific modification and providing versatility for more applications. Its antimicrobial activity can be explained by two main mechanisms. Firstly, positively charged chitosan can interact with negatively charged microbial cell surfaces and will either prevent the transport of essential materials into cells or result in leakage of cellular contents. In the second mechanism, chitosan binds with cellular DNA (*via* protonated amine moieties) and results in microbial RNA synthesis inhibition.[20,21]

Chitosan acts on various types of bacteria and fungi (Table 1.1) and its activity in turn depends on the polymer-related factors (molecular weight, charge density, hydrophilic/hydrophobic character, concentration and chelating capacity), pH, ionic strength, temperature, and the type of microbe.[22] Copolymers with zwitterionic properties were obtained by grafting the mono (2-methacryloyloxyethyl) acid phosphate and vinyl sulfonic acid sodium salt upon chitosan. They showed the optimum antimicrobial activity at 5.75 pH towards *Candida albicans.*[23] The antimicrobial activity of quaternary ammonium salts of chitosan increased with an increase in the alkyl chain length that was attributed to the increased lipophilic properties of the derivatives.[24] Chitosan-based biomedical materials are gaining much attention due to their biodegradability, biocompatibility, non-toxicity and antimicrobial effects. In addition, hydrophilicity and their structural similarity to glycosaminoglycans make them versatile materials for tissue engineering.[25,26] These properties account for wide applications as excipients for drug delivery and gene delivery in wound healing and tissue engineering.[27] Cross-linked, quaternized chitosan/polyvinylpyrrolidone electrospun mats were found to be attractive materials for wound dressings as they were more efficient in inhibiting Gram-positive and Gram-negative bacterial growth.[28] Novel composite scaffolds based on α-chitin/nanosilver[29] and β-chitin/nanosilver[30] exhibited profound antibacterial activity towards *Staphylococcus aureus* and *Escherichia coli.*

Table 1.1 Minimum growth inhibitory concentration for chitosan antibacterial and antifungal activity.

Microorganism	Minimum inhibitory concentration (ppm)	Ref.
Antibacterial action		
Salmonella enterica	2000	31
Bacillus cereus	1000	32
Klebsiella pneumoniae	700	33
Erwinia species	500	33
Xanthomonas campestris	500	34
Erwinia carotovora	200	33
Vibrio cholerae	200	32
Escherichia coli	100	32
Staphylococcus aureus	20	33
Corynebacterium michiganensis	10	33
Antifungal action		
Byssochlamys spp.	1000–5000	35
Trichophyton equinum	2500	33
Trichophyton mentagrophytes	2200	36
Aspergillus fumigatus	>2000	32
Microsporum canis	1100	36
Candida albicans	500	32
Fusarium oxysporum	100	33
Botrytis cinerea	10	33

1.2.1.1.1.2 Heparin. Heparin, a highly sulfated glycosaminoglycan, is widely applicable in the field of hemocompatible biomaterials.[37] The antimicrobial mechanism has not been clearly defined for heparin. However, due to heparin binding with calcium, it seems likely that it acts by chelation of cations that are essential for bacterial growth. Other possible mechanisms might be the inhibition of transport/intracellular utilization of cations. Warren and Graham reported the antimicrobial activity of heparin against *Staphylococcus aureus* and *Erwinia stewartii* at a concentration of 150 U mL^{-1}, when they were grown in protein-free medium.[38,39] Heparin binding or coating has prevented microbial adhesion and colonization *in vitro* and *in vivo* by its ability to favor albumin adsorption and reduced fibrinogen adsorption. In a randomized pilot study, 20 ureteral stents with and without heparin coating were inserted into obstructed ureters for 2–6 weeks and evaluated for encrustation and biofilm formation. About 33% of uncoated stents were colonized by bacteria, while no biofilms were detected on heparin-coated stents. There was a significant decrease in catheter-related infections with heparinized central venous catheters (CVCs) and dialysis catheters that was confirmed by randomized study of heparin-coated and uncoated non-tunnelled CVCs inserted in 246 patients as well as in a retrospective study of coated and uncoated tunnelled dialysis catheters.[37]

1.2.1.1.1.3 ε-Polylysine. ε-Polylysine (ε-PL) is a hydrophilic linear polyamide composed of 25–30 residues of L-lysine with ε-amino and α-carboxyl group linkage.[40] Shoji Shima, Heiichi Sakai, and co-workers first described the production of ε-polylysine by natural fermentation resulting in a compound with wide antibacterial spectrum and lethal effect on bacteria, yeast, mould, viruses *etc.*[41] Its antimicrobial activity depends on the number of L-lysine residues, with >10 residues being necessary to exhibit proper antimicrobial action.[40] It has good antibacterial effect on Gram-negative bacteria that are difficult to control. In addition, it is adsorbed electrostatically to bacterial cell surfaces that have negatively charged lipopolysaccharide, causing the stripping of their outer membrane. This eventually leads to abnormal cytoplasmic distribution and cell death.[41]

Naturally available ε-PL is edible, biodegradable, non-toxic and soluble in water. ε-PL derivatives can be used as emulsifiers, drug carriers, biodegradable fibers, highly water-absorbable hydrogels, biochip coatings, *etc.*[42] ε-PL was studied as an antimicrobial agent in platelet concentrates for the first time in Japan, where it was shown to completely inhibit the growth of *Staphylococcus epidermis*, *Pseudomonas aeruginosa*, *Escherichia coli* and *Staphylococcus aureus* in platelet concentrates after 8 days at 100 $\mu g\,mL^{-1}$. ε-PL and polycaprolactone (PCL) copolymer showed a broad-spectrum antibacterial action towards *Escherichia coli*, *Staphylococcus aureus* and *Bacillus subtilis.*[43]

1.2.1.1.2 Polymers Containing Quaternary Nitrogen Atoms. Most bacterial cells are negatively charged, hence most antimicrobial polymers are positively charged to drive their interaction. In addition, the ones with quaternary ammonium moieties are mostly explored as polymeric biocides. Polycationic biocides act by destructive interaction with the bacterial cell wall.[44] Antimicrobial activity of quaternary ammonium compounds (QAC) depends on the length of their *N*-alkyl chain.[45] For bacteria and fungi, the optimum chain length of QACs is different (14 carbons for Gram-positive bacteria, 16 carbons for Gram-negative bacteria and 12 carbons for yeast and filamentous fungi).[46,47] Antifungal activity of QACs is attributed to their electrostatic interaction with fungal cell membrane resulting in cell lysis. The antifungal activity may also involve the impediment of formation of hyphae.[48] The virucidal mechanism of QACs for enveloped viruses involves disruption or detachment of the viral envelope, with subsequent release of nucleocapsid.[49]

1.2.1.1.2.1 Polymers with Aromatic or Heterocyclic Groups. Cationic polymers with quaternary ammonium groups and aromatic or heterocyclic rings are synthesized from polystyrene and polyvinylpyridine. Imidazole derivatives offer good chemical stability, with resistance to hydrogenation, and undergo numerous substitution reactions for providing functional derivatives. Random and block copolymers containing quaternized poly (4-vinylpyridine) (P4VP) and polystyrene showed good antibacterial action. P4VP possesses reactive pyridine groups that form pyridinium-type antimicrobial polymers.[50,51] The antimicrobial activity and biocompatibility of

N-hexylated P4VP was improved by copolymerization with poly(ethylene glycol) methyl ether methacrylate. Due to increased surface wettability, the antibacterial property of these polymers was found to be 20 times higher than the quaternized homopolymer, without causing any hemolysis.[52]

In general, the antifungal mechanism of action of gemini QACs involves lysis of cell membrane and cell organelles. Gemini QACs contain two pyridinium residues [3,3-(2,7-dioxaoctane) bis(1-decylpyridinium bromide)] per molecule, that cause respiration inhibition and cytoplasmic leakage of adenosine triphosphate as well as magnesium and potassium ions in *Saccharomyces cerevisiae*.[53] D-Glucosamine QA derivatives have potent antifungal activity towards *Coriolus versicolor* and *Poria placenta* by forming complexes with essential elements to block/reduce their fungal growth.[54] Zephiran (alkyl dimethyl benzylammonium chloride) also effectively inactivates enveloped viruses such as vaccinia virus and some non-enveloped viruses such as reovirus and bacteriophages. However, it is not effective against small, non-enveloped viruses such as picorna viruses.[55,56]

1.2.1.1.2.2 Polyacrylamides and Polyacrylates. Among cationic synthetic polymers, polyacrylamides and polyacrylates with tertiary or quaternary amine groups are the most investigated antimicrobial polymers due to their wide versatility and ease of synthesis. Physicochemical properties and antimicrobial activity can be properly modulated by varying the type of monomers, type of counter ion of charged groups, polymer amphiphilicity and alkyl chain length attached to the cationic groups.[57]

Methacrylate polymers with tertiary butylamine groups are considered to be potent antimicrobials.[58] Most of the quaternary polyelectrolytes are obtained from methacrylic monomers such as 2-(dimethylamino) ethyl methacrylate.[59] Antimicrobial properties of modified glycidyl methacrylate polymers with quaternary ammonium and phosphonium groups were tested against Gram-positive bacteria (*Bacillus subtilis* and *Bacillus cereus*), Gram-negative bacteria (*Escherichia coli*, *Pseudomonas aeruginosa*, *Salmonella typhae* and *Shigella* sp.) and fungus (*Trichophyton rubrum*). These polymers showed prominent antimicrobial properties against Gram-negative bacteria and fungi at 24 h contact time.[24] Palermo and Kuroda synthesized copolymers based on polymethacrylate and polymethacrylamides, where the hydrophobic groups, polymer composition and length were varied to get antimicrobials with non-hemolytic properties. Methacrylamides with alkyl pyridinium pendant groups and temperature-responsive *N*-isopropylacrylamides were also synthesized as biocides.[59]

In 1960, Wichterle and Lim first described the use of poly-2-hydroxyethylmethacrylate (PHEMA) for contact lens applications.[60] It got FDA approval in 1971 and was sold by Bausch & Lomb.[61] HEMA/*N*-vinyl-2-pyrrolidone hydrophilic copolymer is used to make soft contact lenses that cover the entire cornea and present good oxygen permeability with great comfort. By contrast, hydrophobic polymers such as poly(methyl methacrylate) (PMMA) and poly(hexa-fluoroisopropyl methacrylate) are widely used for hard contact lenses.[62–64] pH- and thermal-sensitive hydrogels of PHEMA and itaconic acid copolymers have potential biomedical applications, mainly for dermatological treatments and wound dressings. Porphyrin-crosslinked poly(2-hydroxyethyl methacrylate-*co*-methyl methacrylate) copolymers were used for preventing endophthalmitis.[65] Quaternized chitosan-loaded PMMA has been shown to inhibit MRSA and exhibits excellent physical properties and osteogenic activity.[52]

1.2.1.1.2.3 Polysiloxanes. Polysiloxanes with quaternary ammonium and imidazolium groups, as well as polysilsesquioxanes with quaternary ammonium groups, have activity against Gram-positive and Gram-negative bacteria. Polysiloxane polymers with pendant quaternary ammonium salt (QAS) groups show antibacterial action *via* interaction with bacterial membranes.[58] Generally, these polymer systems show the lowest adhesion and best foul release of biofouling, based on a repeating unit (–Si–O–), with saturated organic groups linked to two non-backbone valencies of the silicon. Moreover, the Si–O bond is stronger (108 kcal $mole^{-1}$) than C–C bonds (83 kcal $mole^{-1}$) and is extremely durable, providing long-term control of fouling.[66]

Polysiloxanes with quaternary ammonium groups are gaining interest due to high flexibility of polymer chains that makes the contact of microbe and polymer easier. Their hydrophilic inorganic and hydrophobic organic groups augment the quaternary moiety in the vicinity of the microbial cell wall. Polysiloxanes with *N*,*N*-dialkylimidazolium salt showed higher antibacterial power. Imidazolium-substituted polysiloxane has higher thermal stability compared to alkyl ammonium functionalized polymers.[67] Poly-dimethylsiloxane (PDMS) is the most commonly used silicone polymer. PDMS with QAS moieties will facilitate contact-killing antimicrobial properties of the materials.[68]

1.2.1.1.2.4 Polyionenes. Polyionenes are polymer electrolytes with quaternized nitrogens in the polymer backbone.[69,70] In general, their antimicrobial efficacy depends on chain rigidity, pendant substituents and alkyl chain length.[71–73] When compared to ionenes with flexible spacers, the rigid spacers exhibit greater interaction with lipidic bilayers resulting in their phase separation. Ionenes with low charge density and longer lipophilic chains exhibit effective biocidal activity against yeast, indicating that their hydrophobicity is the predominant factor for cell wall disruption.[9,71] Tiller's group synthesized *N*,*N*,*N*′,*N*′-tetramethyldiamine- and α,ω-dibromoalkane-based polymers and found that they have excellent antimicrobial activity with non-hemolytic properties.[72,73]

1.2.1.1.2.5 Polyoxazolines. Polyoxazolines are pseudopeptides obtained by ring-opening reactions.[74,75] Their properties can be tuned by controlling the end functional groups during initiation and termination chain reactions and by varying the monomer side chain. Due to lower toxicity and functional versatility they are known as biocide end-functionalized polymers.[76]

Polyoxazolines represent a valuable type of macromolecules and are mainly investigated in the biomedical field due to their biocompatibility, blood clearance and protein adsorption. A series of polymethyloxazolines with different satellite groups including hydroxyl-, primary amine- and double bond-containing groups were synthesized. It was found that the functional satellite groups greatly controlled the minimum inhibitory concentration (MIC) towards *Staphylococcus aureus* and *Escherichia coli* at a range of 10–2500 ppm.[77,78]

1.2.1.1.2.6 Hyperbranched and Dendritic Polymers. Branched polyethylene imine (PEI) in its quaternized form adsorbs on the bacterial cell membrane and causes cell death by disrupting the cell membrane and releasing the intracellular contents, thereby showing outstanding antibacterial activity.[79] Its antibacterial properties depend on dendrimer size, length of hydrophobic chains in quaternary ammonium groups and counter anions.[79] The main drawback of branched and hyperbranched polymers is the polydispersity and functional heterogeneity that makes it difficult to rationalize and understand their behavior with microbes. This led to the emergence of dendrimers with compact structure, monodisperse molecular weights and availability of many end groups. Their biocidal properties depend on dendrimer size, hydrophobic chain length, surface porosity and counter anions. Poly(propyleneimine) and poly(amidoamine) (PAMAM) dendrimers are widely used in drug delivery and gene therapy.[80] Polyethylene glycol diacrylate (PEGDA)-based dendrimers are made by reacting PEGDA with ethylene diamine and diethyl amine. Quaternary dendrimer-based copolymers showed antimicrobial action based on the amount of quaternary ammonium moieties and surface porosity.[52,81]

1.2.1.1.3 Polymers with Guanidine Groups. Polyguanidines and polybiguanides possess high water solubility, a broad antimicrobial spectrum and non-toxicity, thereby attracting considerable attention as antimicrobial compounds.[82] They can be synthesized by polycondensation or polyaddition and the starting materials can consist of monomeric guanidines, isocyanide dihalides, guanido acid esters, cyanogen halides or dicyanamides. In the 1940s, the first patent for oligoguanidine compounds as antibacterial agents was filed.[83] Earlier findings suggested that an average molecular weight of 800 Da is required for efficient antimicrobial action.[84] Polyhexamethylene biguanide is a broad-spectrum antimicrobial biocide that kills bacteria, fungi, parasites and certain viruses. It has biguanidine units linked with hexamethylene hydrocarbon chains, thereby providing an amphipathic structure.[85] Its antimicrobial activity is attributed to bacterial cell wall disruption. It binds with lipid membranes, causing increased membrane fluidity and permeability and subsequent microbial death. It has also been reported to bind bacterial DNA, altering the transcription process and causing lethal damage to DNA.[86]

1.2.1.1.4 Polymers Mimicking Natural Peptides. Antimicrobial peptides (AMPs) are the main components of host defense against various infections; they display remarkable activity against bacteria, fungi, viruses and parasites.[87,88] They have roles in immunomodulation and inflammation processes. They kill bacteria by different mechanisms such as cell membrane disruption, interference with metabolism and targeting cytoplasmic components.[89] They usually have hydrophilic and hydrophobic groups that enable the molecule to be solubilized in aqueous environments and pass through the lipidic membranes. However, their utility has been hindered due to high manufacturing costs, susceptibility to proteolysis and poor pharmacokinetic profile.[90] Additionally, the complexity of the native structure imposes difficulty in studying their bioactivity. All these reasons have led to an increase in research interest in synthesizing AMPs.[76]

AMPs can serve as promising candidates for new-generation antimicrobials and are of great interest due to a low risk of bacterial resistance, broader spectra of action, target specificity, high efficacy and synergistic action with classical antibiotics.[91] Extensive research has been done in the area of making synthetic peptides, maintaining the natural peptide skeleton (L-α-amino acids) and non-naturally occurring structures (D-α-amino acids, β-peptides or peptoids). Solid-phase synthesis and solution coupling are the common methods to prepare AMPs. Analogs of idolidicin were synthesized to give less toxic polymers with higher antimicrobial properties.[92] Similarly, grasitin analogs exhibited potent action against Gram-positive and Gram-negative bacteria with significant reduction in hemolysis.[93] β-Peptides are a class of polyamides mimicking AMPs that can show various helical conformations and resistance to degradation by proteases when compared to conventional peptides. Arylamide and phenylene ethynylene oligomers and polymers were made by simple and inexpensive synthetic methods.[94] Polynorbornene derivatives prepared by various synthetic strategies with high molecular mass afford good antibacterial action with minimal cytotoxicity to humans. Ilker *et al.* found that upon increasing the amine groups on such polymers, hemolytic activity decreased significantly.[13] The fungal cell wall contains mannan, chitin and glucans that are absent in other microbes, making them potential targets for therapeutics. Antifungal peptides target fungal cell walls *via* peptide binding to chitin. Moreover, they show lethal effects by disrupting membrane integrity, promoting membrane fluidity or by creating pores.[95] A list of antimicrobial peptides and their antibacterial action is given in Table 1.2.

1.2.1.1.5 Polymers Containing Halogens

1.2.1.1.5.1 Polymers Containing Fluorine. Polymers containing fluorine are most attractive, due to their unique properties such as oil and water

Table 1.2 List of antimicrobial peptides.

Antimicrobial peptide	Chemical ring	Antimicrobial action	Ref.
Magainin	α-Helix	Active against bacteria, fungi and viruses	97–99
Cecropin	α-Helix	Active against bacteria, fungi and viruses	100–102
Brevinin-1	α-Helix	Active against fungi and viruses	103, 104
PMAP-23	α-Helix	Active against fungi	96, 97
Protegrin	β-Sheet	Active against bacteria and viruses	98, 99
Dermaseptin	β-Sheet	Active against viruses	100, 101
Tachyplesin	β-Sheet	Active against viruses	102, 103
Polyphemusin	β-Sheet	Active against viruses	104, 105
Tenecin-3	Extended turn	Active against fungi	106
PR-39	Extended	Active against bacteria	107

repellence due to lower polarizability and high electronegativity of fluorine atoms; higher chemical, thermal and weather resistance; lower dielectric constant and lower surface energy.[76] 2-[(4-Fluorophenyl)amino]-2-oxoethyl-2-methylacrylate was synthesized by free-radical copolymerization. It was found to be more prominent in inhibiting microbial growth due to high fluorine content.[108] By replacing two leucine residues in buforin II with more hydrophobic hexafluoro-leucine residues, antibacterial activity was enhanced without significantly impacting hemolytic activity.[109] Moon *et al.* synthesized a polymer with quinolone and a fluorine atom that proved its capacity to kill bacteria.[110] Guittard's group developed Quaterfluo®, in which perfluoro alkyl groups were incorporated into the gemini structure. The results showed their potent antimicrobial activity after 1 h of contact time.[111]

1.2.1.1.5.2 Polymers Containing Chlorine. Kugel *et al.* modified triclosan with an acrylate functionality followed by copolymerisation with different compositions of ethyl and butyl acrylates. Results showed that antimicrobial properties improved upon increasing triclosan groups without any leaching of triclosan.[112] It acts by deactivation of fatty acid synthesis of bacteria by inhibiting enoylacyl carrier protein reductase.[49]

1.2.1.1.5.3 N-Halamine Compounds. *N*-Halamines are formed by halogenation of amide, imide or amine groups by covalent bonding. They are the most promising candidates as antimicrobials, due to their fast and total killing action against various microbes without any environmental concerns and long-term stability, and it is highly unlikely that microbes will establish resistance to them.[24] They promote the direct transfer of active moiety to the target site or by dissociating into free halogen in aqueous media, resulting in inactivation/inhibition of microbial growth.[76] *N*-halamine acrylamide monomers were copolymerized and used as antimicrobial coatings that exhibited 8-log inactivation of both Gram-positive and Gram-negative bacteria following a 5 min short contact time.[113]

1.2.1.2 Imparting Antimicrobial Activity to Polymer by Chemical Modification

1.2.1.2.1 Covalent Incorporation of Lower Molecular Weight Antimicrobials. Poly(4-vinylphenol) (PVPh) was modified by sulfonation followed by electrospinning and MIC values were measured against a variety of bacteria, where modified polymers exhibited greater antimicrobial action at lower concentration than unmodified PVPh.[114] Worley and coworkers incorporated *N*-chloramine moieties (hydantoins, oxazolidinones and imidazolidinones) into polyester and nylon fabrics by covalent conjugation. The resulting antimicrobial activity suggested that they are highly effective with 7-log reduction in 10 min in case of hydroxymethyl hydantoin functional group incorporation.[115,116] Badrossamay and Sun grafted nitrogen-containing monomers (acrylamide, methacrylamide, *N*-tert-butylacrylamide and *N*-tert-butylmethacrylamide) into polyethylene and polypropylene, where they showed good antibacterial activity at 30 min contact time, even at concentration of bacteria above 10^7 CFU mL^{-1}.[117,118]

1.2.1.2.2 Coupling of Antimicrobial Peptides. Cationic polymers that are hydrophobic can be used as antimicrobial coating materials and they are capable in inhibiting bacteria and human-pathogenic fungi.[119] They interact with bacterial and fungal cell walls, disrupting the integrity of the lipid membrane, and impairing the transportation of compounds, ultimately leading to cell lysis. In fungal cells, cationic amphipathic peptides such as magainin cause membrane lysis and interfere with the DNA integrity of fungi cells.[95,120]

Polyethylene glycol-grafted polystyrene beads were covalently linked to AMPs with specific sequences. The results showed that their antimicrobial action was dependent on exposure time and concentration of modified polystyrene.[121] 2-(2-Methoxyethoxy) ethyl methacrylate and hydroxylated oligoethylene glycol methacrylate copolymer were functionalized with magainin I, and the results showed strong biocidal activity and biofilm prevention, even at low degrees of peptide coupling.[76]

1.2.1.2.3 Grafting Other Antimicrobial Polymers. To infer antimicrobial activity, natural polymers can be grafted to synthetic polymers. Yang *et al.* grafted chitosan onto polypropylene modified with acrylic acid and found that with increasing acrylic acid grafting, cell viability decreased.[122] Lysozyme was immobilized to polyvinyl alcohol cross-linked films and the antimicrobial property was directly proportionate to the amount of enzyme incorporated.[123] Poly(ethylene terephthalate) films were copolymerized and quaternized with hexyl bromide to yield pyridinium groups that were found to be more effective when the surface concentration was larger than 1.5×10^{-4} mol mol^{-1} m^{-2}.[124]

1.2.2 Antimicrobial Nanomaterials

1.2.2.1 Organic Nanoparticles

Polymeric nanoparticles can kill microbes by contact-killing cationic surfaces (quaternary ammonium compounds, quaternary phosphoniums or alkyl pyridiniums) or by releasing antimicrobial agents and antimicrobial peptides. The antibacterial activity of polycations depends on the ability of multiple charges to attach and interact with the bacterial cell wall.[2,125]

1.2.2.1.1 Non-covalent Incorporation of Lower Molecular Weight Antimicrobials. Lu *et al.* incorporated triclosan, a widely used antimicrobial, into cyclodextrin and subsequently into PCL or nylon films. By this modification, the antimicrobial agent was protected against higher temperatures during processing.[126] Sulfamethoxazole was introduced into PAMAM dendrimers as drug carriers in aqueous media. Diuron or 3-(3,4-dichlorophenyl)-1,1-dimethylurea was embedded in poly(ester anhydride) composed of sebacic acid, ricinoleic acid, terephthalic acid and isophthalic acid, by which release of the compound was observed for about 25 days.[76,127]

1.2.2.1.2 Incorporation of Mixture of Antimicrobials Non-covalently. Polymers can be mixed with natural or synthetic antimicrobial polymers. Quaternized PEI at 1% or 2% w/w can be added to the composite resins. Data from antibacterial assays demonstrated that the antibacterial properties were retained up to 3 months with complete growth inhibition of *Enerococcus faecalis*, *Staphylococcus aureus* and *Streptococcus mutans* and a reduced growth of *Staphylococcus epidermis* and *Pseudomonas aeruginosa*.[128,129] Jones *et al.* blended PCL with poly(*N*-vinylpyrrolidone)-iodine which imparted antibacterial properties to the biomaterials without altering mechanical or rheological properties. Moreover, PCL degradation also favored the anti-adherence of *Escherichia coli*.[130]

1.2.2.2 Inorganic Nanoparticles

Organic antibacterials are usually less stable at higher temperatures when compared to inorganic materials, which poses difficulties in designing materials that are stable and able to withstand harsh processing conditions. In order to overcome these problems, inorganic nanosized materials are often used as antimicrobial materials.[2,6] Various coating techniques are listed in Table 1.3. A list of metal and metal oxide nanoparticles and their antimicrobial action is presented in Table 1.4. The antimicrobial mode of action of metal oxide nanoparticles is explained in Figure 1.1 and different approaches for surface modification are shown in Figure 1.2.

Table 1.3 Different methods of surface modification of biomaterials.

Method	Description	Ref.
Sputter deposition	Atoms from the target are ejected from the energized gas ions that travel and bind with the substrate forming a coat	144–146
Electrostatic spray deposition	Process of liquid automization by means of electrical forces	131
Electrophoretic deposition	Involves motion of charged particles towards the oppositely charged electrode and deposit formation under the influence of applied electric field	132
Chemical vapor deposition	Reactive mixture of gas is moved to the coating area from a chemical reaction thus forming a coat on the target substrate	149–151
Pulsed laser deposition	Laser ablate a target material and condense it on the surface of a substrate	133
Photo-chemical deposition	Photo-reduction of a silver precursor results in nanostructured silver coatings	153–155
Sol–gel method	By hydrolysis and condensation, the sol becomes gel. Further drying and heating converts gel into denser particles	134, 135

1.2.3 Antimicrobial Plastics

Bioplastics are biopolymers obtained from proteins and are widely explored for their uses in medicine. They exhibit antimicrobial properties by creating anti-adhesive surfaces, disrupting cell-to-cell communication or leading to cell membrane lysis, thereby killing bacteria.[167] Soya, albumin and whey protein serve as the source of raw materials for producing bioplastics that act as promising materials for fabricating implants. Albumin shows antimicrobial activity by its enzyme lysozyme, which causes cell wall lysis. Albumin from hen egg whites is of particular interest in medical device fabrication due to its inherent antibacterial nature.[168] Albumin-based plastics reduce the growth of *Escherichia coli and Bacillus subtilis* on their surface.[169] Glycomacropeptides and immunoglobulins present in whey protein bind the toxin and prevent microbial infection.[170] Different test methods are available that can be performed to determine whether albumin or whey plastics can be used in medical applications, based on the intended use in areas such as packaging medical products (ASTM F2097-10: *Standard Guide for Design and Evaluation of Primary Flexible Packaging for Medical Products*) and infection testing for medical applications (ASTM F813-07(2012): *Standard Practice for Direct Contact Cell Culture Evaluation of Materials for Medical Devices*).[171]

Currently, silver-based nanoengineered materials are widely applicable in plastic commodities because of their antimicrobial abilities. In medicine and for food safety, titanium-, copper- and zinc-based nanostructures also show promising antimicrobial effects.[172] Liu *et al.* prepared plastics with excellent antibacterial activity by adding Ag/TiO_2 to resins.[173] Matet *et al.*

Table 1.4 Antimicrobial activity of metal oxide nanoparticles.[a]

Metal oxide NPs	Test organism	Antimicrobial action	Ref.
Aluminium oxide (Al_2O_3) NPs	*Escherichia coli*	Growth inhibition of *Escherichia coli*	136
Antimony trioxide (Sb_2O_3) NPs	*Escherichia coli*, *Bacillus subtilis* and *Staphylococcus aureus*	Toxic to all the three microbes	137
Bismuth oxide (Bi_2O_3) NPs	*Pseudomonas aeruginosa*, *Acinetobacter baumannii* and *Escherichia coli*	No effect against all tested microbes	138
Calcium oxide (CaO) NPs	*Lactobacillus plantarum*	Higher bactericidal activity	139
Cerium oxide (CeO) NPs	*Escherichia coli*, *Shewanella oneidensis* and *Bacillus subtilis*	No effect on *Shewanella oneidensis*	140
Cobalt oxide (Co_3O_4) NPs	*Staphylococcus aureus* and *Escherichia coli*	Showed antimicrobial activity on tested bacteria	141
Copper oxide (CuO) NPs	MRSA, *Staphylococcus epidermis*, *Pseudomonas aeruginosa*, *Proteus* sp. *Staphylococcus aureus*, *Bacillus subtilis*, *Escherichia coli*; fish pathogens: *Aeromonas hydrophila*, *Pseudomonas fluorescens*, *Flavobacterium* sp. and *Branchiophilum* sp.	Active against all the tested microbes	142–145
Magnetite (Fe_3O_4) NPs	*Escherichia coli*	Concentration-dependent bacteriostatic action	146
Iron oxide (FeO) NPs	*Staphylococcus aureus*, *Shigella flexneri*, *Escherichia coli*, *Bacillus licheniformis*, *Bacillus subtilis*, *Brevibacillus brevis*, *Vibrio cholerae*, *Pseudomonas aeruginosa*, *Staphylococcus aureus* and *Staphylococcus epidermis*	Moderate antibacterial activity against 6 Gram-positive and 2 Gram-negative bacteria	147
Magnesium oxide (MgO) nanowires	*Escherichia coli* and *Bacillus* spp.	Lower bacteriostatic activity	148
Titanium dioxide (TiO_2) NPs	MRSA	Exhibited antimicrobial effect on tested isolates	149

Zinc oxide (ZnO) NPs	MSSA, MRSA and MRSE, *Streptococcus agalactiae*, *Staphylococcus aureus*, *Escherichia coli*, *Bacillus subtilis*, *Salmonella paratyphi*, *Staphylococcus aureus*, *Pseudomonas aeruginosa*, *Mycobacterium smegmatis*, *Mycobacterium bovis*, *Klebsiella pneumoniae*, *Enterobacter aerogenes*, *Candida albicans*, *Malassezia pachydermatis*, *Bacillus megaterium*, *Bacillus pumilus* and *Bacillus cereus*	Active on tested microbes	150–156
Zinc/iron oxide composite NPs	*Escherichia coli* and *Staphylococcus aureus*	Exhibited greater antibacterial activity with higher Zn/Fe weight ratio	157
ZnO-loaded PA6 nanocomposite	*Staphylococcus aureus* and *Klebsiella pneumoniae*	Dose-dependent antibacterial action	158
Nanosilver-decorated TiO_2 nanofibres	*Staphylococcus aureus* and *Escherichia coli*	Increased antimicrobial effect	159
Hybrid CH-α-Fe_2O_3 nanocomposite	*Staphylococcus aureus* and *Escherichia coli*	Improved antibacterial activity	160
Zinc-doped CuO nanocomposite	*Escherichia coli*, *Staphylococcus aureus* and MRSA	Remarkable biocidal activity	161
PEI-capped ZnO NPs	*Escherichia coli*	Exhibited better antibacterial activity	162
Chitosan-based ZnO NPs	*Candida albicans*, *Micrococcus luteus* and *Staphylococcus aureus*	Showed biofilm inhibition against *Micrococcus luteus* and *Staphylococcus aureus*	163
Carvone functionalized iron oxide	*Staphylococcus aureus* and *Escherichia coli*	Inhibited colonization and biofilm formation	164
Silver-decorated titanium dioxide (TiO_2 : Ag) NPs	MRSA and *Candida* sp.	Conferred antimicrobial effect on tested microbes	165
Graphene oxide modified ZnO NPs	*Escherichia coli*, *Bacillus subtilis*, *Salmonella typhimurium* and *Escherichia faecalis*	Excellent antibacterial activity	166

[a]NPs: nanoparticles; MRSA: methicillin-resistant *Staphylococcus aureus*; MRSE: methicillin-resistant *Staphylococcus epidermidis*; MSSA: methicillin-sensitive *Staphylococcus aureus*; PEI: polyethyleneimine.

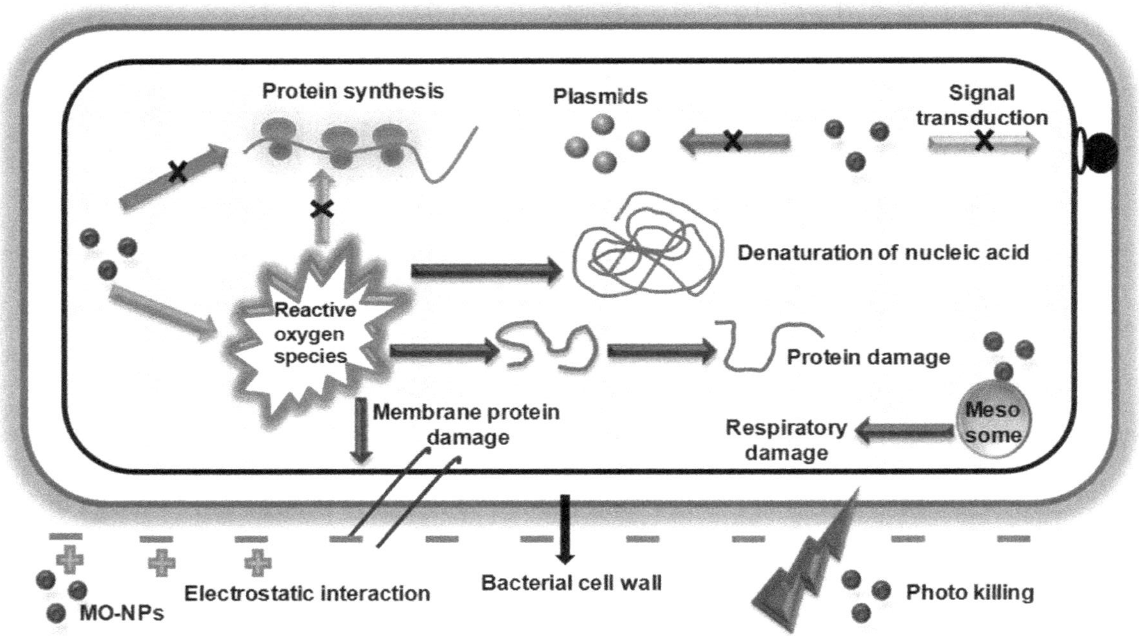

Figure 1.1 Mechanism of antimicrobial action by metal oxide nanoparticles (MO-NPs): MO-NPs cause cell membrane damage by electrostatic interaction. Their accumulation dissipates the proton motive force, disrupting the chemiosmosis process, thereby causing proton leakage. They induce reactive oxygen species generation which damages organic biomolecules (carbohydrates, lipids, proteins and nucleic acids) finally causing microbial death. They bind with mesosomes and alter cellular respiration, cell division and the DNA replication process. Dephosphorylation of phosphotyrosine residues inhibits signal transduction and ultimately obstructs bacterial growth. Protein carbonylation leads to loss of catalytic activity of enzymes, ultimately triggering protein degradation. Photosensitized transition MO-NPs cause alteration in cell membrane, Ca^{2+} permeability, diminution in superoxide dismutase activity, DNA damage and abnormal cell division.

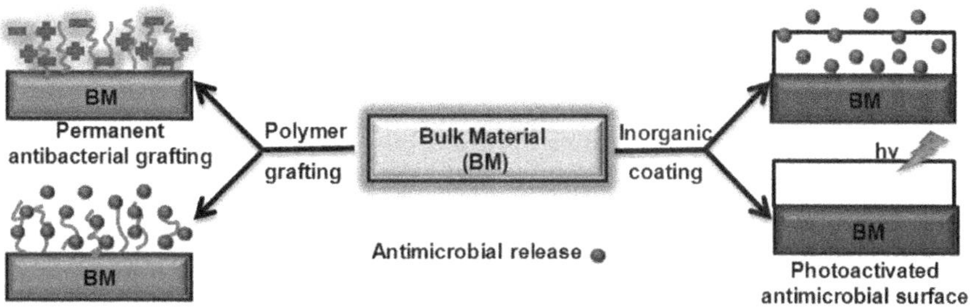

Figure 1.2 Approaches for surface modification in medical devices to impart antimicrobial properties. Polymer coating is preferable for controlled drug release of organic or inorganic antimicrobial compounds, whereas in inorganic coatings both antimicrobial compound release and intrinsic antibacterial activity are possible.

synthesized plasticized chitosan-based polymers containing good antibacterial properties and mechanical strength with easy scale-up.[174] de Olyveira *et al.* developed a polyethylene composite containing silver microparticles.[16,175]

1.2.4 Antimicrobial Ceramics

Hydroxyapatite is a biocompatible and bioactive material in common use as an implant in bone tissue regeneration and as a drug carrier in drug and gene delivery systems. Due to its structural flexibility, various metal ions can be substituted in order to improve solubility, antibacterial activity and mechanical strength for bone implantation.[176,177] In addition, it is a potential candidate for use in cell targeting, fluorescence labeling, imaging and diagnosis materials.[178] Denser hydroxyapatite bioceramics can be used to create middle ear and eye implants, percutaneous device implants and inner dialysis systems. Hydroxyapatite doped with silver, copper oxide and zinc oxide can be used to improve antibacterial properties.[179,180]

1.3 Ideal Features of Antimicrobial Materials

An ideal antimicrobial polymer should have following characteristics:[7]

- highly stable over long periods of time;
- easily and inexpensively synthesized;
- should not decompose or emit toxic products;
- should be water insoluble for disinfection of water;
- should possess broad spectrum of antimicrobial activity;
- should be non-toxic and non-irritating.

1.4 Factors Affecting Antimicrobial Activity

1.4.1 Effect of Molecular Weight

Molecular weight has an important role in determining antimicrobial activity.[7] Chen *et al.* synthesized polypropylenimine dendrimers functionalized with quaternary ammonium groups and found that the antimicrobial properties have parabolic dependence on molecular weight.[79] In the case of polyacrylates and polymethylacrylates with biguanide groups, the optimal range of molecular weight was reported to be from 5×10^4 to 1.2×10^5 Da, with variance above and below this range significantly reducing efficacy.[181] Similarly, poly(tributyl 4-vinylbenzyl phosphonium chloride) also showed optimal antimicrobial action within a range of 1.6×10^4 to 9.4×10^4 Da.[182] However, the bacteriostatic action of fractioned quaternary ammonium salts against *Escherichia coli*, *Pseudomonas aeruginosa* and *Staphylococcus aureus* had little dependence on molecular weight.[7,183]

1.4.2 Effect of Counter Ions

Counter ion effect on antimicrobial properties is not clearly known, except where they change or alter the solubility of host polymers. Kanazawa *et al.* investigated the counter anion dependence of poly[tributyl (4-vinylbezyl) phosphonium] salts where the antimicrobial activity is in the order hexafluorophosphate < perchlorate < tetrafuoride < chloride, which can be correlated with the solubility products of those polymers.[182] Chlorides and bromides exhibit the highest antimicrobial activity in the case of quaternary ammonium compounds. Counter ions with strong binding affinity towards quaternary compounds show lower antibacterial action because of slow and reduced release of free ions in the medium.[79]

1.4.3 Charge Density

Usually, a positive charge density can impart better polymeric electrostatic interaction with negatively charged bacterial cell walls. For chitosan, with increasing degrees of deacetylation, the charge density increase enhances the electrostatic interaction of the polymer and thus antimicrobial property. Higher charge density groups were incorporated in chitosan to form guanidinylated chitosan and asparagine *N*-conjugated chitosan oligosaccharide, which resulted in high antimicrobial action, whereas *N*-carboxyethyl chitosan did not show any antimicrobial action due to a lack of free amino groups.[206–208]

1.4.4 Effect of Spacer Length and Alkyl Chain Length

Spacer length affects the interaction of antimicrobial agents with the bacterial cytoplasmic membrane due to changes in charge density and conformation of the polymer.[184] The antimicrobial activity of quaternary ammonium chlorides depends on the hydrophilic–lipophilic balance. Poly(trialkyl vinyl benzyl ammonium chloride) with the longest carbon chain (C_{12}) showed higher antimicrobial activity.[7]

1.4.5 pH Effect

The pH effect can be seen mostly in amphoteric polymers and chitosan. At acidic pH, chitosan exhibits maximum antimicrobial activity because of polycation formation and better solubility. However, at basic pH, there are no reports of its antimicrobial effect.[185]

1.4.6 Hydrophilicity

Hydrophilic nature is considered an important prerequisite for any antimicrobial agent to show activity. Tailoring of hydrophobic group content and molecular weight in amphiphilic polymethacrylate derivatives showed improvements in antimicrobial activity.[186] In the same manner, compared to the original form, the water-soluble chitosan derivatives synthesized by alkylation, metallization, quaternization and saccharization displayed greater antimicrobial action.[187,188]

1.5 Methods to Evaluate Antimicrobial Properties

Due to the high number of antibiotics in clinical microbiology, sensitivity testing becomes difficult. However, there are two standard testing methods: the serial dilution test and the disc test, by which sensitivity of bacteria to antibiotics can be tested *in vitro.*[189,190] In serial dilution tests, visible microbial growth is tested on a series of agar plates (agar dilution method) or broth (broth dilution method) that contain dilutions of antimicrobial agent.[191] It acts as a reference method for testing antimicrobial susceptibility, which in turn determines the MIC of antimicrobial agents.[192] Determination of MIC has an important role by which the tested microorganism is categorized as clinically susceptible, intermediate or resistant to a tested drug. Antibiotic drug resistance can also be monitored by MIC.[193,194] The disc diffusion method involves the use of different concentrations of antibiotic solutions in paper wells, cups or discs that are placed over the surface or punched into seeded agar plates containing a test bacterial strain.[195] Some of the characterization methods, such as test for microbial count, agar diffusion test and zone of inhibition (ZOI) test are used to determine and evaluate the effectiveness of nanoparticles as antimicrobial agents.[196]

No standard method is advocated in the literature to evaluate the antimicrobial activity of industrial products and medical devices. Moreover, the researchers modify the testing conditions as per their experimental design.[197] There are widely used standardized methods to characterize the antimicrobial materials described in ASTM E-2149 (American Society for Testing and

Materials, 2001), JIS 2801 (Japanese Standards Association, 2000), ZOI method, live–dead fluorescence staining and growth-based methods.[198] The ASTM E-2149-01 test method determines the antimicrobial property of treated specimens under dynamic contact conditions.[199] In the JIS Z 2801 : 2000 (Japanese Industrial Standard) testing method, surfaces (50×50 mm) are inoculated with *Escherichia coli* or *Staphylococcus aureus* suspension in a nutrient broth placed in petri dishes.[200,201] In 1966, Bauer *et al.* performed a test by the measurement of the zone of bacterial growth inhibition, with the testing materials placed on bacteria-inoculated agar plates, through the use of a ruler on the underside of the petri dish.[202,203] Fluorescence methods are based on the detection of intact cell structures and determination of inactive, active, dead and intact cells.[204] A standardized method is reported in Swiss Standard SNV 195929-1992 based on an agar diffusion test, which evaluates the width of bacterial growth inhibition area, around and beneath the samples after incubation with bacteria.[171,205]

1.6 Clinical Trials

Clinical trials for antimicrobial polymers are described in Table 1.5.

1.7 Conclusion and Future Developments

In this chapter, a concise overview on the research and development of novel antimicrobials has been provided. In order to synthesize and incorporate antimicrobial substances in biomaterials, various methods and recent technologies have been stimulated by the need to overcome antibiotic resistance and the risk of infections associated with the clinical use of medical devices.[171] Antimicrobial polymers have various application in the areas of water filtration systems, fibers, food packaging, surgical industries, surfactants and detergents and pharmaceuticals.[6] Nanoantimicrobials can provide new horizons in medical research and it is one of the most interesting areas of development for producing effective antibacterial substrates.[171] Intrinsically antimicrobial polymers represent a promising and novel approach by reducing the drug-resistant bacteria in biofilm.[206] Antimicrobial properties can be incorporated into polymeric materials by chemical modifications or by imparting inorganic/organic antimicrobial agents.[76] There is reduced opportunity for bacterial resistance with antimicrobial polypeptides as they bind with the bacterial cell wall and form pores in the membrane.[207] Short-term activity and environmental toxicity displayed by small molecular weight antimicrobial agents can be overcome by antimicrobial polymers. To obtain materials and products with improved quality and safety, industrial and academic research should come on board to develop innocuous materials that are environmentally friendly and reusable, with a broad range of potent, long-lasting and antimicrobial properties.[6,171]

Table 1.5 Clinical trials for antimicrobial polymers.[a]

Title	Indication	Comments	Phase	Status
Chitosan				
Efficacy and Safety of a Biofunctional Textile in the Management of Atopic Dermatitis	AD	Purpose is to study the use of biofunctional textile coated with chitosan. Shows improved quality of life and diminishes skin colonization with *Staphylococcus aureus* and skin moulds	2	Ongoing
USF Hemostasis: Usage of HemCon for Femoral Hemostasisafter Percutaneous Procedures	Coronary angiography	Used as an adjunct to manual compression for better control of vascular access site bleeding and reduce time to hemostasis after percutaneous coronary angiography	4	Completed
Trial of a Novel Chitosan Hemostatic Sealant in the Management of Complicated Epistaxis	Epistaxis	Purpose is to evaluate applicability of sealant in spontaneous epistaxis and its healing effect on nasal mucosa		Completed
Polyethyleneimine				
The Effects of a Polyethyleneimine-coated Membrane (oXiris™) for Hemofiltration Versus Polymyxin B-Immobilized Fibre Column (Toraymyxin™) for Hemoperfusion on Endotoxin Activity and Inflammatory Conditions in Septic Shock—A Randomized Controlled Pilot Study	Septic shock	It is hypothesized that positively charged inner surface of the membrane allows the absorption of negatively charged bacterial products which leads to activation of pro- and anti-inflammatory mediators at the early stage of sepsis		Not started
A Clinical Study: the Antibacterial Effect of Insoluble Antibacterial Nanoparticles (IABN) Incorporated in Dental Materials for Root Canal Treatment	Endodontic treatment	The effect of antibacterial nanoparticles, incorporated in root canal sealer material and in provisional restoration to be examined	2	Recruiting

Table 1.5 (*Continued*)

Title	Indication	Comments	Phase	Status
Antimicrobial peptides				
Antimicrobial Peptides in Periodontitis: A Pilot Study	Chronic periodontitis	Studied the level of expression of genes coding those peptides by studying periodontal smears		Completed
Analysis of the Response of Subjects with Atopic Dermatitis to Oral Vitamin D3 by Measurement of Antimicrobial Peptide Expression in Skin and Saliva	AD, psoriasis	Examined whether administration of oral vitamin D3 given over 21 days will change the AMP expression in the skin or saliva of subjects with AD		Completed
Role of Antimicrobial peptides in Host Defense Against Vaccinia Virus	AD	Compared smallpox virus replication, number of AMPs and other antiviral molecules in people with AD, as compared to psoriasis or asthma or healthy individuals		Completed
Nanoantimicrobials				
Clinical Study of Antibacterial Nanoparticles Incorporated in Composite Restorations	Oral health	Evaluated the antibacterial effect of alkylated PEI nanoparticles incorporated into flowable and hybrid composite resin disks	2	Completed
Topical Application of Silver Nanoparticles Reduced Oral Pathogens in Mechanically Ventilated Patients: A Randomized Controlled Clinical Trial	Critical illness	Silver nanoparticles are effective to reduce potential pathogen microbial loads in mechanical ventilation patients		Completed
Antibacterial Properties of Silicon Incorporated with Quaternary Ammonium Polyethylenimine Nanoparticles	Head and neck carcinoma	Aim is to evaluate the antibacterial activity of quaternary ammonium PEI nanoparticles (1–2% w/w) when compared to commercial soft liner material	1	Not yet recruiting

[a]AD: atopic dermatitis; AMP: antimicrobial peptide; PEI: polyethylenimine.

Abbreviations

AMP	Antimicrobial peptide
MIC	Minimum inhibitory concentration
PAMAM	Poly(amidoamine)
PCL	Polycaprolactone
PEI	Polyethylenimine
ε-PL	ε-Polylysine
PVPh	Poly(4-vinylphenol)

References

1. H. M. Lode, Clinical impact of antibiotic-resistant Gram-positive pathogens, *Clin. Microbiol. Infect.*, 2009, **15**, 212–217.
2. N. Beyth, Y. Houri Haddad, A. J. Domb, W. Khan and R. Hazan, Alternative antimicrobial approach: nano-antimicrobial materials, *J. Evidence-Based Complementary Altern. Med.*, 2015, 246012.
3. A. N. Neely and M. P. Maley, Survival of enterococci and staphylococci on hospital fabrics and plastic, *J. Clin. Microbiol.*, 2000, **38**, 724–726.
4. A. Jones, A. Mandal and S. Sharma, Protein-based bioplastics and their antibacterial potential, *J. Appl. Polym. Sci.*, 2015, **132**, 41931.
5. F. Siedenbiedel and J. C. Tiller, Antimicrobial polymers in solution and on surfaces: overview and functional principles, *Polymers*, 2012, **4**, 46–71.
6. A. Jain, L. S. Duvvuri, S. Farah, N. Beyth, A. J. Domb and W. Khan, Antimicrobial Polymers, *Adv. Healthcare Mater.*, 2014, **3**, 1969–1985.
7. E. R. Kenawy, S. D. Worley and R. Broughton, The chemistry and applications of antimicrobial polymers: a state-of-the-art review, *Biomacromolecules*, 2007, **8**, 1359–1384.
8. A. Jones, J. Pant, E. Lee, M. J. Goudie, A. Gruzd, J. Mansfield, A. Mandal, S. Sharma and H. Handa, Nitric oxide releasing antibacterial albumin plastic for biomedical applications, *J. Biomed. Mater. Res., Part A*, 2018, **106**, 1535–1542.
9. Y. Xue, H. Xiao and Y. Zhang, Antimicrobial polymeric materials with quaternary ammonium and phosphonium salts, *Int. J. Mol. Sci.*, 2015, **16**, 3626–3655.
10. P. Kaali, Antimicrobial polymer composites for medical applications, KTH Royal Institute of Technology, 2011, 1–89.
11. A. D. Fuchs and J. C. Tiller, Contact-active antimicrobial coatings derived from aqueous suspensions, *Angew. Chem., Int. Ed.*, 2006, **45**, 6759–6762.
12. J. M. Thomassin, S. Lenoir, J. Riga, R. Jerome and C. Detrembleur, Grafting of poly [2-(tert-butylamino) ethyl methacrylate] onto

polypropylene by reactive blending and antibacterial activity of the copolymer, *Biomacromolecules*, 2007, **8**, 1171–1177.
13. M. F. Ilker, K. Nusslein, G. N. Tew and E. B. Coughlin, Tuning the hemolytic and antibacterial activities of amphiphilic polynorbornene derivatives, *J. Am. Chem. Soc.*, 2004, **126**, 15870–15875.
14. C. Dong, Y. Ye, L. Qian, G. Zhao, B. He and H. Xiao, Antibacterial modification of cellulose fibers by grafing β-cyclodextrin and inclusion with ciprofloxacin, *Cellulose*, 2014, **21**, 1921–1932.
15. A. Piozzi and I. Francolini, Editorial of the special issue antimicrobial polymers, Multidisciplinary Digital Publishing Institute, 2013, pp. 18002–18008.
16. D. Sun, M. Babar Shahzad, M. Li, G. Wang and D. Xu, Antimicrobial materials with medical applications, *Mater. Technol.*, 2015, **30**, B90–B95.
17. C. R. Allan and L. A. Hadwiger, The fungicidal effect of chitosan on fungi of varying cell wall composition, *Exp. Mycol.*, 1979, **3**, 285–287.
18. C. Rouget, Des substances amylacees dans les tissus des animaux, specialement des Articules (chitine), *Comptes Rendus*, 1859, **48**, 792–795.
19. M. Rinaudo, Chitin and chitosan: properties and applications, *Prog. Polym. Sci.*, 2006, **31**, 603–632.
20. N. R. Sudarshan, D. G. Hoover and D. Knorr, Antibacterial action of chitosan, *Food Biotechnol.*, 1992, **6**, 257–272.
21. Y. C. Chung and C. Y. Chen, Antibacterial characteristics and activity of acid-soluble chitosan, *Bioresour. Technol.*, 2008, **99**, 2806–2814.
22. E. I. Rabea, M. E. T. Badawy, C. V. Stevens, G. Smagghe and W. Steurbaut, Chitosan as antimicrobial agent: applications and mode of action, *Biomacromolecules*, 2003, **4**, 1457–1465.
23. B. O. Jung, C. H. Kim, K. S. Choi, Y. M. Lee and J. J. Kim, Preparation of amphiphilic chitosan and their antimicrobial activities, *J. Appl. Polym. Sci.*, 1999, **72**, 1713–1719.
24. E. R. Kenawy, I. A. Salem, E. M. Abo-Elghit and A. A. Al-Owais, New trends in antimicrobial polymers: a state-of-the-art review, *Int. J. Chem. Appl. Biol. Sci.*, 2014, **1**, 95–105.
25. L. C. Keong and A. S. Halim, *In vitro* models in biocompatibility assessment for biomedical-grade chitosan derivatives in wound management, *Int. J. Mol. Sci.*, 2009, **10**, 1300–1313.
26. R. A. A. Muzzarelli, Chitosan-based dietary foods, *Carbohydr. Polym.*, 1996, **29**, 309–316.
27. A. Anitha, S. Sowmya, P. T. S. Kumar, S. Deepthi, K. P. Chennazhi, H. Ehrlich, M. Tsurkan and R. Jayakumar, Chitin and chitosan in selected biomedical applications, *Prog. Polym. Sci.*, 2014, **39**, 1644–1667.
28. R. Jayakumar, M. Prabaharan, S. V. Nair and H. Tamura, Novel chitin and chitosan nanofibers in biomedical applications, *Biotechnol. Adv.*, 2010, **28**, 142–150.

29. K. Madhumathi, P. T. S. Kumar, S. Abhilash, V. Sreeja, H. Tamura, K. Manzoor, S. V. Nair and R. Jayakumar, Development of novel chitin/nanosilver composite scaffolds for wound dressing applications, *J. Mater. Sci.: Mater. Med.*, 2010, **21**, 807–813.
30. P. T. S. Kumar, S. Abhilash, K. Manzoor, S. V. Nair, H. Tamura and R. Jayakumar, Preparation and characterization of novel β-chitin/nanosilver composite scaffolds for wound dressing applications, *Carbohydr. Polym.*, 2010, **80**, 761–767.
31. R. C. Goy, D. D. Britto and O. B. G. Assis, A review of the antimicrobial activity of chitosan, *Polimeros*, 2009, **19**, 241–247.
32. G. U. O. Tsai, W. H. Su, H. C. Chen and C. L. Pan, Antimicrobial activity of shrimp chitin and chitosan from different treatments, *Fish. Sci.*, 2002, **68**, 170–177.
33. X. Fei Liu, Y. Lin Guan, D. Zhi Yang, Z. Li and K. De Yao, Antibacterial action of chitosan and carboxymethylated chitosan, *J. Appl. Polym. Sci.*, 2001, **79**, 1324–1335.
34. K. Kurita, Chitin and chitosan: functional biopolymers from marine crustaceans, *Mar. Biotechnol.*, 2006, **8**, 203–226.
35. S. Roller and N. Covill, The antifungal properties of chitosan in laboratory media and apple juice, *Int. J. Food Microbiol.*, 1999, **47**, 67–77.
36. A. Balicka Ramisz, A. Wojtasz Pajak, B. Pilarczyk, A. Ramisz and L. Laurans, Antibacterial and antifungal activity of chitosan, *12th ISAH Congress on Animal Hygiene*, 2005, **2**, 406–408.
37. E. R. Kenawy and H. Xiao, *Polymeric materials with antimicrobial activity: from synthesis to applications*, Royal Society of Chemistry, 2013.
38. (a) J. R. Warren and F. Graham, The effect of heparin on the growth of bacteria and yeasts, *J. Bacteriol.*, 1950, **60**, 171–174; (b) W. Rosett and G. R. Hodges, Antimicrobial activity of heparin, *J. Clin. Microbiol.*, 1980, **11**, 30–34.
39. J. F. Christman and D. G. Doherty, The antimicrobial action of heparin, *J. Bacteriol.*, 1956, **72**, 433.
40. M. Hyldgaard, T. Mygind, B. S. Vad, M. Stenvang, D. E. Otzen and R. L. Meyer, The antimicrobial mechanism of action of epsilon-poly-l-lysine, *Appl. Environ. Microbiol.*, 2014, 02204–02214.
41. S. Shima, Y. Fukuhara and H. Sakai, Inactivation of bacteriophages by ε-poly-L-lysine produced by Streptomyces, *Agric. Biol. Chem.*, 1982, **46**, 1917–1919.
42. S. C. Shukla, A. Singh, A. K. Pandey and A. Mishra, Review on production and medical applications of ε-polylysine, *Biochem. Eng. J.*, 2012, **65**, 70–81.
43. H. T. Naghadeh, Z. Sharifi, S. Soleimani, Z. P. M. Jamaat and S. Ferdowsi, Efficacy of ε-poly-L-lysine as an antibacterial additive for platelets stored at room temperature, *Iran. J. Med. Sci.*, 2017, **42**, 509–511.
44. E. R. Kenawy, F. I. Abdel Hay, R. El Shanshoury, E. R. Abd and M. H. El Newehy, Biologically active polymers. V. Synthesis and antimicrobial activity of modified poly (glycidyl methacrylate-co-2-hydroxyethyl methacrylate) derivatives with quaternary ammonium

and phosphonium salts, *J. Polym. Sci. Part A: Polym. Chem.*, 2002, **40**, 2384–2393.
45. P. Gilbert and A. Al taae, Antimicrobial activity of some alkyltrimethylammonium bromides, *Lett. Appl. Microbiol.*, 1985, **1**, 101–104.
46. M. Kong, X. G. Chen, K. Xing and H. J. Park, Antimicrobial properties of chitosan and mode of action: a state of the art review, *Int. J. Food Microbiol.*, 2010, **144**, 51–63.
47. P. Gilbert and L. E. Moore, Cationic antiseptics: diversity of action under a common epithet, *J. Appl. Microbiol.*, 2005, **99**, 703–715.
48. J. J. H. Oosterhof, K. J. D. A. Buijssen, H. J. Busscher, B. F. A. M. van der Laan and H. C. van der Mei, Effects of quaternary ammonium silane coatings on mixed fungal and bacterial biofilms on tracheoesophageal shunt prostheses, *Appl. Environ. Microbiol.*, 2006, **72**, 3673–3677.
49. Y. Jiao, L. N. Niu, S. Ma, J. Li, F. R. Tay and J. H. Chen, Quaternary ammonium-based biomedical materials: State-of-the-art, toxicological aspects and antimicrobial resistance, *Prog. Polym. Sci.*, 2017, **71**, 53–90.
50. J. C. Tiller, C. J. Liao, K. Lewis and A. M. Klibanov, Designing surfaces that kill bacteria on contact, *Proc. Natl. Acad. Sci. U. S. A.*, 2001, **98**, 5981–5985.
51. J. C. Tiller, S. B. Lee, K. Lewis and A. M. Klibanov, Polymer surfaces derivatized with poly (vinyl-N-hexylpyridinium) kill airborne and waterborne bacteria, *Biotechnol. Bioeng.*, 2002, **79**, 465–471.
52. I. Francolini, G. Donelli, F. Crisante, V. Taresco and A. Piozzi, Antimicrobial polymers for anti-biofilm medical devices: state-of-art and perspectives, *Biofilm-based Healthcare-associated Infections*, Springer, 2015, pp. 93–117.
53. A. Shirai, T. Sumitomo, M. Kurimoto, H. Maseda and H. Kourai, The mode of the antifungal activity of gemini-pyridinium salt against yeast, *Biocontrol Sci.*, 2009, **14**, 13–20.
54. T. Muhizi, V. Coma and S. Grelier, Synthesis of D-glucosamine quaternary ammonium derivatives and evaluation of their antifungal activity together with aminodeoxyglucose derivatives against two wood fungi Coriolus versicolor and Poria placenta: structure-activity relationships, *Pest Manage. Sci.*, 2011, **67**, 287–293.
55. I. F. Tsao, H. Y. Wang and C. Shipman, Interaction of infectious viral particles with a quaternary ammonium chlorid (QAC) surface, *Biotechnol. Bioeng.*, 1989, **34**, 639–646.
56. E. Tuladhar, M. de Koning, I. Fundeanu, R. Beumer and E. Duizer, Different virucidal activities of hyperbranched quaternary ammonium coatings on poliovirus and influenza virus, *Appl. Environ. Microbiol.*, 2012, DOI: 10.1128/AEM.07738-11.
57. W. Siala, F. Van Bambeke, V. Taresco, A. Piozzi and I. Francolini, Synergistic activity between an antimicrobial polyacrylamide and daptomycin versus *Staphylococcus aureus* biofilm, *Pathog. Dis.*, 2016, **74**, ftw042.
58. J. Joca, C. Tukaj, W. Werel, U. Mizerska, W. Fortuniak and J. Chojnowski, Bacterial membranes are the target for antimicrobial

polysiloxane-methacrylate copolymer, *J. Mater. Sci.: Mater. Med.*, 2016, **27**, 55.
59. E. F. Palermo and K. Kuroda, Structural determinants of antimicrobial activity in polymers which mimic host defense peptides, *Appl. Microbiol. Biotechnol.*, 2010, **87**, 1605–1615.
60. O. Wichterle and D. Lim, Hydrophilic gels for biological use, *Nature*, 1960, **185**, 117–118.
61. C. Maldonado-Codina and N. Efron, Hydrogel lenses-materials and manufacture. A review, *Optometry in Practice*, 2003, **4**, 101–113.
62. E. Calo and V. V. Khutoryanskiy, Biomedical applications of hydrogels: A review of patents and commercial products, *Eur. Polym. J.*, 2015, **65**, 252–267.
63. A. W. Lloyd, R. G. A. Faragher and S. P. Denyer, Ocular biomaterials and implants, *Biomaterials*, 2001, **22**, 769–785.
64. J. F. Kunzler and G. D. Friends, Polymer compositions for contact lenses, *Google Pat.*, US5006622A, 1991.
65. S. Li, S. Dong, W. Xu, S. Tu, L. Yan, C. Zhao, J. Ding and X. Chen, Antibacterial Hydrogels, *Adv. Sci.*, 2018, **5**, 1700527.
66. R. L. Townsin and C. D. Anderson, Fouling control coatings using low surface energy, foul release technology, *Advances in Marine Antifouling Coatings and Technologies*, Elsevier, 2009, pp. 693–708.
67. U. Mizerska, W. Fortuniak, J. Chojnowski, R. Haasa, A. Konopacka and W. Werel, Polysiloxane cationic biocides with imidazolium salt (ImS) groups, synthesis and antibacterial properties, *Eur. Polym. J.*, 2009, **45**, 779–787.
68. Q. Zhang, H. Liu, X. Chen, X. Zhan and F. Chen, Preparation, surface properties, and antibacterial activity of a poly (dimethyl siloxane) network containing a quaternary ammonium salt side chain, *J. Appl. Polym. Sci.*, 2015, **132**, 41725.
69. S. R. Williams and T. E. Long, Recent advances in the synthesis and structure-property relationships of ammonium ionenes, *Prog. Polym. Sci.*, 2009, **34**, 762–782.
70. T. Narita, R. Ohtakeyama, M. Nishino, J. P. Gong and Y. Osada, Effects of charge density and hydrophobicity of ionene polymer on cell binding and viability, *Colloid Polym. Sci.*, 2000, **278**, 884–887.
71. T. Ikeda, H. Yamaguchi and S. Tazuke, Phase separation in phospholipid bilayers induced by biologically active polycations, *Biochimica et Biophysica Acta (BBA)-Biomembranes*, 1990, **1026**, 105–112.
72. S. Liu, R. J. Ono, H. Wu, J. Y. Teo, Z. C. Liang, K. Xu, M. Zhang, G. Zhong, J. P. K. Tan and M. Ng, Highly potent antimicrobial polyionenes with rapid killing kinetics, skin biocompatibility and *in vivo* bactericidal activity, *Biomaterials*, 2017, **127**, 36–48.
73. A. Strassburg, F. Kracke, J. Wenners, A. Jemeljanova, J. Kuepper, H. Petersen and J. C. Tiller, Nontoxic, hydrophilic cationic polymers identified as class of antimicrobial polymers, *Macromol. Biosci.*, 2015, **15**, 1710–1723.

74. R. Hoogenboom, Poly (2-oxazoline)s: a polymer class with numerous potential applications, *Angew. Chem., Int. Ed.*, 2009, **48**, 7978–7994.
75. A. Makino and S. Kobayashi, Chemistry of 2-oxazolines: A crossing of cationic ring-opening polymerization and enzymatic ring-opening polyaddition, *J. Polym. Sci., Part A: Polym. Chem.*, 2010, **48**, 1251–1270.
76. A. Munoz Bonilla and M. Fernandez Garcia, Polymeric materials with antimicrobial activity, *Prog. Polym. Sci.*, 2011, **37**, 281–339.
77. S. Bansal and A. K. Halve, Oxazolines: Their synthesis and biological activity, *Int. J. Pharm. Sci. Res.*, 2014, **5**, 4601–4616.
78. B. Guillerm, S. Monge, V. Lapinte and J. J. Robin, How to modulate the chemical structure of polyoxazolines by appropriate functionalization, *Macromol. Rapid Commun.*, 2012, **33**, 1600–1612.
79. C. Z. Chen, N. C. Beck Tan, P. Dhurjati, T. K. van Dyk, R. A. LaRossa and S. L. Cooper, Quaternary ammonium functionalized poly (propylene imine) dendrimers as effective antimicrobials: Structure-activity studies, *Biomacromolecules*, 2000, **1**, 473–480.
80. N. Bourne, L. R. Stanberry, E. R. Kern, G. Holan, B. Matthews and D. I. Bernstein, Dendrimers, a new class of candidate topical microbicides with activity against herpes simplex virus infection, *Antimicrob. Agents Chemother.*, 2000, **44**, 2471–2474.
81. C. K. V. Z. Abid, S. Chattopadhyay, N. Mazumdar and H. Singh, Synthesis and characterization of quaternary ammonium PEGDA dendritic copolymer networks for water disinfection, *J. Appl. Polym. Sci.*, 2010, **116**, 1640–1649.
82. Y. Zhang, J. Jiang and Y. Chen, Synthesis and antimicrobial activity of polymeric guanidine and biguanidine salts, *Polymer*, 1999, **40**, 6189–6198.
83. H. Wang, Preparation and characterization of dual functional antimicrobial (bio) degradable polymers, University Bayreuth, 2016.
84. M. Albert, P. Feiertag, G. Hayn, R. Saf and H. Honig, Structure activity relationships of oligoguanidines influence of counterion, diamine, and average molecular weight on biocidal activities, *Biomacromolecules*, 2003, **4**, 1811–1817.
85. K. Chindera, M. Mahato, A. K. Sharma, H. Horsley, K. Kloc-Muniak, N. F. Kamaruzzaman, S. Kumar, A. McFarlane, J. Stach and T. Bentin, The antimicrobial polymer PHMB enters cells and selectively condenses bacterial chromosomes, *Sci. Rep.*, 2016, **6**, 23121.
86. K. R. Kirker, S. T. Fisher, G. A. James, D. McGhee and C. B. Shah, Efficacy of Polyhexamethylene Biguanide-containing Antimicrobial Foam Dressing Against MRSA Relative to Standard Foam Dressing, *Wounds*, 2009, **21**, 229–233.
87. D. Andreu and L. Rivas, Animal antimicrobial peptides: an overview, *Pept. Sci.*, 1998, **47**, 415–433.
88. V. Teixeira, M. J. Feio and M. Bastos, Role of lipids in the interaction of antimicrobial peptides with membranes, *Prog. Lipid Res.*, 2012, **51**, 149–177.

89. O. G. Travkova, H. Moehwald and G. Brezesinski, The interaction of antimicrobial peptides with membranes, *Adv. Colloid Interface Sci.*, 2017, **247**, 521–532.
90. A. K. Marr, W. J. Gooderham and R. E. W. Hancock, Antibacterial peptides for therapeutic use: obstacles and realistic outlook, *Curr. Opin. Pharmacol.*, 2006, **6**, 468–472.
91. D. Alves and M. Olivia Pereira, Mini-review: Antimicrobial peptides and enzymes as promising candidates to functionalize biomaterial surfaces, *Biofouling*, 2014, **30**, 483–499.
92. R. Halevy, A. Rozek, S. Kolusheva, R. E. Hancock and R. Jelinek, Membrane binding and permeation by indolicidin analogs studied by a biomimetic lipid/polydiacetylene vesicle assay, *Peptides*, 2003, **24**, 1753–1761.
93. M. Tamaki, M. Kokuno, I. Sasaki, Y. Suzuki, M. Iwama, K. Saegusa, M. Shindo, M. Kimura and Y. Uchida, Syntheses of low-hemolytic antimicrobial gratisin peptides, *Bioorg. Med. Chem. Lett.*, 2009, **19**, 2856–2859.
94. G. N. Tew, D. Liu, B. Chen, R. J. Doerksen, J. Kaplan, P. J. Carroll, M. L. Klein and W. F. DeGrado, De novo design of biomimetic antimicrobial polymers, *Proc. Natl. Acad. Sci. U. S. A.*, 2002, **99**, 5110–5114.
95. M. R. Santos, A. C. Fonseca, P. V. Mendonca, R. Branco, A. C. Serra, P. V. Morais and J. F. J. Coelho, Recent developments in antimicrobial polymers: a review, *Materials*, 2016, **9**, 599.
96. D. G. Lee, D. H. Kim, Y. Park, H. K. Kim, H. N. Kim, Y. K. Shin, C. H. Choi and K. S. Hahm, Fungicidal effect of antimicrobial peptide, PMAP-23, isolated from porcine myeloid against *Candida albicans*, *Biochem. Biophys. Res. Commun.*, 2001, **282**, 570–574.
97. K. Park, D. Oh, S. Y. Shin, K. S. Hahm and Y. Kim, Structural studies of porcine myeloid antibacterial peptide PMAP-23 and its analogues in DPC micelles by NMR spectroscopy, *Biochem. Biophys. Res. Commun.*, 2002, **290**, 204–212.
98. W. T. Heller, A. J. Waring, R. I. Lehrer and H. W. Huang, Multiple states of β-sheet peptide protegrin in lipid bilayers, *Biochemistry*, 1998, **37**, 17331–17338.
99. L. Steinstraesser, B. Tippler, J. Mertens, E. Lamme, H. H. Homann, M. Lehnhardt, O. Wildner, H. U. Steinau and K. Eberla, Inhibition of early steps in the lentiviral replication cycle by cathelicidin host defense peptides, *Retrovirology*, 2005, **2**, 2.
100. A. Belaid, M. Aouni, R. Khelifa, A. Trabelsi, M. Jemmali and K. Hani, *In vitro* antiviral activity of dermaseptins against herpes simplex virus type 1, *J. Med. Virol.*, 2002, **66**, 229–234.
101. C. Lorin, H. Saidi, A. Belaid, A. Zairi, F. Baleux, H. Hocini, L. Belec, K. Hani and F. Tangy, The antimicrobial peptide dermaseptin S4 inhibits HIV-1 infectivity *in vitro*, *Virology*, 2005, **334**, 264–275.
102. M. Morimoto, H. Mori, T. Otake, N. Ueba, N. Kunita, M. Niwa, T. Murakami and S. Iwanaga, Inhibitory effect of tachyplesin I on the

proliferation of human immunodeficiency virus *in vitro*, *Chemotherapy*, 1991, **37**, 206–211.
103. T. Murakami, M. Niwa, F. Tokunaga, T. Miyata and S. Iwanaga, Direct virus inactivation of tachyplesin I and its isopeptides from horseshoe crab hemocytes, *Chemotherapy*, 1991, **37**, 327–334.
104. H. Nakashima, M. Masuda, T. Murakami, Y. Koyanagi, A. Matsumoto, N. Fujii and N. Yamamoto, Anti-human immunodeficiency virus activity of a novel synthetic peptide, T22 ([Tyr-5, 12, Lys-7] polyphemusin II): a possible inhibitor of virus-cell fusion, *Antimicrob. Agents Chemother.*, 1992, **36**, 1249–1255.
105. H. Tamamura, A. Otaka, T. Murakami, T. Ishihara, T. Ibuka, M. Waki, A. Matsumoto, N. Yamamoto and N. Fujii, Interaction of an anti-HIV peptide, T22, with gp120 and CD4, *Biochem. Biophys. Res. Commun.*, 1996, **219**, 555–559.
106. (a) D. H. Kim, D. G. Lee, K. L. Kim and Y. Lee, Internalization of tenecin 3 by a fungal cellular process is essential for its fungicidal effect on *Candida albicans*, *Eur. J. Biochem.*, 2001, **268**, 4449–4458; (b) Y. T. Lee, D. H. Kim, J. Y. Suh, J. H. Chung, B. L. Lee, Y. Lee and B. S. Choi, Structural characteristics of tenecin 3, an insect antifungal protein, *IUBMB Life*, 1999, **47**, 369–376.
107. H. G. Boman, B. Agerberth and A. Boman, Mechanisms of action on *Escherichia coli* of cecropin P1 and PR-39, two antibacterial peptides from pig intestine, *Infect. Immun.*, 1993, **61**, 2978–2984.
108. I. Erol, Novel methacrylate copolymers with fluorine containing: synthesis, characterization, reactivity ratios, thermal properties and biological activity, *J. Fluorine Chem.*, 2008, **129**, 613–620.
109. B. Findlay, G. G. Zhanel and F. Schweizer, Cationic amphiphiles, a new generation of antimicrobials inspired by the natural antimicrobial peptide scaffold, *Antimicrob. Agents Chemother.*, 2010, **54**, 4049–4058.
110. W. S. Moon, K. J. Chul, K. H. Chung, E. S. Park, M. N. Kim and J. S. Yoon, Antimicrobial activity of a monomer and its polymer based on quinolone, *J. Appl. Polym. Sci.*, 2003, **90**, 1797–1801.
111. (a) F. Guittard and S. Geribaldi, Highly fluorinated molecular organised systems: strategy and concept, *J. Fluorine Chem.*, 2001, **107**, 363–374; (b) L. Massi, F. Guittard, S. Geribaldi, R. Levy and Y. Duccini, Antimicrobial properties of highly fluorinated bis-ammonium salts, *Int. J. Antimicrob. Agents*, 2003, **21**, 206–212; (c) L. Caillier, E. Taffin de Givenchy, R. Levy, Y. Vandenberghe, S. Geribaldi and F. Guittard, Polymerizable semi-fluorinated gemini surfactants designed for antimicrobial materials, *J. Colloid Interface Sci.*, 2009, **332**, 201–207.
112. A. J. Kugel, L. E. Jarabek, J. W. Daniels, L. J. Vander Wal, S. M. Ebert, M. J. Jepperson, S. J. Stafslien, R. J. Pieper, D. C. Webster and J. Bahr, Combinatorial materials research applied to the development of new surface coatings XII: Novel, environmentally friendly antimicrobial coatings derived from biocide-functional acrylic polyols and isocyanates, *J. Coat. Technol. Res.*, 2009, **6**, 107–121.

113. H. B. Kocer, S. D. Worley, R. M. Broughton and T. S. Huang, A novel N-halamine acrylamide monomer and its copolymers for antimicrobial coatings, *React. Funct. Polym.*, 2011, **71**, 561–568.
114. E. R. Kenawy and Y. R. Abdel Fattah, Antimicrobial properties of modified and electrospun poly (vinyl phenol), *Macromol. Biosci.*, 2002, **2**, 261–266.
115. J. Lin, C. Winkelman, S. D. Worley, R. M. Broughton and J. F. Williams, Antimicrobial treatment of nylon, *J. Appl. Polym. Sci.*, 2001, **81**, 943–947.
116. J. Lin, C. Winkelmann, S. D. Worley, J. Kim, C. I. Wei, U. Cho, R. M. Broughton, J. I. Santiago and J. F. Williams, Biocidal polyester, *J. Appl. Polym. Sci.*, 2002, **85**, 177–182.
117. M. R. Badrossamay and G. Sun, A study on melt grafting of N-halamine moieties onto polyethylene and their antibacterial activities, *Macromolecules*, 2009, **42**, 1948–1954.
118. M. R. Badrossamay and G. Sun, Acyclic halamine polypropylene polymer: Effect of monomer structure on grafting efficiency, stability and biocidal activities, *React. Funct. Polym.*, 2008, **68**, 1636–1645.
119. J. Hoque, P. Akkapeddi, V. Yadav, G. B. Manjunath, D. S. S. M. Uppu, M. M. Konai, V. Yarlagadda, K. Sanyal and J. Haldar, Broad spectrum antibacterial and antifungal polymeric paint materials: synthesis, structure-activity relationship, and membrane-active mode of action, *ACS Appl. Mater. Interfaces*, 2015, **7**, 1804–1815.
120. A. Matejuk, Q. Leng, M. D. Begum, M. C. Woodle, P. Scaria, S. T. Chou and A. J. Mixson, Peptide-based antifungal therapies against emerging infections, *Drugs Future*, 2010, **35**, 197.
121. P. Appendini and J. H. Hotchkiss, Surface modification of poly (styrene) by the attachment of an antimicrobial peptide, *J. Appl. Polym. Sci.*, 2001, **81**, 609–616.
122. J. M. Yang, H. T. Lin, T. H. Wu and C. C. Chen, Wettability and antibacterial assessment of chitosan containing radiation-induced graft nonwoven fabric of polypropylene-g-acrylic acid, *J. Appl. Polym. Sci.*, 2003, **90**, 1331–1336.
123. A. Conte, G. G. Buonocore, M. Sinigaglia and M. A. Del Nobile, Development of immobilized lysozyme based active film, *J. Food Eng.*, 2007, **78**, 741–745.
124. L. Cen, K. G. Neoh and E. T. Kang, Surface functionalization technique for conferring antibacterial properties to polymeric and cellulosic surfaces, *Langmuir*, 2003, **19**, 10295–10303.
125. J. A. Lichter and M. F. Rubner, Polyelectrolyte multilayers with intrinsic antimicrobial functionality: the importance of mobile polycations, *Langmuir*, 2009, **25**, 7686–7694.
126. J. Lu, M. A. Hill, M. Hood, D. F. Greeson, J. R. Horton, P. E. Orndorff, A. S. Herndon and A. E. Tonelli, Formation of antibiotic, biodegradable polymers by processing with Irgasan DP300R (triclosan) and its inclusion compound with β-cyclodextrin, *J. Appl. Polym. Sci.*, 2001, **82**, 300–309.

127. F. Fay, I. Linossier, V. Langlois, K. Vallee-Rehe, M. Y. Krasko and A. J. Domb, Protecting biodegradable coatings releasing antimicrobial agents, *J. Appl. Polym. Sci.*, 2007, **106**, 3768–3777.
128. I. Yudovin-Farber, N. Beyth, A. Nyska, E. I. Weiss, J. Golenser and A. J. Domb, Surface characterization and biocompatibility of restorative resin containing nanoparticles, *Biomacromolecules*, 2008, **9**, 3044–3050.
129. N. Beyth, Y. Houri Haddad, L. Baraness Hadar, I. Yudovin Farber, A. J. Domb and E. I. Weiss, Surface antimicrobial activity and biocompatibility of incorporated polyethylenimine nanoparticles, *Biomaterials*, 2008, **29**, 4157–4163.
130. D. S. Jones, J. Djokic and S. P. Gorman, The resistance of polyvinylpyrrolidone-Iodine-poly (ε-caprolactone) blends to adherence of *Escherichia coli*, *Biomaterials*, 2005, **26**, 2013–2020.
131. A. Chaijaruwanich, Coating techniques for biomaterials: A review, *J. Nat. Sci.*, 2011, **10**(1), 39–50.
132. I. Corni, M. P. Ryan and A. R. Boccaccini, Electrophoretic deposition: From traditional ceramics to nanotechnology, *J. Eur. Ceram. Soc.*, 2008, **28**, 1353–1367.
133. Q. Bao, C. Chen, D. Wang, Q. Ji and T. Lei, Pulsed laser deposition and its current research status in preparing hydroxyapatite thin films, *Appl. Surf. Sci.*, 2005, **252**, 1538–1544.
134. X. Bai, K. More, C. M. Rouleau and A. Rabiei, Functionally graded hydroxyapatite coatings doped with antibacterial components, *Acta Biomater.*, 2010, **6**, 2264–2273.
135. R. Gupta and A. Kumar, Bioactive materials for biomedical applications using sol-gel technology, *Biomed. Mat.*, 2008, **3**, 034005.
136. M. A. Ansari, H. M. Khan, A. A. Khan, S. S. Cameotra, Q. Saquib and J. Musarrat, *Interaction of Al2O3 nanoparticles with Escherichia coli and their cell envelope biomolecules, J. Appl. Microbiol.*, 2014, **116**, 772–783.
137. Y. W. Baek and Y. J. An, Microbial toxicity of metal oxide nanoparticles (CuO, NiO, ZnO, and Sb_2O_3) to *Escherichia coli*, *Bacillus subtilis*, and *Streptococcus aureus*, *Sci. Total Environ.*, 2011, **409**, 1603–1608.
138. A. M. Jassim, S. A. Farhan, J. A. Salman, K. J. Khalaf, M. F. Al Marjani and M. T. Mohammed, Study the antibacterial effect of bismuth oxide and tellurium nanoparticles, *Int. J. Chem. Biomol. Sci.*, 2015, **1**, 81–84.
139. Z. X. Tang, Z. Yu, Z. L. Zhang, X. Y. Zhang, Q. Q. Pan and L. E. Shi, Sonication-assisted preparation of CaO nanoparticles for antibacterial agents, *Quim. Nova*, 2013, **36**, 933–936.
140. D. A. Pelletier, A. K. Suresh, G. A. Holton, C. K. McKeown, W. Wang, B. Gu, N. P. Mortensen, D. P. Allison, D. C. Joy and M. R. Allison, Effects of engineered cerium oxide nanoparticles on bacterial growth and viability, *Appl. Environ. Microbiol.*, 2010, **76**, 7981–7989.
141. T. Ghosh, S. K. Dash, P. Chakraborty, A. Guha, K. Kawaguchi, S. Roy, T. Chattopadhyay and D. Das, Preparation of antiferromagnetic Co_3O_4 nanoparticles from two different precursors by pyrolytic method: in vitro antimicrobial activity, *RSC Adv.*, 2014, **4**, 15022–15029.

142. G. Ren, D. Hu, E. W. C. Cheng, M. A. Vargas Reus, P. Reip and R. P. Allaker, Characterisation of copper oxide nanoparticles for antimicrobial applications, *Int. J. Antimicrob. Agents*, 2009, **33**, 587–590.
143. S. Jadhav, S. Gaikwad, M. Nimse and A. Rajbhoj, Copper oxide nanoparticles: synthesis, characterization and their antibacterial activity, *J. Cluster Sci.*, 2011, **22**, 121–129.
144. Y. Abboud, T. Saffaj, A. Chagraoui, A. El Bouari, K. Brouzi, O. Tanane and B. Ihssane, Biosynthesis, characterization and antimicrobial activity of copper oxide nanoparticles (CONPs) produced using brown alga extract (*Bifurcaria bifurcata*), *Appl. Nanosci.*, 2014, **4**, 571–576.
145. P. V. Kumar, U. Shameem, P. Kollu, R. L. Kalyani and S. V. Pammi, Green synthesis of copper oxide nanoparticles using Aloe vera leaf extract and its antibacterial activity against fish bacterial pathogens, *Bionanoscience*, 2015, **5**, 135–139.
146. S. Chatterjee, A. Bandyopadhyay and K. Sarkar, Effect of iron oxide and gold nanoparticles on bacterial growth leading towards biological application, *J. Nanobiotechnol.*, 2011, **9**, 34.
147. S. S. Behera, J. K. Patra, K. Pramanik, N. Panda and H. Thatoi, Characterization and evaluation of antibacterial activities of chemically synthesized iron oxide nanoparticles, *World J. Nano Sci. Eng.*, 2012, **2**, 196–200.
148. F. Al Hazmi, F. Alnowaiser, A. A. Al Ghamdi, A. A. Al Ghamdi, M. M. Aly, R. M. Al Tuwirqi and F. El Tantawy, A new large scale synthesis of magnesium oxide nanowires: structural and antibacterial properties, *Superlattices Microstruct.*, 2012, **52**, 200–209.
149. A. Jesline, N. P. John, P. M. Narayanan, C. Vani and S. Murugan, Antimicrobial activity of zinc and titanium dioxide nanoparticles against biofilm-producing methicillin-resistant *Staphylococcus aureus*, *Appl. Nanosci.*, 2015, **5**, 157–162.
150. Z. Huang, X. Zheng, D. Yan, G. Yin, X. Liao, Y. Kang, Y. Yao, D. Huang and B. Hao, Toxicological effect of ZnO nanoparticles based on bacteria, *Langmuir*, 2008, **24**, 4140–4144.
151. Y. Liu, L. He, A. Mustapha, H. Li, Z. Q. Hu and M. Lin, Antibacterial activities of zinc oxide nanoparticles against *Escherichia coli* O157: H7, *J. Appl. Microbiol.*, 2009, **107**, 1193–1201.
152. M. A. Ansari, H. M. Khan, A. A. Khan, A. Sultan and A. Azam, Characterization of clinical strains of MSSA, MRSA and MRSE isolated from skin and soft tissue infections and the antibacterial activity of ZnO nanoparticles, *World J. Microbiol. Biotechnol.*, 2012, **28**, 1605–1613.
153. L. Palanikumar, S. N. Ramasamy and C. Balachandran, Size-dependent antimicrobial response of zinc oxide nanoparticles, *IET Nanobiotechnol.*, 2014, **8**, 111–117.
154. P. C. Nagajyothi, T. V. M. Sreekanth, C. O. Tettey, Y. I. Jun and S. H. Mook, Characterization, antibacterial, antioxidant, and cytotoxic activities of ZnO nanoparticles using Coptidis Rhizoma, *Bioorg. Med. Chem. Lett.*, 2014, **24**, 4298–4303.

155. A. B. Patil and B. M. Bhanage, Green methodologies in the synthesis of metal and metal oxide nanoparticles, *Nanomater. Environ. Prot.*, 2014, 293–311.
156. C. Y. Watson, R. M. Molina, A. Louzada, K. M. Murdaugh, T. C. Donaghey and J. D. Brain, Effects of zinc oxide nanoparticles on Kupffer cell phagosomal motility, bacterial clearance, and liver function, *Int. J. Nanomed.*, 2015, **10**, 4173–4184.
157. T. Gordon, B. Perlstein, O. Houbara, I. Felner, E. Banin and S. Margel, Synthesis and characterization of zinc/iron oxide composite nanoparticles and their antibacterial properties, *Colloids Surf., A*, 2011, **374**, 1–8.
158. A. Dural Erem, G. Ozcan and M. Skrifvars, Antibacterial activity of PA6/ZnO nanocomposite fibers, *Text. Res. J.*, 2011, **81**, 1638–1646.
159. C. Srisitthiratkul, V. Pongsorrarith and N. Intasanta, The potential use of nanosilver-decorated titanium dioxide nanofibers for toxin decomposition with antimicrobial and self-cleaning properties, *Appl. Surf. Sci.*, 2011, **257**, 8850–8856.
160. G. P. Halliah, K. Alagappan and A. B. Sairam, Synthesis, characterization of CH- α-Fe_2O_3 nanocomposite and coating on cotton, silk for antibacterial and UV spectral studies, *J. Ind. Text.*, 2014, **44**, 275–287.
161. E. Malka, I. Perelshtein, A. Lipovsky, Y. Shalom, L. Naparstek, N. Perkas, T. Patick, R. Lubart, Y. Nitzan and E. Banin, Eradication of multi-drug resistant bacteria by a novel Zn-doped CuO nanocomposite, *Small*, 2013, **9**, 4069–4076.
162. S. Chakraborti, A. K. Mandal, S. Sarwar, P. Singh, R. Chakraborty and P. Chakrabarti, Bactericidal effect of polyethyleneimine capped ZnO nanoparticles on multiple antibiotic resistant bacteria harboring genes of high-pathogenicity island, *Colloids Surf., B*, 2014, **121**, 44–53.
163. G. S. Dhillon, S. Kaur and S. K. Brar, Facile fabrication and characterization of chitosan-based zinc oxide nanoparticles and evaluation of their antimicrobial and antibiofilm activity, *Int. Nano Lett.*, 2014, **4**, 107.
164. A. M. Holban, E. Andronescu, V. Grumezescu, A. E. Oprea, A. M. Grumezescu, G. Socol, M. C. Chifiriuc, V. Lazar and F. Iordache, Carvone functionalized iron oxide nanostructures thin films prepared by MAPLE for improved resistance to microbial colonization, *J. Sol-Gel Sci. Technol.*, 2015, **73**, 605–611.
165. R. S. Andre, C. A. Zamperini, E. G. Mima, V. M. Longo, A. R. Albuquerque, J. R. Sambrano, A. L. Machado, C. E. Vergani, A. C. Hernandes and J. A. Varela, Antimicrobial activity of TiO_2: Ag nanocrystalline heterostructures: experimental and theoretical insights, *Chem. Phys.*, 2015, **459**, 87–95.
166. L. Zhong and K. Yun, Graphene oxide-modified ZnO particles: synthesis, characterization, and antibacterial properties, *Int. J. Nanomed.*, 2015, **10**, 79–92.
167. Y. Qiu, N. Zhang, Y. H. An and X. Wen, Biomaterial strategies to reduce implant-associated infections, *Int. J. Artif. Organs*, 2007, **30**, 828–841.

168. F. Baron, S. Rehault, Compounds with antibacterial activity, *Bioactive egg compounds*, Springer, 2007, pp. 191–198.
169. A. Jones, A. Mandal and S. Sharma, Protein based bioplastics and their antibacterial potential, *J. Appl. Polym. Sci.*, 2015, **132**, 41931.
170. A. S. Yalcin, Emerging therapeutic potential of whey proteins and peptides, *Curr. Pharm. Des.*, 2006, **12**, 1637–1643.
171. F. Paladini, M. Pollini, A. Sannino and L. Ambrosio, Metal-based antibacterial substrates for biomedical applications, *Biomacromolecules*, 2015, **16**, 1873–1885.
172. A. Llorens, E. Lloret, P. A. Picouet, R. Trbojevich and A. Fernandez, Metallic-based micro and nanocomposites in food contact materials and active food packaging, *Trends Food Sci. Technol.*, 2012, **24**, 19–29.
173. F. Liu, H. Liu, X. Li, H. Zhao, D. Zhu, Y. Zheng and C. Li, Nano-TiO_2@Ag/PVC film with enhanced antibacterial activities and photocatalytic properties, *Appl. Surf. Sci.*, 2012, **258**, 4667–4671.
174. M. Matet, M. C. Heuzey and A. Ajji, Morphology and antibacterial properties of plasticized chitosan/metallocene polyethylene blends, *J. Mater. Sci.*, 2014, **49**, 5427–5440.
175. G. M. de Olyveira, L. M. M. Costa, A. L. Leo, S. F. de Souza, B. M. Cherian, A. J. F. de Carvalho, L. A. Pessan and S. S. Narine, LDPE/EVA composites for antimicrobial properties, *Mol. Cryst. Liq. Cryst.*, 2012, **556**, 168–175.
176. S. Shanmugam and B. Gopal, Copper substituted hydroxyapatite and fluorapatite: synthesis, characterization and antimicrobial properties, *Ceram. Int.*, 2017, **40**, 15655–15662.
177. A. Haider, S. Haider, S. S. Han and I.-K. Kang, Recent advances in the synthesis, functionalization and biomedical applications of hydroxyapatite: a review, *RSC Adv.*, 2017, 7, 7442–7458.
178. M. Mucalo, *Hydroxyapatite (HAp) for biomedical applications*, Elsevier, 2015.
179. J. Kolmas, E. Groszyk and D. Kwiatkowska Rozycka, Substituted hydroxyapatites with antibacterial properties, *BioMed Res. Int.*, 2014, 178123.
180. O. Lukats, P. Bujtar, G. K. Sandor and J. Barabas, Porous hydroxyapatite and aluminium-oxide ceramic orbital implant evaluation using CBCT scanning: a method for *in vivo* porous structure evaluation and monitoring, *Int. J. Biomater.*, 2012, 764749.
181. T. Ikeda, H. Yamaguchi and S. Tazuke, New polymeric biocides: synthesis and antibacterial activities of polycations with pendant biguanide groups, *Antimicrob. Agents Chemother.*, 1984, **26**, 139–144.
182. A. Kanazawa, T. Ikeda and T. Endo, Polymeric phosphonium salts as a novel class of cationic biocides. III. Immobilization of phosphonium salts by surface photografting and antibacterial activity of the surface-treated polymer films, *J. Polym. Sci., Part A: Polym. Chem.*, 1993, **31**, 1467–1472.
183. T. Ikeda, H. Hirayama, H. Yamaguchi, S. Tazuke and M. Watanabe, Polycationic biocides with pendant active groups: molecular weight

dependence of antibacterial activity, *Antimicrob. Agents Chemother.*, 1986, **30**, 132–136.
184. T. Ikeda, H. Hirayama, K. Suzuki, H. Yamaguchi and S. Tazuke, Biologically active polycations, 6. Polymeric pyridinium salts with well defined main chain structure, *Macromol. Chem. Phys.*, 1986, **187**, 333–340.
185. S. H. Lim and S. M. Hudson, Synthesis and antimicrobial activity of a water-soluble chitosan derivative with a fiber-reactive group, *Carbohydr. Res.*, 2004, **339**, 313–319.
186. K. Kuroda and W. F. DeGrado, Amphiphilic polymethacrylate derivatives as antimicrobial agents, *J. Am. Chem. Soc.*, 2005, **127**, 4128–4129.
187. Y. J. Jeon, P. J. Park and S. K. Kim, Antimicrobial effect of chitooligosaccharides produced by bioreactor, *Carbohydr. Polym.*, 2001, **44**, 71–76.
188. Y. Hu, Y. Du, J. Yang, J. F. Kennedy, X. Wang and L. Wang, Synthesis, characterization and antibacterial activity of guanidinylated chitosan, *Carbohydr. Polym.*, 2007, **67**, 66–72.
189. H. Dickert, K. Machka and I. Braveny, The uses and limitations of disc diffusion in the antibiotic sensitivity testing of bacteria, *Infection*, 1981, **9**, 18–24.
190. G. N. Rolinson and E. J. Russell, New method for antibiotic susceptibility testing, *Antimicrob. Agents Chemother.*, 1972, **2**, 51–56.
191. G. Kahlmeter, D. F. J. Brown, F. W. Goldstein, A. P. MacGowan, J. W. Mouton, I. Odenholt, A. Rodloff, C. J. Soussy, M. Steinbakk and F. Soriano, European Committee on Antimicrobial Susceptibility Testing (EUCAST) technical notes on antimicrobial susceptibility testing, *Clin. Microbiol. Infect.*, 2006, **12**, 501–503.
192. J. L. RodriguezTudela, F. Barchiesi, J. Bille, E. Chryssanthou, M. CuencaEstrella, D. Denning, J. P. Donnelly, B. Dupont and W. Fegeler, Method for the determination of minimum inhibitory concentration (MIC) by broth dilution of fermentative yeasts, *Clin. Microbiol. Infect.*, 2003, **9**, i–viii.
193. I. Wiegand, K. Hilpert and R. E. W. Hancock, Agar and broth dilution methods to determine the minimal inhibitory concentration (MIC) of antimicrobial substances, *Nat. Protoc.*, 2008, **3**, 163–175.
194. A. Vipra, S. N. Desai, R. P. Junjappa, P. Roy, N. Poonacha, P. Ravinder, B. Sriram and S. Padmanabhan, Determining the minimum inhibitory concentration of bacteriophages: potential advantages, *Adv. Microbiol.*, 2013, **3**, 181–190.
195. B. Bonev, J. Hooper and J. Parisot, Principles of assessing bacterial susceptibility to antibiotics using the agar diffusion method, *J. Antimicrob. Chemother.*, 2008, **61**, 1295–1301.
196. J. T. Seil and T. J. Webster, Antimicrobial applications of nanotechnology: methods and literature, *Int. J. Nanomed.*, 2012, **7**, 2767.
197. D. Troitzsch, U. Borutzky and U. Junghann, Detection of antimicrobial efficacy in silver-coated medical devices, *Hygiene & Medizin*, 2009, **34**, 80–85.

198. J. B. D. Green, S. Bickner, P. W. Carter, T. Fulghum, M. Luebke, M. A. Nordhaus and S. Strathmann, Antimicrobial testing for surface-immobilized agents with a surface-separated live-dead staining method, *Biotechnol. Bioeng.*, 2011, **108**, 231–236.
199. ASTME, Standard Test Method for Determining the Antimicrobial Activity of Immobilized Antimicrobial Agents Under Dynamic Contact Conditions, 2001.
200. Z. Jis, A. Japanese Standards, Antimicrobial products-Test for antimicrobial activity and afficacy, Minister of International Trade and Industry, 2000.
201. A. E. Madkour and G. N. Tew, Towards self-sterilizing medical devices: controlling infection, *Polym. Int.*, 2008, **57**, 6–10.
202. (a) A. W. Bauer, W. M. M. Kirby, J. C. Sherris and M. Turck, Antibiotic susceptibility testing by a standardized single disk method, *Am. J. Clin. Pathol.*, 1966, **45**, 493–496; (b) H. Zhang, M. Wu and A. Sen, Silver nanoparticle antimicrobials and related materials, *Nano-antimicrobials*, Springer, 2012, 3–45.
203. A. L. Barry, M. B. Coyle, C. Thornsberry, E. H. Gerlach and R. W. Hawkinson, Methods of measuring zones of inhibition with the Bauer-Kirby disk susceptibility test, *J. Clin. Microbiol.*, 1979, **10**, 885–889.
204. F. Joux and P. Lebaron, Use of fluorescent probes to assess physiological functions of bacteriaat single-cell level, *Microbes Infect.*, 2000, **2**, 1523–1535.
205. M. Pollini, M. Russo, A. Licciulli, A. Sannino and A. Maffezzoli, Characterization of antibacterial silver coated yarns, *J. Mater. Sci.: Mater. Med.*, 2009, **20**, 2361–2366.
206. V. Taresco, F. Crisante, I. Francolini, A. Martinelli, L. Dilario, L. Ricci-Vitiani, M. Buccarelli, L. Pietrelli and A. Piozzi, Antimicrobial and antioxidant amphiphilic random copolymers to address medical device-centered infections, *Acta Biomater.*, 2015, **22**, 131–140.
207. S. Pavlukhina, Y. Lu, A. Patimetha, M. Libera and S. Sukhishvili, Polymer multilayers with pH-triggered release of antibacterial agents, *Biomacromolecules*, 2010, **11**, 3448–3456.
208. P. J. Park, J. Y. Je and S. K. Kim, Free radical scavenging activities of differently deacetylated chitosans using an ESR spectrometer, *Carbohydr. Polym.*, 2004, **55**, 17–22.

CHAPTER 2

Introduction to Microbes and Infection in the Modern World

JOSHUA C. DOLOFF[a,b]

[a] Department of Biomedical Engineering, Translational Tissue Engineering Center, Wilmer Eye Institute/Smith Bldg, Johns Hopkins University School of Medicine, 400 N. Broadway, Baltimore, MD 21231, USA; [b] Department of Materials Science and Engineering, Institute of NanoBioTechnology, Johns Hopkins University, 3400 N. Charles St, Baltimore, MD 21218, USA
Email: jcdoloff@jhu.edu

2.1 Introduction

2.1.1 The Many Facets of Microbial Life

Beyond much larger multi-cellular plants and animal species, microbes are largely, but not always, single cell microorganisms that help make up many of the remaining major constituents of the natural ecosystem.[1] They exist everywhere, both in the outside environment in soil, water and air, as well as on and inside of other larger living organisms, including plants and animals. They demonstrate remarkable ability to survive under extreme conditions, including high and low pH and temperature, even those considered unlivable (*e.g.*, boiling, freezing, highly acidic) for many other forms of life. They are quite diverse, being striated into bacteria, protists—including protozoa (free or parasitic), algae, and fungi (including yeast and mold)—viruses, and viroids, as well as prions, known for being protein-based infectious agents (Figure 2.1). Despite the name, which implies that they exist at the micron level, as is the case with unicellular protozoa or bacteria, their

Biomaterials Science Series No. 5
Antimicrobial Materials for Biomedical Applications
Edited by Abraham J. Domb, Konda Reddy Kunduru and Shady Farah

Published by the Royal Society of Chemistry, www.rsc.org

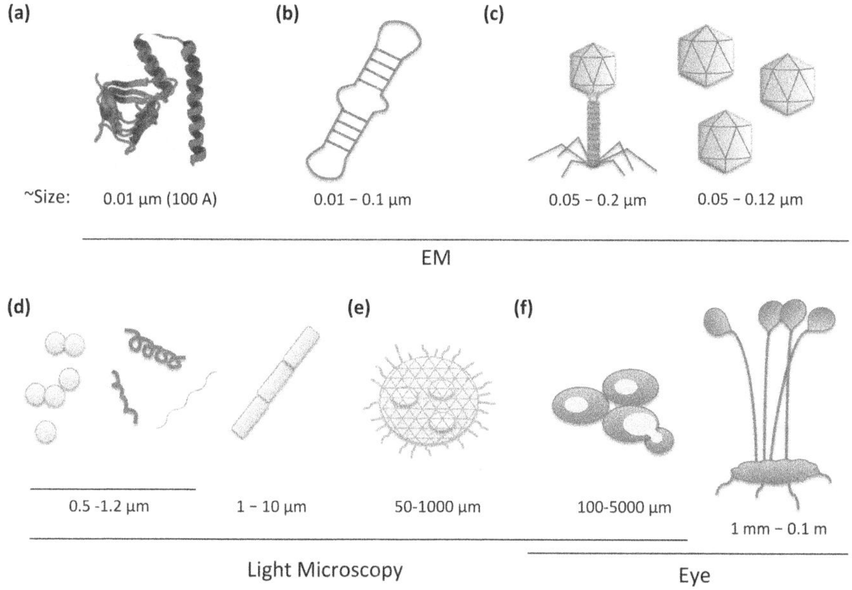

Figure 2.1 Diagrams of different microbes. (a) Prions; (b) viroids; (c) left: bacteriophage, right: viral vectors; (d) left, middle, and right: rod, sphere, and spiral bacteria, respectively; (e) algae; (f) left, middle, and right: fungus, mold, and yeast, respectively. Although not drawn to scale, relative sizing is shown, along with various methods of visualization. EM: electron microscopy; A: angstroms.

scaling can be lower, into the nano, and higher, into the macro ranges, as is the case with subcellular prions and viruses or multi-cellular algae and fungi, respectively. As a consequence, microbes are generally thought of existing at a scale that requires a microscope to properly visualize them as well as their structural complexity. To this point, while some algae and fungi have structures approaching tissue-level organization, most microbes are instead devoid of distinct and complex tissues analogous to internal organs found in humans and other animals. With that said, microbes do contain subcellular or cellular levels of organization, with the most prominent example being the striation between prokaryotes (all bacteria) and eukaryotes (all others except viruses and prions), namely the absence or presence of a membrane-bound nucleus.

2.1.2 Bacteria

In contrast to prokaryotes within archaea (discussed later), bacteria often occupy less caustic environments, and while often beneficial in terms of their utility to society, they also include many known pathogens that cause disease. They can either be producers—creating energy from sunlight or deriving energy from inorganic chemicals to make food for themselves—or consumers—obtaining energy by breaking down dead organisms or absorbing nutrients from other living organisms in a parasitic fashion. In addition, they can be aerobic, with the vast majority living off oxygen-rich environments, or anaerobic, wherein they subsist without the need for oxygen. They can thrive in multiple types of extreme environmental conditions:[2] (i) High temperature (thermophiles)—70–100 °C or more; (ii) low temperature (psychrophiles)—−20 °C or less, such as those found in the Arctic or Antarctic; (iii) high salt (halophiles); (iv) high pressure; (v) low pH; or (vi) acidic. Bacteria are much smaller than plant and animal cells and have low internal complexity containing very few structures. They contain cytoplasm surrounded by both a cell membrane and a cell wall. Bacteria can be rod-shaped (bacillus), spherical (coccus), or spiral (twisted, such as spirilla or spirochaetes). Those that reside in water-laden environments often have motors called flagella. For defense, some bacteria can thicken their cell wall in response to danger, even going into hibernation for hundreds of years as an endospore. However, generally bacteria are identified by the food they consume (*e.g.*, lactose or citric acid) and their waste products (*e.g.*, acetoin).

2.1.3 Archaea

While originally grouped into bacteria (originally being labeled archaebacteria), archaea are now recognized as a distinct group of prokaryotic microbes with their own evolutionary history (now recognized as possibly the oldest forms of life on earth) and biochemistry.[2] While similar in size and shape, major differences differentiating bacteria and archaea biochemically are (i) archaea cell walls do not have peptidoglycan; (ii) bacterial have one ribosomal RNA polymerase while archaea have three (making them more similar to eukaryotic life in this regard); (iii) archaea use ether-linked lipids instead of ester-linked lipids; (iv) bacteria cannot live at temperatures >100 °C, whereas some archaea have been recorded to survive even up to 122 °C; (v) unlike bacteria, archaea cell growth is unaffected by antibiotics; and (vi) while both can reproduce by budding, binary fission, and fragmentation, unlike bacteria, archaea cannot form spores to survive years in dormancy. Euryarchaeota and Crenarchaeota make up much of the archaea and live in extremely harsh conditions. As a consequence, they are commonly called extremophiles. Such conditions include very high or low temperatures and intense pressure, under which very little other life exists. Euryarchaeota include thermophiles; methane-producing methanogens; organotrophs (deriving energy from organic compounds), such as iron, nitrate, sulphate, or sulfur; and aerobes. Crenarchaeota include thermophilic (high temperatures as well <50 °C) and hyperthermophilic (only >80 °C) heterotrophs, which derive their energy from other carbon-based organic matter.

2.1.4 Protists

Protists are single- or many-celled organisms that live in wet environments; however, unlike bacteria they are eukaryotic with a membrane-bound nucleus.[3] Similar to bacteria, protists can also be producers or consumers, as predators, parasites, or saprophytes (scavengers feeding off dead organisms). Fungus-like protists live different phases of their lives as either a single- or multi-cellular organisms. Many are called molds or mildews and are exclusively consumers, either saprophytes or parasites. In addition, animal-like single-celled protists are called protozoans, live in water, earth, and both alive and dead organisms, and are typically known for the way they move, namely *via* flagella, cilia (*e.g.*, *Paramecium*), or pseudopodia (*Amoeba*). In addition to movement, cilia and pseudopodia are also utilized for feeding, as all protozoans are consumers. Lastly, plant-like protists are called algae, being either single- or multi-cellular. All algae, known by the colors they appear, are producers, utilizing chlorophyll to make their own food. Single-celled algae utilize flagella for movement. Seaweed is an example of a multi-celled algae. Ultimately, many protozoans are parasites that cause disease, including toxoplasmosis (*Toxoplasma gondii*),[4] African sleeping sickness, and malaria.

2.1.5 Viruses and Prions

While the validity of whether viruses and those that target bacteria (bacteriophages) are considered to be legitimate organisms capable of all essential functions carried out by most life forms has been questioned, remarkably, they contain subcellular compartmentalization imitating more complex life: DNA or RNA (but never both) into a protein-based capsid coat. Most viruses are considered to be a form of parasite, dependent on operating inside living cells for their propagation and spread (replication). Of note, since they are not capable of replicating on their own, they must first hijack cellular machinery and sources of energy production.

Prions are the most recent newly discovered form of pathogen,[5] with the 1997 Nobel Prize in Physiology or Medicine awarded to Stanley Prusiner in recognition of their significance in biology, physiology, and medicine. While prions are not completely understood, scientists appreciate that despite being comprised of protein, with no DNA or RNA constituents, and even found normally in the brain in harmless contexts, abnormal prions have been shown to convert normally folded proteins into disease-inducing alternatives. As such, they are considered less as microbes and more as aberrant proteins. With that said, infectious variants are responsible for many dementia-type diseases as well as 'mad cow' disease and Creutzfeldt–Jakob disease.

2.1.6 Fungi

Most fungi are many-celled, eukaryotic organisms, including both mold and yeast.[6] While their cells have cell walls, leading scientists to classify them originally as plants, they do not have chlorophyll and therefore cannot produce their own food. Instead, they are popularly known scavengers, most often saprophytes, although some can be parasitic as well. Fungi reproduce through small waterproof spores, which can spread large distances and eventually regrow into a new fungus, and their large multi-cellular bodies are made up of threadlike tubes called hyphae. Despite this, they are considered to not have specialized tissues and organs, and do not have structures such as leaves and roots. While various forms can be used for fermentation or even as food themselves, some produce athlete's foot, ringworm, or other conditions on or in the human body, as well as diseases such as Dutch elm disease or chestnut blight in plants. One infamous water mold was the cause of the Irish potato famine in the 1840s.

2.2 Not All Microbes Are Bad

2.2.1 Microbes Are Utilized in Many Commercial Applications

In many ways, microbes have been successfully utilized for the betterment of mankind, both in the context of general uses in society as well as for commercialized purposes (Figure 2.2). Some of the earliest examples of microbial use in society include food fermentation, such as with lactic acid bacteria,[7] *Lactobacillus*, used to enrich vitamin B12 in milk, as well as in the production of yogurt and curd.[8] Other bacteria and yeast have been used for many food products, such as the expansion of dough through CO_2 released during fermentation. In large quantities, microbes including yeast are used in fermenters or barrels for the large-scale production of wine, beer, and numerous alcoholic beverages including whiskey, rum, and brandy.[9] While many other examples involving fermentation exist, perhaps the most famous examples are often associated with cheese production,[10] wherein the defining smell, flavor or taste, and texture associated with each type are directly the result of the microbes—including bacteria, fungi and mold—used to produce it. Protists such as algae are utilized in many products, such as toothpaste, pudding, and ice cream.

For large-scale industrial applications, microbes have been utilized in many applications including biofertilizer, sewage treatment, and biogas production.[11] Living in a world with an ever-increasing population, the need for appropriate waste disposal or its conversion into other useful products has grown into a huge industry. Municipal waste water sewage systems, often including many potentially dangerous pathogenic microbes, lead into treatment plants that are designed to handle and treat huge volumes of organic waste on a daily basis. Such treatment schemes often involve a two-step process and involve heterotrophic microbes, which must derive nutritional intake and energy from other sources.[12] Following filtration and sedimentation of particulates, microbes are enlisted under mechanical agitation and air infusion which allow for the expansion of aerobic (oxygen-dependent) microbes into large masses called flocs, fungal filament-based mesh-like structures heavily concentrated with bacteria. These microbial systems consume much of the organic waste found in sewage as they grow,

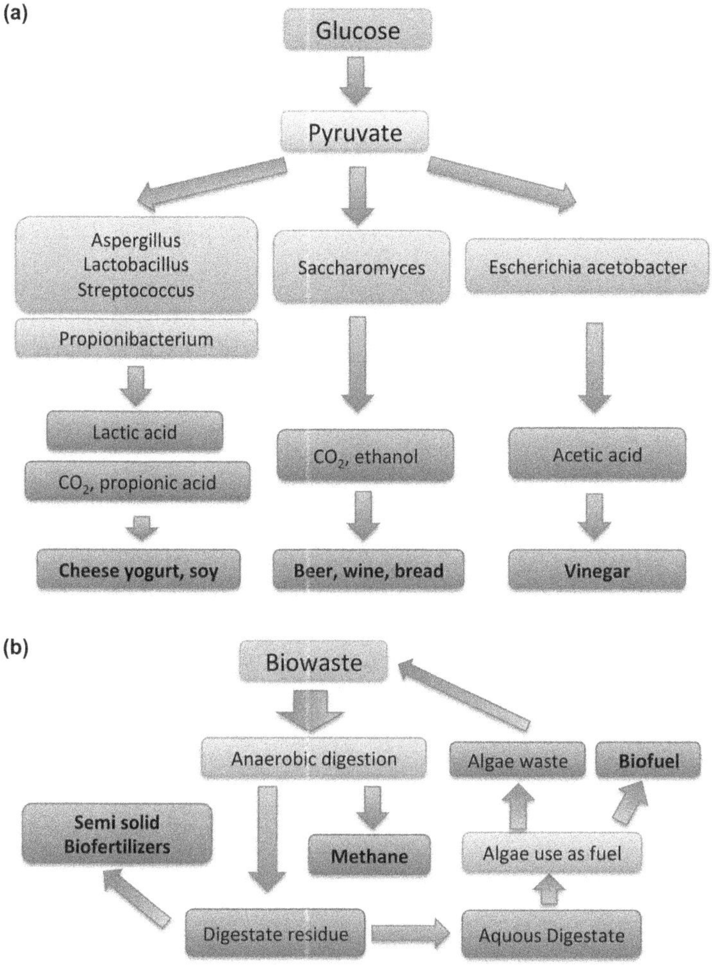

Figure 2.2 Microbial commercial utilities in society outside of medicine. (a) Food and beverage fermentation; (b) biowaste and sewage treatment with benefits in production of biofertilizer (left) and biogas (middle) or biofuel (right).

reducing its biochemical oxygenation and polluting potential. Following this process, small portions of flocs are used to inoculate subsequent batches of water, with the bulk remainder being pumped into large tanks where anaerobic (oxygen-independent) bacteria digest it. Through this last step, many useful gases such as methane, carbon dioxide, and hydrogen sulfide are produced, forming a biogas mixture, which can be later used as an inflammable energy source. In addition, through normal biogas production efforts typically found in rural plants, cellulose-containing materials often involving cow dung are fed to *Methanobacteria*, which produce methane (rather than CO_2) as a heavy byproduct. Methane can be used as burnable energy source for cooking and light, and the spent fecal matter is siphoned off for use as a biofertilizer. Such biofertilizers improve soil quality independent of synthetic chemical treatments and are often derived from bacteria, cyanobacteria, fungi, or even algae that often fix gases such as nitrogen into organic nutrients or help absorb elements such as phosphorus and pass them along to neighboring plant life. Globally, millions of gallons of waste are converted every day, and the importance of microbes is central to this process, as no synthetic technologies have thus far been able to replace these century-plus old methods.

2.2.2 Microbial Uses in Medicine

However, perhaps the most significant microbial contribution to medicinal applications is for production of bioactive molecules such as antibiotics, chemicals, and enzymes. Antibiotics are chemical substances, which kill or slow the growth of disease-causing microbes, whereas other chemical compounds or enzymes can be utilized for chemical reactions or other industrial purposes (Figure 2.3). The most famous example of an antibiotic is penicillin,[13] derived from the mold (or fungus) *Penicillium notatum*, and discovered by Alexander Fleming in 1928, when he observed that *Staphylococcus* bacteria would not grow around a mold also present in one of his unwashed culture plates. Ernst Chain and Howard Florey established its full utility as an antibiotic for use in humans, and it would go on to be used

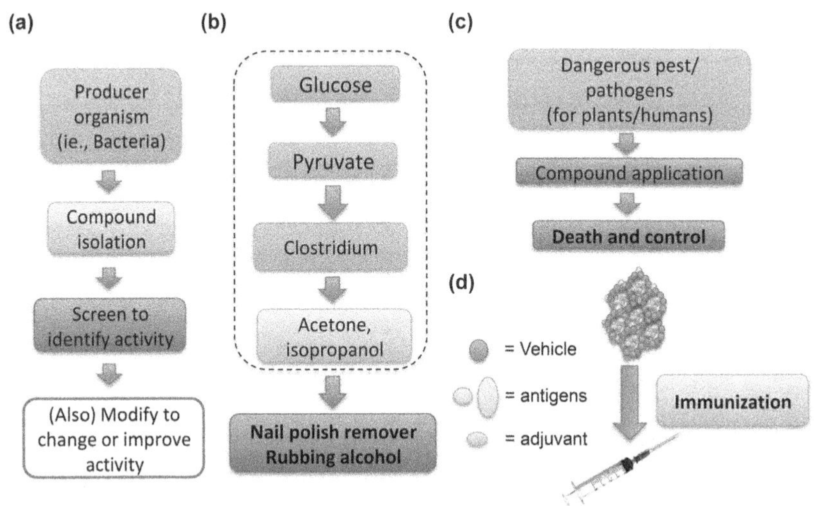

Figure 2.3 Microbial utilities in medicine. Bioactive molecule production of (a) antibiotics and other bioactives such as enzymes—including lipases, streptokinase and proteases, statins and cyclosporine A. (b) Chemicals such as acids/alcohols (dashed line: reaction vessel/incubator). (c) Use as biocontrol for other microbes or pests as well as (d) for education of the immune system (*e.g.*, vaccines).

heavily to save many lives during World War II. For their contributions to advances in human health, Fleming, Chain, and Florey received the Nobel Prize in Physiology or Medicine in 1945. Subsequently, many other antibiotics (*e.g.*, streptomycin, azithromycin, cephalosporins, vancomycin, *etc.*) have been discovered and isolated from additional microbes, and have since been used to save millions of lives by treating many other diseases such as whooping cough, leprosy, plague, and diphtheria.

Other examples of microbe-derived bioactive molecules include organic acids, enzymes, and alcohols. Acid producers include fungi such as *Aspergillus niger* (used for citric acid), bacteria such as *Lactobacillus* (mentioned above in food for lactic acid) and *Acetobacteraceti* (used for acetic acid), as well as *Clostridium butylicum* (for butyric acid). Enzymes produced by and isolated from microbes include lipases,[14] utilized in detergents; streptokinase (from the bacterium *Streptococcus*) has been utilized as an anticlotting agent; and pectinases and proteases used to clear fruit juices. Yeast (*Saccharomyces cerevisiae*) is heavily utilized, not only for alcoholic beverages, as mentioned above, but also for commercial ethanol production. Other organic compounds with bioactivity include cholesterol-combating statins[15] (produced by the yeast *Monascus purpureus*) and the immunosuppressive agent cyclosporin A (produced by the fungus *Trichoderma polysporum*)[16] used in organ transplantation.

2.3 Evolution of Microbes with Humans

It is a testament to their resilience that microbes thrive in so many diverse ecosystems, not only in nature but also in and on our own bodies. Many microbes exist on our skin[17] and in many other spaces inside of or on our bodies: mouth,[18] eyes,[19] ears,[20] vaginal cavity,[21] lungs,[22] and gut.[23] Their small scale (high surface area to volume ratio; about 20 times greater than that of a mammalian cell) makes them well suited for rapid nutrient uptake and therefore more responsive to their surrounding environment.

In fact, the mouth is now considered one of the most diverse bacterial ecosystems,[24] even rivaling the level of biological diversity found in tropical rainforests. In 1999, Stanford scientists discovered 37 new microbes deep in mouth plaque, bringing the then known total to more than 500.[25] Such efforts required utilization of genetic sequencing, because of the inability of traditional culturing methods to grow the vast majority of bacterial species. Scientists are now using this information to try to prevent tooth decay.[26] However, much still isn't known; it is thought that only 2% or so of microbial populations have been identified to date. As a consequence of such realizations, and the acceptance that more information can aid in healthcare efforts, a global survey was initiated by the American Society for Microbiology and the National Science Foundation, and has been extended to form the Human Microbiome Project launched by the National Institutes of Health (NIH) in 2007.[27] To aid in these efforts, these organizations are creating a network of biodiversity research sites termed "microbial observatories".

Complications in such analyses are effects of season, temperature fluctuation, human behavior or demographic backgrounds, food and/or chemical exposure, and even location and exposure to different regional microbes.

2.4 Biocontrol and the Importance of Commensal Microbes

Biocontrol is the manner by which certain biological diseases or pests are kept in check either through predation or the use of chemical treatments, either biological or synthetic in origin.[28] In the modern world, chemical treatments, such as insecticides, pesticides and weedicides, are increasingly relied upon. Heavy use of these agents can lead to saturation and pollution of soil, ground water, and crops, and can affect nearby estuaries through runoff. Unfortunately, many such compounds can be toxic not only to humans, but also to animals and other microbial life that makes up natural ecosystems. In addition, their overuse has led to decreases in efficacy, as many targeted organisms adapt to survive despite ever-increasing levels of use. Such ecosystem imbalance occurs not just on land, but in waterways, where toxic algal blooms[29] can lead to the death of fish and many other endogenous species. As a consequence, this has led many people, including farmers, to revisit the idea of maintaining pests and microbes in balance with one another as a natural method of checks and balances to maintain healthy crop yields without the need for toxic and increasingly ineffective chemical compounds.

With the added issues of microbial resistance (further discussed later), not only farmers, but also clinicians are revisiting more organic methods associated with predation and natural checks and balances that organisms exert over one another. The scientific community has increased its attention towards microbes following additional recent insights into and appreciation of the ability of certain organisms to not only exert control over other harmful pathogens in a manner more akin to predation, but also to regulate or affect other important systems and functions inside the body. Thus, the idea of maintaining a variety of natural microbial life to maintain appropriate ecosystem balance is just as relevant for our bodies and human health practices as it is on a farm or in the environment. In fact, huge coral-like biogeographical organizations and interactions have been observed and well documented among the many different bacteria in the mouth.[18] Furthermore, it has been demonstrated that commensal microbiota play a role in educating the immune system to tolerate microbes normally found in certain tissues, such as the intestine.[30] While such subjects have been the focus of many recent studies as well as the Human Microbiome Project, much attention has been given to traditional sites of commensal microbiota (Figure 2.4), while neglecting respiratory disease (other than in a HIV-related context) and the microbiome of the lungs, which have recently been shown to have a microbial density similar to that of the small bowel.[22] One practical example of a protective commensal microbe is *Lactobacillus*, which can

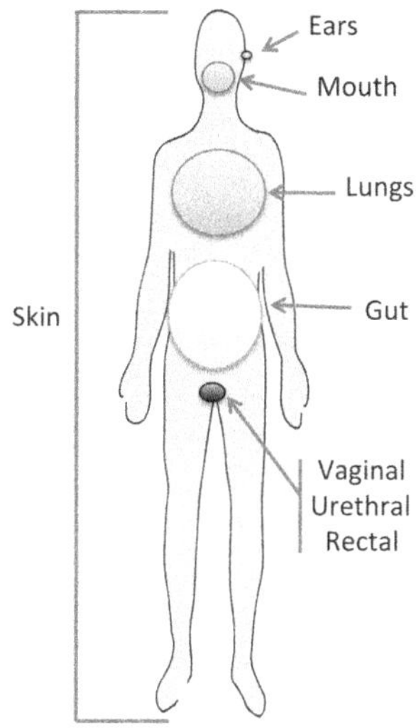

Figure 2.4 Microbiota associated with human patients. There are many ecosystems of microbiota on and in the human body, including the mouth, ears, gut, excretory systems (*e.g.*, vaginal, urethral or rectal cavities), and lungs, as well as in many different regions of the skin (*e.g.*, small white patches show common areas of differential body oil, temperature, humidity, or exposure to open air, all of which have been shown to influence microbial colonization of the skin).

prevent the spread of dangerous disease-causing bacteria in the stomach. Various microbes can be utilized to control non-microbial organisms, for example *Bacillus thuringiensis* is employed to control the caterpillar and stink bug populations to which it is toxic.[31] Such methods are quite targeted, as *B. thuringiensis* is not toxic to most other insect species. Due to its utility, scientists have since isolated genes from this bacterium, incorporating them directly into plants for increased resistance to pest infestation. Moreover, such utilities extend beyond bacteria. The fungus *Trichoderma* has displayed control over various pathogenic risks to plants,[32] and baculoviruses can attack insects in a very species-specific and targeted manner,[33] with no ill effects toward mammalian, plant, fish, bird, or off-target insect species.

Over time, researchers have discovered the importance of microbes in producing nutrients such as vitamin K, necessary for blood clotting,[34] to help break down and absorb the foods we consume, and even to inform and educate our immune systems over time. Scientists have observed that in germ-free mice, the immune system is dysregulated and does not work properly.[35] In addition, it is increasingly understood that the microbiome of an individual is tied to regulating and maintaining human health.[36] While some probiotic supplements have shown benefits in boosting health, poor overall understanding of the indigenous microbiome has resulted in lack of efficacy,[37] or even been shown to be counterproductive in delaying rather than speeding up the reconstitution of the normal gut microbial landscape following antibiotic treatments.[38] In fact, prior use of antibiotics aided the survival of newly introduced microbes and their colonization of the gut following probiotic regimens, most likely due to a reduction in competition by the normal microbiota. Such procedures are desirable in order to transfer various health benefits, such as bacteria aiding in lower rates or nutritional absorption through the lumen of the gut, and often involve fecal microbiota transplantation (FMT) or "poop pill" alternatives, whereby the spores within frozen fecal material from a donor are consumed by a recipient patient.[39] This has followed observations in the field that following FMT from a skinny mouse to a fat mouse (or human-to-human) the recipient subject has exhibited weight loss.[40] The opposite, where patients exhibit weight gain, has also been observed, by accident in some,[41] or perhaps by design in cases of patients being chronically underweight following illness or disorders such as anorexia.[42] Additionally, FMT has recently helped with other issues, such as exerting control over inflammation in the gut;[43] however, applications of FMT are not fully controlled and often can exhibit other adverse side-effects.[44]

2.5 Increases in Emerging Disease

Although natural disasters garner much more media coverage in the modern world, the uncontrolled spread of infectious disease is a far greater human health threat. While numerous wars over the past century are remembered for the numbers of lives lost, since 1945 more than 150 million people have died from AIDS, tuberculosis (TB), and malaria alone; in that same time period, 23 million lives were lost due to war. According to the World Health Organization, there were 219 million malaria cases in more than 90 countries in 2017 alone.[45] In general, infectious diseases constitute the second leading cause of deaths (>25%) worldwide (Figure 2.5),[46] second only to heart disease,

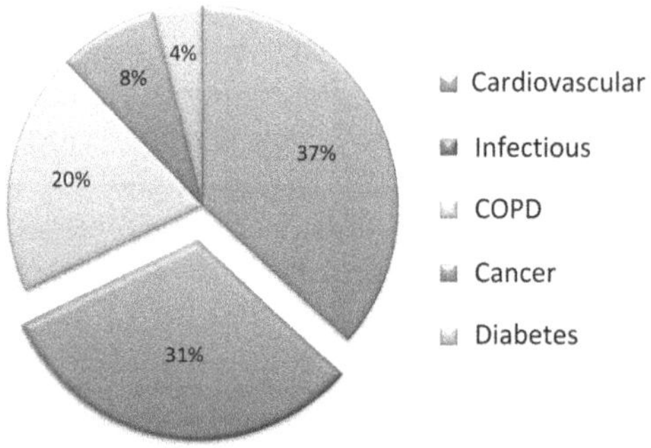

Figure 2.5 Top five global causes of disease-related deaths. Numbers are based on 2017 World Health Organization (WHO) statistics; percentage is normalized to the total number of deaths from the top five causes. Currently, cardiovascular deaths (including stroke) still outrank deaths from infectious diseases, with the largest individual contributing diseases being sepsis, lower respiratory infections, diarrheal diseases, tuberculosis, HIV/AIDS, and malaria. COPD: chronic obstructive pulmonary disease.

which is often reported even in instances of passing in one's sleep from old age. Out of that number, almost 10 million children aged <5 years die from pneumonia/respiratory disease, diarrhea, malaria, measles, and HIV. These numbers are thought to be an underestimate, due to issues of surveillance and reporting throughout the world. However, poor patient outcomes are not simply associated with poverty or lack of education and reporting.

The presence of harmful microorganisms in the area of human health has become a great concern due to a variety of infections and diseases. More than 30 new infectious diseases have been identified and many others previously known have resurged in the past 30 years or so.[47] These include AIDS, Ebola, *Escherichia coli*, flesh-eating *Staphylococcus aureus*, hantavirus, 'mad cow' (prion) disease, measles–mumps–rubella (considered eradicated in 2000, but newly resurgent in part due to the anti-vax (vaccine) movement (discussed later), killed >110 000 people in 2017 alone),[45] and *Salmonella*, among others. TB, also previously thought to be largely obliterated in the United States, resurged in New York City in 1992 in almost 4000 patient cases.[48] In contrast to a change in human behavior (*e.g.*, lack of vaccination), the TB bacterium in this instance instead developed antibiotic resistance. While TB is declining globally (it is decreasing more slowly than hoped, despite educational campaigns),[48] it has increased in other areas with differences in educational or healthcare infrastructure, such as North Korea, which has one of the highest infection rates in the world: ∼500 out of 100 000,[142] and in Africa, where co-infection by HIV has been a complicating factor.[48] In 1993, a much larger infection event (the largest incident of waterborne illness in US history) resulted when the diarrhea-associated protozoan cryptosporidiosis spread through 403 000 people (25% of the city's population) in Milwaukee, Wisconsin.[49] In that same year, hantavirus killed many in New Mexico, Utah, Colorado, and Arizona.[50] Ebola, a hemorrhagic virus, has popped up numerous times in the past few decades, and famously made its way to the United States by flight as an extension of the 2013–2015 West African epidemic, which has been the most geographically extended, long-lasting, and fatal ever.[51] Another example of a major disease-spread event, in 1998 Dengue fever affected 100 million people in more than 50 countries, with 500 000 cases of hemorrhagic Dengue, and >25 000 deaths annually.[52]

2018 marked 25 years since a lack of proper food cleanliness and hygiene contributed to >500 people in the Pacific Northwest and greater Seattle area becoming sick with the bacterium *E. coli* after consuming contaminated burgers across more than 70 Jack-in-the-Box fast-food restaurants.[53] After identification of the contaminating microbe, and in a pre-emptive action to prevent a nationwide outbreak, health officials issued a recall of almost 22 million pounds of ground beef. In another more recent example resulting in a food recall, a $11.2 million fine and a head company executive being convicted to a 28 year sentence, *Salmonella*-contaminated peanut butter made more than 710 people sick in over 46 states in 2007.[54] Many more recent instances of *Salmonella* and *E. coli* have continued to be a health concern in the US, where high manufacture, output and shipment of food products all over the country is now common practice. In fact, a new strain of *E. coli* suspected to come from sprouts on a local farm was identified in Germany in 2011 following illness in 4000 patients, with 50 cases across 16 countries resulting in deaths.[55] Additional resurgences of avian and swine flu, Lyme disease, and severe acute respiratory syndrome (SARS) have emphasized the need to remain vigilant about public health and are a reminder of how easily microbes can spread in the modern world, despite improvements in hygiene, healthcare, and education compared to historical precedents such as in the Middle Ages, where European cities and populations were lacking in all these categories. Unfortunately, given the increased population density globally, the behavior of others is an increasingly important component in the equation to determine the chances of detrimental health consequences for the greater public.

2.6 Identified Medical Threats and Treatments in the Environment

2.6.1 Influences of Environment and Ecological Destruction on Compromised Healthcare

Another major contributor to the emergence of disease is human behavior and how it affects ecological destruction.[47,56] First, bulldozing or slash-and-burn policies for the conversion of forests and jungle into land used for agriculture or construction blunt the ability of microbial-assisted waste elimination due to the reduction of microbes associated with natural land- or water-based ecosystems which help in breaking down potentially hazardous pathogens and organic waste. In addition, as mentioned, heavily used biocontrol agents such as insecticides, pesticides, and weedicides all find their way into the environment and water supply.[31,33] Since they are not specific for larger animal/insect or plant species, microbial wildlife is affected, either leading to the elimination of microbes useful in biowaste digestion, or perhaps unhealthy shifts in microbes that are less useful for waste mitigation.[11] As discussed earlier, this presents an opportunity for more dangerous pathogenic microbes, previously kept in check by competition for resources, to expand.

Another factor contributing to disease emergence resulting from heavy construction projects and the conversion of natural ecosystems into farms and homes is increased proximity to each other, as well as closer contact with other organisms.[47,56] This is especially the case when one considers that deforestation and human population expansion may place us into much closer contact with organisms to which we have had little or no prior contact, and to which we have developed little to no prior immunity. As an example, fungal infections and outbreaks with mortality often ∼50% have been reported in construction crews, who in many cases make first contact while being tasked with disruption of ecosystems for redevelopment.[57] Natural ecosystems not only serve as reservoirs for a large number of insect and animal species, all of which can also act as carriers for infectious microbes, but they also help keep them isolated and contained away from human settlements.

Deforestation also leads to the displacement of organisms (plants and animals) that may be harboring disease, driving them into areas directly around humans that host species and by extension the microbes that they carry might have otherwise considered less ideal than their original habitats. One such example is the rise of rabies in the eastern United States due to infected raccoons surviving on human garbage and food waste. In addition, loss of habitat has been shown to increase avian (*e.g.*, coronaviruses) and bat-associated diseases.[58,59] In Malaysia in the late 1990s, the then new and deadly Nipah virus was thought to have spread from fruit bats as they moved from their destroyed and deforested habitats into human farms; livestock acted as vectors for transmission of disease into humans. Furthermore, leishmaniasis, a once largely forest-isolated protozoan disease, carried by infected flies, has become more urban due to deforestation (and perhaps humans shifting from rural to more urban centers, as discussed later).[60] Another famous example of organismal displacement was not the result of a disturbance of a natural ecosystem, but due to the creation of an unnatural one. The walls of the Aswan High Dam, completed in 1970 after the investment of one billion dollars, created a reservoir of a population of snails, required for a disease caused by a parasitic worm, schistosomiasis, to take hold.[61] In addition, the new body of water contained by the dam resulted in rift valley fever, a mosquito-borne hemorrhagic viral infection.[62] More than 200 000 people became ill and >500 died.

Regardless of the exact route (Figure 2.6) or mechanism of exposure—which can include adherence, invasion, breaching, or compromise of the physiological barriers of the body[63]—carriers of microbial pathogens are referred to as biological vectors. In addition, zoonotic transmission or spillover refers to the transmittance of pathogens and disease, not just across individuals of the same species, but also between members of different species, typically across wild animals, livestock, and humans.[56] Species barriers, which usually limit disease transferal, refer to the understanding that differences between the immune system of one species and another often make it difficult and inhospitable for microbes, that have evolved to operate and thrive in one species, to be capable of survival in another. As a consequence, when microbes do succeed in the rare instances of successfully jumping between species they often exhibit increased resilience and higher levels of morbidity.

Other examples have occurred throughout human history. One of the most significant and modern diseases, AIDS, is thought to have resulted from humans entering deeper into forests and coming into contact with monkey species harboring simian immunodeficiency virus, which in time resulted in patient zero (the first to contract a spreadable form of the disease) developing HIV (human immunodeficiency virus).[64] Another, more recent well-known

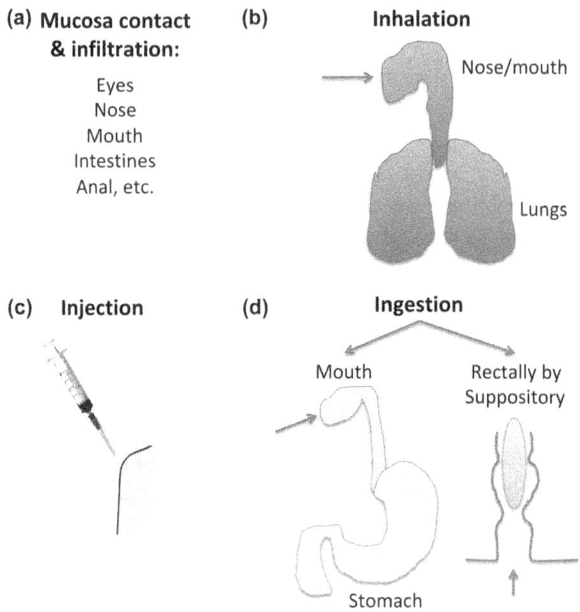

Figure 2.6 **Routes of microbial exposure.** How do microbial infections occur? (a) absorption through mucosal membranes (*e.g.*, kissing, spit, waterborne), simply by having open orifices come in contact with contaminated water, (b) inhalation, (c) injection, or (d) ingestion (orally or by suppository).

example is SARS, thought to have formed into a pathogen that could infect humans by incubating across multiple animals being kept incredibly close together in cages in an open market in China.[65,66] In the case of SARS, it has been suggested that this virus crossed not one, but multiple species barriers. As mentioned, doing so has often resulted in a much more virulent and dangerous infection, one reason why the global media became intensely interested in its coverage when it first emerged. Selection for more robust microbial strains also occurs when a disease moves through multiple patients within the same species, either vertically, from mother to child, or horizontally, between two individuals of the same generation.[67] Combining these concepts, and related to the transmission of disease across individuals of the same species, as well as different species, the germs premise of Jared Diamond's Pulitzer Prize winning book *Guns, Germs and Steel*, in which he postulated that humans of Middle Eastern and European descent that had lived for a few thousand years in close proximity to livestock and farm animals became more resilient to diseases that were very deadly. As a consequence, conquistadors and explorers unknowingly acted as carriers and in turn increased the incidence of diseases among many of the indigenous populations that they met, traded with, conquered and/or colonized, in particular in the Americas[68] and Africa.[69] This was but one factor in what historically led to long-lasting effects on indigenous populations (*e.g.*, the rampant emergence of diseases such as yellow fever and smallpox, and many deaths), who had not traditionally relied upon and lived in close proximity to other animal species for farming or transportation purposes, and as a consequence their immune systems had not yet been challenged in such ways, or at the very least by the germs that early explorers often brought with them.

Closer proximity to animals harboring disease also leads to increased chances of consumption of contaminated meat and/or produce (further discussed later), resulting in increased disease incidence, thought to be a source of some strains of HIV.[64] Many pathogens, even if not deadly, can give off harmful byproducts known as toxins. Botulism is one example of a toxin produced by anaerobic bacteria, which can survive as dormant endospores in canned food, and while such bacteria and endospores can both be destroyed through sterilization processes, illness still occurs.[70] However, exposure to microbes such as prions (*e.g.*, 'mad cow' or Creutzfeldt–Jakob disease), usually through contaminated meat or organs, in some cases infamously ingested in cannibalistic acts, have been notoriously difficult to destroy, even with extreme sterilization temperatures, proteases, ionizing radiation, and fixation (*e.g.*, formaldehyde) treatments.[71] Interestingly, higher exposure among cannibalistic tribes has resulted in prion disease resistance.[72]

2.6.2 Environmental Influences on Improved Health and Healthcare

While there are many environmental factors that can influence the emergence of disease, conversely, and thankfully, there have been many recorded anecdotes over the course of human describing some of the earliest microbial remedies from nature. Antimicrobial peptides are important components of host defense against various infections and display remarkable activity against bacteria, fungi, viruses, and parasites.[73,74] Their activity was observed originally when, for instance, multiple organisms co-mingled in crops or in more modern petri dishes where one was observed to inhibit the growth of the other. Ultimately, such compounds were identified upon subsequent isolation and screening of many different proteins from the suppressing organism. As mentioned earlier, this is how some of the earliest antibiotics, such as penicillin were discovered.

Their modes of action are to kill microbes by various mechanisms such as cell membrane/wall disruption, interference with metabolism, inhibiting DNA or RNA production, or targeting and inhibiting cytoplasmic components involved in translation.[75,76] As an example of microbial specificity, fungal cell walls contain mannan, chitin, and glucans that are absent in other microbes, thus making them potential targets for therapeutics, and antifungal peptides exist that target their cell wall *via* binding to chitin to disrupt membrane integrity.[77] To aid in their efficacy, generally they have evolved to contain hydrophilic and hydrophobic groups that enable them to be solubilized in aqueous environments and pass through lipid membranes. However, characteristics such as susceptibility to proteolysis, high production cost, and poor pharmacokinetic profile have made some consider alternatives to their use.[78] With that said, antimicrobial peptides are of great interest to serve as the basis for next-generation antimicrobial development.[79] Tied into microbial defense, a whole new field of genomic editing has recently spawned out of new CRISPR-Cas9 systems, discovered as a consequence of identifying their role in bacterial innate host defense.[80] Similar to the way intracellular pathogen recognition occurs in mammals through pattern recognition receptors and downstream immune activation,[81–83] CRISPR allows for the targeting of subsequent viral infection events in bacteria, and Cas proteins have been shown to lead to cleavage of DNA,[80] and more recently even RNA-based sequences.[84]

2.7 Increasing Burdens on Healthcare: Population Expansion, Urbanization, and Increasing Age of the General Human Populace

Overpopulation increases difficulties arising from person-to-person transmission (higher probability of contact), an increased rate of global warming, larger numbers of travelers, higher incidence of unrest or war as resources become more scarce, increasing refugees and displaced populations crowding into refugee camps and urban slums, higher absolute numbers living in poverty, increased lack of potable water, and corresponding increases in the volume of dirty contaminated water.[85] Of concern, due to global population expansion and condensation into urban centers, water treatment plant capacity has not sufficiently increased and kept pace with waste production, leading to increased pollution and occurrence of waterborne illnesses.

But what factors effects such emergence? As mentioned earlier, urbanization and concentration of people and waste into condensed areas has dramatically increased the burden on man-made decontamination systems. In 1900, only 13% of the world's population lived in cities; in 2018, that number was 55%, and by 2050, this number is projected to reach almost 70%.[86] Urbanization leads to increased pollution; decreases in available natural resources and clean water, thereby causing additional burdens on public health and city infrastructure such as sanitation, increased populations of vermin, and issues with immunization; and increased poverty, leading to malnutrition and diminished capacity to fight infection.[87] This in turn has led to effluents, some of which may not be properly decontaminated due to waste exceeding treatment capacity in many locations around the world, finding their way into waterways and the consumable water supply.[88] With these problems comes reduced human hygiene, leading to increased outbreaks of disease.

With the increasing average age of the population of the United States, and indeed the world, comes undue effects on healthcare systems as well as the potential for microbial survival and spread throughout the population.[89] In 2010, the percentage of the US population aged >65 years is ~15%, and that figure will only continue to increase, with a prediction that by 2035 adults aged >65 years will for the first time outnumber children (aged <18 years). This trend is occurring not only in the United States but worldwide, with such population growth increasing by as much as two- to five-fold compared to 1960 census information, and depending on the nation.[90] This increased average age allows microbes more permissive survival and spread due to an on average reduced capacity of the immune system in these patients to ward off new illness or successfully eliminate prior infections.[91] Another variable that could affect immunity as well as overall health is that the proportional makeup of the gut microbiota changes with age.[36] While children have traditionally been considered the major source of new pathogen exposure, due to the ongoing education of their immune system until their adaptive immune memory has hit critical mass, as well as regular exposure to other children potentially carrying and spreading sicknesses throughout their younger school years, it is the ever-increasing older geriatric population that is now becoming a larger influence on disease transmission and spread.[89] In addition to their weakened immune systems and the social shift from living with younger relatives to moving into senior, assisted-living centers and homes,[92] it has been well documented that sexually transmitted diseases and other microbes spread *via* unprotected sexual activity have increased dramatically as elderly people deciding that they no longer need to utilize safe practices and contraception due to being past their child-bearing years.

2.8 Approaching Challenges and Perceived Threats

Such instances of spread and transmission have become a reality in a more globalized, interconnected, and modernized world, and has caused the Centers for Disease Control (CDC) to increase their preventative screening policies for those entering the country from high-risk areas, and to improve field team response times and shared patient handling policies for care facilities prior to their arrival, in order to minimize potential pandemic spread scenarios. Table 2.1 details factors driving disease emergence.

2.8.1 Increased Disease Emergence Due to Modern Technology and Human Behavior

Technological advances have developed over time, but also with specific exchange of goods and knowledge over great distances and across different human civilizations. Transmission of information, produce, and living animals, as well as the possibility of carrying rodents, vermin, and insects, as well as blight or fungal/mold contaminants either attached to packaging/crates or living within crops themselves has always increased the chances of microbial pathogen exposure.[85] In the modern world, while travel times have

Table 2.1 Factors influencing global disease emergence. Population expansion, ecological disturbances, technological advances or deficiencies in developed *vs.* developing nations, human behavior (complacency, migration, societal, vaccination, *etc.*), and microbial genetics are all factors in emerging or resilient diseases.

Population expansion	Ecological disturbances	Technological advances or deficiencies in the modern world	Human behavior	Microbial genetics
Urbanization	Deforestation and destruction of natural microbial reservoirs	Globalization and increased contact across different areas of the world	Complacency/lack of appropriate booster shots, or vaccines prior to travel to foreign countries	Mutation emergence—short generation times and ability for high mutational incidence
Aging population	Displacement of microbes and animal carriers into human areas	Faster air travel (shorter trip duration) for people and/or food and goods	Anti-vaccination movement	High probability and incidence of recombination, and gain of resistance genes from plasmid DNA
Microbial evolution—resistance and evasion	Climate change	Improved construction and redevelopment equipment	Migration and displacement (*e.g.*, refugees)	High capacity for sharing of genetic information between individual microbes as well as populations
Overuse of antimicrobials on animals and food	Natural disasters	Lack of access to proper blood or organ pathogen screening	Unsafe tattoos/piercings	
Increased waste production with less management	Building projects creating artificial reservoirs for vectors carrying illness	Poor economics of new drug creation	Lack of food (meat/produce) safety	
Resources and potable water spread thin				

decreased, helping to avoid loss and spoilage of traded goods, the Black Death, while later understood to live within infected mites, was spread heavily throughout Europe by rodents carrying such insects living in close proximity to humans. More related to modern technologies shortening travel times, many disease exposures can now incubate in people on planes, and depending on the type or stage of infection, may or may not spread to many other passengers.[93] Consider also that everyone is essentially located in a contained fuselage, where the chances of human contact as well as exposure to coughed or sneezed pathogens is increased dramatically. One person, flying in 2000 from Israel to New York, died from bacterial meningitis shortly after landing; thankfully their illness did not spread to others, which was not the case when a sick passenger flying from Russia to New York spread TB to 13 others. In 2007, another similar case related to TB occurred when a passenger from Atlanta, who had even previously been diagnosed, traveled outside of the country. In that particular case, thankfully no one else became sick.

While modern technology has improved the safety of blood transfusions and organ transplantations in more advanced areas of the world, lack thereof is a large root cause of the high rates of transmission of infectious diseases in the developing world.[94] In some countries, blood is not even screened and may be positive for hepatitis, HIV, malaria, syphilis, and trypanosomiasis. This problem, and the burden on global healthcare is compounded by additional stresses brought on by population expansion.

Lastly, both long-standing as well as much more recently emergent human behaviors have had dramatic effects on our inability to properly eradicate certain microbial diseases. Tattoos and body piercings have been carried out for hundreds if not thousands of years, the vast majority of which passed without humankind appreciating and adopting hygiene, sterilization, and pasteurization processes.[95] Instances of unlicensed tattoo artists have resulted in infection with *S. aureus*, with a prominent stretch of 44 cases in 2004–2005. Also related to human behavior, if not opting for abstinence from dangerous activities, being open minded to educational programs or at least teaching children at home about being safe and using protection during recreational activities (*e.g.*, sex, drugs, *etc.*) are important in helping to mitigate microbe transmission.

In addition, also related to human behavior, historical practices of inoculation—utilization of pus or debris taken from other patients' open wounds and used for injection or ingestion by another patient, as well as the more modern vaccination—where live or dead vaccines are used to properly educate the immune system, are under attack. Whether vaccination rates are decreasing due to human complacency, religious dogma, or misinformation-fueled anti-vaccine propaganda, campaigns have led to larger percentages of the population, even in developed nations such as the United States, to go unvaccinated. Such populations, while sparse in most cases, can also be found in enriched areas such as Portland, Oregon, where recent unvaccinated rates among higher densities of the populace have dropped dangerously below the ~85–95% required for herd immunity.[96] As a consequence, huge explosions of older diseases once thought largely eradicated in the US, such as measles–mumps–rubella (discussed earlier), are returning and persisting. Related to travel, those seeking to go abroad should always check CDC advisories or with their local travel clinic to determine whether immunizations not standard in the US may be recommended or even required prior to travel to other countries.[97] While Americans currently still have a choice, other countries around the world such as Italy—including the Holy See and Vatican City—have made certain vaccinations mandatory,[98] as they have recognized that not vaccinating children in a globalized, more densely populated modern world puts others at risk.

Climate change, linked to deforestation and human behavior, has also had dramatic effects on microbial disease.[99] Changes in weather as well as average global temperatures have resulted in explosions in mosquito- and other vector-borne illnesses, such as tick-carried Lyme disease. In addition, higher temperatures will create incubating reservoirs with greater activity, from which disease will be much more primed to jump into humans who look to travel through, develop, or just swim and have fun in the moment. Such diseases include but are not limited to malaria, cholera, hepatitis, *E. coli*, dengue, West Nile encephalitis, and leptospirosis. Tying into one of the aforementioned disease incidents in the United States, hantavirus, it is now thought that the increase in humidity in the western United States at the intersection of New Mexico, Utah, Colorado, and Arizona led to an abundant pine kernel crop, causing deer mice numbers to skyrocket 10-fold. Another more recent pathogen carried by mosquitoes, Zika, has also expanded across more tropical areas, and campaigns of increased bug spray usage and utilization of contraception for up to 6 months following potential exposure are being adopted to avoid defects in new birth cases.[100]

2.8.2 Resistance in Patient Care Facilities

It is estimated that 700 000 patients (likely to be underestimated due to poor reporting) die every year from drug-resistant strains and antimicrobial resistance of common bacteria, HIV, TB, and malaria,[101] and of that number, 200 000 or so die from multi-drug resistant and extremely- (or pan)-drug-resistant TB alone.[102] Contributing to this rise, conventional antimicrobials, including antimicrobial peptides, have been shown to easily and in some cases quickly lose efficacy over time due to microbial resistance[103–105] and can lead to environmental contamination and toxicity to humans due to biocidal diffusion.[106,107] As a consequence of the exposure of these agents to microbes in the wild, we have unintentionally induced selection for pathogens that are multi-drug resistant.[108] Resistant microbes were recently shown to exist in nature prior to modern antibiotic use and preserved in permafrost, with resistance genes numbering >170 in some cases[109] and being identified as existing as far back as 30 000 thousand years,[110] in an isolated cave going back 4 million years,[111] and even 1–2 billion years for the emergence of β-lactamase enzymes that deactivate certain antimicrobials.[112]

Therefore, treatment with such agents does not merely induce spontaneous formation of new resistance, but adds a selective pressure to enrich for those that are already resistant. As a consequence, there is now an increased need for exploring new long-lasting, broad-spectrum and more efficient antimicrobial agents or combination strategies[113] that undermine the antibiotic resistome[114]—the collection of resistance genes at play in one environment—as well as the aforementioned unceasing global emergence of new infectious agents.[115] As an example, one group recently found that application of three variant forms of β-lactam not only resulted in the death of methicillin-resistant *S. aureus* (MRSA), but also prevented the emergence of new resistant strains.[116] However, complicating such efforts, findings such pairs or combination drug groupings is difficult and requires a lot of screening.[117] Furthermore, since the 1970s very few new antimicrobial compounds have been discovered, with much more emphasis on modification of already existing agents. And due to the increased effort and time involved in such processes, as well as the issues associated with globalization and human population increase contributing to higher rates of exposure to microbes and subsequent overuse of antimicrobials, bacterial adaptation has kept pace, invalidating drugs soon after they're generated.[118] One such example in 1961 was the quick emergence of resistance to methicillin,[119] and much more recently resistant *S. aureus* just 6 years after the production of aminoglycosides as antimicrobials.[120] These issues are especially important, as other general disinfectants such as hydrogen peroxide, hypochlorite *etc.*, have short periods of effectiveness as well as associated issues with environmental toxicity.[121]

Given these complications with new resistant disease emergence, it is thought that the worldwide number of all antimicrobial resistance-related deaths could even approach 10 million by 2050.[101] To help address such concerns, the emergence of increasingly high levels of drug-resistant pathogens through overuse of chemical compounds led the US Food and Drug Administration (FDA) in 2016 (taking full effect in 2017) to ban use of 19 different ingredients and antimicrobials in soap and hygienic products,[122] many of which for some time had been making their way into our waterways and breeding higher levels of antimicrobial resistance. This just followed McDonalds' 2015 announcement to phase out all meat sources containing antibiotics; the first such declaration ever made by a major fast-food company. Antibiotics used in agriculture are often the same as those used clinically,[123] and food is one of the main routes of introduction and transmission of resistant microbes into the population from animals.[124] While it depends on the host country, even developed nations have supplied antibiotics to animals *via* injection, water or food,[123] even as a cattle growth promoter.[125] In fact, >80% of all antibiotic use is dispensed to food animals, with heavy densities throughout the Midwest and southeast of the United States, parts of Brazil, and much of Europe, India, Eastern China, and southeast Asia.[126]

Another clear area of relevance for the emergence of antimicrobial resistance is in field of biomedical devices, as they are commonly used in hospitals as part of common and more specialized medical practices. They can be a source of microbial infections *via* improper sterilization leading up to surgery or *via* contact with body fluids, tissues, *etc.* leading to nosocomial (hospital-acquired) infections. As an example, out of 150 million intravascular devices used annually in the US, 200 000–400 000 result in nosocomial bloodstream infections. Prevention of these infections becomes necessary to reduce medical complications, patient suffering, and associated medical costs.[127] New molecules with antimicrobial activity and structural modification of polymers to achieve desirable physicochemical and biological properties are being developed.[128] In addition, materials that can provide antimicrobial activity are being explored for biomedical use[129,130] as well as for inhibiting or killing microbes on their surfaces to reduce hospital-acquired infections.[131] Antimicrobial polymers also have the potential for many different applications in biomedical fields, importantly when polymers are in direct contact with human body. Therefore, selecting appropriate polymers against microbes is essential for such applications,[132] and can address these problems by promoting antimicrobial efficacy and reducing residual toxicity.[133,134]

Other hospital-acquired infections have occurred following lack of proper sterilization or decontamination of biopsy equipment. Microbial contamination and colonization of surfaces has become detrimental to health and society. Emphasizing the difficulties encountered in this regard, vancomycin-resistant enterococci and MRSA can survive for a day upon materials used in healthcare systems, with others surviving for >90 days.[135] In addition, while many medical facilities utilize copper as an oxidative surface for self-decontamination, there is evidence that some forms of bacteria started obtaining resistance to copper almost 2000 years ago when copper-based materials and weapons were in high production by the Romans.[136] Lastly, also contributing to insufficient surface decontamination protocols as well as insulation from host immune response in the body,[137] biofilms (also sometimes referred to as mucus) are microbial hydrogel aggregates that are actively secreted by bacteria and aid in their adhesion to different surfaces.[138] They complicate treatment and can result in serious infections and health issues,[139] such as increased patient morbidity, recurrence in 65–80% of cases,[140] and, as a consequence, much higher medical expenses.[141] Therefore, there is a clear need for effective therapeutics that eliminate biofilm.

2.9 Conclusion

In this chapter an overview of microbial infections has been provided, ultimately emphasizing that due to increasing emergence of infection events as well as drug resistance, better adoption of older or newly developed healthcare practices are required to help mitigate risk. Such practices will be needed at multiple levels of society, as multiple contributing factors are at play and involved in microbial risk, such as human behavior, climate change, drug resistance, increasing average age in the global population corresponding to reductions in immune activity, as well as invasion and destruction of natural ecosystems for new development projects. Even with successful implementation of such policies, identification of new therapeutic strategies is crucial for helping to control as well as treat new emergent microbial diseases worldwide. To this effect, synthesis and/or incorporation of antimicrobials into different delivery platforms, such as biomaterial carriers for injection into the body or for direct integration into hospital infrastructure or implanted biomedical devices, are needed. Such advances will lead not only to better retrospective care, but also to improved pre-emptive preventative care in avoiding emergence of dangerous medical events, especially while patients are present in high-risk situations as they are undergoing care or surgical intervention in medical facilities. Many of these topics and more will be explored in the remaining chapters of this text.

Abbreviations

AIDS	acquired immunodeficiency syndrome
CDC	Centers for Disease Control (and Prevention)
CRISPR	clustered regularly interspaced short panlindromic repeats
DNA	deoxyribonucleic acid
FDA	Food and Drug Administration
FMT	fecal microbiota transplantation
HIV	human immunodeficiency virus
MRSA	methicillin-resistant *Staphylococcus aureus*
NIH	National Institutes of Health
RNA	ribonucleic acid
SARS	severe acute respiratory syndrome
TB	tuberculosis

References

1. M. J. McFall-Ngai, *J. Exp. Biol.*, 2015, **218**, 1968–1973.
2. A. Poli, I. Finore, I. Romano, A. Gioiello, L. Lama and B. Nicolaus, *Microorganisms*, 2017, **5**, 25.
3. D. A. Caron, H. Alexander, A. E. Allen, J. M. Archibald, E. V. Armbrust, C. Bachy, C. J. Bell, A. Bharti, S. T. Dyhrman, S. M. Guida, K. B. Heidelberg, J. Z. Kaye, J. Metzner, S. R. Smith and A. Z. Worden, *Nat. Rev. Microbiol.*, 2017, **15**, 6–20.
4. F. Yarovinsky, *Nat. Rev. Immunol.*, 2014, **14**, 109–121.
5. J. Sinkovics, J. Horvath and A. Horak, *Acta Microbiol. Immunol. Hung.*, 1998, **45**, 349–390.
6. J. R. Kohler, B. Hube, R. Puccia, A. Casadevall and J. R. Perfect, *Microbiol. Spectrum*, 2017, **5**, FUNK-0014-2016.
7. F. J. Carr, D. Chill and N. Maida, *Crit. Rev. Microbiol.*, 2002, **28**, 281–370.
8. H. Fang, J. Kang and D. Zhang, *Microb. Cell Fact.*, 2017, **16**, 15.

9. S. Dashko, N. Zhou, C. Compagno and J. Piskur, *FEMS Yeast Res.*, 2014, **14**, 826–832.
10. M. E. Johnson, *J. Dairy Sci.*, 2017, **100**, 9952–9965.
11. H. Daims, M. W. Taylor and M. Wagner, *Trends Biotechnol.*, 2006, **24**, 483–489.
12. S. Chowdhury, *Environ. Monit. Assess.*, 2012, **184**, 6087–6137.
13. E. L. Miller, *J. Midwifery Womens Health*, 2002, **47**, 426–434.
14. R. Gupta, N. Gupta and P. Rathi, *Appl. Microbiol. Biotechnol.*, 2004, **64**, 763–781.
15. M. Manzoni and M. Rollini, *Appl. Microbiol. Biotechnol.*, 2002, **58**, 555–564.
16. T. Anjum, A. Azam and W. Irum, *Indian J. Pharm. Sci.*, 2012, **74**, 372–374.
17. E. A. Grice and J. A. Segre, *Nat. Rev. Microbiol.*, 2011, **9**, 244–253.
18. J. L. Mark Welch, B. J. Rossetti, C. W. Rieken, F. E. Dewhirst and G. G. Borisy, *Proc. Natl. Acad. Sci. U. S. A.*, 2016, **113**, E791–E800.
19. M. Boost, P. Cho and Z. Wang, *Clin. Exp. Optom.*, 2017, **100**, 459–472.
20. R. Lappan, K. Imbrogno, C. Sikazwe, D. Anderson, D. Mok, H. Coates, S. Vijayasekaran, P. Bumbak, C. C. Blyth, S. E. Jamieson and C. S. Peacock, *BMC Microbiol.*, 2018, **18**, 13.
21. J. M. Fettweis, J. P. Brooks, M. G. Serrano, N. U. Sheth, P. H. Girerd, D. J. Edwards, J. F. Strauss, C. The Vaginal Microbiome, K. K. Jefferson and G. A. Buck, *Microbiology*, 2014, **160**, 2272–2282.
22. W. Cookson, M. J. Cox and M. F. Moffatt, *Nat. Rev. Microbiol.*, 2018, **16**, 111–120.
23. J. R. Marchesi, D. H. Adams, F. Fava, G. D. Hermes, G. M. Hirschfield, G. Hold, M. N. Quraishi, J. Kinross, H. Smidt, K. M. Tuohy, L. V. Thomas, E. G. Zoetendal and A. Hart, *Gut*, 2016, **65**, 330–339.
24. D. M. Proctor and D. A. Relman, *Cell Host Microbe*, 2017, **21**, 421–432.
25. I. Kroes, P. W. Lepp and D. A. Relman, *Proc. Natl. Acad. Sci. U. S. A.*, 1999, **96**, 14547–14552.
26. J. L. Baker and A. Edlund, *Front. Microbiol.*, 2018, **9**, 3323.
27. *Human Microbiome Project (HMP)*, National Institute of Health (NIH), 2019, https://commonfund.nih.gov/hmp.
28. A. Kergunteuil, M. Bakhtiari, L. Formenti, Z. Xiao, E. Defossez and S. Rasmann, *Insects*, 2016, 7, 70.
29. D. M. Anderson, A. D. Cembella and G. M. Hallegraeff, *Annu. Rev. Mar. Sci.*, 2012, **4**, 143–176.
30. J. L. Round, R. M. O'Connell and S. K. Mazmanian, *J. Autoimmun.*, 2010, **34**, J220–J225.
31. R. Schunemann, N. Knaak and L. M. Fiuza, *ISRN Microbiol.*, 2014, **2014**, 135675.
32. S. Zhang, Y. Gan and B. Xu, *Front. Plant Sci.*, 2016, **7**, 1405.
33. N. van Beek and D. C. Davis, *Methods Mol. Biol.*, 2007, **388**, 367–378.
34. J. M. Conly and K. Stein, *Prog. Food. Nutr. Sci.*, 1992, **16**, 307–343.

35. J. L. Round and S. K. Mazmanian, *Nat. Rev. Immunol.*, 2009, **9**, 313–323.
36. S. Thomas, J. Izard, E. Walsh, K. Batich, P. Chongsathidkiet, G. Clarke, D. A. Sela, A. J. Muller, J. M. Mullin, K. Albert, J. P. Gilligan, K. DiGuilio, R. Dilbarova, W. Alexander and G. C. Prendergast, *Cancer Res.*, 2017, **77**, 1783–1812.
37. N. Zmora, G. Zilberman-Schapira, J. Suez, U. Mor, M. Dori-Bachash, S. Bashiardes, E. Kotler, M. Zur, D. Regev-Lehavi, R. B. Brik, S. Federici, Y. Cohen, R. Linevsky, D. Rothschild, A. E. Moor, S. Ben-Moshe, A. Harmelin, S. Itzkovitz, N. Maharshak, O. Shibolet, H. Shapiro, M. Pevsner-Fischer, I. Sharon, Z. Halpern, E. Segal and E. Elinav, *Cell*, 2018, **174**, 1388–1405 e1321.
38. J. Suez, N. Zmora, G. Zilberman-Schapira, U. Mor, M. Dori-Bachash, S. Bashiardes, M. Zur, D. Regev-Lehavi, R. Ben-Zeev Brik, S. Federici, M. Horn, Y. Cohen, A. E. Moor, D. Zeevi, T. Korem, E. Kotler, A. Harmelin, S. Itzkovitz, N. Maharshak, O. Shibolet, M. Pevsner-Fischer, H. Shapiro, I. Sharon, Z. Halpern, E. Segal and E. Elinav, *Cell*, 2018, **174**, 1406–1423 e1416.
39. H. P. Browne, B. A. Neville, S. C. Forster and T. D. Lawley, *Nat. Rev. Microbiol.*, 2017, **15**, 531–543.
40. C. A. Marotz and A. Zarrinpar, *Yale J. Biol. Med.*, 2016, **89**, 383–388.
41. N. Alang and C. R. Kelly, *Open Forum Infect. Dis.*, 2015, **2**, ofv004.
42. N. C. de Clercq, M. N. Frissen, M. Davids, A. K. Groen and M. Nieuwdorp, *Psychother. Psychosom.*, 2019, 1–3.
43. C. Burrello, F. Garavaglia, F. M. Cribiu, G. Ercoli, G. Lopez, J. Troisi, A. Colucci, S. Guglietta, S. Carloni, S. Guglielmetti, V. Taverniti, G. Nizzoli, S. Bosari, F. Caprioli, M. Rescigno and F. Facciotti, *Nat. Commun.*, 2018, **9**, 5184.
44. S. Wang, M. Xu, W. Wang, X. Cao, M. Piao, S. Khan, F. Yan, H. Cao and B. Wang, *PLoS One*, 2016, **11**, e0161174.
45. *Malaria Factsheet*, World Health Organization (WHO), 2019.
46. *Top 10 Causes of Death*, World Health Organization (WHO), 2019.
47. N. I. Nii-Trebi, *BioMed Res. Int.*, 2017, **2017**, 5245021.
48. M. W. Borgdorff and D. van Soolingen, *Clin. Microbiol. Infect.*, 2013, **19**, 889–901.
49. K. Vanathy, S. C. Parija, J. Mandal, A. Hamide and S. Krishnamurthy, *Trop. Parasitol.*, 2017, 7, 72–80.
50. C. B. Jonsson, L. T. Figueiredo and O. Vapalahti, *Clin. Microbiol. Rev.*, 2010, **23**, 412–441.
51. L. Baseler, D. S. Chertow, K. M. Johnson, H. Feldmann and D. M. Morens, *Annu. Rev. Pathol.*, 2017, **12**, 387–418.
52. D. J. Gubler, *Clin. Microbiol. Rev.*, 1998, **11**, 480–496.
53. *E. coli infections from Hamburgers in Western US*, Centers for Disease Control (CDC), 1993.
54. *Multistate Outbreak of Salmonella Typhimurium Infections Linked to Peanut Butter, 2008-2009 (FINAL UPDATE)*, Centers for Disease Control (CDC), 2008.

55. *Outbreak of Escherichia coli O104:H4 Infections Associated with Sprout Consumption — Europe and North America, May–July 2011*, Centers for Disease Control (CDC), 2013.
56. R. K. Plowright, C. R. Parrish, H. McCallum, P. J. Hudson, A. I. Ko, A. L. Graham and J. O. Lloyd-Smith, *Nat. Rev. Microbiol.*, 2017, **15**, 502–510.
57. H. Kanamori, W. A. Rutala, E. E. Sickbert-Bennett and D. J. Weber, *Clin. Infect. Dis.*, 2015, **61**, 433–444.
58. A. Afelt, R. Frutos and C. Devaux, *Front. Microbiol.*, 2018, **9**, 702.
59. R. N. Sehgal, *J. Exp. Biol.*, 2010, **213**, 955–960.
60. A. Oryan and M. Akbari, *Asian Pac. J. Trop. Med.*, 2016, **9**, 925–932.
61. P. Steinmann, J. Keiser, R. Bos, M. Tanner and J. Utzinger, *Lancet Infect. Dis.*, 2006, **6**, 411–425.
62. J. M. Drake, A. N. Hassan and J. C. Beier, *J. Vector. Ecol.*, 2013, **38**, 251–259.
63. K. S. Doran, A. Banerjee, O. Disson and M. Lecuit, *Cold Spring Harbor Perspect. Med.*, 2013, **3**, a010090.
64. P. M. Sharp and B. H. Hahn, *Cold Spring Harbor Perspect. Med.*, 2011, **1**, a006841.
65. Y. Guan, B. J. Zheng, Y. Q. He, X. L. Liu, Z. X. Zhuang, C. L. Cheung, S. W. Luo, P. H. Li, L. J. Zhang, Y. J. Guan, K. M. Butt, K. L. Wong, K. W. Chan, W. Lim, K. F. Shortridge, K. Y. Yuen, J. S. Peiris and L. L. Poon, *Science*, 2003, **302**, 276–278.
66. W. Li, S. K. Wong, F. Li, J. H. Kuhn, I. C. Huang, H. Choe and M. Farzan, *J. Virol.*, 2006, **80**, 4211–4219.
67. S. H. Faeth, K. P. Hadeler and H. R. Thieme, *J. Biol. Dyn.*, 2007, **1**, 45–62.
68. K. B. Patterson and T. Runge, *Am. J. Med. Sci.*, 2002, **323**, 216–222.
69. W. H. Schneider, *Med. Hist.*, 2009, **53**, 193–227.
70. P. C. Juliao, S. Maslanka, J. Dykes, L. Gaul, S. Bagdure, L. Granzow-Kibiger, E. Salehi, D. Zink, R. P. Neligan, C. Barton-Behravesh, C. Luquez, M. Biggerstaff, M. Lynch, C. Olson, I. Williams and E. J. Barzilay, *Clin. Infect. Dis.*, 2013, **56**, 376–382.
71. J. Lee, S. Y. Kim, K. J. Hwang, Y. R. Ju and H. J. Woo, *Osong Public Health Res. Perspect.*, 2013, **4**, 57–66.
72. E. A. Asante, M. Smidak, A. Grimshaw, R. Houghton, A. Tomlinson, A. Jeelani, T. Jakubcova, S. Hamdan, A. Richard-Londt, J. M. Linehan, S. Brandner, M. Alpers, J. Whitfield, S. Mead, J. D. Wadsworth and J. Collinge, *Nature*, 2015, **522**, 478–481.
73. D. Andreu and L. Rivas, *Pept. Sci.*, 1998, **47**, 415–433.
74. V. Teixeira, M. J. Feio and M. Bastos, *Prog. Lipid Res.*, 2012, **51**, 149–177.
75. O. G. Travkova, H. Moehwald and G. Brezesinski, *Adv. Colloid Interface Sci.*, 2017, **247**, 521–532.
76. G. Kapoor, S. Saigal and A. Elongavan, *J. Anaesthesiol. Clin. Pharmacol.*, 2017, **33**, 300–305.
77. M. R. Santos, A. C. Fonseca, P. V. Mendonca, R. Branco, A. C. Serra, P. V. Morais and J. F. J. Coelho, *Materials*, 2016, **9**, 599.

78. A. K. Marr, W. J. Gooderham and R. E. W. Hancock, *Curr. Opin. Pharmacol.*, 2006, **6**, 468–472.
79. D. Alves and M. Olivia Pereira, *Biofouling*, 2014, **30**, 483–499.
80. P. D. Hsu, E. S. Lander and F. Zhang, *Cell*, 2014, **157**, 1262–1278.
81. P. Broz and D. M. Monack, *Nat. Rev. Immunol.*, 2013, **13**, 551–565.
82. S. Gordon, *Cell*, 2002, **111**, 927–930.
83. O. Takeuchi and S. Akira, *Cell*, 2010, **140**, 805–820.
84. D. B. T. Cox, J. S. Gootenberg, O. O. Abudayyeh, B. Franklin, M. J. Kellner, J. Joung and F. Zhang, *Science*, 2017, **358**, 1019–1027.
85. J. F. Lindahl and D. Grace, *Infect. Ecol. Epidemiol.*, 2015, **5**, 30048.
86. *2018 Revision of World Urbanization Prospects*, United Nations (UN), 2018.
87. M. K. Ibrahim, M. Zambruni, C. L. Melby and P. C. Melby, *Clin. Microbiol. Rev.*, 2017, **30**, 919–971.
88. A. C. Singer, H. Shaw, V. Rhodes and A. Hart, *Front. Microbiol.*, 2016, **7**, 1728.
89. T. M. Dall, P. D. Gallo, R. Chakrabarti, T. West, A. P. Semilla and M. V. Storm, *Health Aff.*, 2013, **32**, 2013–2020.
90. *Population ages 65 and above (% of total)*, The World Bank, 2018.
91. C. Castelo-Branco and I. Soveral, *Gynecol. Endocrinol.*, 2014, **30**, 16–22.
92. S. A. Tuddenham, K. R. Page, P. Chaulk, E. B. Lobe and K. G. Ghanem, *Int. J. STD AIDS*, 2017, **28**, 330–344.
93. A. J. Tatem, D. J. Rogers and S. I. Hay, *Adv. Parasitol.*, 2006, **62**, 293–343.
94. R. McNerney, *Diagnostics*, 2015, **5**, 200–209.
95. A. Messahel and B. Musgrove, *J. Infect. Public Health*, 2009, **2**, 7–13.
96. M. J. Alam and M. F. Rahman, *Mymensingh Med J.*, 2016, **25**, 392–395.
97. *Need travel vaccines? Plan ahead*, Centers for Disease Control and Prevention, 2019.
98. *Health Information for Travelers to Italy, including Holy See and Vatican City Traveler View*, Centers for Disease Control and Prevention, 2019.
99. E. E. Short, C. Caminade and B. N. Thomas, *Infect. Dis.*, 2017, **10**, 1178633617732296.
100. N. Arora, Y. Sadovsky, T. S. Dermody and C. B. Coyne, *Cell Host Microbe*, 2017, **21**, 561–567.
101. *Antimicrobial resistance: Tackling a crisis for the health and wealth of nations*, Review on AMR, 2016.
102. *Tuberculosis Factsheet*, World Health Organization (WHO), 2019.
103. D. I. Andersson, D. Hughes and J. Z. Kubicek-Sutherland, *Drug Resist. Updates*, 2016, **26**, 43–57.
104. H. M. Lode, *Clin. Microbiol. Infect.*, 2009, **15**, 212–217.
105. M. Frieri, K. Kumar and A. Boutin, *J. Infect. Public Health*, 2017, **10**, 369–378.
106. A. D. Fuchs and J. C. Tiller, *Angew. Chem., Int. Ed.*, 2006, **45**, 6759–6762.
107. J. M. Thomassin, S. Lenoir, J. Riga, R. Jerome and C. Detrembleur, *Biomacromolecules*, 2007, **8**, 1171–1177.
108. S. B. Levy and B. Marshall, *Nat. Med.*, 2004, **10**, S122–S129.

109. M. W. Van Goethem, R. Pierneef, O. K. I. Bezuidt, Y. Van De Peer, D. A. Cowan and T. P. Makhalanyane, *Microbiome*, 2018, **6**, 40.
110. V. M. D'Costa, C. E. King, L. Kalan, M. Morar, W. W. Sung, C. Schwarz, D. Froese, G. Zazula, F. Calmels, R. Debruyne, G. B. Golding, H. N. Poinar and G. D. Wright, *Nature*, 2011, **477**, 457–461.
111. K. Bhullar, N. Waglechner, A. Pawlowski, K. Koteva, E. D. Banks, M. D. Johnston, H. A. Barton and G. D. Wright, *PLoS One*, 2012, 7, e34953.
112. B. G. Hall, S. J. Salipante and M. Barlow, *J. Mol. Evol.*, 2004, **59**, 133–141.
113. M. Baym, L. K. Stone and R. Kishony, *Science*, 2016, **351**, aad3292.
114. T. S. Crofts, A. J. Gasparrini and G. Dantas, *Nat. Rev. Microbiol.*, 2017, **15**, 422–434.
115. Y. Xue, H. Xiao and Y. Zhang, *Int. J. Mol. Sci.*, 2015, **16**, 3626–3655.
116. P. R. Gonzales, M. W. Pesesky, R. Bouley, A. Ballard, B. A. Biddy, M. A. Suckow, W. R. Wolter, V. A. Schroeder, C. A. Burnham, S. Mobashery, M. Chang and G. Dantas, *Nat. Chem. Biol.*, 2015, **11**, 855–861.
117. L. A. Richardson, *PLoS Biol.*, 2017, **15**, e2003775.
118. H. D. Marston, D. M. Dixon, J. M. Knisely, T. N. Palmore and A. S. Fauci, *JAMA*, 2016, **316**, 1193–1204.
119. M. P. Jevons, *Br. Med. J.*, 1961, **1**, 124–125.
120. T. D. Gootz, *Clin. Microbiol. Rev.*, 1990, **3**, 13–31.
121. F. Siedenbiedel and J. C. Tiller, *Polymers*, 2012, **4**, 46–71.
122. *Safety and Effectiveness of Consumer Antiseptics; Topical Antimicrobial Drug Products for Over-the-Counter Human Use*, Food and Drug Administration (FDA), Federal Register, 2016.
123. S. A. McEwen and P. J. Fedorka-Cray, *Clin. Infect. Dis.*, 2002, **34**(Suppl 3), S93–S106.
124. W. Witte, *Science*, 1998, **279**, 996–997.
125. S. B. Levy, *The Antiobiotic Paradox: How Miracle Drugs Are Destroying the Miracle*, Plenum, New York, 1993.
126. T. P. Van Boeckel, C. Brower, M. Gilbert, B. T. Grenfell, S. A. Levin, T. P. Robinson, A. Teillant and R. Laxminarayan, *Proc. Natl. Acad. Sci. U. S. A.*, 2015, **112**, 5649–5654.
127. A. Jones, J. Pant, E. Lee, M. J. Goudie, A. Gruzd, J. Mansfield, A. Mandal, S. Sharma and H. Handa, *J. Biomed. Mater. Res., Part A*, 2018, **106**, 1535–1542.
128. E. R. Kenawy, S. D. Worley and R. Broughton, *Biomacromolecules*, 2007, **8**, 1359–1384.
129. F. Paladini, M. Pollini, A. Sannino and L. Ambrosio, *Biomacromolecules*, 2015, **16**, 1873–1885.
130. A. Jones, A. Mandal and S. Sharma, *J. Appl. Polym. Sci.*, 2015, **132**, 41931.
131. A. Jain, L. S. Duvvuri, S. Farah, N. Beyth, A. J. Domb and W. Khan, *Adv. Healthcare Mater.*, 2014, **3**, 1969–1985.
132. P. Kaali, KTH Royal Institute of Technology, 2011.
133. M. F. Ilker, K. Nusslein, G. N. Tew and E. B. Coughlin, *J. Am. Chem. Soc.*, 2004, **126**, 15870–15875.

134. C. Dong, Y. Ye, L. Qian, G. Zhao, B. He and H. Xiao, *Cellulose*, 2014, **21**, 1921–1932.
135. A. Espert, F. Vilaplana and S. Karlsson, *Composites, Part A*, 2004, **35**, 1267–1276.
136. B. M. Staehlin, J. G. Gibbons, A. Rokas, T. V. O'Halloran and J. C. Slot, *Genome Biol. Evol.*, 2016, **8**, 811–826.
137. J. G. Leid, C. J. Willson, M. E. Shirtliff, D. J. Hassett, M. R. Parsek and A. K. Jeffers, *J. Immunol.*, 2005, **175**, 7512–7518.
138. C. Werlang, G. Carcarmo-Oyarce and K. Ribbeck, *Nat. Rev. Mater.*, 2019, **4**, 134–145.
139. H. Koo, R. N. Allan, R. P. Howlin, P. Stoodley and L. Hall-Stoodley, *Nat. Rev. Microbiol.*, 2017, **15**, 740–755.
140. N. Venkatesan, G. Perumal and M. Doble, *Future Microbiol.*, 2015, **10**, 1743–1750.
141. N. Beyth, Y. Houri Haddad, A. Domb, W. Khan and R. Hazan, *J. Evidence-Based Complementary Altern. Med.*, 2015, 246012.
142. World Health Organization (WHO), https://www.who.int/countries/prk/en/.

CHAPTER 3

Controlled Release of Antimicrobial Small Molecules

MEITAL ZILBERMAN,*[a,b] ELAD KOREN,[b] HELEN GUEZ[a] AND LIOR MATSLIAH[a]

[a] Department of Biomedical Engineering, Faculty of Engineering, Tel-Aviv University, Tel-Aviv 69978, Israel; [b] Department of Materials Science and Engineering, Faculty of Engineering, Tel-Aviv University, Tel-Aviv 69978, Israel
*Email: meitalz@post.tau.ac.il

3.1 Introduction

The world of drug delivery systems includes six main types of systems: nanoparticles, fibers, dendrimers, liposomes, nanotubes, and films. The main characteristics of each one is presented shortly below.

Although each one of those polymeric platforms has its unique set of characteristics, they all must obtain a suitable drug release profile in order to treat the specific infection on the target site. Drug dosage is a critical parameter in infection treatment. It is necessary to control the therapeutic window in order to prevent toxicity while still allowing full recovery. The ability to control release kinetics is a main tool in biomedical applications.[1]

The physical and chemical properties of the drug have a tremendous influence on the release rate. The solubility of the drug affects its release rate and its permeability through the biological membrane. In research done on the subject, it was found that an increase in drug solubility and a decrease in its molecular weight increases the drug release rate.[2]

Biomaterials Science Series No. 5
Antimicrobial Materials for Biomedical Applications
Edited by Abraham J. Domb, Konda Reddy Kunduru and Shady Farah

Published by the Royal Society of Chemistry, www.rsc.org

The parameters that influence the nano-device drug release profile are polymer type, molecular weight, crystallinity, and glass transition temperature. Low molecular weight accelerates the degradation of the polymer and creates a faster drug release rate, in comparison to polymers with higher molecular weights.[3] A polymer with high crystallinity can absorb less water, as opposed to amorphic polymers. This results in a slower drug release rate.[4]

3.1.1 Nanoparticles

Nanoparticles, sometimes referred to as nanocarriers or nanospheres, are nanometric polymeric colloidal particles made of either natural or synthetic polymers (or a combination of both), and can contain therapeutic agents within the polymer matrix. Nanoparticles have high intracellular uptake, they can penetrate through the submucosal layers, can carry both hydrophobic and hydrophilic bioactive molecules, may improve the availability of hydrophobic medications in the surrounding tissue, and can deliver high drug dosages. Nanoparticles have been extensively investigated for sustained and targeted/localized delivery of different agents. For all these reasons and more, they have become very popular systems to deliver many kinds of antibiotics for bacterial infection treatments.

3.1.2 Nanofibers

Nanofibers are drug carriers with diameters in the nanometer range. They can be produced by different types of polymers. Owing to their special geometry, they can be used for both systemic and local drug delivery. Their surface-to-volume ratio, in comparison to other nanoparticles, enables high drug loadings.[5] During processing, a variety of drugs can be incorporated into nanofibers.

The drug release profile from nanofibers can be controlled by various parameters, such as the polymer concentration and its molecular weight. The atmospheric conditions, such as temperature and humidity, as well as the manufacturing process itself, also affect the drug release rate. By creating the optimal combination of all of the above components, it is possible to develop nanofibers with various properties.

3.1.3 Dendrimers

Dendrimers are a family of synthetic polymers with three-dimensional, highly branched constructions. Dendrimers have three parts: a core with reactive groups, middle layers attached to the core with covalent bonds, and terminal functional groups on the outer side. Their physiochemical and biological properties depend on the terminal groups.[6]

Due to their nanoscale size, their excellent geometry and the ability to create various kinds of dendrimers, the dendrimers have a significant advantage as a drug-releasing material in the biomedical world. Their excellent antibacterial activity is influenced by their surface. By manipulating the types and quantities of the functional groups on the surface, it is possible to incorporate different kinds of antibacterial materials and help treat a wide range of infections.

3.1.4 Liposomes

Liposomes are spherical nanometric particles characterized by diameters of 20 nm to 1 μm and may contain one or more lipid bilayers surrounding aqueous cores. In controlled drug release, liposomes can be used as bioactive molecules for several reasons:[7]

1. High biocompatibility
2. They can be integrated with both hydrophobic and hydrophilic medicines
3. They can carry bioactive molecules in an efficient manner to the target site.

Liposomes are mainly used as nanometric devices for controlled antibiotic injection for the treatment of intracellular infections. The use of liposomes allows the expansion of the therapeutic window and enhances the accessibility to the intracellular infections. Additionally, their unique structure enables drug supply with high concentration over long periods of time in order to treat infections optimally.

The main disadvantages of liposomes are instability of vesicles, low drug capacity compared to other drug-carrying systems, and sensitivity of the lipids to radiation, heat, and other chemical materials, which is a significant downside for the sterilization process.[8]

3.1.5 Nanotubes

Nanotubes are a class of nanoparticles shaped as tubes, which are able to contain different materials. In the biomedical world, nanotubes are widely used because of their excellent ability to serve as controlled drug delivery systems. Most industries use nanotubes made of titanium dioxide (TiO_2), but there are additional kinds of nanotubes, such as carbon nanotubes (CNT).

3.1.6 Films

A thin film is a layer of material ranging from micrometers to hundreds of nanometers in thickness. Thin films are made of thin and flexible layers of polymer. Film-based platforms are excellent for drug release in cases in which there is no accessibility to the target site through tablets or liquid formulations.[9]

Films have many advantages: improved bioavailability, low production cost, ease of manufacture, and they are user-friendly. These factors make films very useful for a wide range of biomedical applications;[10] they can improve drug effectiveness and reduce the needed dosage frequency. Ideal films need to have the ability to be loaded with a sufficient amount of the medicine; they should be non-toxic, biocompatible, and biodegradable.[11] It is possible to use various types of polymers to control a variety of properties, such as the drug release rate, adhesion, mechanical strength, degradation rate, *etc.*[3] As a result, it is possible to provide solutions for a wide range of infections and diseases.

3.2 Nanoparticles

Nanoparticles are nanometric polymeric colloidal particles that may contain therapeutic materials within the polymer matrix.[12] They may be made of natural polymers, such as albumin, collagen, gelatin, chitosan, haemoglobin, and alginate, or synthetic polymers such as poly(amides), poly(amino acids), poly(alkyl-α-cyanoacrylates), poly(esters), and poly(*ortho* esters).[13,14]

3.2.1 Design Characteristics of Nanoparticles

Nanoparticles are applied in many biomedical applications. In the field of drug delivery systems they have significant advantages in comparison to other drug delivery systems. They can improve the solubility of hydrophobic medications, inhibit and control the release rate of the drug in an optimal way, and deliver higher drug concentrations due to the effect of enhanced permeation and retention.[15] Nanoparticles can be prepared *via* emulsion polymerization, emulsification/solvent evaporation, and nanoprecipitation.[8]

Nanoparticles can carry both hydrophobic and hydrophilic bioactive molecules to various parts of the body, from the lymphatic system to the arteries.[8] An efficient platform is considered to be one that can contain a large amount of the drug in order to decrease the number of carriers needed for the treatment.[8] The characteristics of the carrier influence the solubility and encapsulation quality of the drug. It is possible to add the drug during the polymerization process or at the end of the production process of the nanoparticle.[7] It is even possible to encapsulate several active molecules within the nanoparticles, to improve the penetration ability of the antibiotics and to provide better treatment of various infections.

3.2.2 Examples for Specific Nanoparticle-based Systems

3.2.2.1 Systems Releasing Silver Ions

Silver nanoparticles (AgNPs) are characterized by excellent antibacterial activity.[16]

Silver ions have excellent antibacterial qualities, and the incorporation of silver ions into nanoparticles enables the incorporation of these qualities with enhanced penetration to the site in need of treatment. In fact, silver has an excellent ability to kill hundreds of types of disease-causing organisms.[17] The ability of AgNPs to penetrate cell membranes allows the treatment of diseases without enhancing the resistance of bacteria to the treatment, as opposed to treatment methods using conventional antibacterial agents.[18] AgNPs present antibacterials for Gram-negative bacteria as well as for Gram-positive bacteria. Many researchers in the field have tried to develop models describing their antibacterial action, but full understanding of this is yet to be achieved. The release of silver ions from the nanoparticle affects the antibacterial activity.[19] The size of the silver ions influences the number of released ions; release of smaller ions is greater than those of larger sizes.[19]

Each bacterium has a minimal concentration of silver ions at which it will stop growing. This minimal concentration is called minimum inhibitory concentration (MIC). MIC is different to minimal bacterial concentration, which defines the concentration at which the bacteria will be completely eradicated. *Fusobacteriun nucleatum* is required to have 0.003 mg mL^{-1} of AgNPs to reach MIC, whereas for *Streptococcus mutans* and *Actinomyces oris*, the concentration must be 0.04 mg mL^{-1} and 0.5 mg mL^{-1}, respectively.[20,21] AgNPs present antifungal action against 44 different fungal species.[22] AgNPs are more toxic to the human body than are antibiotics. The use of different formulas allows the creation of a variety of attributes, resulting in the ability to give treatment to various types of diseases. Moreover, the use of AgNPs allows the world of medicine to deal with antibiotic-resistant microbes.[23]

3.2.2.2 Systems Releasing Ampicillin

Ampicillin is an antibiotic drug from the amino penicillin family. The drug is used for the treatment of a variety of bacterial infections, such as respiratory tract infections, urinary tract infections, and more.[24,25] Research done on ampicillin-loaded polyisohexylcyanoacrylate nanoparticles found that 90% of the ampicillin was tightly bound to the nanoparticles. Moreover, the size of the nanoparticles increased dramatically with increasing ampicillin payload. It was proved that an increase in the ampicillin concentration in the polymerization process increases the size of the polymeric nanoparticle.[24] Studies on mice have shown that loading ampicillin into nanoparticles can increase the chances of survival, even though some of the infections were not eliminated entirely.[25]

Great improvement in antimicrobial activity can be achieved by incorporating all sorts of antibiotics, such as ciprofloxacin with ampicillin, into nanoparticles. It should be noted that it is not always possible to achieve a complete eradiation of the infection in such cases. This could be caused by several reasons, the most probable relating to some non-dividing intracellular bacteria being less sensitive to antibiotic treatment.[26]

3.2.2.3 Systems Releasing Rifampicin

Great efforts have been made using nanoparticles as nanodevices for delivering anti-tubercular drugs.[26] One study has shown the applicability of rifampicin containing 1,3-β-glucan functionalized poly(lactic-*co*-glycolic acid) (PLGA) nanoparticles for quicker uptake of the drug by macrophages, and as a result, improved targeting of the intracellular *Mycobacterium tuberculosis* effectively.[26] In another study, nanoparticles were developed with a novel hydrophobic chitosan derivative, octanoyl, containing rifampicin for pulmonary delivery. Over a period of 2 months, the nanoparticles were found to be stable. Moreover, the nanoparticles inhibited the drug release for 72 h, and improved patient compliance.[26]

In another study, rifampicin was loaded into chitosan nanoparticles. The *in vitro* experiment demonstrated orally sustained release of the drug with an initial burst effect.[27] In another study, the group tried to make nanoparticles that enclosed sufficient rifampicin to efficiently clear macrophages of infection with *Mycobacterium bovis* bacillus Calmette–Guérin. In the experiment, PLGA copolymers were used as biodegradable vehicles to selectively deliver the drug. The experiment focused on using 25 and 50 $\mu g\,mL^{-1}$ PLGA. Analysis showed that the PLGA nanoparticles encapsulating only 1% rifampicin had little effect on the colony forming unit (CFU) levels over 15 days, relative to the untreated controls or infected cells treated with empty nanoparticles. In contrast, nanoparticles loaded with 31% rifampicin led to a total killing of the bacillus Calmette–Guérin by 15 days post-infection, with the 50 $\mu g\,mL^{-1}$ showing a faster rate of eradication than the 25 $\mu g\,mL^{-1}$ nanoparticles. The results showed that one dose of nanoparticles loaded with sufficient rifampicin concentration given after infection can efficiently clear the bacillus Calmette–Guérin infection within 9–12 days of treatment.[28]

A different example of notable improvement due to antibiotic release from nanoparticles is the combination of rifampicin, isoniazid and pyrazinamide, encapsulated in PLGA. This combination resulted in complete sterilization of pigs' lungs in 10 days, as opposed to 46 days of daily doses of the free drug without the use of nanoparticles.[29]

3.3 Nanofibers

Nanofibers are fibers with nanometric diameters that can be produced from different types of polymers. By doing so, the physical and chemical properties of the nanofiber can be controlled. The diameter is determined by the type of polymer used to make the fiber, as well as by the production technique of the fiber. The biomedical world uses nanofibers in many applications of tissue engineering and in wound healing, thanks to the great resemblance to the morphological structure of the extracellular matrix.[15]

The main characteristics of nanofibers are the surface-to-volume ratio in comparison to other nanoparticles, high porosity, and substantial mechanical strength.[30] The efficient proportion of area-to-volume of the fiber increases the availability of the medicine in an aqueous environment and enhances the drug's effectiveness. By controlling different parameters, such as temperature and moisture, it is possible to determine the surface morphology of the fiber, thus influencing the release rate of the drug.[15]

3.3.1 Methods of Preparation

There are several methods of nanofiber manufacture: bicomponent extrusion, phase separation, template synthesis, drawing, electrospinning, and more. For bicomponent extrusion, two kinds of polymers extrude together to a spinneret hole. The polymers are split by a septum that feeds the two segment components into side-by-side arrangements. The main technique to manufacture bicomponent fibers is called the pipe-to-pipe method.[30]

Phase separation is a process in which the polymers are blended with solvents, followed by gelation and solvent extraction. As a result, the solvent phase leaves the other residual phase. The next stages of production involve freezing and freeze-drying. An example of production by phase separation method is polylactic acid (PLLA) nanofibers.[30] During the template synthesis method, the final geometric shape of the desired material is achieved by using a nano-porous membrane template composed of cylindrical pores. This method is commonly used to produce inorganic nanofibers, such as carbon nanofibers.[30]

Drawing is a process that uses tensile force to stretch the raw material. Similar to what is done for metal, during drawing processes for polymeric fibers, the polymer is pulled and becomes thinner, to the required diameter. The process may be described as dry spinning at a molecular level. During the drawing process, a high degree of deformation is applied to the polymer. Only viscoelastic materials can be drawn by this process and still remain intact after the harsh pulling.[31]

The most common method of fabricating nanofibers is electrospinning.[32] This method was developed in the early 1920s, gaining popularity in the biomedical world in its first years of use, especially as a manufacturing method for drug-carrying systems, biosensors, tissue engineering scaffolds, and more. The advantages of this method include the ability to easily mass-produce fibers of several diameters from a variety of polymers. By using this method, it is possible to manipulate the fiber's geometry for its intended application.

The typical electrospinning device contains at least one syringe pump, a polymer solution with or without drug, a high voltage supply, a source electrode, and a collector electrode. During the electrospinning process, an electrical field is applied to the polymer solution in order to charge the polymer and the needle. Once the electrical rejection power overcomes the surface tension, the polymer is injected from the tip of the needle in a manner that creates microscopically-thin drops with a relatively broad surface. The solvent evaporates rapidly from the drops, thus creating nanofibers charged with the opposite charge of the collector electrode, on which the fibers accumulate. This electrode could be static, moving in a direct line, or in circular motion.[32]

The electrospinning process produces fibers with diameters varying from 2 nm up to several micrometers. During the process, the fiber properties may be influenced by a number of parameters, such as polymer concentration and molecular weight, solution density, polymer conductivity, amplitude of the electrical field, type of electrodes, distance between the needle and the collector electrode, and more. By controlling these parameters along with temperature and moisture, one can determine the diameter of the fiber and its morphology, and adjust the fiber to the desired application.[15]

3.3.2 Antibacterial Activity

Antibacterial materials and antibiotics can easily be incorporated into nanofibers to prevent microbial colonization. The antimicrobial activity of nanofibers is directly influenced by their drug-release rate, and the release rate is higher for bigger surfaces. The drug release rate defines its concentration, which has to be above the specified MIC.[33]

Nanofibers that are used as platforms for controlled drug release must have good biocompatibility with the human body. The degradation rate of the polymer that makes up the nanofibers is one of the factors that influences the drug-release profile.[4]

Many antibacterial materials prevent the synthesis of enzymes that are vital for bacteria. One example is the use of penicillin as an antibiotic, causing cell lysis. In one study, the researchers developed penicillin-integrated antibacterial polyurethane nanofiber networks, which have shown better antibacterial activity and bioavailability compared to topical antibiotic usage.[34] Nanofibers are also used as platforms for carrying metal ions. For biomedical applications, silver ions are mainly used, due to their excellent antibacterial properties, compared to those of other metals. The antibacterial activity of silver ions is generated from their strong bond with the thiolate groups of cellular enzymes and proteins.

In dealing with bacterial resistance to antimicrobial drugs or agents, another method of bacterial extermination has been developed in which bacterial colonization is prevented ahead of time. There are two techniques for this method: killing the bacteria at first contact, and preventing bacterial adhesion to the nanofiber.[4]

3.3.2.1 Antibacterial Contact-killing

Killing bacteria at first contact is based on charging the exposed surface of the nanofiber. At the moment of contact, the membrane cells are destroyed, and as a result are completely exterminated. The surface that comes into contact with the bacteria is the crucial factor when using the contact-killing method.[4] Contact-active surfaces are created by the formation of a covalent bond between them and the biocides—contact-active antibacterial agents that may be attached to substrates. The nearby environment of the surface is not contaminated and therefore does not contribute to the development of drug resistance in the bacteria, as opposed to the damage of bacteria cells in physical ways.[4]

There are studies about various contact-active antibacterial agents, the most popular being quaternary ammonium compounds. Quaternary ammonium compounds are positively charged organic molecules that contain four alkyl groups covalently attached to a central nitrogen atom. A common quaternary ammonium compound is alkyl pyridinium. In a study that examined its contact-active antibacterial efficiency, poly(vinyl-*N*-pyridinium) bromide was bonded covalently to the following polymers: high-density polyethylene, low-density polyethylene, nylon, polypropylene, and polyethylene terephthalate. Results showed that each of these polymers managed to kill Gram-negative and Gram-positive bacteria with 99% success.[35]

The surface of the fiber may be coated with specific materials that allow charging.[4] The most common materials for this technique include cationic chitosan, polyethyleneimine, and ammonium salts. Research on the subject has examined the influence of chitosan coating on the antibacterial activity of poly(ε-caprolactone) nanofibers. Results showed that the coating significantly improves the antibacterial activity of the nanofiber, compared to an uncoated reference nanofiber.[36] Another study of chitosan-coated nanofibers showed that the use of smaller fibers could improve antibacterial properties. This may be due to improved surface-to-volume ratio (of all the fibers combined), and as previously mentioned, the surface has a crucial influence on antibacterial activity.[37]

3.3.2.2 Anti-biofouling Performance

In order to prevent initial adhesion and bacterial growth on the surface of the nanofibers, there are several methods that can be used to decrease friction and lower surface energy. By controlling the topography of the surface, we can influence the adhesion, as smoother surfaces decrease adhesion, compared to rough surfaces. In accordance with the specific application, hydrophobic or hydrophilic materials could be used to decrease bacterial adhesion.[4] An example of this can be found in a study that used electrospun fibers derived from novel polysulfobetaine and polysulfabetaines. Polysulfabetaines are types of zwitterionic polymers that are characterized by strong hydrophilicity and exhibit excellent antifouling properties. When using such materials, the bacterial adhesion could be decreased due to its hydrophobia and its surface hydrate ion nature. In the study, the antifouling ability of the electrospun polysulfabetaine nanofibers was evaluated by researching the adhesion and the bacteria's growth in sea water and natural environments. The polysulfabetaine nanofiber-coated surfaces were stable in sea water, and they prevented bacterial growth without showing any biocidal nature.[38]

3.3.3 Drug-release Kinetics of Antibacterial Nanofibers

Many parameters influence the drug-release kinetics of nanofibers. The release kinetics can be classified into two categories: the release rate and the amount. In order to treat infection efficiently, it is necessary to find the appropriate release kinetics for the site at hand.[1]

Nanofibers have many advantages as drug-release devices. They can enhance dissolution and permeation, and they can control the drug release rate locally or systematically. Furthermore, nanofibers have a high ratio of surface-to-volume, short diffusion distance of the drug, allow high control over the surface morphology, can be used with a wide variety of drugs, and many more advantages. The mechanisms by which drugs may be released from nanofibers include diffusion, desorption, or polymer degradation.[1]

As previously mentioned, the hydrophilic/hydrophobic nature of the drug molecules also significantly influences the drug release rate. Hydrophilic drugs do not dissolve in a non-polar solvent, which must be taken into consideration when designing a nanofiber that must include a hydrophobic medicine. In this case, there would be selective distribution of the drug: the drug would tend to aggregate at the surface of the fiber, which would lead to a burst of quick drug release in the first hours of the application followed by a dramatic decrease in the release rate. In many cases, this is the desired behavior for effective infection treatment. Therefore, depending on the site and the type of infection, accurate examination of all materials in the drug-release system must be undertaken in order to optimize the device.[39]

Aside from the issue of quick initial release as a vital component of infection treatment, the drug release rate following the burst is also a critical consideration in the correct design of nanofibers. At this point, the polymer type and its degradability have a tremendous effect on the release rate of the remaining drug. In certain cases, two types of polymers are used in order to regulate the drug release in the desired profile.[40]

For example, core/sheath nanofibers were produced to provide adjustable biphasic drug release. The nanofiber structures include model, sheath, and core matrix. Ketoprofen was used as the model drug, polyvinylpyrrolidone as the sheath polymer, and ethyl cellulose as the core matrix. The research demonstrated that different parameters such as drug concentrations in the sheath or core fluids can tailor the amount of drug released at different phases.[27]

As previously mentioned, hydrophobic or hydrophilic attributes are a crucial parameter in the correct design of drug-releasing systems. In a research that examined nanofibers as oral drug carriers, gelatin nanofibers were loaded with the hydrophobic drug, piperine. Results showed that controlling the density of the cross-links affected the porosity of the polymer matrix, and therefore controlled the drug release rate over time.[41]

The degree of ionization of the loaded drug affects the relative location in the nanofiber's surface. For higher degrees of ionization, the medicine would be placed closer to the surface, thus creating a more significant burst effect. For example, cefoxitin sodium was found to localize on the surface of

PLGA-based nanofibers. The reason for that was related to the high ionic strength of the drug. The introduction of an amphiphilic PEG-*b*-polylactic acid block copolymer in the polymer matrix inhibited the drug release up to 1 week.[4]

The relevant properties for controlled drug release are physical, chemical, therapeutic, and microbial. The electrospinning process enables the preservation of these properties, since the drug is protected from heat exposure or exposure to radiation, which might damage it. By using this method, it is possible to preserve the properties of the drug.

Although the basic antibacterial delivery system can deliver a wide range of drugs, advanced antibacterial systems are still being developed to prevent initial burst release and maintain a controlled drug release for longer periods of time for the treatment of chronic infections.[4]

Cloxacillin benzathine (CLOX) is a semi-synthetic penicillin and β-lactam antibiotic, which has good activity against most common infections caused by Gram-positive bacteria such as *Staphylococcus aureus*. One study reported a fabrication of electrospun nanofibrous membranes based on a biodegradable polymer for efficient release of CLOX. For that purpose, nanofibrous membranes of Ecovio® loaded with CLOX were successfully produced *via* the electrospinning technique. Ecovio® is a polymer blend composed of poly(lactic acid) and poly(butylene adipate-*co*-terephthalate). Ecovio® nanofiber loaded with CLOX 20% w/w was chosen to perform the drug-release experiments in solutions of pH 7.3 and 5.5. The drug-release profile at pH 7.3 starts with an initial burst release at a relatively rapid rate within the first 30 minutes, compared to the release profile at pH 5.5, which is characterized by a typical biphasic pattern; in the beginning there is an initial burst release similar to pH 7.3, and subsequently a more sustained drug release is seen. In the biphasic pattern, the initial burst release can suppress the bacterial activity, and the sustained release of the remaining drug can prolong the therapeutic time. As a result, an optimized antibacterial effect could be achieved. The conclusions of the research were that the cumulative release of drug from Ecovio® nanofibers containing 20% CLOX was pH-dependent, in which the antibiotic release rate was much faster for pH 7.3 than for pH 5.5. The researchers succeeded in developing antimicrobial nanofibrous membranes, which is promising for drug-delivery carriers in general, and more specifically, gives another method for inhibiting *S. aureus* growth.[42]

3.4 Dendrimers

3.4.1 Characteristic Features

Dendrimers are three-dimensional homogenous nanometric constructions that are classified as macromolecules with distinct separation between their core and periphery.[43] They have a spherical, organized structure, which completely distinguishes them from linear structures—the more common structure in the polymer world.[8] Unlike conventional polymers, dendrimers have become very popular in biomedical applications due to their biocompatibility, water solubility, precise molecular weight, and polyvalency.[44] The characteristics of dendrimers are different from those of regular polymers. Considering their microscopic proportions, they are used for many nano-based medical applications, mostly as drug carriers for various treatments.[44]

The dendrimer contains inner void spaces, which are used as encapsulants of guest molecules. The surface of the dendrimer contains functional groups that can interact with external molecules. The physical and chemical attributes are affected by the functional groups and the branching unit. The number of layers of branching units is called the generation number. G-2 represents two layers of branching units, G-3 means three layers, and so on. A dendrimer's biocompatibility is a critical criterion for biomedical applications. The dendrimer must be non-toxic and non-immunogenic in the case of drug carriers. The cytotoxicity of the dendrimer is influenced by its size—the more layers the dendrimer has, the more cytotoxic it is. The biocompatibility of a dendrimer can be improved by controlling its surface.[45]

As previously mentioned, the attributes dendrimers are widely affected by their surface. The reason for this is the interactions of the dendrimer's surface with the guest molecules. The structure of the dendrimer determines the number of guest molecules that can be incorporated. By correctly matching the terminal groups to the guest molecules of the dendrimer, and by correctly controlling the generation number, the type and amount of surface interactions can be manipulated, which eventually dictates the chemical and physical attributes of the dendrimer.[46]

Correct utilization of the three-dimensional structure of the dendrimers and their terminal groups at the surface allows encapsulation of different drugs within the dendrimers, by creating covalent or electrostatic bonds between the medicine and the terminal functional groups. In the literature, there are reports of two mechanisms for drug delivery that are derived from the breaking of bonds between terminal functional groups of dendrimers and the drug that interacts with them. The first mechanism is the breaking of these covalent bonds by using the appropriate enzymes. The second mechanism for bond breaking is the altering of pH and temperature conditions.[47]

The surface of the dendrimers is the key to their excellent antibacterial activity, compared to traditional drug-carrying devices. It is possible to load the surface of the dendrimer with a large amount of antibacterial materials, thus creating a strong bond to surfaces of organisms. Initially, an electrostatic absorption occurs between the dendrimer's cations and the negatively charged bacteria. As a result, the permeability of the bacteria membrane rises, thus allowing a penetration of more and more dendrimers.[48]

3.4.2 Synthesis of Dendrimers

During the synthesis of the dendrimer, its molecular weight is doubled with every new layer. This has a tremendous influence on the surface of the dendrimer, and therefore on the amount or type of functional group on the dendrimer's surface. The amount has an integral part in determining the chemical and physical attributes of the dendrimer, as well as the biological interaction of the dendrimer with its environment. By controlling the molecular weight and the number of functional groups of the dendrimer, the dendrimer can be designed to fit a variety of needs and purposes.[45]

Dendrimers are synthesized by one of two different methods: the divergent method or the convergent method. In the divergent method, the formation begins from the core; more molecules are formed over the core in steps, and in each step a new layer of molecules is formed over the previous layer. To each new layer is added a branching unit, which increases the molecular weight. In each step, the monomers are doubled, and the end of the new monomer constitutes the new surface of the dendrimer. This method has some drawbacks: the number of reactions increases rapidly, which could cause defects in the structure of the dendrimer, and the ability to separate the final product from the reactors is compromised due to the similar molecular weight.[45]

In the convergent method, the process of creating the dendrimer starts from the periphery inwards, towards the core; every two monomers in the outer layer are linked to a single monomer in an inner layer, and so forth, while decreasing the number of branches. At the end of the process, the core is created by linking two segments together. Using this method, it is possible to avoid defects, since the number of layers is constant. The drawback of this method is that it is very slow for creating dendrimers with large molecular weight. Therefore, this method is preferable when relatively small dendrimers are needed.[44]

3.4.3 Main Types of Antibacterial Dendrimers

The main types of dendrimers include polyamidoamine (PAMAM), peptide, and poly (propyleneimine) (PPI).

3.4.3.1 Polyamidoamine

This kind of dendrimer is characterized by more branches than other types of dendrimers. They have the ability to be used as carriers of antimicrobial agents. In the case of antimicrobial delivery, these dendrimers improve the solubility, the efficiency of the treatment and the cell bioavailability.[49] Research into antibacterial activity of PAMAM with PEG have shown that at relatively low concentrations of 6% PEG, it is possible to reach MICs of *Pseudomonas aeruginosa* without harming the corneal epithelial cells. Additionally, it has been proven that as the generation level decreases, the effectiveness of the antibacterial activity grows.[48]

A study on the subject examined the effect of three surface functional groups on the antibacterial activity of the PAMAM dendrimer against Gram-negative *Escherichia coli*. The functional groups examined were hydroxyl, amino, and carboxyl. The research showed that the G-4 dendrimer with amino terminal groups has the best antibacterial activity. However, it is toxic to epithelial cells, even at concentrations as low as 10 μg mL^{-1}. In contrast, a dendrimer with hydroxyl terminal groups becomes toxic at a concentration 100 times higher, but is characterized with low antibacterial properties in comparison to the other functional groups.[50]

The high toxicity of amino-terminated PAMAM dendrimers harms mammalian cells, which is a crucial factor when choosing the necessary dendrimer for treatment. In research that examined ways to deal with septicemia, it was found that treatment through G-2 dendrimers with amino groups prevents the growth of bacteria without causing toxicity effects, in contrast to dendrimers above G-3. Furthermore, it has been shown that the antibacterial activity decreases when using dendrimers with higher generation numbers.[51]

3.4.3.2 Peptide Dendrimers

These dendrimers are hyper-branched macromolecules that have either a core or a surface that contains peptide bonds. Their molecular weight varies between 2 kDa and 100 kDa.

Peptide dendrimers can be classified into three types, depending on their branching units and surface functional sites. The first type is the most common, and it includes unnatural amino acids or organic groups at the core, and peptides or proteins attached as surface functional groups; the second type is the smallest, and it includes polyamino acids as branching units and as surface functional sites; the third type includes amino acid branching units and surface peptidyl chains.[52]

Chloroquine is employed for the suppression and treatment of malaria. The formulation of chloroquine phosphate (CP) as a therapeutic drug for sustained release can prevent acute chloroquine toxicity. In a study on this subject, peptide dendrimers based on lysine amino acid were synthesized up to the fourth generation, and coated with D-galactose. Galactose-coated *vs.* uncoated, and third- *vs.* fourth-generation poly-L-lysine dendrimers (UPD3-CP, GCPD3-CP, UPD4-CP, and GCPD4-CP) were designed, characterized, and compared as novel drug carriers for delivery of CP. It was found that for the same generation number, the CP release rate from coated dendrimers was observed to be slower, in comparison to uncoated formulations. For different generation numbers, G-4 demonstrated sustained drug release compared to G-3. This may be explained by the difference in the compact and sealed architecture between fourth and third generations. The conclusion of the research is that galactose-coated poly-L-lysine dendrimers can be utilized for controlled delivery of CP more safely than the uncoated formulation, and are therefore suitable for treating malaria.[53]

3.4.3.3 Poly (Propylene Imine)

PPI dendrimers were developed commercially in the 1990s. In fact, this is the oldest type of dendrimer in the industry. This dendrimer has a spherical, symmetrical shape, and is mainly used as a drug-carrying material.[54] A rise in the generation number increases the number of the cationic groups, while also increasing the toxicity level of the dendrimer.[21] PPIs contain poly-alkylamines with primary amine terminal groups and tertiary tris-propyleneamine cores.[45]

A study that examined the differences in the antibacterial activity of G-3 PAMAM-NH2 and of G-3 PPI against *E. coli* showed that, in contrast to PAMAM, the activity of PPI was not sufficient.[55] Another study compared the effect of attaching maltose to G-4 PPI dendrimers on the antibacterial activity against different bacteria. The research examined the selectivity of the maltose-modified dendrimers. The antibacterial activity of G-4 PPI against Gram-positive strains and the yeast *Candida albicans* was better, in comparison to the modified dendrimers.[56]

In another study, the loading of ceftazidime into G-1 PPI dendrimers was carried out to produce an appropriate drug-delivery platform for battling *P. aeruginosa*. The amine groups of dendrimers interacted with the carboxylic groups of ceftazidime. PPI-G1 was dissolved in dry tetrahydrofuran, and ceftazidime was added to the solution. The release profile of ceftazidime from PPI-G1 dendrimer was analyzed using UV-spectrophotometry. Results showed an initial rapid drug release in the first 20 hours, during which 50% of the total loaded ceftazidime was released. Subsequently, a gradually increasing drug release continued over the next ~60 hours. Overall, a gradual drug release was observed, within 3 days, of up to 92% of the loaded ceftazidime.

3.5 Liposomes

Liposomes are small spheres composed of an aqueous core entrapped by one or more phospholipids that form closed, bi-layered systems.[57] They are widely used as an advanced technology for delivering bioactive molecules due to their high biocompatibility and their ability to incorporate hydrophilic and hydrophobic drugs, as well as their ability to deliver bioactive molecules directly to the desired site.[58–61] Hydrophobic drugs are usually encapsulated in the lipid bilayers of liposomes, whereas hydrophilic drugs may be either encapsulated inside the aqueous cores of liposomes or located in the external aqueous phase.[62]

An important application for liposomes in drug delivery is their combination with antibiotic agents for different ailments. Liposomal systems have been widely studied, either to target antibiotics to the surface of bacterial biofilms, or by virtue of their property of being absorbed by cells of the reticuloendothelial system, to target antibiotics towards intracellular bacteria.[63] One example is a study done by Meers *et al.*, in which they investigated biofilm penetration, mechanism of drug release, and *in vivo* antimicrobial activity of a unique nanoscale liposomal formulation of amikacin designed specifically for nebulization and inhaled delivery for cystic fibrosis patients.[64] They found that fluorescent liposomes penetrated readily into biofilms and infected mucus, whereas the largest fluorescent beads did not. Inhaled liposomal antibiotic was released in a slow, controlled manner in normal rat lungs, as opposed to inhaled free amikacin. Another such study engineered a liposome formulation with a lipid composition sensitive to bacterium-secreted phospholipase A2 (PLA2) and adsorbed chitosan-modified gold nanoparticles (AuChi) onto the liposome surface.[65] These complexes showed prohibited fusion activity and negligible drug leakage. When exposed to either purified PLA2 enzyme or PLA2 secreted by *Helicobacter pylori* bacteria in culture, the AuChi-liposomes rapidly released the encapsulated contents, but this rate was decreased by adding a PLA2 inhibitor. When the antibiotic doxycycline was loaded into the liposomes, there was effective inhibition of *H. pylori* growth. This displays the ability of smart "on-demand" antibiotic delivery by liposomes—as more enzymes or bacteria are present at infection site, more drugs may be released for treatment, showing yet another benefit in using liposomes for drug delivery systems. Another study regarding antibiotic release from liposomes aimed to create a localized antibiotic delivery for chronic non-healing wound infection therapy. This team developed low temperature sensitive liposomes (LTSLs) containing an antimicrobial agent (ciprofloxacin) for induced release at mild hyperthermia ($\sim$42 °C).[66] In addition, this team characterized *in vitro* ciprofloxacin release and efficacy against *S. aureus* plankton and biofilms, and determined the feasibility of localized ciprofloxacin delivery in combination with hyperthermia in rat models. They found that at body temperature, $<$5% of ciprofloxacin was released from the LTSL, whereas more than $>$95% was released at 42 °C. Additionally, triggered release at 42 °C from LTSL resulted in significantly higher bacterial killing and induced membrane deformation and changes in biofilm matrix compared to free ciprofloxacin or LTSL at body temperature. Finally, a different study showed that the incorporation of rifabutin in liposomes resulted in a significant enhancement of activity against *M. avium* infection compared to free rifabutin.[67] Moreover, the antitubercular action of rifampin was significantly increased when encapsulated in egg phosphatidylcholine liposomes.

Many antibacterial drugs have difficulty passing through the bacterial cell membrane, especially those with high molecular weights or high-volume structures. Therefore, many studies have been done in order to incorporate these molecules into liposomes.[68] Another study incorporated fusidic acid (FUS), a bacteriostatic steroidal compound with activity limited to Gram-positive bacteria, into fusogenic liposomes.[69] These compounds were then evaluated for their effect on improving cell penetration and antibacterial activity. Results showed that encapsulation of FUS in a liposome can improve antimicrobial efficacy and reduce the effective concentration required, most likely due to increased diffusion through the bacterial cell membrane. Furthermore, the FUS-loaded fusogenic liposomes inhibited growth of both Gram-positive and Gram-negative bacterial strains, as opposed to free FUS, which is active only against Gram-positive strains. Finally, a different team developed a novel liposomal linolenic acid formulation, which was found to have effective bactericidal activity against several antibiotic-resistant strains of *H. pylori*.[70] Results showed that liposomal linolenic acid complexes had the most effective bactericidal effect, killing *H. pylori* within 5 minutes, as opposed to liposomal stearic acid and oleic acid. Additionally, the permeability of the outer membrane of the bacteria increased when treated with liposomal oleic acid and liposomal linolenic acid, with the latter providing significantly higher permeability. Each of the above examples, along with many more ongoing studies, shows the beneficial value of using liposomes for antibacterial drug delivery in various biomedical applications.[70]

In conclusion, it is clear that liposomes are worthy candidates for drug-delivery vehicles, either when injected independently, or when incorporated into scaffolds and other platforms. A wide variety of drugs and other molecules may be encapsulated within liposomes, for more efficient, long-lasting release; examples include anti-cancer agents, nucleic acids, and antibiotic/antibacterial agents. Due to their biocompatibility, biodegradability, ability to incorporate both hydrophobic and hydrophilic molecules, and ability to deliver bioactive molecules to a specific target site, liposome-based drug-delivery systems show promising efforts for a variety of biomedical applications.

3.6 Nanotubes

Nanotubes, a class of nanomaterials, have increasingly gained popularity due to their excellent physical properties and close resemblance to native structures in tissues. Due to their hollow tubular structure, they are great candidates for drug and growth factor delivery.[71] Nanotubes are self-assembling, organic or inorganic sheets of atoms arranged in tubes with single- or multi-walled structures, with large internal volume. The external surface of nanotubes can be functionalized easily.[72] One of the most popular nanotubes is CNT. CNTs are graphene sheets rolled up into tubular form, in which each layer consists of hexagonal networks of carbon atoms. CNTs exhibit high target specificity and drug delivery potential due to their high aspect ratio, flexible surface chemistry, and relatively easy control of their structure and morphology.[72] TiO_2 nanotubes are another popular type;[73] they are widely used—both within and outside biomedical applications—due to their excellent structural, optical, catalytic, and electrochemical properties. Typical TiO_2 nanotube arrays are composed of millions of densely packed nanotubes, which can serve as nano-reservoirs with the ability to incorporate substantial amounts of active agents, such as insoluble drugs, antibiotics, proteins, genes, and drug carriers.[73]

Many studies have been conducted into carbon nanotubes. For example, a nano-hybrid made of silver nanoparticles within third-generation dendritic PAMAM grafted onto multi-walled carbon nanotubes was used as an antimicrobial agent in solution.[74] Antimicrobial properties were analyzed and results showed that dendrimer-modified nano-hybrids had highest antimicrobial activity. Many CNT–organic polymer composites require the use of organic solvents in order to synthesize them; however, this leads to limited polymer functionalization. For this reason, one study generated multi-walled

CNT–chitosan nanoparticle hybrids.[75] Results showed that the hybrids are biocompatible at concentrations up to 100 μg mL^{-1} for 24 hour incubation, and they improve protein immobilization efficiency 0.8 times while decreasing cellular toxicity by ~50% in comparison to carboxylated multi-walled CNTs. With regards to antibiotics, although amphotericin B is a very potent and effective antibiotic for treatment of chronic fungal infections, it's not widely used due to its cytotoxicity within mammals. However, a study was done in which amphotericin B was incorporated into CNTs in order to increase solubility and decrease aggregation.[76] Results showed a decrease in toxicity and increased antifungal activity of the CNT–amphotericin B complex in Jurkat cells. Furthermore, modulation of antibiotic activity was displayed against three types of fungi. A similar study investigated antimicrobial activity for iron oxide nanoparticle-decorated CNTs.[77] The iron oxide nanoparticle CNTs were easily generated by pulsed laser ablation of carbon and iron targets in de-ionized water. Results showed that anti-bacterial activity was higher for the nanoparticles prepared at lower energy, most likely due to their reduction in size, and was higher for Gram-positive bacteria than Gram-negative bacteria. The healing activity of CNTs was enhanced by doping with iron oxide nanoparticles.

In regards to TiO_2 -based technologies, many studies have been completed. For example, one-dimensional TiO_2 nanotubes doped with silver were synthesized in a study, then characterized, and their ability for anti-bacterial activity was analyzed.[78] Results showed promising antibacterial action against both Gram-negative and Gram-positive bacteria by the silver-doped TiO_2 nanotubes. To generate titanium materials with anticancer and antibacterial properties, TiO_2 nanotube arrays loaded with the anticancer agent selenium were synthesized onto titanium substrates and then covered with a chitosan layer.[79] The selenium-deposited and chitosan-coated TiO_2 nanotube substrates showed great potential for promoting proliferation of healthy osteoblasts while inhibiting growth of cancerous ones. Furthermore, results showed that the substrates had long-term antibacterial ability.[79]

Another example of the use of TiO_2 nanotubes for antimicrobial delivery is their incorporation onto implants. Medical-grade titanium alloy is widely used for bone and dental implants, but the material alone has no innate antimicrobial properties that would reduce infection risk following surgery. As AgNPs are known to be antibacterial, this study investigated the growth of AgNPs on TiO_2 nanotubes on Ti-6Al-4V discs.[80] After *S. aureus* bacteria was cultured on the composite coatings for 24 hours, results showed that both the micron- and nano-sized clusters of the AgNPs were found to be anti-bacterial. A similar study aimed to give antibacterial traits to Ta2O5 nanotube-coated Ti6Al4V substrates by incorporation of silver oxide nano-particles.[81] The silver oxide nanoparticle-decorated TiO_2 nanotubes promoted the formation of bone-like apatite layer and significantly reduced the viability of *E. coli* cells. The coating also increased density and dispersion of osteoblast cells in comparison to the bare implant. Another study has evaluated the ability of super-hydrophobic TiO_2 nanotube surfaces to reduce bacterial adhesion regardless of whether the bacteria are Gram-positive or Gram-negative.[82] Although the super-hydrophobic surfaces did not repel bacteria completely, they had negligible bacteria attached after 24 hours, and there was no evidence of biofilm formation, which is a significant factor in antibiotic delivery. Another study regarding titanium implants synthesized a nanostructured film composed of one-dimensional titanate nanowires and used it as a carrier of AgNPs and chitosan in order to improve the surface antibacterial activity and biocompatibility of titanium implants.[83] They found that the AgNPs were successfully carried by titanate nanowires and distributed homogeneously on the entire surface. They also successfully deposited a chitosan nanofilm on the AgNPs. Antibacterial analysis tests displayed better antibacterial activity for the samples with a higher initial concentration of $AgNO_3$ solution while chitosan nanofilms could enhance antibacterial activity of the titanate nanowires. All the above results help illustrate the benefit of using nanotubes in conjunction with implants for antibacterial release as well as for osseointegration.

A variety of other types of nanotubes, other than carbon and TiO_2, are also widely used for antimicrobial delivery. For example, a study deals with sustained drug delivery using drug-loaded halloysite clay nanotubes doped into poly(caprolactone)/gelatin microfibers.[84] The metronidazole-loaded halloysite nanotubes incorporated in the microfibers allowed for extended release of the drugs over 20 days, as opposed to the 4 day release resulting from direct release of the drug from microfibers. The sustained antibiotic release prevented colonization of Fusobacteria, while eukaryotic cells could still adhere to and proliferate on the drug-loaded composite membranes. Another application that uses halloysite nanotubes is the food industry.[85] This study used halloysite nanotubes as nanocontainers for salycilic acid (SA), and investigated the system for its ability to stabilize halloysite suspensions, its release kinetics in water, and its antibacterial activity. Results proved halloysite/SA stability of halloysite suspension in water and controlled release of SA over 50 hours. Furthermore, SA released by halloysite/SA demonstrated antibacterial activity at lower concentrations than free SA.

In conclusion, the use of nanotubes as carriers for a variety of elements has great potential, both in the biomedical world, and for external applications. Due to their excellent physical properties, close similarity to native structures in tissues, and ease of surface functionalization, nanotubes provide an efficient platform for local and systemic drug delivery. All types of nanotubes (*i.e.* carbon, TiO_2, halloysite, *etc.*) may be used for a variety of different applications, whether used as independent vehicles or coated onto implants, and show promising antimicrobial action regardless of the mechanism.

3.7 Films

Drug-eluting films are a novel platform attracting increasing interest from pharmaceutical scientists, as a convenient, cost-effective technology for localized drug delivery. Whether administered orally, through ocular and transdermal paths, or even coating surfaces to prolong the shelf life of food, films loaded with antimicrobial agents have been marked as a versatile, promising system that can provide several crucial advantages over pre-existing drug delivery methods.

Films are a thin, flexible layer (or multi-layers) of polymer that may contain plasticizer.[86] The polymeric matrix must be capable of releasing drugs effectively. The films have shown capabilities to reduce dose frequency, enhance drug efficiency and improve onset of drug action.[87] Other crucial requirements of the films, as any other drug-eluting platform, are:

1. Sufficient drug loading capacity
2. Non-toxic, biocompatible, and biodegradable
3. Preserve the bioactivity of the drug or drugs
4. Have a sufficient shelf life
5. Must be comfortably administered.[88,89]

Films often contain inactive agents such as plasticizers (to prolong shelf life) and fillers (to improve mechanical and physical properties), in order to modify key characteristics of the film. These agents may vary according to the application purpose.[90]

Various types of wounds result in tissue loss. These include burn wounds, wounds caused as a result of trauma, diabetic ulcers, and pressure sores. In the United States, more than 1.25 million people experience burns every year and 6.5 million experience various chronic skin ulcers. In burns, infection is the major complication after the initial period of shock and it is estimated that about 75% of the mortality following burn injuries is related to infection. Wound dressings aim to restore the milieu required for skin regeneration by protecting the wound from environmental threats, including penetration of bacteria, and by maintaining a moist healing environment. A wide variety of wound dressing products targeting various types of wounds and different aspects of the wound healing process are currently available on the market. Most wound dressings are based on films and therefore here we focus on antibiotic-eluting films as wound dressings. Ideally, a dressing should be easy to apply and remove, and its design should meet both physical and mechanical requirements, *i.e.* water absorbance and transmission rate, handleability and strength. Novel concepts of antibiotic-eluting films for wound healing applications are described in the final part of this chapter.

3.7.1 Advantages of Drug-eluting Films

Thin films, which are based on water soluble polymers, are known to dissolve rapidly, thus releasing a maximal dose in a rapid manner.[91] When used as a transdermal patch, thin films are less likely to cause skin irritations, thanks to their similarity to the skin tissue in water vapor permeability.[92] Thin films are easy to administer and are more appealing to less cooperative patients such as pediatric, geriatric, and psychiatric patients.[90] Antimicrobial agents can be incorporated into biodegradable films, which can be used as an active packaging compound, prolonging the shelf life of fresh or processed food. Antimicrobial peptides have been proven as effective tools in the fight against biofilms, which are resilient, hard-to-treat structures of bacteria, with the ability to share resistance to antibiotics and the body's immune system. The creation of antimicrobial peptide-containing films could prevent the establishment of biofilms on medical equipment, such as catheters.[93]

3.7.2 Preparation and Characterization of Antimicrobial Films

The most widely used method of preparing films for medical purposes is solvent casting.[94] Other methods include hot melt extrusion and inkjet printing, which is a relatively new way of manufacturing thin films. Solvent casting is usually preferred due to its low cost of processing and easy preparation process.[95] The process is described in Figure 3.1 and the

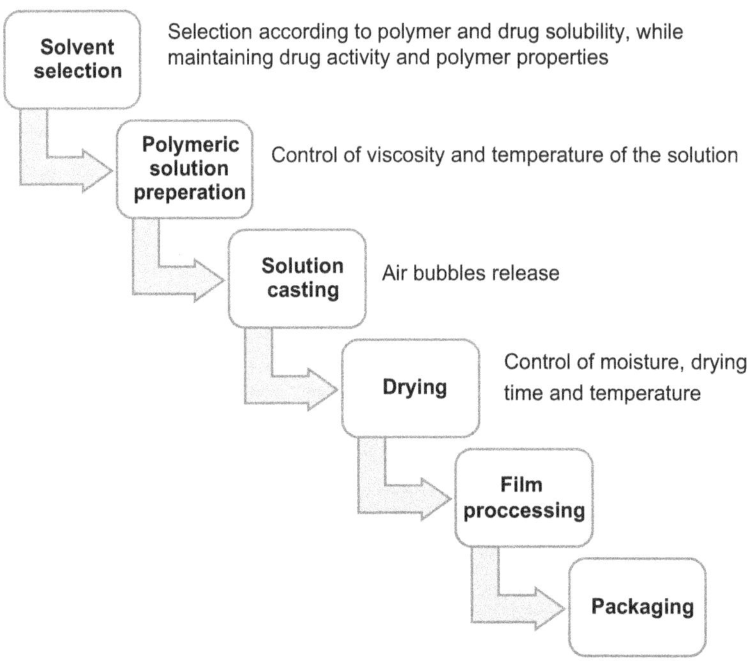

Figure 3.1 The solvent casting process for preparation of polymeric films. Drug molecules can be incorporated in the second stage of solution preparation.

parameters affecting the product are mentioned at each stage. According to this method of preparation, film-forming solutions are prepared by dissolving the desired polymer (and bioactive agent) in a suitable solvent, while stirring the solution (sometimes heating is used) to ensure complete dissolution of the polymer, resulting in a homogenous solution. Optional agents such as cross-linker and plasticizer can then be added under further stirring.[93] The following issues should be addressed: toxicity of the cross-linker and the use of organic solvents in some formulations, which might also be a health hazard, and the tendency to get brittle upon storage. Adding a plasticizer may help eliminate the latter, or reduce its magnitude. When drug molecules are incorporated it is crucial to choose solvents that will not affect the activity of the drug.

While attempting to evaluate the drug release rate, it is important to simulate the conditions of the human environment. There are different tests to evaluate drug release, such as high-performance liquid chromatography and immunoassay. Other widely used tests examine the ability of the antimicrobial films to inhibit the growth of pathogens; one example is the zone of inhibition test. The Clinical and Laboratory Standards Institute instructs that the developing disk diffusion tests consist of inoculum preparation, inoculation of test plates, disk application on inoculated agar plates, incubation at specific conditions, and the measurement of inhibition zone surrounding of the disk.[96] The inoculant should contain a turbidity standard corresponding to the number 0.5 of the McFarland scale, which represents a total of $1\text{–}2 \times 10^8$ CFU mL^{-1}. The zones of film inhibition discs on the plates are then measured by their diameter and by visual observations through photographs at different times of inhibition.[97,98]

Water vapor transmission rate (permeability) of a film (WVTR) is defined as the quantity of water transmitted through a unit area of film in a unit of time. The test shows the resemblance between the film and the human skin, in terms of water permeability, which affects hydration, skin temperature, and blood flow.[99] Films are placed on a permeability cup, pre-filled with water and sealed with a metal ring. The device is then weighed and kept at a desired temperature and low moisture, then weighed again at every time point. A graph is formed using the weight loss *vs.* time and then WVTR is calculated according to the following equation:[100]

$$\mathrm{WVTR} = \frac{\mathrm{slope} \times 24}{\mathrm{area}} \left[\frac{\mathrm{g}}{\mathrm{m}^2 \times \mathrm{day}}\right]$$

It has been claimed that burn wound dressings should ideally possess a WVTR in the range of 2000–2500 $g\,m^{-2}\,d^{-1}$.[101] As antimicrobial films for transdermal usage replace or cover layers of the skin until it is healed, they should imitate the properties of the human skin, including water vapor permeability.

A swelling test is performed to measure the polymer hydration, as it affects drug release.[102] With the increase of hydrogen bonds and the strength between polymer chains, the penetration and movement of water through the films is decreased. The swelling properties are measured by evaluating the percentage of hydration. It is done by weighing the film, immersing it in a medium solution and weighing it again after a predetermined period of time. The difference between the initial and final weight, divided by the initial weight gives the hydration percentage.[103] Adhesion begins shortly following the initial swelling. Bioadhesion increases as hydration continues, until a drop in adhesive strength occurs, due to disentanglement at the polymer–tissue interface. The rate and extent of film hydration and swelling affects the film adhesion and consequently the drug release from the film,[104] including antimicrobial films.

The average tensile strength and strain of normal human skin are 7.7 MPa and 100%, respectively.[105] Adequate mechanical properties are crucial for routine handling and functional stability of biomedical products, especially for tissue engineered porous constructs, as they need to provide an initial biomechanical profile for the cells before new tissue can be formed.[93,100]

The topography of the film can give an insight regarding the homogeneity of the film, as well as the distribution of the drug. Ensuring a uniform distribution of the drugs without drug–polymer interactions is of great importance, as aggregates might lead to a wrinkled surface, and a rough surface might occur as a result of interactions between a crystalline drug and the polymer.[106] Scanning electron microscopy, transmission electron microscopy, and other imaging methods can show the films' texture, thickness, and drug distribution.[107]

3.7.3 Examples of Antibacterial Films

Various studies have been conducted regarding antibacterial release from films, for a variety of applications. In regards to attaining the ideal wound dressing, studies used a variety of formulations, materials, and drugs/other agents. In one study, hydrogels based on the natural polysaccharide, pullulan, were synthesized by chemical cross-linking.[108] The hydrogel was loaded with gentamycin, which resulted in effective suppression of bacterial proliferation, ultimately protecting the wound from bacterial infiltration. Furthermore, the hydrogels had good swelling capacity, high water absorption, excellent mechanical strength, and sustainable antimicrobial release ability. Another study used the solvent casting method in order to prepare chitosan films incorporated with thyme oil, and checked its antimicrobial properties on various Gram-positive and Gram-negative bacteria.[109] Results showed that the minimum thyme oil concentration displaying antimicrobial activity on both types of bacteria was 1.2% (v/v). This concentration also showed the highest antioxidant activity. Another group developed a kind of antibacterial bacterial cellulose dry film and characterized its potential to act as a functional wound dressing for acute trauma treatment.[110] This process consisted of immersing a freeze-dried bacterial cellulose film in a benzalkonium chloride solution, which is a cationic surfactant type of antimicrobial agent, and freeze-drying again. Results showed stable and sustained antimicrobial activity for at least 24 hours, especially against *S. aureus* and *Bacillus subtilis*. Another useful biomaterial for antibiotic-releasing platforms is silk, as it has unique characteristics for antibiotic delivery, such as biocompatibility, tunable biodegradation, stabilizing effects, water-based processing, and diverse material formats. A study involving silk films investigated the release of penicillin and ampicillin from bulk-loaded silk films.[111] Results showed that, when prepared with sufficient drug concentrations, silk films can be used to prevent infection and suppress bacterial growth completely. These coatings can attain high local concentrations that cannot be administered systemically due to negative side effects.

Regarding infections surrounding prosthetic implants, many studies have been trying to find the optimal solution for targeted antibiotic delivery. One such study demonstrated that a vancomycin-containing sol–gel film on titanium alloy rods can successfully treat infections in rat models.[112] Results showed that these films significantly decreased quantities of *S. aureus*. Furthermore, while there was bone degradation present in the control group, coated rods gave minimal signs of infection. A similar study showed that films consisting of tannic acid combined with one of several cationic antibiotics (tobramycin, gentamicin, and polymyxin B) can be generated using the layer-by-layer technique, in order to act as highly efficient, bioresponsive, controlled-release antibacterial coatings on implanted biomedical devices.[113] Coatings as thin as 40 nm strongly inhibited *S. epidermidis* and *E. coli* growth both at surfaces and in surrounding medium, while supporting adhesion and proliferation of murine osteoblast cells. Another research group fabricated polyelectrolyte multilayers incorporating gentamicin using the layer-by-layer technique.[114] The films showed efficient antibacterial activity against *S. aureus* and proved non-toxic toward murine osteoblasts MC3T3. This fabrication technique allows incorporation of a wide variety of different active species within the structures of multilayered films—a huge benefit in drug-delivery vehicles.

In conclusion, the popularity of drug-eluting films is on the rise, due to their cost-effective production, easy application, versatile properties, and various ways of administration. Films can be used as sublingual, ocular, and transdermal drug-delivery platforms, and can coat food packages to prolong their shelf life. The films can be well adjusted for each purpose due to the wide range of polymer formulations available and different production techniques. Further research is required to reach an optimal drug release while maintaining good mechanical properties and keeping high biocompatibility. Antimicrobial films should maintain the balance between potent dosage, biocompatibility and constant drug release throughout the entire term of application. The following section focuses on unique film structure, developed for controlling the release profile of the bioactive agents.

3.8 Novel Concepts in Antibiotic-loaded Bioresorbable Films

Bioresorbable polymeric films can be used for various biomedical applications, such as coatings for orthopedic implants, wound dressings, and treating periodontal infections. Here several novel antibiotic-eluting film-based structures are presented. These include synthetic bioresorbable dense and porous structures, hybrid structures, and soy protein films. In these unique systems structuring effects of the films and the relative quantities of their components enabled excellent control of the drug release profile, as well as the mechanical and physical properties.[115]

3.8.1 Dense Structured Synthetic Films with Controlled Drug Location/Dispersion

Bacterial adhesion to biomaterials and the ability of many microorganisms to form biofilms on foreign bodies are well established as major contributors to the pathogenesis of implant-associated infections. The major problems in treating osteomyelitis include poor distribution of the antimicrobial agent at the site of infection due to limited blood circulation to infected skeletal tissue, and inability to directly address the biofilm pathogen scenario. Controlled antimicrobial release systems inside orthopedic devices thus represent alternatives to conventional systemic treatments.[115] Antibiotic-loaded bioresorbable films that can be "bound" to orthopedic implants (by slightly dissolving their surface before attaching them to the implant surface) were developed and studied.[116] These films are designed to prevent bacterial infections by a gentamicin-controlled release phase for at least 1 month. In addition to providing desired drug-delivery profiles, these films prevent the need for an additional implant for antibiotic release to the implantation site. Such films can also be used as "stand alone" structures and thus serve as wound dressings or films for treatment of periodontal infections.[116]

PLLA and poly(DL-lactic-*co*-glycolic acid) (PDLGA) films containing antibiotic drugs were prepared by solution processing accompanied by a post-preparation isothermal heat treatment. In the process of film preparation, the solvent evaporation rate determines the kinetics of drug and polymer solidification and thus the drug dispersion/location in the film. The resulting drug-eluting systems are therefore termed "structured films". In general, two types of polymer/drug film structures were created and studied for all matrix polymer types:

1. A polymer film with drug particles located on its surface. This structure, which is derived from a dilute solution, was obtained using a slow solvent evaporation rate which enables prior drug nucleation and growth on the polymer solution surface. This skin formation is

accompanied by a later polymer core formation/solidification. This structure was named the "A-type".

2. A polymer film with most of the drug particles distributed within the bulk. This structure, which is derived from a concentrated solution, was obtained using a fast solvent evaporation rate and resulted from drug nucleation and segregation within a dense polymer solution. Solidification of drug and polymer occurred concomitantly. This structure was named the "B-type".

Both types of antibiotic drugs, water-soluble and partially water-soluble, were incorporated in these films and their release kinetics was studied. The effects of polymer type, initial molecular weight, film morphology (drug location/dispersion) and drug loading on the gentamicin release profile were examined. The cumulative gentamicin release profiles from films of various polymer/gentamicin systems are presented in Figure 3.2. All release profiles exhibited a burst release followed by a relatively slow release phase. Such profiles are desirable for applications such as fracture fixation, where a burst release is needed in order to prevent infection and kill the microorganisms found in the implant area before they settle and create a biofilm which antibiotics cannot easily penetrate. A second phase of slow drug release is necessary to prevent microbial infections at the implant site during healing. The burst effect is obtained due to diffusion of drug molecules located on the surface and in polymer layers close to the surface, while the continuing release is obtained due to diffusion of drug molecules from the bulk and is affected by the host polymer's degradation rate. Gentamicin's therapeutic level in serum is 4–8 $\mu g\,mL^{-1}$ and its toxic level is 12 $\mu g\,mL^{-1}$.[117] All studied films released gentamicin at levels higher than the MIC. As expected, polymers with lower molecular weights exhibited higher burst effects and higher release rates, due to a higher quantity of hydroxylic and carboxylic edge groups, which make them more hydrophilic. Furthermore, a lower molecular weight results in a lower glass transition temperature, which facilitates faster drug release from the polymer.

Processing conditions strongly affect the release profile through morphology. Thus, dilute solutions and slow evaporation rates resulted in A-type films with the drug located on the surface. These films exhibited a relatively high burst effect followed by a slow release rate. In contradistinction, concentrated solutions and fast evaporation rates resulted in B-type films, in which most of the drug is located in the polymeric film and some is located on the surface. These films exhibited a relatively low burst effect followed by a lower release rate. Only the PDLGA with relatively low molecular weight films exhibited similar release profiles for A- and B-type films. This behavior is obtained due to the polymer's relatively high degradation rate and gentamicin's extremely hydrophilic nature. It was concluded that the gentamicin release profiles from the various systems is determined by the host polymer, its initial molecular weight and the processing conditions, which

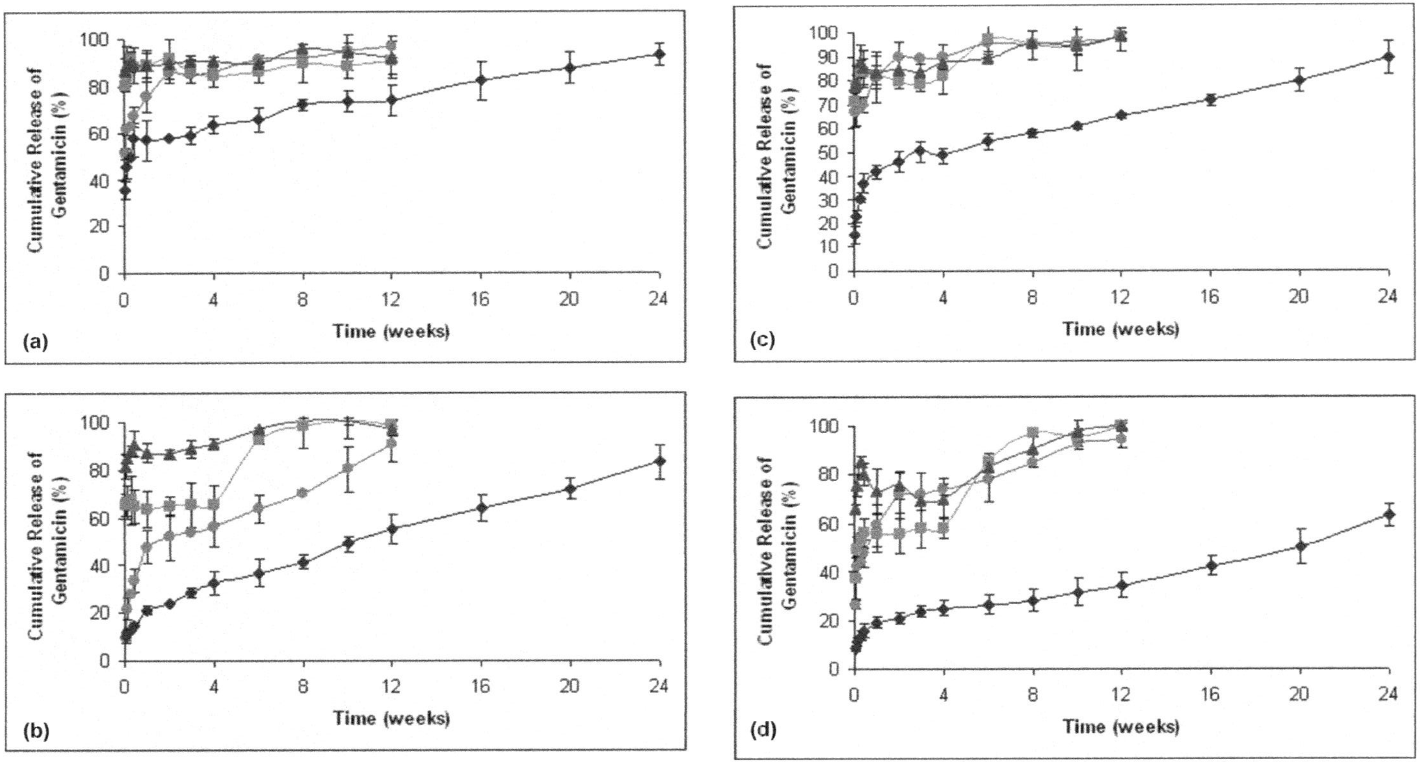

Figure 3.2 *In vitro* cumulative gentamicin release from polymer/gentamicin dense films. (a) A-type films containing 10% w/w gentamicin; (b) B-type films containing 10% w/w gentamicin; (c) A-type films containing 30% w/w gentamicin; (d) B-type films containing 30% w/w gentamicin. ◆ PLLA1; ● PLLA2; ■ PDLGA1; ▲ PDLGA2.
Reproduced from ref. 115 with permission from John Wiley and Sons, Copyright © 2007 Wiley Periodicals, Inc.

affect the drug location/dispersion in the film. Drug loading has a minor effect on the release profile.

Microbiological evaluation of the effect of gentamicin release on bacterial viability was performed. These experiments were carried out in order to monitor the effectiveness of various concentrations of the antibiotic released from the films in terms of the residual bacteria compared with the initial bacterial concentration. Bacteria present in PBS only served as the control. The bacteria were added at the beginning of the films' release, in order to simulate contamination at the time of implantation. The results are presented in Figure 3.3. No bacteria were left after 1–3 days in comparison to the control, where all bacteria survived even after 7 days in the presence of a very high concentration of the starter (1×10^8 CFU mL^{-1}). All films exhibited marked gentamicin release, which was responsible for the dramatic decrease in bacterial survival (10^3 CFU mL^{-1} after 1 day). Moreover, the polymer/gentamicin film preparation did not affect gentamicin's activity as an antimicrobial agent. This study enabled in-depth understanding of gentamicin-loaded structured films and as a result, the production of systems with the desired controlled gentamicin release profiles, *i.e.* with the desired burst effect and continuing release rate (within the therapeutic window) for several weeks. The developed systems can be applied on the surface of any metallic or polymeric fracture fixation device, and can therefore make a significant contribution to the field of orthopedic implants.

For the application of treating periodontal infections metronidazole-loaded 50/50 PDLGA, 75/25 PDLGA and PLLA films were developed and studied, all of them were B-type with most of the drug located within the bulk. These films were designed to be inserted into periodontal pockets and treat infections during the metronidazole controlled release phase, for at least 1 month. The effects of copolymer composition and drug content on the release profile, on cell growth, and on bacterial inhibition were investigated. The metronidazole release profiles from films containing 10% drug showed that although the 50/50 PDLGA film degraded faster than the 75/25 PDLGA and PDLLA films, the rate of drug release from the latter two films loaded with 10% metronidazole was faster than from the former, due to differences in drug location/dispersion within the film. The drug crystals appear to be located mainly on the surface of the PDLLA and 75/25 PDLGA films, whereas in the 50/50 PDLGA films the drug was located in the bulk and also on the surface. These results indicate that the copolymer composition affects the release profile, while the drug content did not show any significant effect on the shape of the release curves. Human gingival cells and rat mesenchymal bone marrow cells have demonstrated normal *in vitro* growth on the drug-eluting films. The released drug also exhibited effectiveness against *Bacteroides fragilis*. The microbiological inhibition kinetics showed that metronidazole cumulative release over 3 days succeeded in totally inhibiting bacterial growth after 2 days.[118]

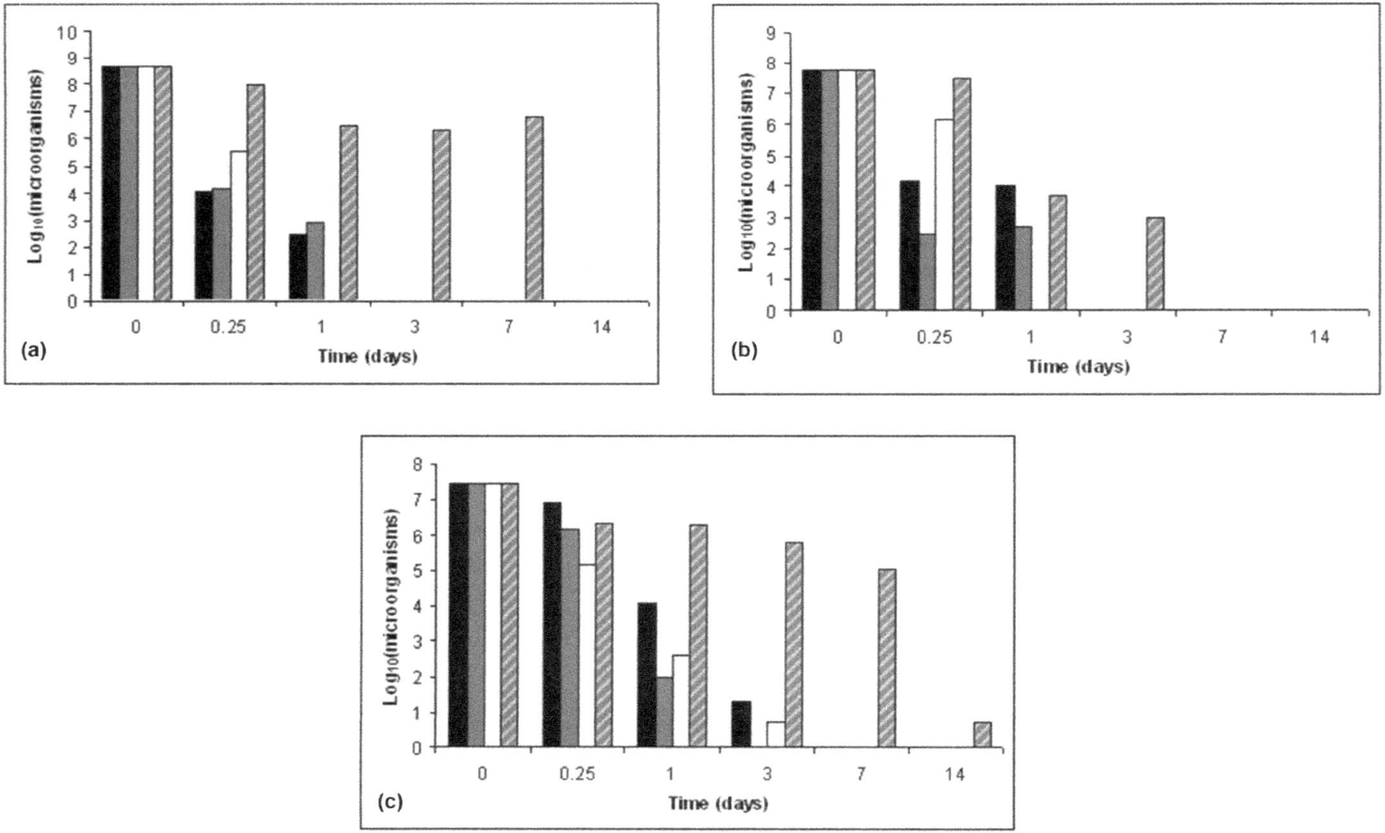

Figure 3.3 Number of colony forming units (CFU) *vs.* time for (a) *Pseudomonas aeruginosa*; (b) *Staphylococcus epidermidis*; (c) *Staphylococcus aureus*. The releasing films are ■ A-type PLLA film containing 30% w/w gentamicin; ■ B-type PLLA film containing 30% w/w gentamicin; □ B-type PDLGA film containing 10% w/w gentamicin; ▨ (control) A-type PLLA film without gentamicin. Reproduced from ref. 115 with permission from John Wiley and Sons, Copyright © 2007 Wiley Periodicals, Inc.

3.8.2 Porous Synthetic Film Structures

A special technique has been developed and studied, termed freeze-drying of inverted emulsions. This technique can be used to create drug-loaded porous films with better control of the drug release profile, compared to dense films. When loaded with antibiotic drugs, these porous film structures can be successfully used as wound dressings and skin substitutes. The effects of process and formulation parameters on the obtained microstructure and on the resulting drug-release profile and other properties that are relevant for the application were thoroughly studied.[119–124]

The inverted emulsions were prepared by homogenization of two immiscible phases: an organic solution containing a known amount of PDLGA in chloroform, and an aqueous phase containing double-distilled water. Homogenization of the two phases is usually performed for the duration of 90 seconds at an average rate of 16 000 RPM using a homogenizer. Both process parameters and formulation parameters are controllable and affect the microstructure and properties. The "process parameters" are the homogenization rate and duration and are termed as kinetic parameters, and the "formulation parameters" are the polymer content of the organic phase, the polymer's molecular weight, the copolymer composition (glycolic acid : lactic acid), the organic : aqueous (O : A) phase ratio, the drug content and incorporation of surfactants. These are termed "thermodynamic parameters", due to their strong effect on the microstructure through the emulsion's stability, as explained in the details and examples below. The formulation parameters were found to be more important than the process parameters in determining the microstructure. After preparing the inverted emulsions they are poured into a dish, followed by immediate freezing in a liquid nitrogen bath so as to form a porous drug-loaded film. The following freeze-drying process enables the preservation of the micro-/nano-structure of the inverted emulsion and results in a solid implant encapsulated with drug molecules.

The freeze-drying of inverted emulsions technique is unique in being able to preserve the liquid structure in solids and was employed in our studies in order to produce highly porous micro and nano-structures. This fabrication process enables the incorporation of both water-soluble and water-insoluble drugs into the film and be released to the surrounding in a controlled manner. Water-soluble bioactive agents are incorporated in the aqueous phase of the inverted emulsion, whereas water-insoluble drugs are incorporated in the organic (polymer) phase. Sensitive bioactive agents, such as proteins, can also be incorporated in the aqueous phase. This prevents their exposure to harsh organic solvents and enables the preservation of their activity.

Many antibiotic drugs are water soluble and therefore are incorporated in the aqueous phase of the inverted emulsion, and as a result are located on the pore walls of the highly porous solid structures created during the freeze-drying process. In such structures relatively high burst release can be obtained when immersed in aqueous surrounding, due to the high water solubility of these drugs. Their location in the pores (rather than in the polymeric domains), as a result of the process of preparation, also tend to increase the burst release. Therefore, it is extremely important to be able to control the release profile of such drugs through structuring of the porous matrix. Such structuring effects are obtained by choosing the appropriate formulation parameters.

The controlled release of antibiotics from wound dressings is challenging, since various related design considerations need to be addressed. Specifically, porosity, which is desired to provide adequate gaseous exchange and absorption of wound exudates may act as a double-edged sword; allowing rapid water penetration which typically leads to a rapid release of the water soluble active agent within several hours to several days. Structural effects on the controlled release of gentamicin and ceftazidime from the porous structures were extensively studied. Samples with a relatively low emulsion O : A phase ratio (up to 8 : 1) typically demonstrated much pore connectivity and their *in vitro* release patterns displayed a burst release of approximately 95%. In contradistinction, porous structures derived from higher O : A phase ratios (for example 12 : 1), displayed reduced pore connectivity and a lower pore fraction, resulting in a significant half-fold decrease in the burst release of antibiotics to approximately 45%. An increase in the polymer's molecular weight from 100 kDa to 240 kDa resulted in a tremendous effect on the microstructure, with porosity of only 16%. Since high viscosity increases the shear forces during the process of emulsification and reduces the tendency of droplets to move, it is expressed in significantly smaller pores and a relatively thick polymeric domain between them. These changes in microstructure reduced the burst release of the encapsulated antibiotics to approximately 30% and enabled a continuous moderate release over a period of 1 month. Finally, an increase in the emulsion's polymer content to 20% w/v also resulted in a dramatic decrease in the burst release. A higher polymer content in the organic phase results in denser polymer walls between pores after freeze-drying and therefore poses better constraint on the release of drugs out of pores. It is important to note that the porous structure of the film, in terms of pore size and porosity, not only affected the drug release profile from the film wound dressing, but also strongly affected its physical properties. This is important because properties such as water uptake and WVTR strongly affect the healing process, and therefore they should be adjusted to the wound type.[121]

3.8.3 Hybrid Synthetic-natural Films for Wound Healing Applications

In an advanced version of these highly porous structured films, hybrid wound dressings were developed and studied. These combined a porous drug-eluting PDLGA top layer with a spongy collagen sublayer in contact with the damaged skin. The porous synthetic PDLGA top layer is designed to control moisture transmission and prevent bacterial penetration as well as to act as a drug reservoir. PDLGA is a mechanically reliable polymer that has been proven to perform well in various implants and long-term drug-delivery systems. Taken together, both materials synergistically produce properties that are not available in the individual constituent materials. The microstructure of the upper PDLGA layer was observed and its effect on the release profiles of gentamicin was elucidated. The effects of the gentamicin release profile on the healing process of burn wounds were also studied.[123,124]

As mentioned earlier, the main challenge in designing a device for the release of low molecular weight hydrophilic bioactive agents (such as the antibiotics used in this study) is to overcome their rapid discharge from the device. A drug-eluting bilayer structure is even more challenging, especially when the drug is incorporated within the top layer and its discharge from the device also depends on the swelling rate of the lower layer. The non-cross-linked collagen sponge with high porosity and swelling rate was used, so that it would not decrease the drug release rate from the upper layer to the wound bed.

The freeze-drying of inverted emulsions technique was used to produce the drug-eluting top layer. Three different emulsion formulations loaded with gentamicin were used. These three formulations yielded different resultant upper layer microstructures. The results show that the highly porous (63%) upper layer (based on the formulation with an O : A phase ratio of 6 : 1) exhibited a relatively high burst release of antibiotics (38%) and 80% release of the encapsulated drug within 4 days. Lower porosity, achieved by employing an emulsion with a higher 12 : 1 O : A phase ratio, reduced the burst release and the antibiotic release rate. A finer microstructure with thicker polymer walls between pores thus enables slower diffusion of the hydrophilic antibiotic molecules to the surrounding area (Figure 3.4).

Finally, it is important to mention that one of the challenges in fabricating a bi-layered structure so that it can fulfill its function is to ensure adhesion between the two distinct layers. Integration between a synthetic and a natural polymer is challenging due to their different structural and chemical properties. Contrary to previously described methodologies for chemically combining natural and synthetic polymers, a study conducted by M. Zilberman *et al.* reported a simple dip-coating technique for physically binding between the natural polymer collagen and the synthetic polymer PDLGA, which enables the penetration of the inverted emulsion into the collagen pores when vacuum is used.[123] This results in an interface between the collagen and PDLGA porous layer in the solid state, which actually behaves like an interphase in which both materials are mechanically mixed and therefore the two layers are well held together. Superior mechanical properties such as tension, as well as physical properties (water uptake, WVTR) were reported.

In vivo results using second-degree burns on a guinea-pig model showed that the hybrid bilayer wound dressings, which combine a drug-loaded porous PDLGA top layer with a spongy collagen sublayer, are very promising. The hybrid dressing, with relatively slow gentamicin release, enabled the highest degree of wound healing, which is at least double that obtained

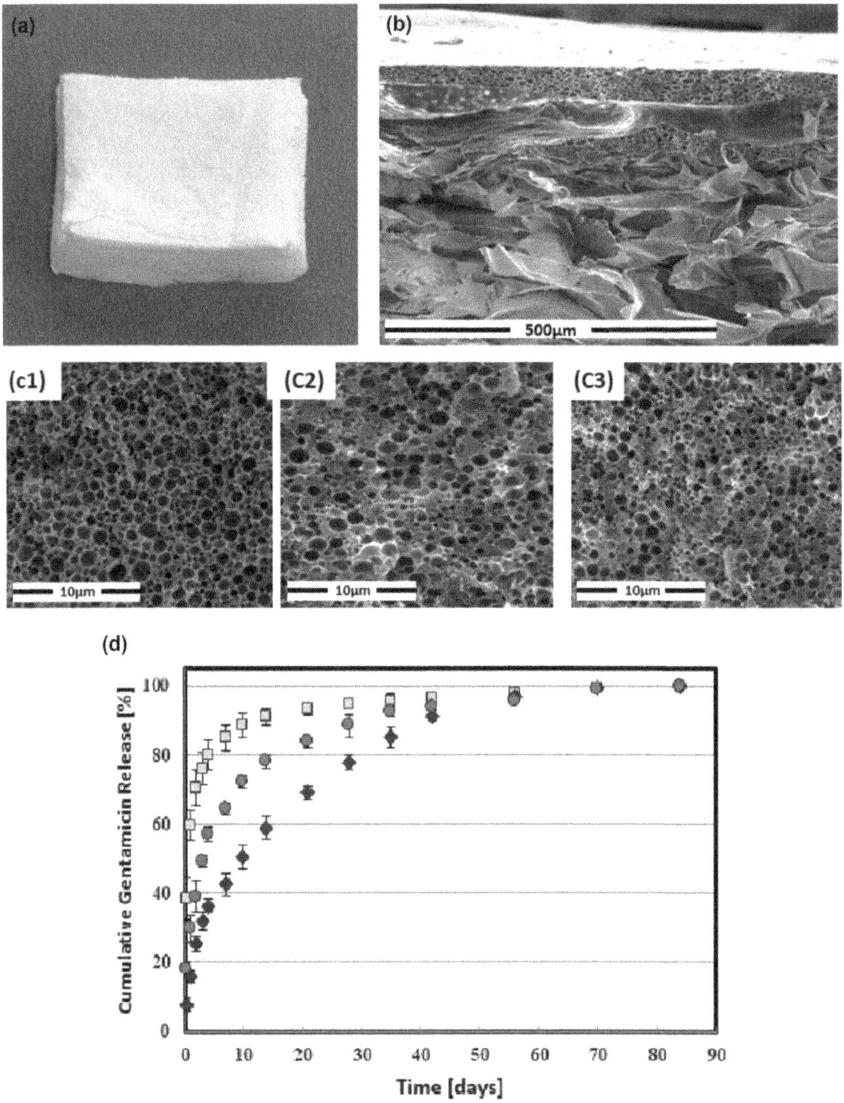

Figure 3.4 (a) Photograph of the hybrid wound dressing, composed of a PDLGA upper layer loaded with gentamicin and a collagen lower layer; (b) cross-sectional cryo-fractured SEM image demonstrating the two layers and their interface; (c) examples for the microstructure of the porous upper layer; (d) cumulative release profiles of gentamicin from the structures presented in (c): C1▭, C2 -●, C3 ◆.
Reproduced from ref. 123 with permission from Elsevier, Copyright 2019.

by the other dressing formats. It resulted in the lowest degree of wound contraction and a relatively low number of inflammatory cells compared to the controls. This dressing was found to be superior to hybrid wound dressings with fast gentamicin release and the neat hybrid dressing without drug release. Since this dressing exhibited promising results and does not require frequent bandage changes, it offers a potentially valuable concept for treating large infected burns.[124]

3.8.4 Soy Protein Films

Interest in developing biodegradable materials from renewable biopolymers for biomedical applications has been on the rise in recent years. Soybean is one of the most important and widely consumed legume crops in the world. It is composed of approximately 38% protein, 30% carbohydrate, 18% oil, and 14% minerals, other materials and moisture.[125] Soybeans are widely consumed in the food industry in the form of inexpensive high-protein soy meals, soy oil, soy milk, and tofu. Soybean protein has been explored mainly in the polymer, food, and agriculture fields. The use of soybean protein as a food source is still increasing, due to its functional and nutritional value, availability, and low price.[126] The carbohydrate and oil components of the soybean can be removed for obtaining soy protein (at least 90%) that can be used for various applications. As an abundant plant protein, soy protein has the advantages of being economically competitive, biodegradable and biocompatible. It presents good water resistance, storage stability, and its properties can be tailored by physical, chemical, or enzymatic treatments such that it can provide diverse requirements for different biomedical applications.

Soy protein has been investigated for wound coverage applications, due to these advantages as a biomaterial. Special emphasis is given to novel antibiotic-eluting soy protein films used as wound dressings.[127,128] These films were prepared from aqueous solutions and their investigation focused on the effects of the formulation and process parameters on the mechanical and physical properties and on the release profile of the antibiotic drug gentamicin from the dressings. It was found that the initial mechanical properties of the wound-dressing structures are affected mainly by the plasticizer and cross-linking agent and can also be controlled by the cross-linking method. Glycerol was chosen as the preferred plasticizer for our soy protein films and glyoxal was chosen as the preferred cross-linking agent. The pH and temperature of the starting solution showed only a slight effect on the tensile properties of the films. Films that were cross-linked by a combination of cross-linking agent and thermal treatment were found to be superior in combining relatively high resistance to tear (tensile strength of 17 MPa) and ductility (maximal strain of 160%). It was also shown that the physical properties of the wound dressing (water absorbance and weight loss profile) can be controlled by the cross-linking process. Films that were cross-linked by thermal treatment or addition of cross-linking agent exhibited lower water uptake and weight loss rate than non-cross-linked films. A combination of both cross-linking methods resulted in higher trends of these results. The WVTR of the films was in the desired range for wound dressings (~2300 $g\,m^{-2}\,day^{-1}$) and was not affected by the cross-linking method.

The gentamicin release profile from these soy protein films exhibited a moderate burst effect followed by a decreasing release rate that lasted for at least 4 weeks. The dominant release mechanism of gentamicin from cross-linked soy protein films is diffusion. Cross-linking by a combination of glyoxal and thermal treatment resulted in a lower burst release and lower total released drug, compared to cross-linking by glyoxal only.[127]

The gentamicin release profiles demonstrated great efficiency towards the relevant bacteria, which are abundant on human skin. The films could effectively inhibit *S. aureus* and *S. albus* infections for at least 2 weeks and *P. aeruginosa* for 3 days. The inhibition profiles obtained during 2 weeks of incubation fit the *in vitro* release profiles. Film extracts released during the first 24 hours from all samples showed no cytotoxic effect on cells, except for a slight inhibition in cell proliferation after 3 days of culture in the presence of some film extracts containing relatively high concentrations of glycerol. In general, the films were found to be biocompatible.[128] In addition, an *in vivo* study was performed to elucidate the function of these gentamicin-loaded soy protein films as wound dressings.[129] The dressing material was tested in contaminated deep second-degree burn wounds in a guinea-pig ($n = 20$) model and the results demonstrated its ability to accelerate epithelialization with 71% epithelial coverage compared to an unloaded format of the soy material (62%) and a significant improved epithelial coverage as compared to the conventional dressing material (55%). Thus, this new platform of antibiotic-eluting wound dressings is advantageous over currently used popular dressing materials that provide controlled release of silver ions, due to its gentamicin release profile, which is safer. Another advantage of this novel concept is that it is based on a biodegradable natural polymer and therefore does not require bandage changes and offers a potentially valuable and economic approach for treating burn-related infections. In conclusion, the versatile antibiotic release profiles from the soy protein films, their effect on bacterial growth and the relatively high biocompatibility achieved, together with the *in vivo* study results and the desired mechanical and physical properties, indicate that these new drug-eluting films are potentially useful as wound dressings.

An additional advancement in the soy protein films for wound-healing applications was achieved when porous blend structures with antibiotic release were studied.[130,131] Such structures are beneficial because the porosity enables tissue growth and penetration inside, thus making them attractive as scaffolds for skin regeneration. The physical properties, microstructure and the clindamycin release profile from selected soy protein–gelatin and soy protein–alginate films were studied, as well as the effect of drug-release profile on bacterial inhibition. The release of clindamycin from the soy protein-based scaffolds was relatively short. It exhibited an initial burst effect followed by a second phase of slow release rates, up to 4 days. Clindamycin demonstrated activity as well as efficiency towards the relevant bacterial strains, which are abundant on human skin. The zone of inhibition results showed that soy protein-based porous structures loaded with clindamycin could effectively inhibit *S. aureus* and *S. albus* infections for at least 4 days. The viable count results demonstrated a significant decrease (by three orders of magnitude) in the bacterial viability after 4 days. The authors therefore concluded that their novel soy protein porous blend films may provide a potentially safer, lower-cost biomaterial system for various tissue regeneration applications.

In conclusion, synthetic dense and porous films, natural (soy) films and hybrid (synthetic-natural) films were presented. In these systems, structuring effects were demonstrated for controlling the drug release profile and the mechanical and physical properties, *i.e.* controlling the microstructure through formulation and process parameters, enabled to achieve desired release profiles of various antibacterial agents. These systems can be used for many applications in the medical field, where combating infections is crucial.

References

1. A. L. Anthony, *Science and Technology of Polymer Nanofibers*, Wiley, 2008.
2. K. T. Savjani, A. K. Gajjar and J. K. Savjani, Drug solubility: importance and enhancement techniques, *ISRN Pharm*, 2012, **2012**, 195727.
3. S. Feng, L. Nie, P. Zou and J. Suo, Effects of drug and polymer molecular weight on drug release from PLGA-mPEG microspheres, *J. Appl. Polym. Sci.*, 2015, **132**(6), 41431.
4. S. Calamak, R. Shahbazi, I. Eroglu, M. Gultekinoglu and K. Ulubayram, An overview of nanofiber-based antibacterial drug design, *Expert Opin. Drug Discovery*, 2017, **12**(4), 391–406.
5. M. Foox and M. Zilberman, Drug delivery from gelatin-based systems, *Expert Opin. Drug Delivery*, 2015, **12**(9), 1547–1563.
6. N. Abed and P. Couvreur, Nanocarriers for antibiotics: A promising solution to treat intracellular bacterial infections, *Int. J. Antimicrob. Agents*, 2014, **43**(6), 485–496.
7. A. Deepak, A. K. Goyal and G. Rath, Nanofiber in transmucosal drug delivery, *J. Drug Delivery Sci. Technol.*, 2018, **43**, 379–387.
8. D. Astruc, E. Boisselier and C. Ornelas, Dendrimers Designed for Functions: From Physical, Photophysical, and Supramolecular Properties to Applications in Sensing, Catalysis, Molecular Electronics, Photonics, and Nanomedicine, *Chem. Rev.*, 2010, **110**(4), 1857–1959.
9. D. Sharma, *et al.*, Fast dissolving oral films technology: A recent trend for an innovative oral drug delivery system, *Int. J. Drug Delivery*, 2015, 7(2), 60–75.
10. D. Rokaya, V. Srimaneepong, J. Sapkota, J. Qin, K. Siraleartmukul and V. Siriwongrungson, Polymeric materials and films in dentistry: An overview, *J. Adv. Res.*, 2018, **14**, 25–34.

11. S. Karki, H. Kim, S.-J. Na, D. Shin, K. Jo and J. Lee, Thin films as an emerging platform for drug delivery, *Asian J. Pharm. Sci.*, 2016, **11**(5), 559–574.
12. M. Goldberg, R. Langer and X. Jia, Nanostructured materials for applications in drug delivery and tissue engineering, *J. Biomater. Sci., Polym. Ed.*, 2007, **18**(3), 241–268.
13. J. K. Kim, H. J. Kim, J.-Y. Chung, J.-H. Lee, S.-B. Young and Y.-H. Kim, Natural and synthetic biomaterials for controlled drug delivery, *Arch. Pharm. Res.*, 2014, **37**(1), 60–68.
14. R. A. Jain, The manufacturing techniques of various drug loaded biodegradable poly(lactide-co-glycolide) (PLGA) devices, *Biomaterials*, 2000, **21**(23), 2475–2490.
15. R. Goyal, L. K. Macri, H. M. Kaplan and J. Kohn, Nanoparticles and nanofibers for topical drug delivery, *J. Controlled Release*, 2016, **240**, 77–92.
16. X.-F. Zhang, Z.-G. Liu, W. Shen and S. Gurunathan, Silver Nanoparticles: Synthesis, Characterization, Properties, Applications, and Therapeutic Approaches, *Int. J. Mol. Sci.*, 2016, **17**(9), 1534.
17. S. H. Jeong, S. Y. Yeo and S. C. Yi, The effect of filler particle size on the antibacterial properties of compounded polymer/silver fibers, *J. Mater. Sci.*, 2005, **40**(20), 5407–5411.
18. K. Chaloupka, Y. Malam and A. M. Seifalian, Nanosilver as a new generation of nanoproduct in biomedical applications, *Trends Biotechnol.*, 2010, **28**(11), 580–588.
19. M. C. Roco, Nanoscale Science and Engineering: Unifying and Transforming Tools, *AIChE J.*, 2004, **50**(5), 890–897.
20. A. Jain, K. Jain, P. Kesharwani and N. K. Jain, Low density lipoproteins mediated nanoplatforms for cancer targeting, *J. Nanoparticle Res.*, 2013, **15**(9), 1888.
21. D. Kaur, K. Jain, N. K. Mehra, P. Kesharwani and N. K. Jain, A review on comparative study of PPI and PAMAM dendrimers, *J. Nanoparticle Res.*, 2016, **18**(6), 146.
22. K.-J. Kim, W. S. Sung, S.-K. Moon, J.-S. Choi, J. G. Kim and D. G. Lee, Antifungal effect of silver nanoparticles on dermatophytes, *J. Microbiol. Biotechnol.*, 2008, **18**(8), 1482–1484.
23. K. Niska, N. Knap, A. Kędzia, M. Jaskiewicz, W. Kamysz and I. Inkielewicz-Stepniak, Capping Agent-Dependent Toxicity and Antimicrobial Activity of Silver Nanoparticles: An In Vitro Study. Concerns about Potential Application in Dental Practice, *Int. J. Med. Sci.*, 2016, **13**(10), 772–782.
24. S. Henry-Michelland, M. J. Alonso, A. Andremont, P. Maincen, J. Sauzières and P. Couvreur, Attachment of antibiotics to nanoparticles: preparation, drug-release and antimicrobial activity in vitro, *Int. J. Pharm.*, 1987, **35**(1–2), 121–127.
25. E. M. Gipps, R. Arshady, J. Kreuter, P. Groscurth and P. P. Speiser, Distribution of polyhexyl cyanoacrylate nanoparticles in nude mice bearing human osteosarcoma, *J. Pharm. Sci.*, 1986, **75**(3), 256–258.

26. L. G. Garcia, *et al.*, Antibiotic activity against small-colony variants of Staphylococcus aureus: review of in vitro, animal and clinical data, *J. Antimicrob. Chemother.*, 2013, **68**(7), 1455–1464.
27. D. G. Yu, X. Wang, X. Y. Li, W. Chian, Y. Li and Y. Z. Liao, Electrospun biphasic drug release polyvinylpyrrolidone/ethyl cellulose core/sheath nanofibers, *Acta Biomater.*, 2013, **9**(3), 5665–5672.
28. R. Kalluru, *et al.*, Poly(lactide-co-glycolide)-rifampicin nanoparticles efficiently clear *Mycobacterium bovis* BCG infection in macrophages and remain membrane-bound in phago-lysosomes, *J. Cell Sci.*, 2013, **126**(Pt 14), 3043–3054.
29. R. Pandey, A. Sharma, A. Zahoor, S. Sharma, G. K. Khuller and B. Prasad, Poly (DL-lactide-co-glycolide) nanoparticle-based inhalable sustained drug delivery system for experimental tuberculosis, *J. Antimicrob. Chemother.*, 2003, **52**(6), 981–986.
30. E.-R. Kenawy, F. I. Abdel-Hay, M. H. El-Newehy and G. E. Wnek, Processing of polymer nanofibers through electrospinning as drug delivery systems, *Mater. Chem. Phys.*, 2009, **113**(1), 296–302.
31. A. A. Almetwally, M. El-Sakhawy, M. H. Elshakankery and M. H. Kasem, Technology of nano-fibers: Production techniques and properties – Critical review, *J. Text. Assoc*, 2017, 5–14.
32. A. M. Moydeen, M. S. Ali Padusha, E. F. Aboelfetoh, S. S. Al-Deyab and M. H. El-Newehy, Fabrication of electrospun poly(vinyl alcohol)/dextran nanofibers via emulsion process as drug delivery system: kinetics and *in vitro* release study, *Int. J. Biol. Macromol.*, 2018, **116**, 1250–1259.
33. M. Gopiraman, *et al.*, Silver coated anionic cellulose nanofiber composites for an efficient antimicrobial activity, *Carbohydr. Polym.*, 2016, **149**, 51–59.
34. Y. Sathish Kumar, A. R. Unnithan, D. Sen, C. S. Kim and Y. S. Lee, Microgravity biosynthesized penicillin loaded electrospun polyurethane–dextran nanofibrous mats for biomedical applications, *Colloids Surf. A Physicochem. Eng. Asp*, 2015, **477**, 77–83.
35. R. Kaur and S. Liu, Antibacterial surface design – Contact kill, *Prog. Surf. Sci.*, 2016, **91**(3), 136–153.
36. F. Croisier, P. Sibret, C. C. Dupont-Gillain, M. J. Genet, C. Detrembleur and C. Jérôme, Chitosan-coated electrospun nanofibers with antibacterial activity, *J. Mater. Chem. B*, 2015, **3**(17), 3508–3517.
37. A.-F. M. Seyam, S. M. Hudson, H. M. Ibrahim, A. I. Waly, Nabil and Y. Abou-Zeid, Healing performance of wound dressing from cyanoethyl chitosan electrospun fibres, *Indian J. Fibre Text. Res.*, 2012, **37**, 205–210.
38. V. A. Vasantha, *et al.*, Antibacterial, electrospun nanofibers of novel poly(sulfobetaine) and poly(sulfabetaine)s, *J. Mater. Chem. B*, 2016, **4**(15), 2731–2738.
39. S.-F. Chou, D. Carson and K. A. Woodrow, Current strategies for sustaining drug release from electrospun nanofibers, *J. Controlled Release*, 2015, **220**, 584–591.

40. D. G. Yu, X. Wang, X. Y. Li, W. Chian, Y. Li and Y. Z. Liao, Electrospun biphasic drug release polyvinylpyrrolidone/ethyl cellulose core/sheath nanofibers, *Acta Biomater.*, 2013, **9**(3), 5665–5672.
41. A. Laha, S. Yadav, S. Majumdar and C. S. Sharma, In-vitro release study of hydrophobic drug using electrospun cross-linked gelatin nanofibers, *Biochem. Eng. J.*, 2016, **105**, 481–488.
42. R. Schneider, L. A. Mercante, R. S. Andre, H. de, M. Brandão, L. H. C. Mattoso and D. S. Correa, Biocompatible electrospun nanofibers containing cloxacillin: Antibacterial activity and effect of pH on the release profile, *React. Funct. Polym.*, 2018, **132**, 26–35.
43. E. Abbasi, *et al.*, Dendrimers: synthesis, applications, and properties, *Nanoscale Res. Lett.*, 2014, **9**(1), 247.
44. D. Pandita, N. Poonia, S. Kumar, V. Lather and K. Madaan, Dendrimers in drug delivery and targeting: Drug-dendrimer interactions and toxicity issues, *J. Pharm. BioAllied Sci.*, 2014, **6**(3), 139.
45. A. P. Sherje, M. Jadhav, B. R. Dravyakar and D. Kadam, Dendrimers: A versatile nanocarrier for drug delivery and targeting, *Int. J. Pharm.*, 2018, **548**(1), 707–720.
46. D. G. Mullen, *et al.*, Design, Synthesis, and Biological Functionality of a Dendrimer-Based Modular Drug Delivery Platform, *Bioconjug. Chem.*, 2011, **22**(4), 679–689.
47. S. Tripathy and M. K. Das, Dendrimers and their Applications as Novel Drug Delivery Carriers, *J. Appl. Pharm. Sci.*, 2013, **3**(09), 142–149.
48. U. H. Sk, Nanosize Dendrimers: Potential Use as Carriers and Antimicrobials, *Antimicrob. Nanoarchitecton.*, 2017, 323–355.
49. S. J. Strydom, W. E. Rose, D. P. Otto, W. Liebenberg and M. M. de Villiers, Poly(amidoamine) dendrimer-mediated synthesis and stabilization of silver sulfonamide nanoparticles with increased antibacterial activity, *Nanomed. Nanotechnol.*, Biol. Med., 2013, **9**(1), 85–93.
50. B. Wang, *et al.*, Inhibition of bacterial growth and intramniotic infection in a guinea pig model of chorioamnionitis using PAMAM dendrimers, *Int. J. Pharm.*, 2010, **395**(1–2), 298–308.
51. X. Xue, *et al.*, Amino-Terminated Generation 2 Poly(amidoamine) Dendrimer as a Potential Broad-Spectrum, Nonresistance-Inducing Antibacterial Agent, *AAPS J.*, 2013, **15**(1), 132–142.
52. K. Sadler and J. P. Tam, Peptide dendrimers: applications and synthesis, *Rev. Mol. Biotechnol.*, 2002, **90**(3–4), 195–229.
53. P. Agrawal, U. Gupta and N. K. Jain, Glycoconjugated peptide dendrimers-based nanoparticulate system for the delivery of chloroquine phosphate, *Biomaterials*, 2007, **28**(22), 3349–3359.
54. D. He, Studies of PPI Dendrimers: Structures, Properties, and Potential Applications, Theses, *Dissertations and Capstones*, Paper 631, 2002.
55. C. Zhisheng Chen, S. L. Cooper and N. C. Beck Tan, Incorporation of dimethyldodecylammonium chloride functionalities onto poly(propylene imine) dendrimers significantly enhances their antibacterial properties, *Chem. Commun.*, 1999, (16), 1585–1586.

56. A. Felczak, *et al.*, Antimicrobial activity of poly(propylene imine) dendrimers, *New J. Chem.*, 2012, **36**(11), 2215.
57. A. D. Bangham, M. W. Hill and N. G. A. Miller, Preparation and use of liposomes as models of biological membranes, *Methods Membr. Biol.*, 1974, 1–68.
58. E. Mastrobattista, G. A. Koning and L. van Bloois, *et al.*, Functional characterization of an endosomedisruptive peptide and its application in cytosolic delivery of immunoliposomeentrapped proteins, *J. Biol. Chem.*, 2002, **277**(30), 27135–27143.
59. A. Schnyder and J. Huwyler, Drug transport to brain with targeted liposomes, *NeuroRx*, 2005, **2**(1), 99–107.
60. M. L. Immordino, F. Dosio and L. Cattel, Stealth liposomes: review of the basic science, rationale, and clinical applications, existing and potential, *Int. J. Nanomed.*, 2006, **1**(3), 297–315.
61. C. Chen, D. Han, C. Cai and X. Tang, An overview of liposome lyophilization and its future potential, *J. Controlled Release*, 2010, **142**(3), 299–311.
62. M. Foox and M. Zilberman, Drug delivery from gelatin-based systems, *Expert Opin. Drug Delivery*, 2015, **12**(9), 1547–1563.
63. S. Thamphiwatana and W. Gao, *et al.*, Phospholipase A2-responsive antibiotic delivery via nanoparticle-stabilized liposomes for the treatment of bacterial infection, *J. Mater. Chem. B*, 2014, **2**(46), 8201–8207.
64. P. Meers, *et al.*, Biofilm penetration, triggered release and in vivo activity of inhaled liposomal amikacin in chronic Pseudomonas aeruginosa lung infections, *J. Antimicrob. Chemother.*, 2008, **61**, 859–868.
65. R. Wardlow, C. Bing, J. VanOsdol, D. Maples and M. Ladouceur-Wodzak, *et al.*, Targeted antibiotic delivery using low temperature-sensitive liposomes and magnetic resonance-guided high-intensity focused ultrasound hyperthermia, *Int. J. Hyperthermia*, 2016, **32**(3), 254–264.
66. H. Daraee, A. Etemadi, M. Kouhi, S. Alimirzalu and A. Akbarzadeh, Application of liposomes in medicine and drug delivery, *Artif. Cells,* Nanomed. Biotechnol., 2016, **44**(1), 381–391.
67. M. T. Aytekin Aydin, H. L. Hosgun, A. Dede and K. Guven, Synthesis, characterization, and antibacterial activity of silver-doped TiO_2 nanotubes, *Spectrochim. Acta, Part A*, 2018, **205**, 503–507.
68. D. Nicolosi, S. Cupri, C. Genovese, G. Tempera, R. Mattina and R. Pignatello, Nanotechnology approaches for antibacterial drug delivery: Preparation and microbiological evaluation of fusogenic liposomes carrying fusidic acid, *Int. J. Antimicrob. Agents*, 2015, **6**, 622–626.
69. S. W. Jung, S. Thamphiwatana, L. Zhang and M. Obonyo, Mechanism of Antibacterial Activity of Liposomal Linolenic Acid against *Helicobacter pylori*, *PLoS One*, 2015, **10**(3), e0116519.
70. M. Maniruzzaman, J. S. Boateng, M. J. Snowden and D. Douroumis, A review of hot-melt extrusion: process technology to pharmaceutical products, *ISRN Pharm*, 2012, **2012**, 436763, 9 pages.

71. P. E. Mikael, J. A. Wallace and S. P. Nukavarapu, Nanotubes for tissue engineering, *Nanomedicine*, 2012, 460–489.
72. H. B. Ruttalaand and Y. T. Ko, Liposomal co-delivery of curcumin and albumin/paclitaxel nanoparticle for enhanced synergistic antitumor efficacy, *Colloids Surf., B*, 2015, **128**, 419–426.
73. X. Chen, K. Cai, J. Fang, M. Lai, Y. Hou, J. Li, Z. Luo, Y. Hu and L. Tang, Fabrication of selenium-deposited and chitosan-coated titania nanotubes with anticancer and antibacterial properties, *Colloids Surf., B*, 2013, **103**, 149–157.
74. J. Xue, Y. Niu, M. Gong, R. Shi, D. Chen, L. Zhang and Y. Lvov, Electrospun Microfiber Membranes Embedded with Drug-Loaded Clay Nanotubes for Sustained Antimicrobial Protection, *ACS Nano*, 2015, **9**(2), 1600–1612.
75. W. Wu, S. Wieckowski, G. Pastorin, M. Benincasa, C. Klumpp and J. P. Briand, *et al.*, Targeted delivery of amphotericin B to cells by using functionalized carbon nanotubes, *Angew. Chem. Int. Ed. Engl.*, 2005, **44**(39), 6358–6362.
76. K. S. Khashan, G. M. Sulaiman and R. Mahdi, Preparation of iron oxide nanoparticles-decorated carbon nanotube using laser ablation in liquid and their antimicrobial activity, *Artif. Cells Nanomed. Biotechnol.*, 2017, **45**(8), 1699–1709.
77. N. Monteiro, M. Martins, A. Martins, N. A. Fonseca, J. Moreira, R. Reis and N. Neves, Antibacterial activity of chitosan nanofiber meshes with liposomes immobilized releasing gentamicin, *Acta Biomater.*, 2015, **18**, 196–205.
78. Z. Xu, M. Li, X. Li, X. Liu, F. Ma, S. Wu, K. W. K. Yeung, Y. Han and P. K. Chu, Antibacterial activity of silver doped titanate nanowires on Ti Implants, *ACS Appl. Mater. Interfaces*, 2016, **8**(26), 16584–16594.
79. W. Yuan, G. Jiang, J. Che, X. Qi, R. Xu, M. W. Chang, Y. Chen, S. Y. Lim, J. Dai and M. B. Chan-Park, Deposition of Silver Nanoparticles on Multiwalled Carbon Nanotubes Grafted with Hyperbranched Poly(amidoamine) and Their Antimicrobial Effects, *J. Phys. Chem. C*, 2008, **112**(48), 18754–18759.
80. U. F. Gunputh, H. Le, R. D. Handy and C. Tredwin, Anodised TiO_2 nanotubes as a scaffold for antibacterial silver nanoparticles on titanium implants, *Mater. Sci. Eng., C*, 2018, **91**, 638–644.
81. M. Sarraf, *et al.*, Silver oxide nanoparticles-decorated tantala nanotubes for enhanced antibacterial activity and osseointegration of Ti6Al4V, *Mater. Des.*, 2018, **154**, 28–40.
82. K. Bartlet, S. Movafaghi, L. P. Dasi, A. K. Kota and K. C. Popat, Antibacterial activity on superhydrophobic titania nanotube arrays, *Colloids Surf., B*, 2018, **166**, 179–186.
83. D. Losic, M. S. Aw, A. Santos, K. Gulati and M. Bariana, Titania nanotube arrays for local drug delivery: recent advances and perspectives, *Expert Opin. Drug Delivery*, 2015, **12**(1), 103–127.

84. C. Li, K. Yang, Y. Zhang, H. Tang, F. Yan, L. Tan, Q. Xie and S. Yao, Highly biocompatible multi-walled carbon nanotube-chitosan nanoparticle hybrids as protein carriers, *Acta Biomater.*, 2011, 7(8), 3070–3077.
85. L. Ghezzi, A. Spepi, M. Agnolucci, C. Cristani, M. Giovannetti, M. R. Tine and C. Duce, Kinetics of release and antibacterial activity of salicylic acid loaded into halloysite nanotubes, *Appl. Clay Sci.*, 2018, **160**, 88–94.
86. A. F. Borges, C. Silva, J. F. J. Coelho and S. Simões, Oral films: Current status and future perspectives: I—Galenical development and quality attributes, *J. Controlled Release*, 2015, **206**, 1–19.
87. D. Achouri, K. Alhanout, P. Piccerelle and V. Andrieu, Recent advances in ocular drug delivery, *Drug Dev. Ind. Pharm.*, 2013, **39**(11), 1599–1617.
88. E. Barbu, L. Verestiuc, T. G. Nevell and J. Tsibouklis, Polymeric materials for ophthalmic drug delivery: trends and perspectives, *J. Mater. Chem.*, 2006, **16**(34), 3439.
89. R. P. Dixit and S. P. Puthli, Oral strip technology: Overview and future potential, *J. Controlled Release*, 2009, **139**(2), 94–107.
90. S. C. Prabhu, S. D. Parsekar, A. Shetty, S. S. Monteiro, M. Azharuddin and A. R. Shabaraya, A Review on Fast Dissolving Sublingual Films for Systemic Drug Delivery, In*t. J. Pharm. Chem. Sci.*, 2014, **3**(2), 501–511.
91. R. Guo, X. Du, R. Zhang, L. Deng, A. Dong and J. Zhang, Bioadhesive film formed from a novel organic–inorganic hybrid gel for transdermal drug delivery system, *Eur. J. Pharm. Biopharm.*, 2011, **79**(3), 574–583.
92. S. Kunte and P. Tandale, Fast dissolving strips: A novel approach for the delivery of verapamil, *J. Pharm. BioAllied Sci.*, 2010, **2**(4), 325–328.
93. M. Rocha, F. A Ferreira, M. M. Souza and C. Prentice, Antimicrobial films- a review, *Microb. Pathog. Strateg. Combat. them Sci. Technol. Educ.*, 2013, 23–31.
94. S. Karki, H. Kim, S. J. Na, D. Shin, K. Jo and J. Lee, Thin films as an emerging platform for drug delivery, *Asian J. Pharm. Sci.*, 2016, **11**(5), 559–574.
95. I. Zurdo Schroeder, P. Franke, U. F. Schaefer and C. M. Lehr, Development and characterization of film forming polymeric solutions for skin drug delivery, *Eur. J. Pharm. Biopharm.*, 2007, **65**(1), 111–121.
96. S. Rivero, L. Giannuzzi, M. A. García and A. Pinotti, Controlled delivery of propionic acid from chitosan films for pastry dough conservation, *J. Food Eng.*, 2013, **116**(2), 524–531.
97. Ó. L. Ramos, *et al.*, Features and performance of edible films, obtained from whey protein isolate formulated with antimicrobial compounds, *Food Res. Int.*, 2012, **45**(1), 351–361.
98. M. Bharkatiya, *et al.*, Development and characterization of transdermal patches of metoprolol tartrate, *Asian J. Pharm. Clin. Res.*, 2010, **3**(2), 130–134.

99. H. Barkay-Olami and M. Zilberman, Novel porous soy protein-based blend structures for biomedical applications: Microstructure, mechanical, and physical properties, *J. Biomed. Mater. Res., Part B*, 2016, **104**(6), 1109–1120.
100. D. Queen, J. D. Gaylor, J. Evans, J. Courtney and W. Reid, The preclinical evaluation of the water vapour transmission rate through burn wound dressings, *Biomaterials*, 1987, **8**(5), 367–371.
101. M. H. Aburahma and A. A. Mahmoud, Biodegradable Ocular Inserts for Sustained Delivery of Brimonidine Tartarate: Preparation and In Vitro/In Vivo Evaluation, *AAPS PharmSciTech*, 2011, **12**(4), 1335–1347.
102. M. C. Bottino, R. A. Arthur, R. A. Waeiss, K. Kamocki, K. S. Gregson and R. L. Gregory, Biodegradable nanofibrous drug delivery systems: effects of metronidazole and ciprofloxacin on periodontopathogens and commensal oral bacteria, *Clin. Oral Investig.*, 2014, **18**(9), 2151–2158.
103. R. Kumria, A. B. Nair, G. Goomber and S. Gupta, Buccal films of prednisolone with enhanced bioavailability, *Drug Deliv.*, 2016, **23**(2), 471–478.
104. H. Yamada and F. G. Evans, *Strength of Biological Materials*, Baltimore, Williams & Wilkins, 1970.
105. K. Hermans, *et al.*, Development and characterization of mucoadhesive chitosan films for ophthalmic delivery of cyclosporine A, *Int. J. Pharm.*, 2014, **472**(1–2), 10–19.
106. P. Baranowski, B. Karolewicz, M. Gajda and J. Pluta, Ophthalmic drug dosage forms: characterisation and research methods, *Sci. World J.*, 2014, **2014**, 861904.
107. H. Li, J. Yang, X. Hu, J. Liang, Y. Fan and X. Zhang, Superabsorbent polysaccharide hydrogels based on pullulan derivative as antibacterial release wound dressing, *J. Biomed. Mater. Res., Part A*, 2011, **98A**, 31–39.
108. D. Altiok, E. Altiok and F. Tihminlioglu, Physical, antibacterial and antioxidant properties of chitosan films incorporated with thyme oil for potential wound healing applications, *J. Mater. Sci.: Mater. Med.*, 2010, **21**(7), 2227–2236.
109. B. Wei, G. Yang and F. Hong, Preparation and evaluation of a kind of bacterial cellulose dry films with antibacterial properties, *Carbohydr. Polym.*, 2011, **84**(1), 533–538.
110. E. M. Pritchard, T. Valentin, B. Panilaitis, F. Omenetto and D. L. Kaplan, Antibiotic-releasing silk biomaterials for infection prevention and treatment, *Adv. Funct. Mater.*, 2013, **23**, 854–861.
111. C. S. Adams, V. Antoci, G. Harrison, P. Patal, T. A. Freeman, I. M. Shapiro, J. Parvizi, N. J. Hickok, S. Radin and P. Ducheyne, Controlled release of vancomycin from thin sol-gel films on implant surfaces successfully controls osteomyelitis, *J. Orthop. Res.*, 2009, **27**, 701–709.
112. I. Zhuk, F. Jariwala, A. B. Attygalle, Y. Wu, M. R. Libera and S. A. Sukhishvili, Self-Defensive Layer-by-Layer Films with Bacteria-Triggered Antibiotic Release, *ACS Nano*, 2014, **8**(8), 7733–7745.

113. H. F. Chuang, R. C. Smith and P. T. Hammond, Polyelectrolyte Multilayers for Tunable Release of Antibiotics, *Biomacromolecules*, 2008, **9**(6), 1660–1668.
114. M. Baro, *et al.*, In vitro-in vivo characterization of gentamicin bone implant, *J Controlled Release*, 2002, **83**(3), 353–364.
115. M. Aviv, I. Berdicevsky and M. Zilberman, Gentamicin-Loaded Bioresorbable Films for Prevention of Bacterial Infections Associated with Orthopedic Implants, *J. Biomed. Mater. Res., Part A*, 2007, **83**(1), 10–19.
116. http://www.healthdigest.org/drugs/gentamicinsulfate.html.
117. Y. Levy and M. Zilberman, Novel Bioresorbable Fiber Structures Loaded with Proteins for Tissue Regeneration Applications: Microstructure and Protein Release, *J. Biomed. Mater. Res., Part A*, 2006, **79A**, 779–787.
118. O. Grinberg, I. Binderman, H. Bahar and M. Zilberman, Highly Porous Bioresorbable Scaffolds with Controlled Release of Bioactive Agents for Tissue Regeneration Applications, *Acta Biomater.*, 2010, **6**, 1278–1287.
119. A. Kraitzer, L. Ofek, R. Schreiber and M. Zilberman, Long-Term In-Vitro Study of Paclitaxel Eluting Bioresorbable Core/Shell Fiber Structures, *J. Controlled Release*, 2008, **126**, 139–148.
120. J. J. Elsner, I. Berdicevsky and M. Zilberman, *In-vitro* Microbial Inhibition and Cellular Response to Novel Biodegradable Composite Wound Dressings with Controlled Release of Antibiotics, *Acta Biomater.*, 2011, **7**(1), 325–336.
121. J. J. Elsner, A. Shefy-Peleg and M. Zilberman, Novel Biodegradable Composite Wound Dressings with Controlled Release of Antibiotics: Microstructure, Mechanical and Physical Properties, *J. Biomed. Mater. Res., Part B*, 2010, **93**(2), 425–435.
122. M. Shemesh and M. Zilberman, Structure-Property Effects of Novel Bioresorbbale Hybrid Structures with Controlled Release of Bioactive Agents for Wound healing Applications, *Acta Biomater.*, 2014, **10**, 1380–1391.
123. M. Zilberman, D. Egozi, M. Shemesh, E. Mazor, M. Baranes, A. Keren and Y. Ullmann, Novel Hybrid Dressings with Controlled Release of Antibiotics: Results in a Guinea Pig Burn Model, *Acta Biomater.*, 2015, **22**, 155–163.
124. M. Santin and L. Ambrosio, Soybean-based biomaterials: preparation, properties and tissue regeneration potential, *Expert Rev. Med. Devices*, 2008, **5**(3), 349–358.
125. S. K. Park, Z. Y. Ju, A. Gennadios, Formation and properties of soy protein films and coatings, *Protein-Based Films & Coatings*, CRC Press, New York, 2002, pp. 123–137.
126. Z. Peles and M. Zilberman, Novel Soy Protein Wound Dressings with Controlled Antibiotic Release: Mechanical and Physical Properties, *Acta Biomater.*, 2011, **8**, 209–217.
127. Z. Peles, I. Binderman, I. Berdicevsky and M. Zilberman, Soy Protein Films for Wound Healing Applications: Antibiotic Release, Bacterial

Inhibition and Cellular Response, *J. Tissue Eng. Regener. Med.*, 2013, 7(5), 401–412.
128. D. Egozi, M. Baranes-Zeevi, Y. Ullmann, A. Keren, E. Matanes and M. Zilberman, Biodegradable Soy Wound Dressings with Controlled Release of Antibiotics: Results in a Guinea Pig Burn Model, *Burns*, 2015, **41**, 1459–1467.
129. H. Barkay-Olami, I. Berdicevsky and M. Zilberman, Novel Soy Protein Blend Scaffolds Loaded with Antibiotics: Drug Release Profile - Bacterial Inhibition Effects, *Advanced Biomaterials and Devices in Medicine*, 2015, **2**, 23–31.
130. H. Barkay-Olami and M. Zilberman, Novel Soy-Protein Based Porous Blend Structures for Biomedical Applications: Microstructure, Mechanical and Physical Properties, *J. Biomed. Mater. Res., Part B*, 2016, **104**(6), 1109–1120.
131. H. Barkay-Olami and M. Zilberman, Microstructure and In Vitro Cellular Response to Novel Soy Protein-Based Porous Structures for Tissue Regeneration Applications, *J. Biomater. Appl.*, 2016, **30**(7), 1004–1015.

CHAPTER 4

Biomimetic Antimicrobial Polymers

UPAYAN BAUL*[a] AND SATYAVANI VEMPARALA*[b]

[a] Institute of Physics, Albert-Ludwigs-University of Freiburg, Hermann-Herder-Strasse 3, 79104 Freiburg, Germany; [b] The Institute of Mathematical Sciences, C.I.T. Campus, Taramani, Chennai 600113, India
*Email: upayan.baul@physik.uni-freiburg.de; vani@imsc.res.in

4.1 Introduction

The discovery and use of antibiotics has been a very effective tool for humanity in our continual fight against infectious pathogens, and has saved millions of lives.[1,2] However, indiscriminate usage of antibiotics in agricultural and food industries coupled with rapid ability of the microorganisms to mutate and develop resistance against the antibiotics, has forced us to use higher dosages of antibiotics, setting up of a vicious cycle and resulting in emergence of multidrug-resistant bacteria.[3,4] Dramatically increased levels of such antibiotic resistance is a global public health threat and has initiated much research into shifting the focus from developing stronger antibiotics to the development of alternative and new types of antimicrobial agents to replace conventional antibiotics. In this direction, the discovery of naturally occurring antimicrobial peptides (AMPs), which have played a critical role in the survival of higher organisms against pathogens for thousands of years, has attracted much attention.[5,6]

AMPs are an integral part of innate immunity system in higher organisms.[7–10] The immunity system in higher complex organisms can be broadly divided into adaptive and innate mechanisms (systems). Much focus has

Biomaterials Science Series No. 5
Antimicrobial Materials for Biomedical Applications
Edited by Abraham J. Domb, Konda Reddy Kunduru and Shady Farah

Published by the Royal Society of Chemistry, www.rsc.org

been on the highly specific adaptive immunity mechanisms, but increasingly the other facet of the immunity system, the innate immunity system, is gaining importance in its crucial role in the defense against several microbial infections.[11–13] AMPs, also referred to as host defense peptides, are known to act against a wide variety of pathogens, albeit with low specificity, which is reflected in the large structural diversity of this class of peptides. AMP structural classes range from α-helices and β-sheets to more extended or looped structures, lacking in specific secondary structural elements.[14,15] The divergence of structures has been explained in terms of divergence of the microbial families against which such peptides evolved. The afore mentioned non- (low) specific way in which the AMPs interact with pathogens is probably at the heart of reduced antimicrobial resistance exhibited against them by the pathogens. Naturally occurring AMPs are typically 15–50 amino acids long and are mostly positively charged, along with hydrophobic content.[4,16] Many of these cationic peptides also exhibit a property called facial amphiphilicity, which is the spatial separation of charged and hydrophobic elements about the peptide backbone.[8] Such facially amphiphilic conformations of the AMPs facilitate simultaneous interactions with both hydrophobic lipid tail groups and hydrophilic lipid head groups and water, and has been recognized as a hallmark of antimicrobial mechanisms. A cationic peptide generally recognizes a microbe *via* several structural and chemical elements of a microbial membrane, including presence of negatively charged lipids, curvature, specific head groups, lipid packing defects, *etc.*, which are different from host cells whose outer cell membranes consist of lipid molecules that are predominantly neutral in their charge, with lower packing defects and curvature.[13,17,18] The antimicrobial mechanism of AMPs includes both direct targeting of microbial membrane integrity (generically termed as membranolytic mode) and/or orchestration of immunomodulatory responses triggered by the presence of pathogenic species (for this reason they are also called host defense peptides).[4,10,16,19–22] There have been many research efforts in exploring the possibility of usage of naturally occurring AMPs in clinical use and there are several AMPs already in clinical trials or development.[23–25]

In recent years, there also has been increased focus in developing biomimetic synthetic polymers that exhibit antimicrobial mechanisms. This is because the therapeutic usage of natural antimicrobial peptides is fraught with issues not limited to high cost of manufacturing, vulnerability to proteolysis and low bioavailability.[10,19,21,26–30] Most such polymers are cationic in nature and much of the design aspects of these biomimetic polymers comes from structural information from cationic AMPs, including small number of monomers, presence of charged and hydrophobic moieties, and in some cases in-built facial amphiphilicity.[31–37] For example, the common presence of lysine residues as charged moieties in many natural AMPs has encouraged the inclusion of primary amines when designing antimicrobial polymers. Also in the design of antimicrobial polymers with intended facial amphiphilicity, aromatic rings were used as a part of backbone of biomimetic polymers to confer rigidity. In addition, the charged and hydrophobic groups were explicitly designed to be on either side of this rigid backbone. In terms of in-built secondary structure conformations of AMPs, while many AMPs exhibit well-defined amphiphilic secondary structures such as α-helices and β-sheets,[8] much of the literature suggests that this may not be an essential ingredient for antimicrobial mechanisms.[16,38,39] However, the ability to adopt functional amphiphilic structure near the bacterial membranes, even if unstructured in solution, has emerged as an important ingredient in determining the efficacy of the antimicrobial agents.[4,16,34,40–42] For this reason, although many examples of lab-designed biomimetic polymers include structures that have in-built backbone rigidity, such as polyacrylamides,[43,44] polynorbornenes,[45] and phenylene ethynylenes,[46,47] a class of polymers such as polymethacrylate copolymers and nylon-derivatives[34,48–51] have flexible backbone conformations. These biomimetic polymers have many advantages over their natural counterparts, including enhanced stability under various physiological conditions, ease of preparation methods, and high antimicrobial potency.[4,52,53] In addition to being biomimetic counterparts to naturally occurring AMPs with possibilities of clinical applications, the antimicrobial polymer (AMPoly) systems with their tunable properties can also be used as a platform to understand the nuances of antimicrobial mechanisms. This includes the effect of antimicrobial agents on the structural properties of the membrane such as the bilayer phase, membrane hydration and thickness, order in the packing of lipid acyl tails, local curvature and pressure profile across the bilayer, *etc.*

Computer simulations play an important role in probing the structural and dynamical aspects of antimicrobial agent–membrane interactions at various length and time-scale resolutions. Although demonstration of full lysis of a bacterial cell by antimicrobial agents is beyond the current capabilities of numerical simulations, many molecular dynamics simulation studies have elucidated the initial agent–membrane interactions: roles played by the type of headgroups, composition of the microbial membranes, distribution of charged and hydrophobic groups of antimicrobial agents, backbone rigidity, *etc.*, at resolutions that experiments cannot probe.[43,44,51,54–78] The resolution of many of these molecular dynamics simulations can vary from fully atomistic to coarse-grained simulations and can span time scales of the order of hundreds of nanoseconds to microseconds.

The focus of this chapter is on biomimetic antimicrobial polymers in general and on our work on methacrylate-based polymers in particular. There are many excellent reviews on the more general topic of AMPs, the antimicrobial mechanism and their role in immune system. Following a brief summary of the physiochemical features of methacrylate AMPoly, and the most significant developments in their design as potent antimicrobial agents, special emphasis is laid on recent insights gained using computer simulations on their mechanism of microbial membrane recognition, invasion, and subsequent destabilization. For a detailed description of the development, and antimicrobial activity of methacrylate AMPoly, interested readers are referred to insightful reviews.[78–82] The rest of the chapter is organized as follows. In Section 4.2, we describe the well-known modes of antimicrobial action and in Section 4.3 we describe our work on AMPolys based on methacrylate copolymers with special emphasis on the role of computer simulations.

4.2 Models of Antimicrobial Action

The basis of antimicrobial action by antimicrobial agents largely lies in the ability of these agents to detect the differences[8] in the lipid composition of the mammalian and microbial cell membranes, although many disruption models are suggested in the literature[18] in addition to other modes of antimicrobial mechanism.[4,16,83] The microbial membranes typically have negatively charged lipids and zwitterionic lipids with smaller head groups like phosphatidylethanolamine (PE), compared to mammalian lipids. The presence of negatively charged lipids in the outer leaflet of microbial cell membranes is suggested as a reason for the presence of positively charged groups in AMPs. These cationic groups in AMPs can detect and thereby differentiate between the mammalian and microbial cells, *via* electrostatic interactions, a first step in the antimicrobial mechanism. It has been well established that the subsequent cell disruption mechanism is largely driven by the hydrophobic groups present in the antimicrobial agents which interact strongly with lipid hydrophobic tails. Although structure–activity studies have indicated that the antimicrobial activity of many α-helical AMPs strongly depends on the content and composition of hydrophobic residues,[84–87] there are also studies which show that increases in hydrophobic content above an optimal balance can potentially lead to toxicity to host cells, a highly undesirable trait.[4] Therefore, the relative proportions of cationic and hydrophobic residues is of critical importance in optimal antimicrobial activity: highly effective against microbial cells and displaying low cytotoxicity to host cells. Recent works have suggested that in addition to detecting the oppositely charged lipid groups in microbial membranes, the antimicrobial agents are also capable of detecting lipid head group defects, and curvature of the microbial membranes, making the initial process of selective recognition a multi-mode mechanism. It has also been observed that presence of facially amphiphilic conformations is essential to the antimicrobial mechanism, and this could be either built into the agents or the agents themselves have the ability to acquire such functionally relevant structures at the water–membrane interface. The facially amphiphilic conformations of antimicrobial agents will facilitate favorable interactions of the cationic groups with both the water and polar/anionic lipid head groups, while the hydrophobic constituents would be buried deeper into the hydrophobic membrane interior.

Regarding the mechanisms of cell disruption, several models have been proposed, including barrel-stave, toroidal-pore and carpet mechanisms.[5,16,17,88–92] The first two mechanisms involve the penetration of antimicrobial agents into the cell membrane and are evoked in a concentration-dependent manner. They also require some form of co-operative self-assembly of antimicrobial agents outside or inside the membrane phase. Helical AMPs like alamethicin are predicted to disrupt the microbial membranes *via* the barrel-stave model,[93–95] while more toxic AMPs like magainin and melittin have been shown to disrupt the membrane in a toroidal mode.[96,97] The most general model of antimicrobial mechanism is the 'carpet' model, wherein the antimicrobial agents interact predominantly with the lipid headgroups at multiple sites and typically at high concentrations of antimicrobial agents. In this mechanism, typically the antimicrobial agents are aligned parallel to the cell membrane, and do not permeate as deeply into the cell membrane as in the other two mechanisms. There have been variants of the carpet mechanism, in which prior to the disruption of the cell membranes, the membrane-partitioned antimicrobial agents can recruit the negatively charged lipids, creating defective phase domain boundaries affecting the integrity of the microbial membranes.[98,99] There have been examples of α-helical AMPs that have been shown to induce such phase demixing of anionic and zwitterionic lipids.[4] Laterally demixed phases in cell membranes occur naturally, yet they are different from the antimicrobial agent-induced demixing of the lipids which occur on a much smaller time scale, making it difficult for the membrane to respond to the emergent defects.[98,100–105] These phase boundary defects can manifest in differences in curvature, membrane thickness, mismatch in lipid head packing and eventually as increased leakage of cell contents. While experimental studies have shown that such phase demixing is induced by antimicrobial polymer agents with flexible backbones and randomly distributed charged groups,[51,105,106] simulation studies of helical antimicrobial agents from the latarcin family (with a more rigid backbone secondary structure)[75] have also demonstrated such a mechanism. Regardless of the specific mode of antimicrobial mechanism, most of the antimicrobial agents exhibit broad-spectrum action largely involving the microbial cell membrane as the primary target and do not interact with any specific membrane proteins and this general non-specific mode of action is at the heart of increased difficulty for microbes to develop resistance against these antimicrobial agents, though there have been reports of resistance in the literature.

4.3 Antimicrobial Polymers with Flexible Backbones

Many different classes of synthetic and biomimetic polymers have been designed, inspired by the natural AMPs, which demonstrate antimicrobial potency as well as selectivity towards microbial cells while exhibiting low toxicity to host cells.[34,43,45,48,79,107] There is significant effort in designing polymers with rigid backbones with amphiphilicity built into the molecular structure such as phenylene ethynylenes,[46,47,107] arylamides,[43,44,107] and polynorbornenes.[45,107] The biomimetic design strategies have also been extended to polymers based on flexible backbone conformations such as polymethacrylate copolymers[34] and nylon derivatives.[48,49] The flexible polymers have advantages over polymers with rigid backbones, especially in terms of relative ease of preparation; they also have been instrumental in being used as molecular platforms to understand the general class of membrane-active agents including intrinsically disordered proteins.[108]

As a natural consequence of the observed secondary structure-forming propensities of AMPs, early studies probed the significance of secondary structural scaffolds such as an α-helix or a β-sheet in the mechanism of antimicrobial action.[107,109–113] Initially involving agent molecules with only peptide backbones, these studies were extended to molecules with rigid polymeric backbones and associated built-in facially amphiphilic conformations, including arylamides and polynorbornenes.[38,43,45,114–118] However, systematic and exhaustive research led to an unexpected finding. While the essential significance of facially amphiphilic conformations in the membrane permeation and disruption was abundantly evident, the same did not hold true for the correlation between antimicrobial efficacy and the stability, or rigidity, of the secondary structural scaffolds such as α-helices.[38,39,79,119] In fact, molecules with stable helical structures even in the dissolved aqueous phase were often attributed greater toxicity to host cells, and hence reduced selectivity to microbial membranes.[4]

In hindsight, the apparently negative, and at the time counterintuitive, results led to a revolutionary development in the design strategy of biomimetic AMPolys. With the ability to adopt facially amphiphilic conformations at the membrane–water interface emerging as the necessary and, applied to the same context, sufficient criterion for antimicrobial activity,[79,119] the design of antimicrobial agent molecules was extended to amphiphilic molecules with flexible polymer backbones.[12,36,50,78–82,108,120–125] Such molecules (prominent examples including methacrylate polymers and nylon derivatives[34,48]) are expected to be unstructured in solution. In a membranous environment, however, the intrinsic flexibility of their backbones enable them to adopt to energetically favorable facially amphiphilic conformations where the hydrophobic and charged/polar moieties are preferentially sequestered to two sides of the polymeric backbone.[36,51,76–78,82,126] Such adaptive changes to the polymeric conformations, coined as *acquired facial amphiphilicity*, and its significance in the antimicrobial activity is the theme of this section. In the following discussions, we consider methacrylate based (co-)polymers as the prototypical example of such flexible AMPolys. However, the general physiochemical features, and the principles governing antimicrobial activity through acquired facial amphiphilicity can be envisaged to hold for the more general class of flexible AMPoly.

Historically, the potency for membrane disruption of an antimicrobial agent has been attributed to the content and composition of its hydrophobic residues, while the selectivity has been envisaged to arise from favorable long-range electrostatic interactions between its cationic moieties and the anionic lipid heads groups present in the exterior leaflets of microbial membranes.[8] We especially hope to draw the reader's attention to additional complexities beyond such phenomenological description, and in the process question the oft-cited clear demarcations of roles of individual moieties.[89,90,127] Specifically, we discuss that the mechanism of antimicrobial action of methacrylate AMPoly is cooperative, and also responsive to the composition of lipid membranes.[51,126] Through adaptive responses of both the polymers and the lipid membrane, these molecules are intrinsically capable of distinguishing a microbial membrane from a mammalian one, beyond the favorable electrostatic interactions. Further, methacrylate AMPoly can induce membrane destabilization through multiple modes of action, depending on the lipid profile of the microbial membrane.

The chemical composition of a representative methacrylate AMPoly is shown in Figure 4.1. These copolymers consist of monomers with hydrophobic and cationic side chains randomly distributed along the polymer backbone. Methacrylate AMPolys have a broad spectrum of antimicrobial activity, including both Gram-positive and Gram-negative bacteria. They are also highly resistant to the development of bacterial resistance, and have been observed to be equally efficacious against *Escherichia coli* and *Staphylococcus aureus* in both the stationary and exponential growth phases of the bacteria.[50,77,79] As their chemical structure would suggest, these polymers are not designed with any built-in facially amphiphilic structural motifs. In computer simulations, their aqueous conformations have been observed to be random, with the degree of compaction dictated by the relative ratio of hydrophobic to cationic monomers.[76] In experiments, this compositional attribute, and more generally the *hydrophobic balance*, defined as the relative content of hydrophobic to hydrophilic groups, has been observed to critically influence the antimicrobial efficacy of methacrylate AMPoly.[77,128] Generally speaking, enhanced hydrophobicity increases the potency for microbial membrane disruption by the molecules, but at the expense of greater toxicity to host cells.

The antimicrobial efficacy of methacrylate AMPoly is also responsive to the chemical composition of cationic side chains, both through the basicity of terminal ammonium groups, and the length of the spacer alkyl chain separating the amines from the backbone (*m* in Figure 4.1).[77,129] Methacrylate AMPolys comprising primary amines have been observed to be more potent antimicrobial agents compared to those containing secondary and tertiary amines.[129] These observations indicate that specific interactions, such as hydrogen bonding between cationic moieties and lipid head groups can also contribute to the antimicrobial activity of methacrylates, and the role of

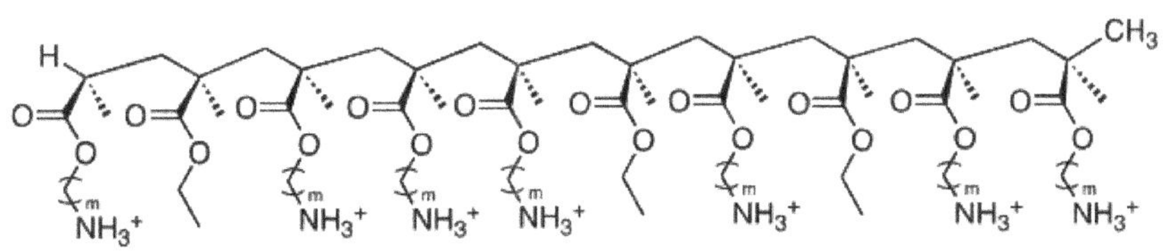

Figure 4.1 Chemical structure of methacrylate AMPoly.

cationic groups is not limited to imparting specificity through electrostatic interactions. The experimental study of the effect of length of spacer arms, *m*, on the antimicrobial efficacy of methacrylates was aided by computer simulations probing the membrane-bound conformations of the polymers.[77] In simulations of 100 ns duration, the membrane partitioning propensities for the polymers were systematically observed to increase from polymers with aminoethylene side chains ($m = 2$) to aminobutylene ($m = 4$) and aminohexylene side chains ($m = 6$). For the latter two, the partitioned polymers were further observed to adopt the well-known *snorkeling* conformations observed with α-helical AMPolys.[77,130,131] In the snorkeling conformation, the partitioned polymers adopt facially amphiphilic conformations about the polymer backbones, which lie parallel to the membrane–water interface, and the cationic groups are anchored to the locations of lipid head groups. In experiments, the polymers with butylene spacer arms were observed to be optimal considering the balance between antimicrobial activity and hemolytic toxicity, with ethylene and hexylene spacer arms leading to low potency and high toxicity, respectively.[77] The subsequent simulation studies thus exclusively probed the membrane interactions of methacrylate polymers with aminobutylene side chains, denoted here as E4 AMPoly.

Selective disruption of microbial membranes by antimicrobial agents is a collective process, involving the cooperative presence of multiple agent molecules, as well as lipid membranes of varying lipid compositions, including the mammalian membranes of the host cells. Following the identification of a methacrylate polymer composition that is optimal through considerations of both antimicrobial potency and selectivity to microbial membranes, namely E4 AMPoly, subsequent computer simulation studies have thus focused on the collective conformations of multiple E4 AMPoly, and their influence on the structural properties of membranes of varying lipid compositions.[51,126] In an aqueous environment, an isolated E4 AMPoly is characterized by dynamically fluctuating random conformations. Simulations of multiple E4 AMPolys in 0.15 M NaCl solution showed that the E4 AMPolys are aggregated in physiological salt solutions. The aggregates have a micellar structure, with cationic groups lining the surface, and the hydrophobic groups buried towards the interior of the aggregates. Within the aggregates, the individual polymers retain the conformational disorder, with no signatures of any facial amphiphilicity.[51] It can be envisaged, that the E4 AMPolys encounter the lipid membranes while in this aggregated state.

The events leading to antimicrobial action by agent molecules can be broadly categorized into (a) selective recognition and partitioning into microbial membranes by agent molecules over host cell membranes; and (b) induced destabilization/disruption of the microbial membranes. The former is a predominantly interfacial phenomenon, in which the lipid head group compositions are of critical import. The latter further involves the composition of the interior of the membrane, or the lipid acyl tails. Using fully atomistic molecular dynamics simulations of 300 ns or longer in each case, the interactions of an aggregate of four E4 AMPolys with microbial membranes of varying lipid compositions have been studied.[51,126] Owing to the abundance of zwitterionic ethanolamine (PE) and anionic glycerol (PG) head groups in microbial membranes,[75] the microbial membrane models studied included doPE-PG (1,2-dioleoyl-*sn*-glycero-3 phosphoethanolamine and phosphoglycerol), poPE-PG (1-palmitoyl-2-oleoyl phosphatidylethanolamine and phosphatidylglycerol) and dpPE-PG (1,2-dipalmitoyl-*sn*-glycero-3 phosphorylethanolamine and phosphoglycerol). The head group compositions remaining equivalent, the three model membranes differed in the degree of saturation of their acyl tails. Specifically, all the acyl tails of doPE-PG are unsaturated, all tails of dpPE-PG are saturated, while only half the lipid tails of poPE-PG are saturated. An additional membrane composition has also been studied, doPC-PG (1,2-dioleoyl-*sn*-glycero-3 phosphocholine and phosphoglycerol), which can be described as the mammalian equivalent of the microbial membrane model doPE-PG, but with anionic lipid content identical to the microbial counterpart.

In simulations, the favorable electrostatic interactions among anionic lipid head groups and cationic E4 moieties were observed to facilitate the immediate recognition of the membranes by the aggregate. However, following the initial approach of the aggregate to the membrane–water interface, the mechanism of partitioning was observed to differ categorically between the microbial membrane models and doPC-PG.[126] Figure 4.2 presents a pictorial description of the differences, using snapshots from system trajectories for doPE-PG and doPC-PG. In the presence of all microbial membranes, namely doPE-PG, poPE-PG, and dpPE-PG, upon contact the aggregate of E4 polymers was observed to adopt extended conformations, maximizing the contacts with lipid head group atoms. Subsequently, individual polymers were released from the aggregate in a time-phased manner, which readily partitioned into the membrane interior, as shown in Figure 4.2(A). In the proximity of a doPC-PG membrane, however, the aggregate retained its compact conformations even after multiple transient contacts with lipid atoms were initiated (Figure 4.2(B)). The eventual anchoring of the aggregate was observed to be facilitated by a hydrophobic side arm of an E4 polymer penetrating beyond the head group region of the membrane. Subsequently, the entire aggregate was observed to gradually partition into the membrane as a single entity, and three of the four partitioned polymers remained loosely bound through the duration of the simulation.

Such marked differences in membrane affinities cannot be interpreted in terms of membrane–polymer energetics alone. Simulation results showed that not only were the doPE-PG and doPC-PG membranes equivalent in terms of electrostatic interactions with the E4 AMPoly, the energetic favorability for polymer partitioning was also equivalent for all membrane models simulated.[126] The greater partitioning propensities for E4 AMPoly into microbial membranes has been qualitatively mapped to their ability to sense interfacial lipid packing defects. Interfacial lipid packing defects are transient exposures of lipid hydrophobic groups at the membrane–water interface, arising from the dynamic fluctuations of the lipid head groups.[132–135]

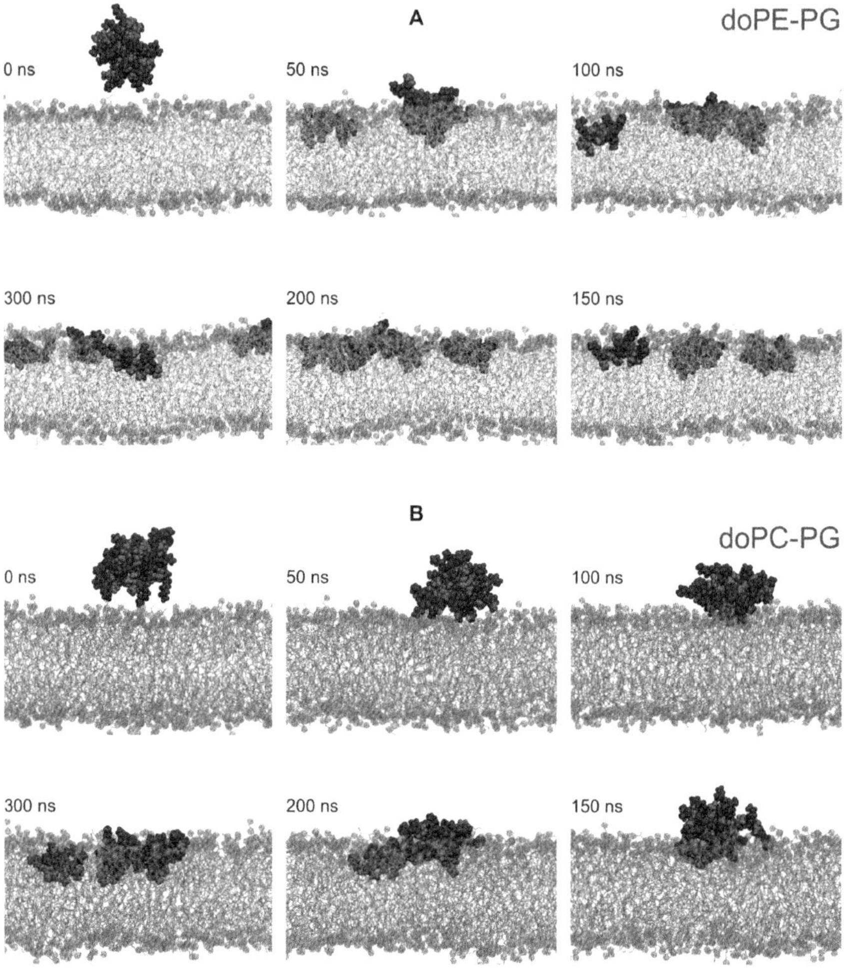

Figure 4.2 Snapshots of the evolution of (A) E4 – doPE-PG and (B) E4 – doPC-PG membrane systems. Water and ions are not shown for clarity. The cationic and hydrophobic groups of E4 AMPoly are shown in darker and lighter shades of gray, respectively.
Adapted from ref. 126 with permission from the Royal Society of Chemistry.

It has been suggested that membrane permeation of amphiphilic molecules, including AMPs such as *smp24*, can be facilitated by the presence of such defects.[134,136,137] In the simulations, the conical (PE) lipid head groups of the microbial membranes were shown to result in greater abundance of the lipid packing defects, compared to the more cylindrical (PC) head groups of the mammalian membrane model.[126] These results highlighted that the selectivity of E4 AMPoly for microbial membranes is not limited to favorable electrostatic interactions with the anionic lipids of microbial membranes.

E4 AMPolys are known to adopt snorkeling conformations following the partition into lipid membranes. In this acquired facially amphiphilic conformation, the cationic groups localize close to the lipid head groups, and the hydrophobic moieties are buried deeper within the membrane interior.[51,77,126] The extent of facial amphiphilicity in the snorkeling conformations has been strongly linked to the antimicrobial activity of E4 AMPoly.[77,78] In the density profiles for a representative E4 AMPoly shown in Figure 4.3; this can be estimated from the overlap in the density profiles for cationic and hydrophobic moieties, with minimal overlap representing robust acquired facial amphiphilicity. Clearly, the acquired facial amphiphilicity in partitioned polymer conformations is more pronounced upon partitioning into model microbial membranes, and weak when partitioned into the doPC-PG membrane studied. The extent of facial amphiphilicity within microbial membranes is also responsive to the composition of lipid acyl tails. The robust signatures of facial amphiphilicity, and the extended nature of the conformations observed inside microbial membranes asserts that the interactions of E4 AMPoly are stronger with microbial membranes, compared to the mammalian membrane model. In terms of the bilayer responses too, the E4 AMPolys were observed to maintain a passive presence within the doPC-PG membrane. However, their influences on the structural and morphological properties of microbial membranes were strong, and

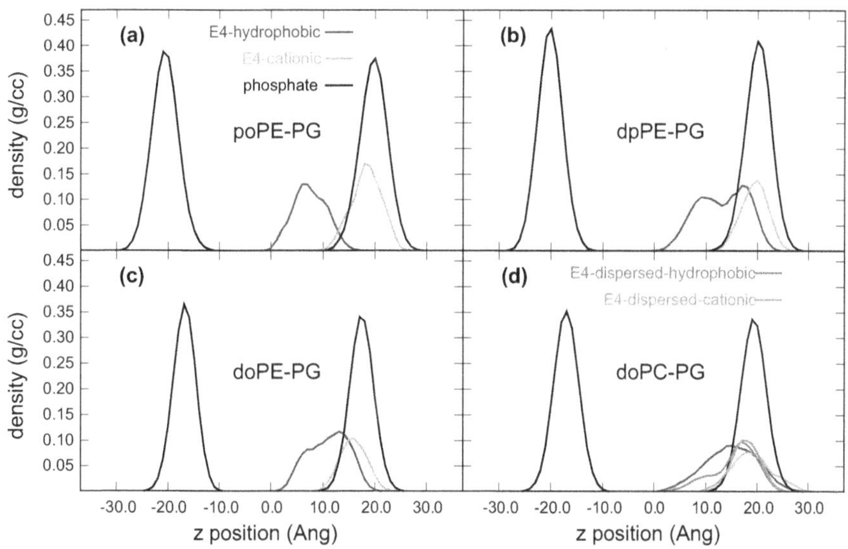

Figure 4.3 Density profiles of various system components along the direction of membrane normal (z-axis). Results for E4 – poPE-PG, E4 – dpPE-PG, E4 – doPE-PG, and E4 aggregate – doPC-PG systems are shown in (a), (b), (c), and (d), respectively.
Reproduced from ref. 126 with permission from the Royal Society of Chemistry.

specific to the lipid compositions of the membrane models.[51,126] A hallmark of efficacious antimicrobial agents is that they can assert their antimicrobial activity using varied modes of action based on the specific physiochemical properties of the embedding membranes.[4,16,83] Along with pedagogical pore-forming mechanisms such as the barrel-stave, toroidal- and disordered toroidal-pore, and the carpet modes of action,[16,88–90] recent studies have shown that induced lateral inhomogeneities in membrane properties can lead to efficient membrane destabilization and eventual disruption.[98,138–140] Such inhomogeneities are often associated with the demixing of lipid species in a mixed lipid membrane. In simulations, evidence for the onset of such phase demixing, and associated lateral inhomogeneities in local membrane thicknesses was clearly observed for the poPE-PG model lipid membrane, induced by the cooperative presence of multiple partitioned E4 AMPoly.[51] While no such evident demixing of charged and zwitterionic lipid species was observed with the other microbial membrane models studied, namely dpPE-PG and doPE-PG, the induced inhomogeneities in membrane thicknesses was also prominently observed for the dpPE-PG membrane. The computed local thickness profiles for all the discussed membranes are shown in Figure 4.4, along with the thickness profiles for the respective control membranes (in the absence of E4 AMPoly). For the doPE-PG membrane, substantial lateral heterogeneities were observed also for the control membrane, which is restrictive for such comparisons.[126]

For the doPE-PG membrane, the presence of the partitioned E4 AMPoly was observed to reduce the average membrane thickness, while the reverse was observed for poPE-PG and dpPE-PG. The thickness of a lipid membrane has strong correlations with the order in lipid tails, quantifiable through the deuterium order parameter (S_{cd}), defined as $S_{cd} = \frac{1}{2} \langle 3\cos^2\theta - 1\rangle$. In the equation, θ is the angle between a C–H bond and the membrane normal. Increased S_{cd} is reflective of increased order in the packing of lipid tails, reduced fluidity of the membrane, and enhanced membrane thickness. The S_{cd} values computed from the simulations were generally observed to be consistent with the observed thickness profiles of the membranes shown in Figure 4.4. No consistent effects were observed for the doPC-PG membrane, with high dispersion in results attributed to localized effects due to instability in E4–lipid tail interactions. Lipid tail composition specific effects for the microbial membrane models further highlighted the adaptive influences of E4 AMPoly on microbial membrane structure and dynamics, and indicated the likely existence of multiple modes of antimicrobial action. In presence of saturated lipid tails, E4 AMPoly were observed to promote further order in packing, and a consequent stratification of lipid tails. On the contrary, they enhanced the fluidity of a microbial membrane comprising of only unsaturated lipid tails. In summary, the selectivity of flexible antimicrobial AMPoly E4 towards microbial membranes over mammalian ones is robust, and observable at every stage of membrane–polymer interactions.

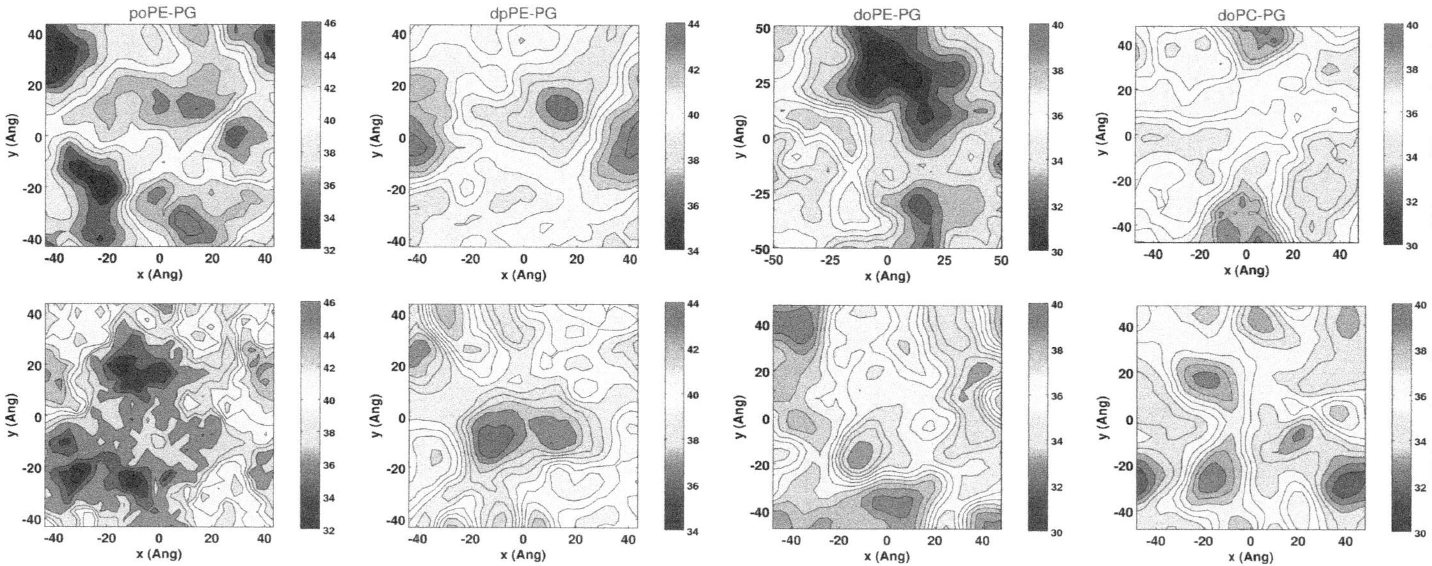

Figure 4.4 Two-dimensional thickness profiles for E4 – membrane (upper row) and control (lower row) systems. For comparisons across models, the limits of the heatmaps (thickness in ångström) representing magnitude of local fluctuations should be taken into consideration. The observations for different membrane models are described in the text.
Adapted from ref. 51 with permission from AIP Publishing, Copyright 2014, and from ref. 126 with permission from the Royal Society of Chemistry.

4.4 Conclusion

In this chapter, a focused summary of biomimetic antimicrobial polymers with special emphasis on atomistic simulations of antimicrobial polymers based on methacrylate copolymers that we have been working on for the last few years is presented. There have been renewed efforts in the past few years to design better antimicrobial agents for possible therapeutic usage in the context of the alarming rise in microbial resistance to even the strongest of antibiotics. Hence there is an urgent need to understand the design principles of such antimicrobial agents, since incorporating both the nuanced selectivity to microbial cells and low toxicity to host cells in the laboratory is a daunting task. It is also of crucial importance to understand the atomistic details of interactions of antimicrobial agents with different cell membranes in order to design more effective antimicrobial agents. Concerted efforts to design polymers, inspired from naturally occurring AMPs, have been made in recent years, both to provide effective alternatives to natural AMPs and also to understand the design principles behind the evolution of such peptides. The synthetic polymers can also be used as platforms to explore the structure–function relationship of biopolymers and can play a pivotal role in probing the acquisition of functional structures in right environments in systems such as intrinsically disordered proteins. Computer simulations can play a vital role in all these efforts, as they can probe the mechanics of antimicrobial agents and their interaction with cell membranes. With increasing availability of fine-tuned forcefields, simulation methodologies, and ever-increasing availability of computing power, simulations can systematically probe the systems belonging to different structural classes, different side chain groups, and can aid significantly in rational drug design.

References

1. K. C. Nicolaou and S. Rigol, A brief history of antibiotics and select advances in their synthesis, *J. Antibiot.*, 2018, **71**, 153.
2. R. I. Aminov, A brief history of the antibiotic era: lessons learned and challenges for the future, *Front. Microbiol.*, 2010, **1**, 134.
3. G. D. Wright, The antibiotic resistome: the nexus of chemical and genetic diversity, *Nat. Rev. Microbiol.*, 2007, **5**, 175–186.
4. L. T. Nguyen, E. F. Haney and H. J. Vogel, The expanding scope of antimicrobial peptide structures and their modes of action, *Trends Biotechnol.*, 2011, **29**, 464–472.
5. J. M. Ageitos, A. Sánchez-Pérez, P. Calo-Mata and T. G. Villa, Antimicrobial peptides (amps): ancient compounds that represent novel weapons in the fight against bacteria, *Biochem. Pharmacol.*, 2017, **133**, 117–138.
6. L.-J. Zhang and R. L. Gallo, Antimicrobial peptides, *Curr. Biol.*, 2016, **26**, R14–R19.

7. J. B. Mcphee and R. E. W. Hancock, Function and therapeutic potential of host defence peptides, *J. Pept. Sci.*, 2005, **11**, 677–687.
8. M. Zasloff, Antimicrobial peptides of multicellular organisms, *Nature*, 2002, **415**, 389–395.
9. A. Tossi, L. Sandri and A. Giangaspero, Amphipathic, -helical antimicrobial peptides, *Pept. Sci.*, 2000, **55**, 4–30.
10. R. E. W. Hancock and H.-G. Sahl, Antimicrobial and host-defense peptides as new anti-infective therapeutic strategies, *Nat. Biotechnol.*, 2006, **24**, 1551–1557.
11. H. G. Boman, Peptide antibiotics and their role in innate immunity, *Annu. Rev. Immunol.*, 1995, **13**, 61–92.
12. G. J. Gabriel, A. Som, A. E. Madkour, T. Eren and G. N. Tew, Infectious disease: Connecting innate immunity to biocidal polymers, *Mater. Sci. Eng., R*, 2007, **57**, 28–64.
13. M. Pasupuleti, A. Schmidtchen and M. Malmsten, Antimicrobial peptides: key components of the innate immune system, *Crit. Rev. Biotechnol.*, 2012, **32**, 143–171.
14. S. Thomas, S. Karnik, R. S. Barai, V. K. Jayaraman and S. Idicula-Thomas, Camp: a useful resource for research on antimicrobial peptides, *Nucleic Acids Res.*, 2009, **38**, D774–D780.
15. G. Wang, X. Li and Z. Wang, Apd2: the updated antimicrobial peptide database and its application in peptide design, *Nucleic Acids Res.*, 2008, **37**, D933–D937.
16. H. Jenssen, P. Hamill and R. E. W. Hancock, Peptide antimicrobial agents, *Clin. Microbiol. Rev.*, 2006, **19**, 491–511.
17. O. G. Travkova, H. Moehwald and G. Brezesinski, The interaction of antimicrobial peptides with membranes, *Adv. Colloid Interface Sci.*, 2017, **247**, 521–532.
18. L. T. Nguyen, E. F. Haney and H. J. Vogel, The expanding scope of antimicrobial peptide structures and their modes of action, *Trends Biotechnol.*, 2011, **29**, 464–472.
19. M.-D. Seo, H.-S. Won, J.-H. Kim, T. Mishig-Ochir and B.-J. Lee, Antimicrobial peptides for therapeutic applications: A review, *Molecules*, 2012, **17**, 12276–12286.
20. V. Teixeira, M. J. Feio and M. Bastos, Role of lipids in the interaction of antimicrobial peptides with membranes, *Prog. Lipid Res.*, 2012, **51**, 149–177.
21. S. Rotem and A. Mor, Antimicrobial peptide mimics for improved therapeutic properties, *Biochim. Biophys. Acta, Biomembr.*, 2009, **1788**, 1582–1592.
22. R. E. W. Hancock and G. Diamond, The role of cationic antimicrobial peptides in innate host defences, *Trends Microbiol.*, 2000, **8**, 402–410.
23. L. Zhang and T. J. Falla, Cationic antimicrobial peptides–an update, *Expert Opin. Invest. Drugs*, 2004, **13**, 97–106.
24. A. Lewies, J. F. Wentzel, G. Jacobs and L. H. Du Plessis, The potential use of natural and structural analogues of antimicrobial peptides in the

fight against neglected tropical diseases, *Molecules*, 2015, **20**, 15392–15433.
25. J. L. Fox, Antimicrobial peptides stage a comeback, *Nat. Biotechnol.*, 2013, **31**, 379–382.
26. A. K. Marr, W. J. Gooderham and R. E. W. Hancock, Antibacterial peptides for therapeutic use: obstacles and realistic outlook, *Curr. Opin. Pharmacol.*, 2006, **6**, 468–472.
27. S. Riedl, D. Zweytick and K. Lohner, Membrane-active host defense peptides–challenges and perspectives for the development of novel anticancer drugs, *Chem. Phys. Lipids*, 2011, **164**, 766–781.
28. A. Rozek, J.-P. S. Powers, C. L. Friedrich and R. E. W. Hancock, Structure-based design of an indolicidin peptide analogue with increased protease stability, *Biochemistry*, 2003, **42**, 14130–14138.
29. I. H. Lee, Y. Cho and R. I. Lehrer, Effects of ph and salinity on the antimicrobial properties of clavanins, *Infect. Immun.*, 1997, **65**, 2898–2903.
30. H.-L. Chu, H.-Y. Yu, B.-S. Yip, Y.-H. Chih, C.-W. Liang, H.-T. Cheng and J.-W. Cheng, Boosting salt resistance of short antimicrobial peptides, *Antimicrob. Agents Chemother.*, 2013, 00252.
31. Y. Zhang, A. Algburi, N. Wang, V. Kholodovych, D. O. Oh, M. Chikindas and K. E. Uhrich, Self-assembled cationic amphiphiles as antimicrobial peptides mimics: role of hydrophobicity, linkage type, and assembly state, *Nanomed.: Nanotechnol. Biol. Med.*, 2017, **13**, 343–352.
32. G. N. Tew, R. W. Scott, M. L. Klein and W. F. DeGrado, De novo design of antimicrobial polymers, foldamers, and small molecules: from discovery to practical applications, *Acc. Chem. Res.*, 2009, **43**, 30–39.
33. R. Liu, X. Chen, Z. Hayouka, S. Chakraborty, S. P. Falk, B. Weisblum, K. S. Masters and S. H. Gellman, Nylon-3 polymers with selective antifungal activity, *J. Am. Chem. Soc.*, 2013, **135**, 5270–5273.
34. K. Kuroda and W. F. DeGrado, Amphiphilic polymethacrylate derivatives as antimicrobial agents, *J. Am. Chem. Soc.*, 2005, **127**, 4128–4129.
35. J. Haldar, D. An, L. A. de Cienfuegos, J. Chen and A. M. Klibanov, Polymeric coatings that inactivate both influenza virus and pathogenic bacteria, *Proc. Natl. Acad. Sci. U. S. A.*, 2006, **103**, 17667–17671.
36. C. Ergene, K. Yasuhara and E. F. Palermo, Biomimetic antimicrobial polymers: recent advances in molecular design, *Polym. Chem.*, 2018, **9**, 2407–2427.
37. M. M. Konai, B. Bhattacharjee, S. Ghosh and J. Haldar, Recent progress in polymer research to tackle infections and antimicrobial resistance, *Biomacromolecules*, 2018, **19**, 1888–1917.
38. C. M. Goodman, S. Choi, S. Shandler and W. F. DeGrado, Foldamers as versatile frameworks for the design and evolution of function, *Nat. Chem. Biol.*, 2007, **3**, 252–262.
39. M. A. Schmitt, B. Weisblum and S. H. Gellman, Interplay among folding, sequence, and lipophilicity in the antibacterial and hemolytic activities of α/β-peptides, *J. Am. Chem. Soc.*, 2007, **129**, 417–428.

40. K. Kuroda and G. A. Caputo, Antimicrobial polymers as synthetic mimics of host-defense peptides, *Wiley Interdiscip. Rev.: Nanomed. Nanobiotechnol.*, 2013, **5**, 49–66.
41. M. S. Ganewatta and C. Tang, Controlling macromolecular structures towards effective antimicrobial polymers, *Polymer*, 2015, **63**, A1–A29.
42. P. M. Hwang and H. J. Vogel, Structure-function relationships of antimicrobial peptides, *Biochem. Cell Biol.*, 1998, **76**, 235–246.
43. G. N. Tew, D. Liu, B. Chen, R. J. Doerksen, J. Kaplan, P. J. Carroll, M. L. Klein and W. F. DeGrado, De novo design of biomimetic antimicrobial polymers, *Proc. Natl. Acad. Sci. U. S. A.*, 2002, **99**, 5110–5114.
44. D. Liu, S. Choi, B. Chen, R. J. Doerksen, D. J. Clements, J. D. Winkler, M. L. Klein and W. F. DeGrado, Nontoxic membrane-active antimicrobial arylamide oligomers, *Angew. Chem., Int. Ed.*, 2004, **43**, 1158–1162.
45. M. F. Ilker, K. Nusslein, G. N. Tew and E. B. Coughlin, Tuning the hemolytic and antibacterial activities of amphiphilic polynorbornene derivatives, *J. Am. Chem. Soc.*, 2004, **126**, 15870–15875.
46. L. Arnt, K. Nusslein and G. N. Tew, Nonhemolytic abiogenic polymers as antimicrobial peptide mimics, *J. Polym. Sci., Part A: Polym. Chem.*, 2004, **42**, 3860–3864.
47. L. Arnt and G. N. Tew, New poly(phenyleneethynylene)s with cationic, facially amphiphilic structures, *J. Am. Chem. Soc.*, 2002, **124**, 7664–7665.
48. T. Ikeda and S. Tazuke, Biologically active polycations: Antimicrobial activities of poly[trialkyl (vinylbenzyl) ammonium chloride]-type polycations, *Die Makromol. Chem., Rapid Commun.*, 1983, **4**, 459–461.
49. M. A. Gelman, B. Weisblum, D. M. Lynn and S. H. Gellman, Biocidal activity of polystyrenes that are cationic by virtue of protonation, *Org. Lett.*, 2004, **6**, 557–560.
50. I. Sovadinova, E. F. Palermo, M. Urban, P. Mpiga, G. A. Caputo and K. Kuroda, Activity and mechanism of antimicrobial peptide-mimetic amphiphilic polymethacrylate derivatives, *Polymers*, 2011, **3**, 1512–1532.
51. U. Baul, K. Kuroda and S. Vemparala, Interaction of multiple biomimetic antimicrobial polymers with model bacterial membranes, *J. Chem. Phys.*, 2014, **141**, 084902.
52. X.-Z. Lai, Y. Feng, J. Pollard, J. N. Chin, M. J. Rybak, R. Bucki, R. F. Epand, R. M. Epand and P. B. Savage, Ceragenins: Cholic acid-based mimics of antimicrobial peptides, *Acc. Chem. Res.*, 2008, **41**, 1233–1240.
53. B. E. Haug, W. Stensen, M. Kalaaji, Ø. Rekdal and J. S. Svendsen, Synthetic antimicrobial peptidomimetics with therapeutic potential, *J. Med. Chem.*, 2008, **51**, 4306–4314.
54. E. Matyus, C. Kandt and D. P. Tieleman, Computer simulation of antimicrobial peptides, *Curr. Med. Chem.*, 2007, **14**, 2789–2798.
55. B. Orioni, G. Bocchinfuso, J. Y. Kim, A. Palleschi, G. Grande, S. Bobone, Y. Park, J. I. Kim, K-soo Hahm and L. Stella, Membrane perturbation by

the antimicrobial peptide pmap-23: a fluorescence and molecular dynamics study, *Biochim. Biophys. Acta, Biomembr.*, 2009, **1788**, 1523–1533.
56. G. Bocchinfuso, A. Palleschi, B. Orioni, G. Grande, F. Formaggio, C. Toniolo, Y. Park, K.-S. Hahm and L. Stella, Different mechanisms of action of antimicrobial peptides: insights from fluorescence spectroscopy experiments and molecular dynamics simulations, *J. Pept. Sci.*, 2009, **15**, 550–558.
57. A. Stavrakoudis, I. G. Tsoulos, Z. O. Shenkarev and T. V. Ovchinnikova, Molecular dynamics simulation of antimicrobial peptide arenicin-2: β-hairpin stabilization by noncovalent interactions, *Pept. Sci.*, 2009, **92**, 143–155.
58. L. Thogersen, B. Schiøtt, T. Vosegaard, N. C. Nielsen and E. Tajkhorshid, Peptide aggregation and pore formation in a lipid bilayer: A combined coarse-grained and all atom molecular dynamics study, *Biophys. J.*, 2008, **95**, 4337–4347.
59. W. Soliman, S. Bhattacharjee and K. Kaur, Interaction of an antimicrobial peptide with a model lipid bilayer using molecular dynamics simulation, *Langmuir*, 2009, **25**, 6591–6595.
60. J. Li, R. Lakshminarayanan, Y. Bai, S. Liu, L. Zhou, K. Pervushin, C. Verma and R. W. Beuerman, Molecular dynamics simulations of a new branched antimicrobial peptide: A comparison of force fields, *J. Chem. Phys.*, 2012, **137**, 215101.
61. X. Zhao, H. Yu, L. Yang, Q. Li and X. Huang, Simulating the antimicrobial mechanism of human β-defensin-3 with coarse-grained molecular dynamics, *J. Biomol. Struct. Dyn.*, 2015, 1–34.
62. Y. Wang, D. E. Schlamadinger, J. E. Kim and J. A. McCammon, Comparative molecular dynamics simulations of the antimicrobial peptide cm15 in model lipid bilayers, *Biochim. Biophys. Acta, Biomembr.*, 2012, **1818**, 1402–1409.
63. T. D. Romo, L. A. Bradney, D. V. Greathouse and A. Grossfield, Membrane binding of an acyl-lactoferricin b antimicrobial peptide from solid-state {NMR} experiments and molecular dynamics simulations, *Biochim. Biophys. Acta, Biomembr.*, 2011, **1808**, 2019–2030.
64. M. Mihajlovic and T. Lazaridis, Charge distribution and imperfect amphipathicity affect pore formation by antimicrobial peptides, *Biochim. Biophys. Acta, Biomembr.*, 2012, **1818**, 1274–1283.
65. J. Li, S. Liu, R. Lakshminarayanan, Y. Bai, K. Pervushin, C. Verma and R. W. Beuerman, Molecular simulations suggest how a branched antimicrobial peptide perturbs a bacterial membrane and enhances permeability, *Biochim. Biophys. Acta, Biomembr.*, 2013, **1828**, 1112–1121.
66. H. Khandelia and Y. N. Kaznessis, Structure of the antimicrobial β-hairpin peptide protegrin-1 in a dlpc lipid bilayer investigated by molecular dynamics simulation, *Biochim. Biophys. Acta, Biomembr.*, 2007, **1768**, 509–520.

67. J. Mondal, X. Zhu, Q. Cui and A. Yethiraj, Sequence-dependent interaction of β-peptides with membranes, *J. Phys. Chem. B*, 2010, **114**, 13585–13592.
68. J. N. Horn, A. Cravens and A. Grossfield, Interactions between fengycin and model bilayers quantified by coarse-grained molecular dynamics, *Biophys. J.*, 2013, **105**, 1612–1623.
69. K. P. Santo, S. J. Irudayam and M. L. Berkowitz, Melittin creates transient pores in a lipid bilayer: Results from computer simulations, *J. Phys. Chem. B*, 2013, **117**, 5031–5042.
70. C. I. E. von Deuster and V. Knecht, Competing interactions for antimicrobial selectivity based on charge complementarity, *Biochim. Biophys. Acta, Biomembr.*, 2011, **1808**, 2867–2876.
71. G. Bocchinfuso and S. Bobone, Claudia Mazzuca, Antonio Palleschi, and Lorenzo Stella, Fluorescence spectroscopy and molecular dynamics simulations in studies on the mechanism of membrane destabilization by antimicrobial peptides, *Cell. Mol. Life Sci.*, 2011, **68**, 2281–2301.
72. D. Sengupta, H. Leontiadou, A. E. Mark and S. J. Marrink, Toroidal pores formed by antimicrobial peptides show significant disorder, *BBA – Biomembr.*, 2008, **1778**, 2308–2317.
73. H. Leontiadou, A. E. Mark and S. J. Marrink, Antimicrobial peptides in action, *J. Am. Chem. Soc.*, 2006, **128**, 12156–12161.
74. A. J. Rzepiela, D. Sengupta, N. Goga and S. J. Marrink, Membrane poration by antimicrobial peptides combining atomistic and coarse-grained descriptions, *Faraday Discuss.*, 2010, **144**, 431–443.
75. A. A. Polyansky, R. Ramaswamy, P. E. Volynsky, I. F. Sbalzarini, S. J. Marrink and R. G. Efremov, Antimicrobial peptides induce growth of phosphatidylglycerol domains in a model bacterial membrane, *J. Phys. Chem. Lett.*, 2010, **1**, 3108–3111.
76. I. Ivanov, S. Vemparala, V. Pophristic, K. Kuroda, W. F. DeGrado, J. A. McCammon and M. L. Klein, Characterization of nonbiological antimicrobial polymers in aqueous solution and at waterlipid interfaces from all-atom molecular dynamics, *J. Am. Chem. Soc.*, 2006, **128**, 1778–1779.
77. E. F. Palermo, S. Vemparala and K. Kuroda, Cationic spacer arm design strategy for control of antimicrobial activity and conformation of amphiphilic methacrylate random copolymers, *Biomacromolecules*, 2012, **13**, 1632–1641.
78. E. F. Palermo, S. Vemparala, and K. Kuroda, Antimicrobial polymers: Molecular design as synthetic mimics of host-defense peptides, in *Tailored Polymer Architectures for Pharmaceutical and Biomedical Applications*, American Chemical Society, 2013, chap. 20, pp. 319–330, http://pubs.acs.org/doi/pdf/10.1021/bk-2013-1135.ch019.
79. K. Kuroda and G. A. Caputo, Antimicrobial polymers as synthetic mimics of host-defense peptides, *Wiley Interdiscip. Rev.: Nanomed. Nanobiotechnol.*, 2013, **5**, 49–66.

80. H. Takahashi, E. F. Palermo, K. Yasuhara, G. A. Caputo and K. Kuroda, Molecular design, structures, and activity of antimicrobial peptide-mimetic polymers, *Macromol. Biosci.*, 2013, **13**, 1285–1299.
81. E. F. Palermo and K. Kuroda, Structural determinants of antimicrobial activity in polymers which mimic host defense peptides, *Appl. Microbiol. Biotechnol.*, 2010, **87**, 1605–1615.
82. U. Baul and S. Vemparala, *Chapter Four – Membrane-bound Conformations of Antimicrobial Agents and Their Modes of Action*, Academic Press, 2015, pp. 97–128.
83. A. Peschel and H.-G. Sahl, The co-evolution of host cationic antimicrobial peptides and microbial resistance, *Nat. Rev. Microbiol.*, 2006, **4**, 529–536.
84. I. Zelezetsky, U. Pag, H.-G. Sahl and A. Tossi, Tuning the biological properties of amphipathic α-helical antimicrobial peptides: Rational use of minimal amino acid substitutions, *Peptides*, 2005, **26**, 2368–2376.
85. S. Kim, S. S. Kim and B. J. Lee, Correlation between the activities of α-helical antimicrobial peptides and hydrophobicities represented as rp hplc retention times, *Peptides*, 2005, **26**, 2050–2056.
86. H. Yan, S. Li, X. Sun, H. Mi and B. He, Individual substitution analogs of mel(12–26), melittin's c-terminal 15-residue peptide: their antimicrobial and hemolytic actions, *FEBS Lett.*, 2003, **554**, 100–104.
87. T. Wieprecht, M. Dathe, M. Beyermann, E. Krause, W. L. Maloy, D. L. MacDonald, and M. Bienert, Peptide hydrophobicity controls the activity and selectivity of magainin 2 amide in interaction with membranes, *Biochemistry* **36**, 6124–6132 (1997).
88. K. Matsuzaki, K. Sugishita, N. Ishibe, M. Ueha, S. Nakata, K. Miyajima and R. M. Epand, Relationship of membrane curvature to the formation of pores by magainin 2, *Biochemistry*, 1998, **37**, 11856–11863.
89. K. A. Brogden, Antimicrobial peptides: pore formers or metabolic inhibitors in bacteria? *Nat. Rev. Microbiol.* **3**, 238–250 (2005).
90. M. R. Yeaman and N. Y. Yount, Mechanisms of antimicrobial peptide action and resistance, *Pharmacol. Rev.*, 2003, **55**, 27–55.
91. Y. Huang, J. Huang and Y. Chen, Alpha-helical cationic antimicrobial peptides: relationships of structure and function, *Protein Cell*, 2010, **1**, 143–152.
92. N. Papo and Y. Shai, Can we predict biological activity of antimicrobial peptides from their interactions with model phospholipid membranes? *Peptides*, 2003, **24**, 1693–1703.
93. E. S. Salnikov, M. D. Zotti, F. Formaggio, X. Li, C. Toniolo, J. D. J. ONeil, J. Raap, S. A. Dzuba and B. Bechinger, Alamethicin topology in phospholipid membranes by oriented solid-state nmr and epr spectroscopies: a comparison, *J. Phys. Chem. B*, 2009, **113**, 3034–3042.
94. S. J. Archer, J. F. Ellena and D. S. Cafiso, Dynamics and aggregation of the peptide ion channel alamethicin. measurements using spin-labeled peptides, *Biophys. J.*, 1991, **60**, 389–398.

95. P. Yang, F.-G. Wu and Z. Chen, Lipid fluid–gel phase transition induced alamethicin orientational change probed by sum frequency generation vibrational spectroscopy, *J. Phys. Chem. C*, 2013, **117**, 17039–17049.
96. K. Matsuzaki, O. Murase, N. Fujii and K. Miyajima, An antimicrobial peptide, magainin 2, induced rapid flip-flop of phospholipids coupled with pore formation and peptide translocation, *Biochemistry*, 1996, **35**, 11361–11368.
97. R. Mani, S. D. Cady, M. Tang, A. J. Waring, R. I. Lehrer and M. Hong, Membrane-dependent oligomeric structure and pore formation of a-hairpin antimicrobial peptide in lipid bilayers from solid-state nmr, *Proc. Natl. Acad. Sci. U. S. A.*, 2006, **103**, 16242–16247.
98. R. M. Epand and R. F. Epand, Lipid domains in bacterial membranes and the action of antimicrobial agents, *BBA – Biomembr.*, 2009, **1788**, 289–294.
99. R. M. Epand, C. Walker, R. F. Epand and N. A. Magarvey, Molecular mechanisms of membrane targeting antibiotics, *Biochim. Biophys. Acta, Biomembr.*, 2016, **1858**, 980–987.
100. Z. Zerrouk, S. Alexandre, C. Lafontaine, V. Norris and J.-M. Valleton, Inner membrane lipids of escherichia coli form domains, *Colloids and Surfaces B: Biointerfaces*, 2008, **63**, 306–310.
101. R. M. Epand and R. F. Epand, Domains in bacterial membranes and the action of antimicrobial agents, *Mol. BioSyst.*, 2009, **5**, 580–587.
102. R. F. Epand, M. A. Schmitt, S. H. Gellman and R. M. Epand, Role of membrane lipids in the mechanism of bacterial species selective toxicity by two α/β-antimicrobial peptides, *Biochim. Biophys. Acta, Biomembr.*, 2006, **1758**, 1343–1350.
103. R. M. Epand, S. Rotem, A. Mor, B. Berno and R. F. Epand, Bacterial membranes as predictors of antimicrobial potency, *J. Am. Chem. Soc.*, 2008, **130**, 14346–14352.
104. P. Joanne, C. Galanth, N. Goasdoue, P. Nicolas, S. Sagan, S. Lavielle, G. Chassaing, C. E. Amri and I. D. Alves, Lipid reorganization induced by membrane-active peptides probed using differential scanning calorimetry, *Biochim. Biophys. Acta, Biomembr.*, 2009, **1788**, 1772–1781.
105. R. F. Epand, B. P. Mowery, S. E. Lee, S. S. Stahl, R. I. Lehrer, S. H. Gellman and R. M. Epand, Dual mechanism of bacterial lethality for a cationic sequence-random copolymer that mimics host-defense antimicrobial peptides, *J. Mol. Biol.*, 2008, **379**, 38–50.
106. B. Kwon, A. J. Waring and M. Hong, A 2h solid-state nmr study of lipid clustering by cationic antimicrobial and cell-penetrating peptides in model bacterial membranes, *Biophys. J.*, 2013, **105**, 2333–2342.
107. F. Sgolastra, B. M. deRonde, J. M. Sarapas, A. Som and G. N. Tew, Designing mimics of membrane active proteins, *Acc. Chem. Res.*, 2013, **46**, 2977–2987.
108. H. Takahashi, G. A. Caputo, S. Vemparala and K. Kuroda, Synthetic random copolymers as a molecular platform to mimic host-defense antimicrobial peptides, *Bioconjugate Chem.*, 2017, **28**, 1340–1350.

109. T. L. Raguse, E. A. Porter, B. Weisblum and S. H. Gellman, Structure-activity studies of 14-helical antimicrobial-peptides: probing the relationship between conformational stability and antimicrobial potency, *J. Am. Chem. Soc.*, 2002, **124**, 12774–12785.
110. E. A. Porter, B. Weisblum and S. H. Gellman, Mimicry of host-defense peptides by unnatural oligomers: antimicrobial -peptides, *J. Am. Chem. Soc.*, 2002, **124**, 7324–7330.
111. P. S. Farmer and E. J. Ariëns, Speculations on the design of nonpeptidic peptidomimetics, *Trends Pharmacol. Sci.*, 1982, **3**, 362–365.
112. R. P. Cheng, S. H. Gellman and W. F. DeGrado, β-peptides: from structure to function, *Chem. Rev.*, 2001, **101**, 3219–3232.
113. N. P. Chongsiriwatana, J. A. Patch, A. M. Czyzewski, M. T. Dohm, A. Ivankin, D. Gidalevitz, R. N. Zuckermann and A. E. Barron, Peptoids that mimic the structure, function, and mechanism of helical antimicrobial peptides, *Proc. Natl. Acad. Sci. U. S. A.*, 2008, **105**, 2794–2799.
114. B. Mensa, Y. H. Kim, S. Choi, R. Scott, G. A. Caputo and W. F. DeGrado, Antibacterial mechanism of action of arylamide foldamers, *Antimicrob. Agents Chemother.*, 2011, **55**, 5043–5053.
115. D.-W. Zhang, X. Zhao, J.-L. Hou and Z.-T. Li, Aromatic amide foldamers: Structures, properties, and functions, *Chem. Rev.*, 2012, **112**, 5271–5316.
116. I. Huc, Aromatic oligoamide foldamers, *Eur. J. Org. Chem.*, 2004, **2004**, 17–29.
117. Z. M. AL-Badri, A. Som, S. Lyon, C. F. Nelson, K. Nüsslein and G. N. Tew, Investigating the effect of increasing charge density on the hemolytic activity of synthetic antimicrobial polymers, *Biomacromolecules*, 2008, **9**, 2805–2810.
118. A. Al-Ahmad, D. Laird, P. Zou, P. Tomakidi, T. Steinberg and K. Lienkamp, Nature-inspired antimicrobial polymers–assessment of their potential for biomedical applications, *PLoS One*, 2013, **8**, e73812.
119. E. Marie, S. Sagan, S. Cribier and C. Tribet, Amphiphilic macromolecules on cell membranes: From protective layers to controlled permeabilization, *J. Membr. Biol.*, 2014, **247**, 861–881.
120. D. S. S. M. Uppu, M. M. Konai, U. Baul, P. Singh, T. K. Siersma, S. Samaddar, S. Vemparala, L. W. Hamoen, C. Narayana and J. Haldar, Isosteric substitution in cationic-amphiphilic polymers reveals an important role for hydrogen bonding in bacterial membrane interactions, *Chem. Sci.*, 2016, 7, 4613–4623.
121. E.-R. Kenawy, S. D. Worley and R. Broughton, The chemistry and applications of antimicrobial polymers: a state-of-the-art review, *Biomacromolecules*, 2007, **8**, 1359–1384.
122. P. Li, X. Li, R. Saravanan, C. M. Li and S. S. J. Leong, Antimicrobial macromolecules: synthesis methods and future applications, *RSC Adv.*, 2012, **2**, 4031–4044.
123. L. Timofeeva and N. Kleshcheva, Antimicrobial polymers: mechanism of action, factors of activity, and applications, *Appl. Microbiol. Biotechnol.*, 2011, **89**, 475–492.

124. A. Muñoz-Bonilla and M. Fernández-García, Polymeric materials with ntimicrobial activity, *Prog. Polym. Sci.*, 2012, **37**, 281–339.
125. M. S. Ganewatta and C. Tang, Controlling macromolecular structures towards effective antimicrobial polymers, *Polymer*, 2015, **63**, A1–A29.
126. U. Baul and S. Vemparala, Influence of lipid composition of model membranes on methacrylate antimicrobial polymermembrane interactions, *Soft Matter*, 2017, **13**, 7665–7676.
127. K. Glinel, P. Thebault, V. Humblot, C. M. Pradier and T. Jouenne, Antibacterial surfaces developed from bio-inspired approaches, *Acta Biomater.*, 2013, **8**, 1670–1684.
128. K. Kuroda, G. A. Caputo and W. F. DeGrado, The role of hydrophobicity in the antimicrobial and hemolytic activities of polymethacrylate derivatives, *Chem. - Eur. J.*, 2009, **15**, 1123–1133.
129. E. F. Palermo and K. Kuroda, Chemical structure of cationic groups in amphiphilic polymethacrylates modulates the antimicrobial and hemolytic activities, *Biomacromolecules*, 2009, **10**, 1416–1428.
130. G. A. Caputo and E. London, Cumulative effects of amino acid substitutions and hydrophobic mismatch upon the transmembrane stability and conformation of hydrophobic -helices, *Biochemistry*, 2003, **42**, 3275–3285.
131. S. Jaud, M. Fernández-Vidal, I. M. Nilsson, N. M. Meindl-Beinker, N. C. Hübner, D. J. Tobias, G. von Heijne and S. H. White, Insertion of short transmembrane helices by the sec61 translocon, *Proc. Natl. Acad. Sci. U. S. A.*, 2009, **106**, 11588–11593.
132. S. Vanni, H. Hirose, H. Barelli, B. Antonny and R. Gautier, A subnanometre view of how membrane curvature and composition modulate lipid packing and protein recruitment, *Nat. Commun.*, 2014, **5**, 4916.
133. L. Vamparys, R. Gautier, S. Vanni, W. F. D. Bennett, D. P. Tieleman, B. Antonny, C. Etchebest and P. F. J. Fuchs, Conical lipids in flat bilayers induce packing defects similar to that induced by positive curvature, *Biophys. J.*, 2013, **104**, 585–593.
134. P. L. Harrison, G. R. Heath, B. R. G. Johnson, M. A. Abdel-Rahman, P. N. Strong, S. D. Evans and K. Miller, Phospholipid dependent mechanism of smp24, an α-helical antimicrobial peptide from scorpion venom, *Biochim. Biophys. Acta, Biomembr.*, 2016, **1858**, 2737–2744.
135. B. Antonny, Mechanisms of membrane curvature sensing, *Annu. Rev. Biochem.*, 2011, **80**, 101–123.
136. G. Drin and B. Antonny, Amphipathic helices and membrane curvature, *FEBS Lett.*, 2010, **584**, 1840–1847.
137. J. Bigay, J.-F. Casella, G. Drin, B. Mesmin and B. Antonny, Arfgap1 responds to membrane curvature through the folding of a lipid packing sensor motif, *EMBO J.*, 2005, **24**, 2244–2253.
138. A. A. Polyansky, R. Ramaswamy, P. E. Volynsky, I. F. Sbalzarini, S. J. Marrink and R. G. Efremov, Antimicrobial peptides induce growth of phosphatidylglycerol domains in a model bacterial membrane, *J. Phys. Chem. Lett.*, 2010, **1**, 3108–3111.

139. W. Kopéc, J. Telenius and H. Khandelia, Molecular dynamics simulations of the interactions of medicinal plant extracts and drugs with lipid bilayer membranes, *FEBS J.*, 2013, **280**, 2785–2805.
140. A. Arouri, M. Dathe and A. Blume, Peptide induced demixing in PG/PE lipid mixtures: a mechanism for the specificity of antimicrobial peptides towards bacterial membranes? *Biochim. Biophys. Acta, Biomembr.*, 2009, **1788**, 650–659.

CHAPTER 5

Synthetic Cationic Water-soluble Antimicrobial Polymers: An Alternative to Conventional Small-molecule Antibiotics

STEVEN MANKOCI, CHAO PENG AND ABRAHAM JOY*

Department of Polymer Science, The University of Akron, Akron, Ohio, 44325 USA
*Email: abraham@uakron.edu

5.1 Introduction

The world is on the verge of a global health crisis due to the rise of antimicrobial resistance among pathogenic bacteria. There have been various recent reports and case studies of outbreaks and individual incidents of bacterial infections that were impossible or very difficult to treat because none of the antimicrobials available worked to control the infection.[1–4] Within the USA alone, there have been multiple high-profile cases where sick individuals were unable to be treated because all available antimicrobials did not work. However, it is the poorer parts of the world that are affected disproportionately by antimicrobial resistance because of the limited access to all possible treatment options and the rapid spread of resistant organisms due to the lack of sanitary practices. It is estimated that approximately 700 000 people worldwide die annually as a result of antimicrobial

Biomaterials Science Series No. 5
Antimicrobial Materials for Biomedical Applications
Edited by Abraham J. Domb, Konda Reddy Kunduru and Shady Farah

Published by the Royal Society of Chemistry, www.rsc.org

resistance.[5] Yet, this number is expected to rise to a staggering 10 million deaths annually, and to become the greatest single cause for morbidity worldwide if no drastic actions are taken to address this problem.[5] Bacteria have developed many different, sophisticated mechanisms of resistance that are capable of combatting the action of antimicrobials. These mechanisms of resistance may be inherent to the species of the organism, since it provides an advantage to the fitness of the organism, or it may be generated spontaneously due to adaptation to environmental pressures. Regardless of the reason for the development of resistance, there are various common mechanisms that bacteria can adopt as a strategy to attenuate the effects of antimicrobials.[6]

One widespread mechanism of action is the synthesis of an enzyme that is capable of either modifying and inactivating the antimicrobial or destroying the antimicrobial. A typical example of this mechanism is the production of the enzyme β-lactamase by certain bacteria. To date, there have been more than 800 different β-lactamase variants identified to destroy the β-lactam ring structure, which is an important moiety for the function of penicillin and all of its derivatives.[7] Historically, in order to get around the action of β-lactamase, different chemical variants that contain the same β-lactam chemical structure motif have been synthesized and developed as antimicrobials. However, this strategy has been largely unsuccessful because bacteria have been able to adapt and develop resistance to these new antimicrobials rather quickly and it has been shown that it is typically a matter of months that bacteria become resistant to a new β-lactam antimicrobial after it is introduced to clinics.

Another common mechanism of resistance to antimicrobials is developed by modifying the permeability of the antibiotic molecules through the cytoplasmic membrane and the outer membrane, in the case of Gram-negative bacteria. For many small-molecule antimicrobials to function, they need to permeabilize the bacterial cell in order to reach their specific target site. For example, quinolones and fluoroquinolones must enter the cell to inhibit the action of topoisomerases, which are enzymes that are important for controlling the coiling of DNA during replication and transcription. Aminoglycosides and tetracyclines require entrance into the cell in order to inhibit protein synthesis. To counter the action of these antimicrobials, bacteria can either change the permeability of antimicrobials through the membrane by modifying the pores in the membrane or by the use of efflux pumps, which are active transport transmembrane proteins that expel the antimicrobial outside the cytoplasm or the periplasm.

Bacteria have also been documented to resist the action of antimicrobials by modifying the target sites of the antimicrobials. Since many antimicrobials function by affecting specific target sites, such as a specific catalytic site of enzyme(s) vital for metabolism, bacteria can either protect this target site using special proteins or modifying the target sites themselves. A very well studied example of target site protection originally identified among *Streptococcus* spp. involves the use of a protein known as TetM, which functions to remove tetracycline from its binding site within the ribosome. TetM can change the conformation of the target site that tetracycline binds, which prevents the binding of tetracycline while simultaneously allowing the binding of ternary complex that contains the aminoacyl-tRNA for protein synthesis.[8] In addition, there are ways in which bacteria can modify the target site of antimicrobials *via* genetic mutations or enzymatic modification so that the antimicrobial doesn't work anymore, yet the target site is still able to function properly.

In order to address the threat of antimicrobial resistance among pathogenic bacteria, the Centers for Disease Control and Prevention (CDC) has identified four core actions that are necessary to help combat resistance. These actions are (1) prevent the transmission of infections; (2) track the spread of resistant bacteria; (3) utilize the antimicrobial tools at our disposal more effectively; and (4) promote the development of novel antimicrobials and diagnostics.[9] Each of these four actions is a very worthy endeavor; however, the action that is perhaps most appealing and suitable for the scientific research community to address is the need for novel antimicrobials and diagnostics. One class of molecules that has been gaining more and more attention is water-soluble synthetic cationic polymers that are capable of controlling bacterial infections. Cationic antimicrobial macromolecules (CAMs) are different from conventional small molecule antibiotics in their mode of action. Instead of acting on a very specific target sites on bacteria, antimicrobial polymers target bacteria in a way that is much broader and less specific, so it is much more difficult for bacteria to adapt and resist their action. Broadly, most CAMs target the bacterial membranes and do not require a specific protein target to function. The purpose of this chapter is to provide a brief review of what is currently known about CAM chemistry, function, and mechanisms of action, while providing insight into possible future investigations.

5.2 Biocidal Polymers

There are many CAMs that have been discovered and developed since the concept of polymers was established in the early 1900s. Broadly, CAMs can be sorted based on the original purposes to which they were applied and the means through which they are synthesized. In this section, compounds referred to as "biocidal polymers" are discussed. Historically, this class of CAMs has been known for a long time and includes materials that are completely synthetic in nature. The broad term "biocide" has been selected to describe the action of these polymers, because many of them were synthesized with a broad objective of inhibiting the growth or killing bacteria, often with significantly less regard for the compatibility or survival of other cell types, such as mammalian cells, that may come into contact with the material. Many of these materials were originally synthesized to be used as disinfectants.

5.2.1 Polyhexanide

The oldest synthetic material that has been used as a water-soluble CAM is polyhexanide, which is also known as polyhexanide and poly(hexamethylene biguanide). It is a synthetic polymer in which biguanide groups are linked together *via* a six-carbon hydrocarbon chain along its backbone. Polyhexanide was originally synthesized and investigated in the early 1950s by Rose and Swain, who worked for Imperial Chemical Industries (ICI).[10] The objective of the original research was to synthesize novel antimalaria therapeutics, since other biguanide-containing molecules were found to be active against *Plasmodium* spp., the causative agent for malaria. Even though polyhexanide is not very effective in the treatment of malaria, it was soon discovered that polyhexanide and chlorhexidine, another frequently used disinfectant that has a similar chemical structure and was discovered around the same time, are both rather effective against bacteria. In the original work, it was noted that in order for both chemicals to have their antimicrobial properties, it is essential for them to contain at least two biguanide groups and for the hydrocarbon chain between the biguanide groups to contain either 5, 6, or 7 carbons.[10]

Traditionally, polyhexanide has been produced by step-growth polymerization of hexamethylenediamine and sodium dicyanamide to yield number-average molecular weights (M_n) in the range of approximately 1–2 kDa or 2–12 repeat units (Scheme 5.1).[11,12] Interestingly, resulting polymers (or oligomers) have been shown to have four different end group functionalities. As expected from the polymerization, polyhexanide has been shown to have a mixture of cyanoguanidine groups and amines. However, it has also been shown that during the synthesis of polyhexanide, there exists a dynamic state of polymerization/depolymerization due to the instability of the biguanide functional group at elevated temperatures.[13] As a result, polyhexanide has been shown to contain guanidine and cyanoamine terminal groups as well.

Polyhexanide is very soluble (>20% w/v) in polar solvents like water and alcohols and minimally soluble in less polar solvents.[14] It has been documented that polyhexanide exhibits a critical micelle concentration of between 20 and 50 mM in aqueous solution as determined by surface tension and electrical impedance measurements.[12] In addition, the biguanide group is very basic and shows two pKa values for the two imido groups in each repeat unit: $pK_{a1} = pH\ 2.0–3.0$ and $pK_{a2} = pH\ 10.5–11.5$.[14] Therefore, at physiological pH, each biguanide group is monoprotonated and carries a charge of 1+ on each repeat unit.

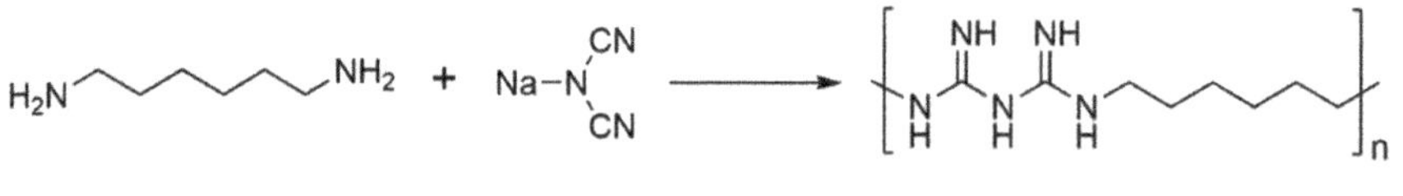

Scheme 5.1 Step growth polymerization of hexamethylenediamine and sodium dicyanamide.

After the initial discovery of polyhexanide in the 1950s, it went on to be used in non-medical consumer products, such as in the preservation of leather, the treatment of pools, a preservative in cosmetics, and in the disinfection of food products.[14] In the 1980s and 1990s, polyhexanide regained interest and it was investigated for biomedical applications. It was in the early 1990s that polyhexanide was first used as an antiseptic during surgical procedures.[14] Since then, there have been various medical products in which polyhexanide has been incorporated, such as mouthwashes and wound dressings,[15–17] as polyhexanide is relatively non-toxic, with observed rat oral LD_{50} values in the range ~1–5 $g\,kg^{-1}$ and no apparent ability to cause significant toxicity upon chronic exposure to small amounts.[14,18] Also, polyhexanide has been shown to have a broad-spectrum of effectiveness against many different potentially pathogenic bacteria, fungi, and protozoans. Under controlled laboratory conditions, it has been shown that polyhexanide can effectively control the growth and kill microbial pathogens at concentrations ranging from 0.1 to 100 $\mu g\,mL^{-1}$, depending on the pathogen and the culture conditions.[18] Additionally, it has been shown to be fast acting and is capable of achieving a 99.999% killing of *Escherichia coli*, *Staphylococcus aureus*, *Enterococcus faecalis*, *Pseudomonas aeruginosa*, and *Candida albicans* in less than 5 min when exposed to a concentration of 200 $\mu g\,mL^{-1}$.[19]

The primary killing mechanism of polyhexanide, as well as most other CAMs, has been believed to be the disruption of bacterial membranes. The membranes of most bacteria are composed of lipid moieties that contain an abundance of anionic charges on the surface presented to the outside environment, especially when compared to mammalian cells. For example, the cytoplasmic membranes of most mammalian cells are composed of phospholipids that have a phosphatidyl choline head group on the side of the bilayer that is presented to the outside environment, which results in a net neutral membrane surface, while many bacteria present lipids such as cardiolipin, phospholipids with phosphatidyl glycerol head groups, and, in the case of Gram-negative bacteria, lipopolysaccharide, all of which carry a net negative charge across the molecule. Therefore, polyhexanide can interact with the bacterial membrane surface by electrostatic interaction. A significant piece of early evidence that polyhexanide disrupts bacterial membranes was provided by Davies *et al.*[20] In their work, they grew *E. coli* spheroplasts, or bacteria that have had their outer membrane and peptidoglycan layers removed, in a medium containing ^{32}P- and ^{13}C-labeled nutrients and measured the release of the radio-labeled components from the inside of the cells. It was observed that the addition of polyhexanide was disrupting the membrane at a range of concentrations and, when the concentration was high enough, eventually led to the formation of biomolecule–polyhexanide aggregates.[20] Further investigation with liposomes that contained net neutral and anionic head groups has indicated that polyhexanide interacts more selectively towards membranes that contain anionic head groups.[21] This was determined using liposomes that incorporated 1,6-diphenyl-l,3,5-hexatriene, a dye which exhibits changes in the polarization of its fluorescence with changes in the fluidity of the membrane, and through changes in the phase transition temperature of the lipids measured by differential scanning calorimetry.

However, it has been observed recently that polyhexanide may affect bacteria through a completely different mechanism of action. In contrast to previous experiments that indicate that the primary mechanism of action of polyhexanide is through the disruption of bacterial membranes, work performed by Good and co-workers has provided intriguing evidence that the primary mechanism of action is by being taken into the bacterial cell through transmembrane transport and binding to chromosomes, which causes them to precipitate.[22,23] The precipitation of chromosomes then leads to an arrest of cell division and the eventual death of the cell. In experiments with the dye SYTOX green, which is frequently used to determine the integrity of cell membranes, it was shown that the dye is incapable of entering bacteria exposed to polyhexanide, which is in contradiction to various other reports that indicate membrane disruption. Instead, it was observed that fluorescein isothiocyanate (FITC)-tagged polyhexanide accumulates in the cytoplasm of bacterial cells and condensed the chromosomes at various concentrations, as shown using fluorescence microscopy (Figure 5.1).

5.2.2 Quaternary Ammonium Functionalized Polymers

Another class of water-soluble polymeric biocides that have been quite heavily investigated is polymers that functionalize with quaternary ammonium (QA) functional groups, both in pendant groups and along the backbone, which impart a permanent cationic charge on the molecule no matter the changes in pH. The use of small-molecule QA salts in antimicrobial applications can be dated back to the early 20th century when they were originally identified as antimicrobials in 1916 by Jacobs and co-workers.[24–26] However, it wasn't until the 1930s, when Domagk investigated QA salts having at least one long aliphatic chain, that this class of molecules began to gain widespread attention and began to be used as antimicrobials.[27] Today, the small-molecule QA salts are ubiquitous throughout consumer products and are the main active ingredient in many disinfectant products, such as Lysol. Such effectiveness in controlling the growth of bacteria leads to efforts to make synthetic polymers that are functionalized with QAs on pendant groups to (1) impart antimicrobial properties to solid materials; and (2) help localize the effects of the QAs to a certain area and prevent the spread of the biocide away from the point of application. Therefore, there has been much work in applying polymeric QA salts to coatings and grafting them to surfaces. However, in this section, small subclass of these polymers that were synthesized and investigated as water-soluble polymers is discussed (Figure 5.2).

Water-soluble cationic polymers that incorporate QA functional groups into the pendant group can be synthesized in a variety of ways. In most

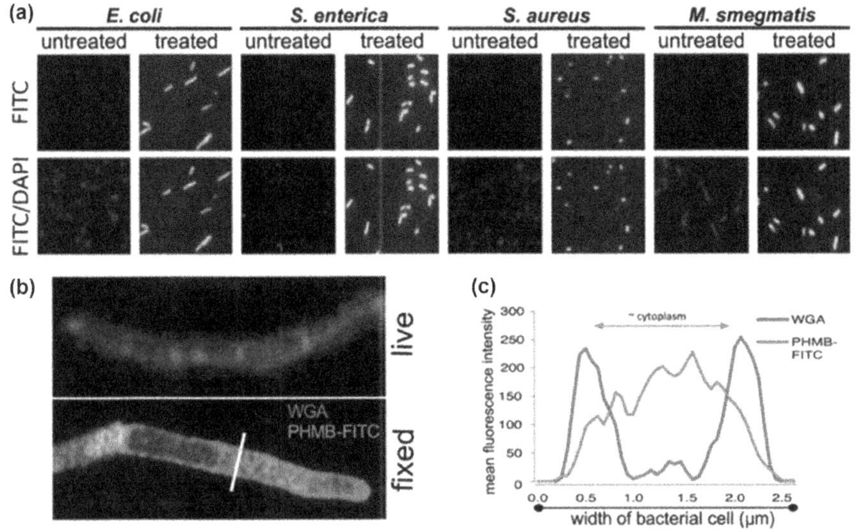

Figure 5.1 PHMB effects on cell membrane permeability and entry into bacteria. (a) Fluorescence microscopy of PHMB-FITC entry into diverse bacteria; PHMB-FITC (2 µg mL^{-1}) was added to bacterial cultures and the cells were counterstained with DAPI; (b) confocal image showing localization of PHMB-FITC in *Bacillus megaterium*; bacteria were counterstained with the membrane localizing probe wheat germ agglutinin (WGA) conjugated to Alexa Fluor-555 and visualized as live (top) and fixed (bottom) cells; bar = 5 µm; (c) fluorescence intensity profile plot analysis of cellular localization of PHMB-FITC and WGA fluorescence (the white line in (b) indicates the cross-section used for analysis).

cases, these polymers are synthesized by chain growth polymerizations of vinyl monomers such as acrylates, methacrylates, functionalized styrenes, and vinyl pyridines. The formation of the QA can be achieved either before or after polymerization of the monomer(s).[28] If the monomer is functionalized with the QA prior to polymerization, care must be taken to ensure that the presence of the functional group does not have any detrimental effects on polymerization or destabilize the monomer in any way. While, if the polymer is functionalized with the QA after polymerization, it can be rather difficult to reach complete conversion due to unfavorable steric and electrostatic interactions introduced by neighboring groups.[29] To form QA moieties, there are two synthetic strategies that are usually employed. One approach to synthesize a QA is through the Menshutkin reaction of a tertiary amine and an alkyl halide, an S_N2 reaction where the nucleophilic lone pair of electrons on the nitrogen forms a bond with the electrophilic carbon of the alkyl halide. Also, QAs are frequently formed from another S_N2 reaction that

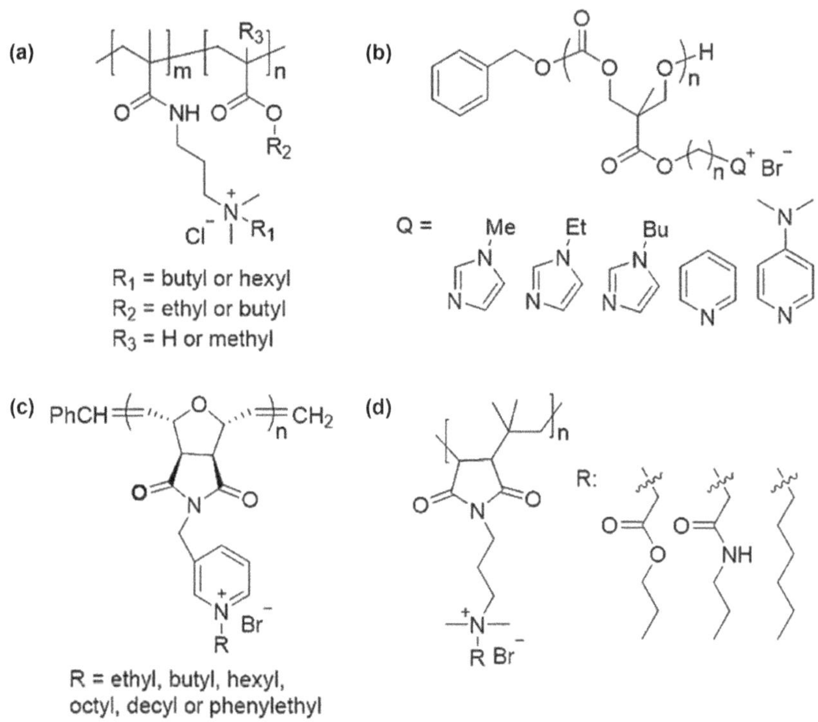

Figure 5.2 Typical examples of quaternary ammonium functionalized antimicrobial polymers; (a) acrylate/acrylamide based copolymers containing quaternary ammonium salts (b) antimicrobial polycarbonates with nitrogen-containing heterocycles (c) amphiphilic polyoxanorbornenes with various quaternary alkyl pyridinium side chains (d) *N*-alkyl maleimide based antimicrobial polymers with quaternary ammonium salts.

involves the reaction of a primary amine with an alkyl halide through exhaustive alkylation. This method of synthesis involves the stepwise synthesis of a secondary amine, then a tertiary amine, and finally, a QA through the stepwise nucleophilic attack of the alkyl halide by the lone pair of electrons on the nitrogen of the amine.

When making a QA functionalized polymer, there are various important parameters that drive their rational design to enable them to exhibit antimicrobial properties. For example, polymerization of an antimicrobial monomer does not necessarily lead to an antimicrobial polymer, as is the case with 4-vinyl-*N*-benzylpyridinium chloride and its resulting polymer.[30] While, at the same time, the opposite is true; the monomer itself doesn't necessarily need to exhibit antimicrobial activity for the polymer to have antimicrobial properties. This can be observed in the behavior of a methacrylate monomer that contains a pendant QA based on a DABCO ring that has either a hexyl or a butyl group; the monomer is not antimicrobial, while the resulting polymer is antimicrobial.[31]

Molecular weight has also been indicated in modulating the behavior of QA polymers. It is necessary for the QA polymer to reach a specific molecular weight, especially in cases where the monomer itself is not antimicrobial, to have an optimal antimicrobial effect. Yet, there also exists a maximum molecular weight at which the antimicrobial properties will begin to decrease. It is believed that molecular weight affects the ability of the molecule to bind to and disrupt the membrane.[32,33] For instance, when comparing the ability of a charged small molecule species and a polycation, the polycation will bind to anionic surfaces, such as that of a bacterial membrane, more tightly than that of a small molecule. However, at the same time, with increasing molecular weight, the solubility of the polymer as well as the ability for the polymer to transcend the outer most layers of the bacteria, *i.e.* the thick peptidoglycan layer of Gram-positive bacteria, the outer membrane of Gram-negative bacteria, and the glycocalyx of many bacteria, will decrease with increasing molecular weight. Therefore, to achieve the most desirable antimicrobial activity, there is typically a range of molecular weights that work most effectively and this range of molecular weights can vary between different polymers. Some of the earliest work of this was demonstrated by Ikeda *et al.*, in which they fractionated poly(vinylbenyl ammonium chloride) into molecular masses of narrow dispersity.[34] When testing the effectiveness of the polymer fractions, as well as the monomer, at a fixed concentration in killing *Bacillus subtilis*, the polymers of an intermediate molecular mass of approximately 16 kDa were most effective.

Another important factor in the synthesis of successful water-soluble antimicrobial QA polymers is the balance of hydrophobic and cationic character. Even with the small-molecule QA salts, there exists a balance in achieving optimal activity through modification of the length of the alkyl chains attached to the nitrogen. In many instances, it has been found that a small-molecule QA salt has optimal effectiveness against bacteria when it has a ~8–12-carbon alkyl chain attached to the nitrogen. Analogously, there have been various investigations into the effect of multiple parameters on QA functionalized polymers, such as length of the hydrophobic segment within the backbone, length of the hydrophobic segment connecting the pendant QA, and the length of the alkyl chains attached to the nitrogen of the QA.[32,33] In many instances, optimal antimicrobial activity of QA functionalized polymers can be achieved by modifying the length of the hydrophobic segment(s). In a study of poly(trialkylbenzylammonium chlorides) by Ikeda *et al.*, there were a variety of polymers synthesized that varied the length of the substituents on the nitrogen of the pendant QA.[35] It was found that by increasing the length of one of the alkyl substituents, a greater antimicrobial effect can be achieved. Also, more recently, Fernández-Garcia and co-workers demonstrated various important relationships in the tuning of the properties of the biocidal polymethacrylates based on quaternized 1,3-thiazole and 1,2,3-triazole side-chain groups.[36] In their investigations, they found that length and identity of the spacer between the backbone and the quaternized moiety is important; a succinate spacer is not as effective as an aliphatic hydrocarbon spacer and an aliphatic hydrocarbon spacer with the intermediate length tested works best. Also, with these polymers, it was observed that the polymers quaternized with methyl and butyl groups showed more desirable properties over the analogues that were synthesized with longer hydrocarbon chain substituents.

Some of the best-characterized and most frequently studied QA functional polymers are those functionalized with pyridinium groups, which are the quaternized form of the nitrogen-containing heterocycle pyridine. Originally, it was discovered that insoluble, cross-linked poly(4-vinyl pyridinium bromide) resin was capable of very effectively binding and retaining various Gram-negative and Gram-positive bacteria, such as *E. coli*, *S. aureus*, and *P. aeruginosa*.[37] Interestingly, it was found that this resin was capable of irreversibly binding the bacteria in a non-lethal manner. However, further investigation of the same group discovered that water-soluble poly(4-vinyl pyridinium) worked well to kill both Gram-positive and Gram-negative bacteria.[38] In particular, it was determined that poly(4-vinyl pyridinium) exhibited the greatest antimicrobial activity with an *N*-benzyl substituent, as opposed to an aliphatic alkyl chain, and demonstrated antimicrobial activity that was comparable to that of widely used small molecule antimicrobial disinfectants benzalkonium chloride and chlorohexidine. Copolymers of poly(4-vinyl pyridinium-*co*-styrene) have also been investigated for their antimicrobial properties as well as for their toxicity to mammalian organisms.[39] It was observed with this series of copolymers that greater incorporation of the 4-vinyl pyridinium group resulted in more efficacious killing of *E. coli*, *S. aureus*, and *Pseudomonas geniculate*. Also, it was demonstrated that the copolymer is a non-irritant with low oral toxicity. Pyridinium functional groups have additionally been added to polymers synthesized *via* ring-opening metathesis polymerization and characterized for their antimicrobial structure–property relationships.[40] The study included a series of polymers that were synthesized with two molar masses, ~3 kDa and ~10 kDa, and the pyridinium group formed using a variety of alkyl bromides and phenylethyl bromide. Antimicrobial testing revealed that there was a very clear difference in performance between the polymers that contained *N*-alkyl substituents of 4 carbons or fewer and those that contained the more hydrophobic substituents of 6 carbons or more, with the polymers that had the more hydrophobic substituents being more potent antimicrobials. However, with the increase in antimicrobial activity in the polymers with the more hydrophobic substituents, there was also a noticeable decrease in the compatibility of the polymers with human blood cells, indicating that the increased antimicrobial performance of the antimicrobial polymers also results in increased activity against non-bacterial cells. For example, the polymer with ethyl substituent (3 kDa) exhibited a HC_{50} of 4030 $\mu g\,mL^{-1}$ and a selectivity (HC_{50}/MIC) of 20 for *E. coli* and *B. subtilis*, while the HC_{50} and selectivity decreased to 8 $\mu g\,mL^{-1}$ and 1.7, respectively, when the polymer has an octyl substituent (3 kDa).

QA functional groups have also been incorporated into the backbones of polymer chains and investigated for their antimicrobial properties. These polymers are a special class of QA polymers, which are frequently referred to as ionene polymers, that have been studied for a variety of applications in addition to their use as antimicrobials, such as for use as a flocculating agent in the treatment of clay suspensions and waste water, as antistatic agents, and as hair conditioning agents. These polymers were first synthesized in the 1930s, but they were not very widely investigated until after Renbaum *et al.* determined the mechanism through which they are formed and demonstrated a useful means to characterize the resulting polymer's molecular weight.[41–44] The popularity of ionene polymers is in large part due to their rather simple synthesis. The most frequently used method to synthesize ionene polymers is through the step-growth polymerization of di-tertiary amines with di-alkyl halides through successive Menshutkin reactions. This method of polymerization is rather efficient and polymers with molecular weights in the range of a few thousand Daltons all the way up into the tens of thousands of Daltons can easily be synthesized. In these polymerizations it has been observed that (1) the rate of polymerization increases with the increase polarity of the solvent used in polymerization; and (2) the reactivity of the of the halogen follows $I > Br > Cl$.[45]

In terms of investigation of antimicrobial properties, ionene polymers have not been as extensively investigated within the literature as polymers that have QA functional groups in their pendant groups. However, within the past few years, there have been multiple high-profile publications that have stimulated the resurgence in interest in these polymers.[46–48] These polymers hold great potential for antimicrobial applications because of their demonstrated effectiveness against a broad-spectrum of microbial pathogens as well as compatibility with mammalian cells. Also, there have been various patents filed that claim the superior effectiveness of these QA polymers. Originally, ionene polymers were investigated for their antimicrobial properties by Renbaum *et al.* in the early 1970s.[44,49] Since then, there have been multiple investigations on tuning their hydrophobicity to obtain the most ideal antimicrobial effect as well as studies that have been performed to better understand their mechanism of action. Ikeda *et al.* synthesized a series of ionene polymers that varied the size of aliphatic and aromatic spacers between the QAs of the backbone.[50] It was observed that the polymers were significantly more effective in controlling the growth of the Gram-positive bacteria *S. aureus* and *B. subtilis*, with concentrations in the range of 5–66 $\mu g\,mL^{-1}$, than against various Gram-positive bacteria and fungi, with effective concentrations in the range of 100 $\mu g\,mL^{-1}$ to >1000 $\mu g\,mL^{-1}$. When comparing the various alkyl chain lengths and the variations in the number of aromatic rings per repeat unit, it was observed that the ionene polymer containing two successive *o*-xylene spacers performed the best in controlling bacterial growth, while the ionene polymer containing shorter alkyl chains or hydrophilic hydroxyl groups within the spacer performed the poorest. In a mechanistic investigation of these polymers, it was shown that these polymers are capable of disrupting phospholipid membranes and the ability of the ionene polymers to induce disruption is dependent on the identity of the spacers between the QAs along the backbone.[50] A complementary investigation set out to determine the relationship between charge density and hydrophobicity with the ability of the ionene polymer to bind to cells and impact their viability.[51] Within the study, it was determined that the ionene polymers with short alkyl spacers bind to the cells without causing disruption while ionene polymers with long alkyl spacers cause disruption after binding to the cells. Also, the effectiveness of the ionene polymers can be modulated though the length of the *N*-alkyl substituents. It was observed that methyl and octyl substituents showed greater antimicrobial efficacy when compared to that of butyl and hexyl substituents. It is believed that this interesting behavior of the effect of the length of the *N*-alkyl substituents is due to changes in the conformation of the chain around the QA.

5.2.3 Quaternary Phosphonium Functionalized Polymers

Besides QAs, another important class of biocidal polymers that has been investigated is polymers that contain quaternary phosphonium (QP) functional groups. They appear rather similar to QAs in that they have a permanent positive charge present across the functional group due to bonding of the central phosphate atom with four organic substituents. In fact, the most frequent means in which QPs are synthesized appears to be very similar to the Menshutkin reaction that is used frequently to form QAs, where a tertiary phosphine reacts with an alkyl halide. However, in terms of properties of the functional groups, QAs and QPs have some rather distinct differences. Most significantly, the difference in electronegativity of the nitrogen and phosphorus atoms leads to a very different distribution of electrons and partial charges. In QAs, it has been calculated *ab initio* that the more electronegative nitrogen atom has a partial negative charge, while the surrounding carbon atoms have a partial positive charge.[52] While, in contrast, when the calculations are performed on QPs, the phosphorus atom has a partial positive charge and the surrounding carbon atoms have a partial negative charge. In addition, a significant difference between ammonium and phosphonium moieties is their stability. Phosphonium species tend to be significantly more stable and resistant to degradation than ammonium species, which has led to their utilization in anionic exchange membranes and fuel cells.[53–55]

As biocidal polymers, QP polymers have been rather well studied. In many respects, the same variables that can modulate the properties of QA polymers can also have similar effects in modulating the properties of QP polymers. This was demonstrated in a series of papers by Kanazawa, Ikeda, and Endo, where monomers synthesized from chloroalkylstyrene and tertiary phosphines were polymerized and the effects of various changes in chemical structure on antimicrobial properties were measured.[56–61] It was observed from a series of polymers with either ethyl, *n*-butyl, phenyl, or *n*-octyl substituents attached to the QP that the more hydrophobic substituents lead to more potent antimicrobial activity.[56] However, when the effect of the length of the alkyl spacer between the QP and the backbone of the polymer chain was varied as either a propyl or an ethyl chain, the polymer with the shorter spacer proved to be a better antimicrobial.[60] In terms of molecular weight, when QP polymers that ranged in weight-average molecular mass between 16 000 and 94 000 Da, it was observed that increasing molecular weight resulted in greater antimicrobial efficacy.[57] Also, there have been ionene polymers synthesized to contain a QP group along the backbone of the polymers.[58] Again, just as with the polymers that incorporated with the QP on the pendant group, these QP ionene polymers showed similar changes in antimicrobial properties when hydrophobicity was adjusted. In a study that measured the effect of a wide range of alkyl chain spacers, it was observed that the polymers that had the greatest separation of QP groups along the backbone, which was a polymer with alternating hexyl and butyl chain lengths, possessed the greatest activity against *S. aureus*, with all the polymers with fewer carbons in the backbone showing lower activity. There have additionally been investigations into the mechanism of action of QP polymers and they indicate that the polymers function to disrupt membranes of cells.[61]

When compared head-to-head, QP polymers have been found to work better as antimicrobials than QA polymers. Although it is not known exactly why QP polymers are stronger antimicrobials, it has been speculated that it has to do with how tightly anions bind to the quaternary species; since phosphorus is has more electrons than nitrogen, the ionic radius of a phosphonium species is larger than that of an ammonium.[28] Nonetheless, whatever the reason, there is significant experimental evidence that shows greater activity of QP polymers. Kenawy *et al.* have shown using methacrylate-based polymers functionalized with QA and QP functional groups that the polymers incorporated with QP functional group demonstrated greater antimicrobial effectiveness in killing various Gram-negative and Gram-positive bacteria as well as fungi when compared to a QA-functionalized polymer.[62] Also, in the work of Kanazawa, Ikeda, and Endo, there was a comparison of the ability of a QP polymer they synthesized to an analogous QA polymer. The QP polymer was shown to be able to kill *S. aureus* more quickly than the QA polymer.[56] There have been various investigations using water-insoluble materials that show QP functional groups to have a greater antimicrobial effectiveness than QA functional groups.[63,64]

5.3 Synthetic Mimics of Antimicrobial Peptides

The final category of CAMs is known as synthetic mimics of antimicrobial peptides (SMAPs), which form a class of polymers that have been synthesized to intentionally mimic CAPs. Superficially, this class of polymers appears to be no different to the biocidal polymers discussed in Section 5.2. However, there do exist some important differences that differentiate these two types of synthetic antimicrobial polymers. Unlike biocidal polymers, which in many ways can be considered unintentionally biomimetic, this class of synthetic polymers was developed with the intention to determine whether synthetic polymers that are designed to mimic CAPs can function in a way that matches or exceeds the ability of CAPs to control the growth of bacteria and demonstrate desirable interactions with the cells of the organism that is being treated for infection. To further differentiate this class of molecules from biocidal polymers, there are differences in the types of cationic functional groups used by each class of CAM. SMAPs can be characterized by their use of biologically utilized cationic functional groups, such as the primary amine found on the amino acid lysine, the guanidine group found on the amino acid arginine, and the imidazole group found on the amino acid histidine. Whereas biocidal polymers, which draw inspiration from biguanide antimalaria drugs and small-molecule QA salts, typically use functional groups that are not naturally found on CAPs, such as biguanide, ammonium, and phosphonium moieties. These differences between polymeric biocides and SMAPs can also be viewed as two different generations of synthetic cationic polymers. Biocidal polymers are the first generation of synthetic polymers that were developed concurrently yet independently from CAPs. SMAPs, in contrast, are a newer generation of synthetic polymers that are the result of a greater understanding of how similar CAPs and biocidal polymers truly are in terms of how they affect bacteria.

Investigations of SMAPs have revealed that proper design of any synthetic polymer can result in the creation of polymers that are antimicrobial when they are synthesized in a way that mimics the basic design features of CAPs. As discussed earlier, CAPs come in many shapes, conformations, and lengths of amino acid residues, but what really ties them all together is the fact that they are amphiphilic and contain an abundance of cationic amino acids and hydrophobic amino acids. By using these simple specifications that unite all CAPs as design parameters, successful SMAPs can be synthesized that get around some of the disadvantages of CAPs, such as their high cost of production and short *in vivo* lifetime. This section explores the best-studied SMAPs categorized by the chemistry upon which they are synthesized and discuss what is known about their antimicrobial properties and mechanisms of action.

5.3.1 Polyamides

Containing perhaps the most intuitive selection of backbone chemistry for the synthesis of SMAPs, polyamides have been thoroughly investigated for their antimicrobial properties. Since the terms "polypeptide", "peptide", and "polyamide" vary depending on the branch of science and field of study, the use of the term polyamide to describe a molecule that mimics a naturally occurring peptide may seem somewhat redundant. There are indeed many synthetic peptides that are synthesized to make a better CAP that have been investigated.[65] However, here we do discuss these molecules, since they should be considered as CAPs. Instead, the focus is on polyamide-type polymers that have been assembled from monomers other than α-amino acids. To get around some of the limitations of CAPs, various polyamide SMAPs have been innovated, such as β-peptides, which are constructed from β-amino acids, and peptoids, which are polyamides constructed from *N*-functionalized glycines (Figure 5.3).[66] Polyamides synthesized in this manner are capable of evading the action of enzymes and, in some cases, still capable of achieving a range of folded structures in solution.

β-peptides are among the earlier types of SMAPs that have been investigated for their antimicrobial properties. Originally, interest in β-peptides was sparked by the broad need identified by peptide researchers for the development of a more robust means to create molecules that exhibit the ability to fold into the bioactive shapes of peptides, but without the detrimental features of peptides, *e.g.* susceptibility to protease degradation. β-peptides that are synthesized using a solid-phase peptide synthesizer, so that they have a well-defined sequence and monodisperse molecular weight distribution, have been well characterized to fold into many peptide-like secondary structures that have been shown to have peptide-like behavior in various applications. DeGrado and co-workers were among the first to investigate whether β-peptides are able to mimic the structures of CAPs and preserve the antimicrobial properties.[67] Inspired by the α-helical magainins and cecropsins, a series of helical β-peptides were synthesized that incorporated β-amino acids that resembled the α-amino acids lysine, leucine, and valine. Their results showed that the β-peptides were indeed antimicrobial, but also quite toxic towards mammalian cells. It was speculated that the reason for this was that the peptides were too hydrophobic, leading to a decrease in the selectivity of the peptides to disrupt bacterial cells over mammalian cells.[67] In a follow-up investigation, the composition of the helical β-peptides was changed: the β-amino acid that mimicked leucine was exchanged with a more hydrophilic β-amino acid that mimicked the amino acid alanine.[68] By swapping out these monomers in the sequence of the β-peptide, a greater selectivity of the killing of bacteria over mammalian cells was achieved; one of the β-peptides showed a minimum inhibitory

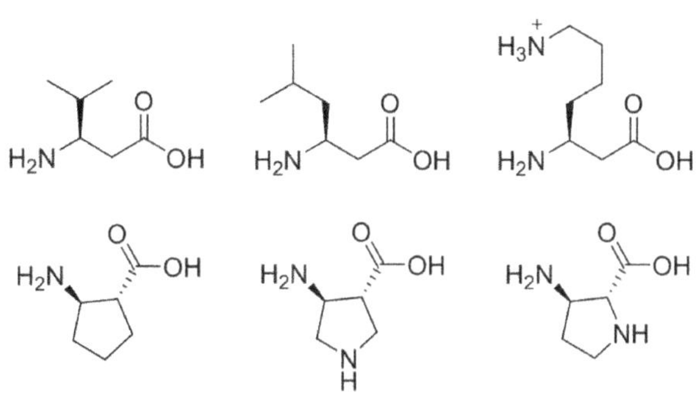

Figure 5.3 Building blocks used for β-peptide synthesis.

concentration (MIC) of 26 μg mL^{-1} and a HC_{50} of 910 μM. Additionally, using model phospholipid membranes, it was demonstrated that these β-peptides had a greater affinity to interact with and disrupt membranes that have a greater amount of anionic phospholipids. In a similar approach, Gellman and co-workers also designed a successful series of β-peptides.[69–71] However, in their approach, β-amino acids imparted a greater degree of rigidity to the synthetic peptide. The peptides were designed to form helical structures in solution and they were able to demonstrate a range of antimicrobial effectiveness depending on the functionality of the β-amino acid. The most noteworthy aspect of this work is that they synthesized a β-peptide that they named β-17 (also referred to as APC40 in later works), that has superior effectiveness against a broad spectrum of bacteria and a selectivity that other β-peptides have been incapable of matching.

Directly as a result of the success of β-peptides being applied as SMAPs, peptoids, another type of polyamide SMAP, were investigated for their antimicrobial effectiveness. Peptoids, unlike other polyamide SMAPs, are unable to form hydrogen-bonded secondary structures, but are still capable of acquiring other less stable secondary structures through interactions of the *N*-substituted groups if the sequence of monomers in the chain allows for steric or electrostatic interactions that promote formation (Figure 5.4).[72] The ability of peptoids to function as successful SMAPs was first observed by Patch and Barron, when they tested peptoids that were constructed to mimic the helical structure, cationic charge, and size of CAP magainin-2.[73] In a subsequent study, it was determined that not only did the antimicrobial peptoids resemble CAPs in structure and function, but also in their mechanism of action.[74] Evidence that antimicrobial peptoids interact and insert themselves into the membranes of phospholipids was obtained through X-ray reflectivity measurements. Even though it could not be determined from these data if the structures that are formed on the surface of the phospholipids are the same as those formed by CAPs, the results obtained were consistent with what was obtained for the CAPs pexiganan and LL-37. It was also noticed within the study that other characteristics of the peptoids were rather peptide-like as well, such as antimicrobial properties that are independent of the chirality of the helix and dependent on the proper balance of hydrophobic and cationic character.[74] Peptoids have also received attention for having antibiofilm properties.[75] The performance of some peptoids in preventing the growth of planktonic *P. aeruginosa* as well

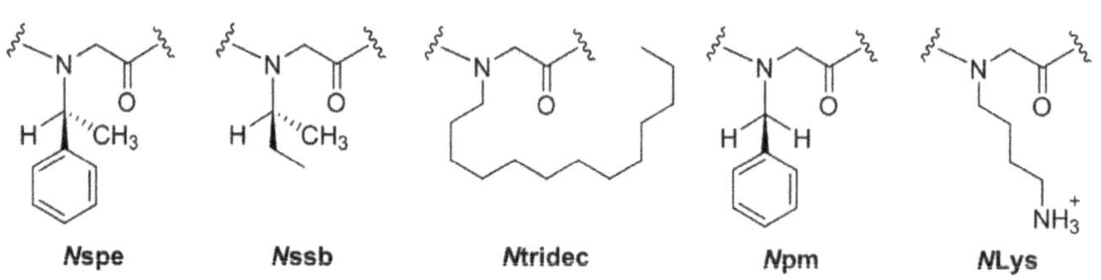

Figure 5.4 Building blocks used for the synthesis of antimicrobial peptoids.

as the formation of *P. aeruginosa* biofilms is comparable to or even exceeds the ability of currently available antimicrobial therapeutics (Table 5.1).

In much of the earlier work where polyamide SMAPs were investigated for their antimicrobial properties, the SMAPs were designed to not only mimic the chemical structure of the peptides, *i.e.* primary structure, but also the secondary structure of the peptides. However, it was soon realized that it was not necessary for the polyamides to mimic the secondary structures of peptides to have desirable antimicrobial properties. The basic design features that are most important in dictating the antimicrobial properties of SMAPs was outlined very well within the work of Mor and co-workers.[76] Within their work, they developed a class of SMAPs known as oligoacyllysines (OAKs), which are oligomers that are composed of lysine amino acids linked together by hydrocarbon spacers *via* amide bonds. OAKs demonstrated that there is not one specific hydrophobic amino acid nor is there any specific secondary structure that a peptide forms that is absolutely necessary for imparting antimicrobial properties. Instead, the most important feature for a SMAP molecule to have to function as an antimicrobial is the correct balance of cationic and hydrophobic character. This was observed with OAKs when they were synthesized with various different lengths of hydrocarbon spacers and different lengths of the hydrocarbon on the *N*-terminal end of the molecule. It was shown that the OAK that contained 8-carbon hydrocarbon spacers, seven lysine residues between each spacer, and a 12-carbon hydrocarbon tail at the *N*-terminal end, which was named $C_{12}K\text{-}7_{\alpha 8}$, functioned as a very potent antimicrobial that killed bacteria at therapeutically relevant concentrations while at the same time demonstrating very good compatibility with mammalian cells. Different to traditional antimicrobial peptides, $C_{12}K\text{-}7_{\alpha 8}$ showed very low toxicity as demonstrated by hemolysis assay using human red blood cells, which did not show dose-dependent toxicity in the tested range (0–100 μM). Additionally, at around the same time, Gellman and co-workers showed similar results that demonstrated that the hydrophobic and cationic features of the primary structures of peptides are the most important features in imparting antimicrobial

Table 5.1 Sequences of peptoids and peptides, and their antimicrobial activities against planktonic *Pseudomonas aeruginosa*.

Sequence of peptoid or peptide	MIC for *P. aeruginosa* μM	MIC for *P. aeruginosa* $mg\,L^{-1}$
H-(NLys-Nspe-Nspe)$_4$-NH$_2$	12.5	22.7
H-(NLys-Nspe-Nspe)$_3$-NLys-Nspe-NH$_2$	12.5–25	20.7–41.4
H-(NLys-Nspe-Nspe)$_2$-NLys-Nspe-L-Pro-NLys-Nspe-Nspe-NH$_2$	25–50	43.8–87.6
H-(NLys-Npm-Npm)$_4$-NH$_2$	25	42.6
H-Ntridec-NLys-Nspe-Nspe-NLys-NH$_2$	12.5–25	10.4–20.8
H-NLys-Nspe-Nspe-NLys-NH$_2$	>100	>59.5
H-(NLys-Nssb-Nssb)$_4$-NH$_2$	>100	>143.4
LLGDFFRKSKEKIGKEFKRIVQRIKDFLRNLVPRTES (LL-37)	25–50	112.3–224.6
GIGKFLKKAKKFGKAFVKILKK (Pexiganan)	12.5–25	30.9–61.8

properties to the polymer (Scheme 5.2).[77,78] Within their work, they synthesized random/statistical copolymers of nylon-3, which were synthesized *via* anionic ring opening polymerization of β-lactam monomers functionalized with hydrophobic and cationic functional groups to yield polyamides that had the same repeat units of the β-peptides they had previously investigated. The results showed that the lack of regular structure found within the random/statistical copolymers did not reduce the ability of the polymers to affect the bacteria when compared to CAPs, but instead led to the discovery that one of the copolymers actually performed better than all of the CAPs against the clinically isolated strains of *S. aureus* and *Enterococcus faecium* that were used in the study.[77] These findings led to the proposal that these polymers that do not have any structure in solution take on what was referred to as a "globally induced amphiphilic conformation", which means that the repeat units of the random/statistical copolymers are flexible enough to segregate into an unorganized amphiphilic structure that segregates into cationic and hydrophobic domains when interacting with the surface of the bacteria.

Since it has been demonstrated that no exact secondary structure is needed for polyamides to acquire antimicrobial properties, there have been many studies that have investigated the structure/property relationships and suitability of polyamides for specific applications. Further investigation of nylon-3 SMAPs has yielded various interesting observations about how to modulate the antimicrobial properties of the polymers. In a comparison of the nylon-3 random/statistical copolymers that were composed of a cationic repeat unit and one of four different hydrophobic, cyclic, and non-aromatic repeat units, it was observed that none of the tested ring sizes, which varied between five- and eight-membered rings, had a substantial effect on controlling the growth of four different types of bacteria when compared at analogous monomer ratios and molecular masses.[78] However, the changes in ring size had a more profound effect on the compatibility of the polymer with mammalian cells, with the incorporation of larger rings into the repeat unit of the polymer leading to an increase in lysis of red blood cells. A similar effect on both bacteria and blood cells was observed when comparing the effect of molecular weight on the cells, with molecular weight having very little effect on the activity of the polymers against bacteria and a significant drop off in compatibility of the polymers with red blood cells being observed as the molecular weight was increased over the range of ~10 to ~60 repeat units.[78] Subsequently, it was observed that the use of non-cyclic hydrophobic repeat units can result in the formation of polyamides that have comparable effectiveness against bacteria and superior compatibility with mammalian cells.[79] However, it appears that just as with their cyclic counterparts, if the non-cyclic hydrophobic repeat unit becomes too hydrophobic, there is a decrease in red blood cell compatibility. When comparing the polymers in terms of the distribution of the repeat units along the backbone, it was determined that the more ideal copolymers were synthesized from random/statistical distributions than from copolymers that have a blockier

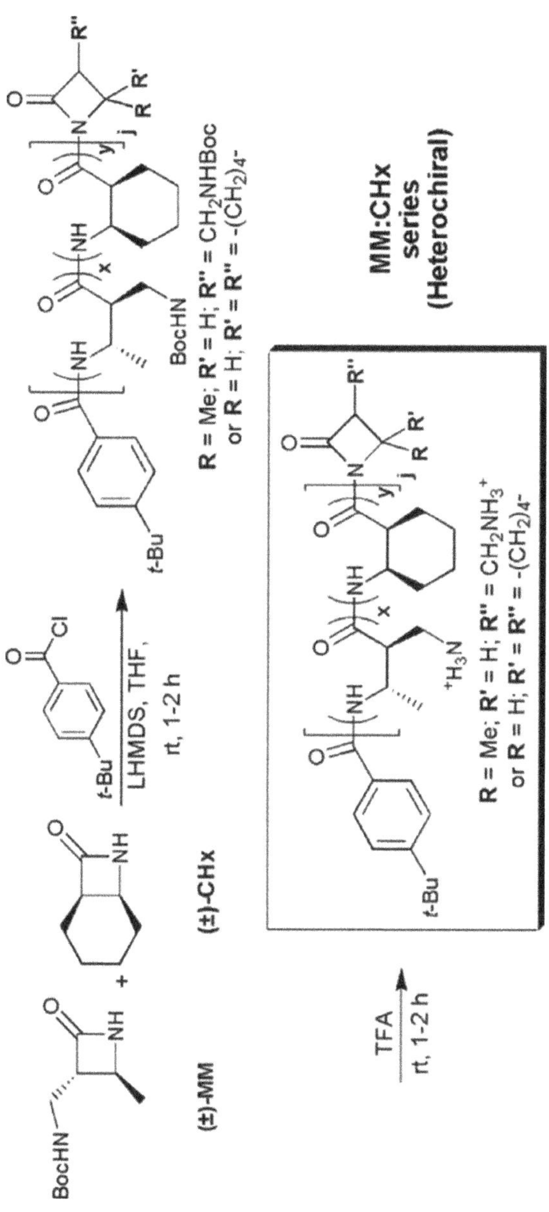

Scheme 5.2 Synthesis of antimicrobial random β-amino acid copolymers using β-lactam monomer units. Reproduced from ref. 78 with permission from American Chemical Society, Copyright 2009.

distribution, such as a diblock copolymer.[78] For each of the copolymers synthesized, the random/statistical copolymers had a significantly lower MIC against all tested bacteria than the diblock copolymers. Nylon-3 terpolymers have also been investigated, where a third hydrophilic monomer is added to the random/statistical sequence of the polymers. The addition of polar and uncharged groups to the sequence of the nylon-3 polymers, no matter whether they replace the hydrophobic repeat unit or the cationic repeat unit or both, resulted in polymers that had significantly greater mammalian cell compatibility without sacrificing antimicrobial activity.[80] For example, $MM_{60}CO_{30}HG_{10}$ exhibited a HC_{10} of 3.13 $\mu g\,mL^{-1}$. While $MM_{60}CO_{30}HS_{10}$ that replaced a small proportion of the hydrophobic subunit with uncharged polar unit exhibited much higher HC_{10} value around 50–100 $\mu g\,mL^{-1}$. In terms of applications of polyamide SMAPs, they have been shown to have a broad spectrum of activity against both Gram-negative and Gram-positive bacteria. However, the best-performing polyamide SMAPs have been shown to have potent activity against the difficult to treat pathogen *Clostridium difficile*, which surpasses the *in vitro* performance of vancomycin and LL-37 in being able to affect all strains tested.[81]

Mechanistically, there have been various studies to determine how exactly the nylon-3 SMAPs work to kill bacteria. In experiments where phospholipids were used to form vesicles that mimic the cytoplasmic membranes of *E. coli*, *S. aureus*, and red blood cells, it was observed that nylon-3 SMAPs disrupted the *E. coli*-like and the *S. aureus*-like vesicles to a greater degree than the red blood cell-like vesicles, which is consistent with a CAP-like mechanism.[82,83] Evidence from single-cell time-resolved microscopy also indicated that the polymers disrupt the cytoplasmic membrane.[84] The microscope images indicated that the polymers worked very rapidly to halt growth of the bacteria by inducing severe osmotic shock after translocating across the membrane. Yet, it has been observed with actual cells that membrane disruption is only part of what is actually occurring on the molecular level. It has been shown that the membrane disruption mechanism of nylon-3 SMAPs does indeed occur at low concentrations of the polymers near MIC. However, at very high concentrations of polymer, bacteria are effectively killed through a different mechanism of action.[82] Once a high enough concentration was achieved with *E. coli*, it was observed that the cytoplasmic membrane was not disrupted, but rather that the cell became unable to transport any solutes into or out of the cell because the cell was encapsulated in an impermeable polymer layer that coated the outer membrane of the bacteria. Evidence for this mechanism of action was indicated in part by live/dead staining that showed that bacteria at high concentrations of nylon-3 SMAP were not permeabilized by the dead stain, even though it was determined that the bacteria were not viable.[82] In addition, evidence for the formation of a polymer coating was indicated by the use of the water-soluble molecule *ortho*-nitrophenyl-β-galactoside (ONPG), which is colorless in solution and can be hydrolyzed by the intracellular enzyme β-galactosidase into galactose and the chromophore *ortho*-nitrophenol.[82]

In experiments where *E. coli* was treated with nylon-3 SMAP, the ability of ONPG to gain entrance into the cell and generate *ortho*-nitrophenol was progressively reduced with increasing concentrations of polymer.

5.3.2 Polyurethanes

Antimicrobial polyurethanes are synthesized by step-growth polymerization of functionalized diol and diisocyanate, which contain both cationic and hydrophobic moieties. One representative example is the water-soluble antimicrobial polyurethanes developed by Joy and co-workers.[85] As shown in Scheme 5.3, the antimicrobial polyurethanes were synthesized by co-polymerization of *N*-substituted diol monomers containing cationic or hydrophobic groups and hexamethylene diisocyanate catalyzed by tin(II) octoate. As demonstrated by MIC and time-kill assays, these antimicrobial polyurethanes show narrow spectrums of activity, which are effective against Gram-negative *E. coli* (MIC = 16 μg mL^{-1}) but are not effective against Gram-positive *S. aureus* (MIC = 250 μg mL^{-1}). Inner membrane permeability assays showed that these polymers were able to disrupt the cytoplasmic membrane of *E. coli*. In addition, these antimicrobial polyurethanes exhibit low toxicity toward mammalian cells even at high concentrations. According to the results from hemolysis assays using sheep blood cells, some of the examples exhibited less than 10% hemolysis even at 625 μg mL^{-1}.

5.3.3 Chain Growth Polymers

Besides polyamide SMAPs, many chain growth polymer (CGP) SMAPs have also been investigated. After it became widely recognized that a well-defined secondary structure is not necessary for a successful SMAP, many

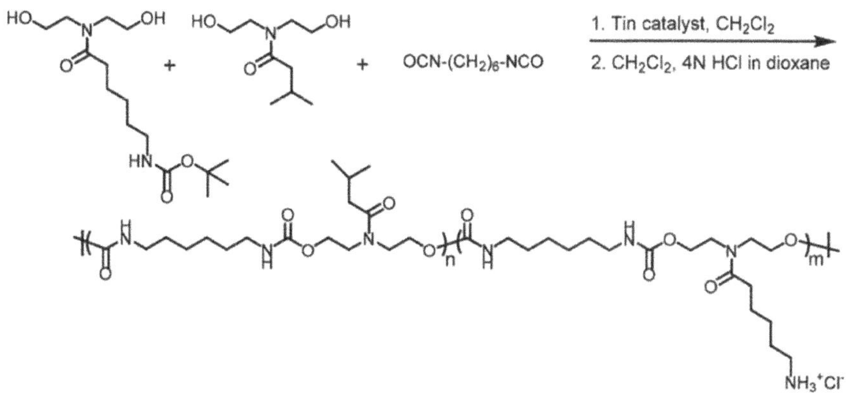

Scheme 5.3 Synthesis of peptidomimetic polyurethanes with mLys and mVal pendant groups.
Reproduced from ref. 85 with permission from American Chemical Society, Copyright 2017.

conventional CGPs synthesized from readily available libraries of acrylate, methacrylate, and acrylamide monomers were investigated to determine whether they could be used to construct SMAPs that can be effective in controlling the growth of bacteria while having a minimal effect on mammalian cells. The appeal of these types of polymers as SMAPs is driven by the fact that many of these polymers can be made very economically, especially when compared to CAPs. These CGPs are distinguished from the previously investigated biocidal polymers, which incorporated many similar repeat units, by the fact that these polymers contain primary ammonium in the repeat unit, which was previously dismissed as a functional group that is less biocidal than other cationic groups, and that these polymers are designed to draw inspiration from naturally occurring peptides. The earliest investigation of CGPs that were inspired by CAPs was performed by Kuroda and DeGrado using polymers that were synthesized from aminoethyl methacrylate and butyl methacrylate.[86] Even though their results were not very extraordinary, the SMAPs synthesized had decent antimicrobial activity and poor compatibility with red blood cells; the investigation began a long series of studies to better understand the structure/property relationships as they relate to antimicrobial activity for CGPs as well as mechanistic investigations into how exactly they affect bacteria.

There have been many investigations into the rational design of CGP SMAPs and their applications. In a subsequent investigation by Kuroda, Caputo, and DeGrado, many different methacrylate-based SMAP random/statistical copolymers were synthesized that had different ratios of cationic and hydrophobic methacrylate repeat units, differently functionalized hydrophobic groups, and different molecular weights to determine how all of these variables affect the antimicrobial properties.[87] Just as with polyamide SMAPs, these CGP SMAPs were observed to have antimicrobial activity and mammalian cell compatibility that are greatly dependent on achieving a balance of cationic and hydrophobic functionality. Also, polymers that had a greater molecular weight resulted in greater antimicrobial activity, but at the expense of compatibility with mammalian cells. In a similar investigation of polyacrylamide SMAPs, the importance of cationic/hydrophobic balance was also observed.[88] However, since the amide groups found in these polymers are much more polar than the analogous esters of polymethacrylates, the polyacrylamides showed greater effectiveness when copolymerized with hydrophobic acrylamides that had more hydrophobic pendant groups than that of the analogous polymethacrylates. Furthermore, not only have there been investigations into the effects of hydrophobic groups, but also changes in other functionalities. In a head-to-head comparison of copolymers formed from primary, tertiary, and quaternized ammonium functional methacrylates and a hydrophobic methacrylate, it was actually observed that primary ammonium and tertiary ammonium worked better to kill bacteria than quaternized ammonium groups (Scheme 5.4).[89] Guanadinylated cationic functionalized polymethacrylate SMAPs have also been investigated for their antimicrobial properties (Scheme 5.5).[90] Polymers that are functionalized

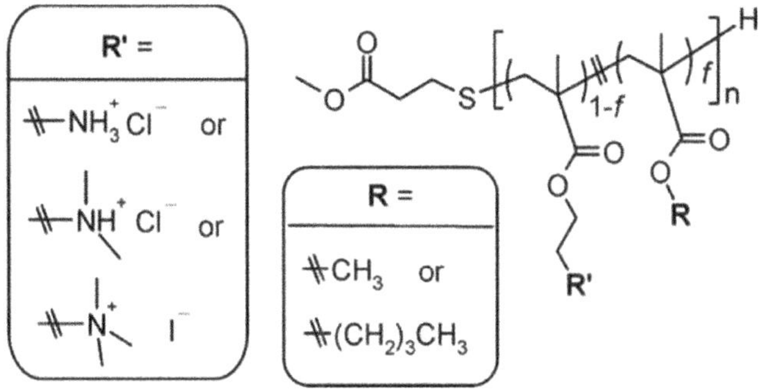

Scheme 5.4 Synthesis of methacrylate copolymers containing different cationic groups and alkyl groups in the side chains.
Reproduced from ref. 89 with permission from American Chemical Society, Copyright 2009.

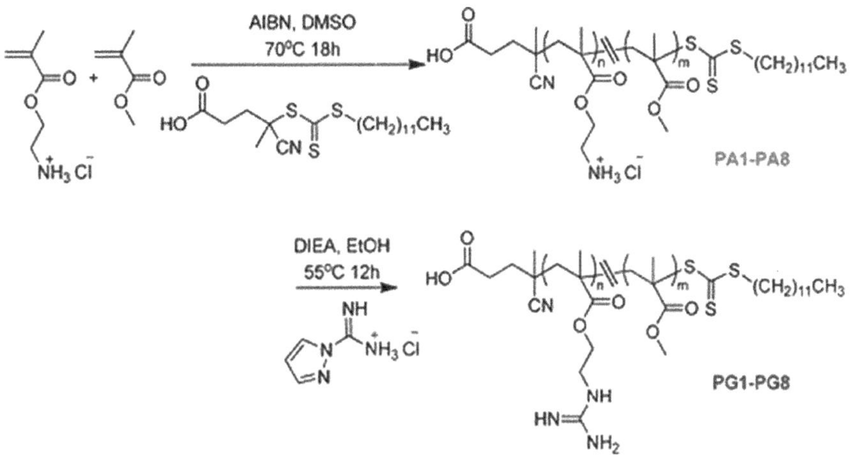

Scheme 5.5 Synthesis of random copolymers containing either amine or guanidine side chains.
Reproduced from ref. 90 with permission from American Chemical Society, Copyright 2013.

with guanidine functional groups are interesting because they mimic the cationic amino acid arginine. These polymers were shown to be more effective against Gram-positive bacteria *S. aureus* and *S. epidermidis*, efficacious in protein-rich medium, and less hemolytic than comparable amine functionalized polymethacrylates. For example, guanidine containing polymers showed lower hemolytic activity compared to amine containing polymers at the same polymer chain length. Also, when making a copolymer with methyl methacrylate, the guanidine polymers showed lower hemolytic activity than the amine-containing polymers when the percentage of methyl methacrylate in the polymer is the same. Moreover, there have been CGP

SMAPs that replace the hydrophobic repeat unit with one that has a PEG chain.[91] With a cationic polyacrylate that is sufficiently hydrophobic, the addition of PEG chains can lead to the creation of SMAPs that are potent antimicrobials while having a very limited effect on blood cells if the right amount of PEG functionality is incorporated. With such wide availability of the monomers used to synthesize these polymers, there have been multiple specific applications for which these polymers have been investigated. Some of the polymethacrylate-based SMAPs have been shown to be effective against many Gram-positive bacteria and have shown promising in controlling staphylococcal infections.[92] In addition, it has been shown that poly(dimethylaminoethyl methacrylate), a polymer that has been functionalized with a tertiary amine pendant group is rather effective in controlling the growth of *Mycobacteria* spp., which indicates that it may be useful in the treatment of tuberculosis.[93]

Various studies have been performed to determine the mechanism of action of random/statistical CGP SMAPs. There have been multiple experiments performed that have indicated through the use of model phospholipid membranes encapsulated with fluorescent dyes that these polymers have an affinity for phospholipid membranes that are anionically charged.[87,88] However, beyond probing for rudimentary electrostatic interactions, there have been multiple reports that shed further light on what is exactly occurring on the molecular level to cause disruption of the membranes of the cells. Small-angle X-ray scattering has indicated that CGP SMAPs actually behave very similarly to CAPs in the way they destabilize model bacterial membranes.[94] Both CGP SMAPs and CAPs show that they disrupt membranes by generating similar degrees of negative gaussian curvature across the membranes. However, for the CGP SMAPs to induce the same amount of negative gaussian curvature as CAPs, the CGP SMAPs involved in this process need to be significantly more cationic and hydrophobic than the CAPs because they are less effective in binding and inserting themselves into the membranes due to their unordered sequence.[94] Also, molecular dynamics simulations have been performed using a similar CGP SMAP that shows that the polymer forms micelle-like aggregates in solution and that the ability of these polymers to function as a SMAP is dependent upon the ability of the polymer chains to dissociate from the aggregate and bind to the surface of the membrane.[95] Upon binding to the surface of the membrane, the flexible polymer chain then adopts a conformation that results in the formation of a hydrophobic domain that is inside the membrane and a cationic domain that is present on the exterior surface. The presence of the polymers within the membrane results in the electrostatic attraction of anionic phospholipids, which causes a change in the distribution of lipids in the membrane, and leads to inconsistencies in the thickness of the phospholipid bilayer, which lowers the membrane permeability barrier. However, these investigations have not only been limited to the effects of CGP SMAPs on bacterial cells; the mechanism of action against mammalian cells has also been investigated.[96] Evidence indicates that once a CGP SMAP achieves a certain degree of hydrophobicity, the polymer causes the formation of nano-sized holes in the cytoplasmic membrane of red blood cells that ultimately results in colloid-osmotic lysis of the cell, which is lysis of the cell due to an imbalance of osmolarity caused by unregulated diffusion of low molecular weight solutes across the membrane.

5.3.4 Other Polymers

In addition to polymers formed *via* chain growth polymerizations, there have been various other synthetic polymers that have been investigated to determine their suitability as SMAPs. One such family of polymers is those that are formed *via* ring-opening metathesis polymerization (ROMP) of norbornene derivatives. Just as with the other SMAPs, the application of polymers using this method of polymerization provide a useful means to determine how different structures can affect antimicrobial efficacy. However, what makes these polymers different from those have been previously investigated is that careful monomer design can lead to the creation of a polymer with a sequence of alternating functional groups and a more rigid polymer backbone. In the first investigation of the antimicrobial properties of ROMP SMAPs, a series of homopolymers that contained a norbornene-derived repeat unit that had a cationic pendant group as well as a hydrophobic pendant group.[97] The polymers were synthesized to determine how molecular weight and different hydrophobic pendant groups affected antimicrobial performance and compatibility with red blood cells. Unlike previous studies, the SMAPs of these studies were synthesized to molecular weights over a broad range from ~1600 Da to ~140 000 Da. Over this range of molecular weights, it was observed that polymers in the range ~1600 Da to ~15 000 Da actually had better antimicrobial activity than those were of greater molecular weights. Also, none of the homopolymers synthesized exhibited an ideal high effectiveness in controlling bacterial growth while simultaneously having little effect on red blood cells. So, random/statistical copolymers of the ROMP SMAPs were synthesized and, in doing so, a SMAP with an ideal high antimicrobial efficacy with low hemolytic properties was obtained. Subsequent studies demonstrated that ROMP SMAPs can have potent antimicrobial activity and low hemolytic activity by using guanidinylated cationic groups and changing the location of the hydrophobic groups on the repeat unit.[98,99] As shown in Figure 5.5, the amine containing poly(oxanorbornene) was not effective against bacteria such as *E. coli*, *S. marcescens*, *S. aureus*, and *B. subtilis* (MIC = 200–400 $\mu g\,mL^{-1}$), while the guanidine-containing poly(oxanorbornene) exhibited much lower MIC values (6–50 $\mu g\,mL^{-1}$) without greatly increasing hemolytic activity (HC_{50} = 1500 $\mu g\,mL^{-1}$).However, the most interesting effect of changing pendant groups was obtained when ROMP SMAPs had two pendant primary ammoniums on each repeat unit. These polymers were capable of selectively affecting Gram-positive *S. aureus* over Gram-negative *E. coli*.[100] Further investigation revealed that this polymer is selective towards *S. aureus* because it is incapable of penetrating the outer membrane of *E. coli*.

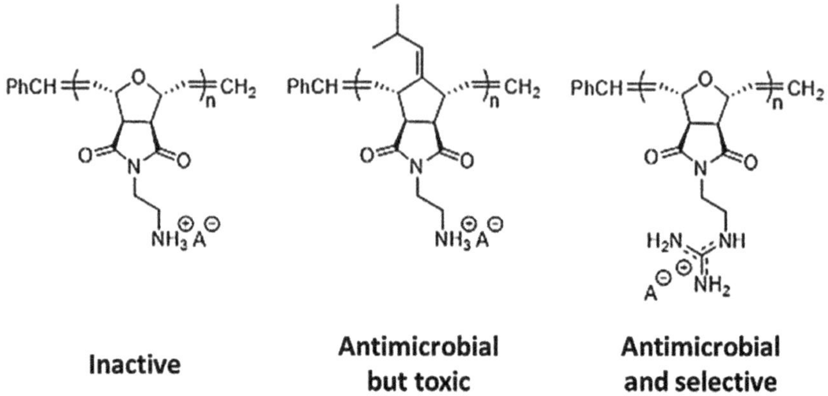

Figure 5.5 Oxanorbornene-based antimicrobial polymers with different cationic groups and hydrophobic groups provide different activity and selectivity against bacteria.

Various other polymers have been investigated for their use as SMAPs. Polymers that have been synthesized *via* a simultaneous step-growth and chain-growth polymerization have been studied to determine their usefulness as antimicrobials.[101] As a result of this polymerization mechanism, the synthesized polymer incorporated hydrolyzable ester bonds interspersed along an aliphatic backbone that had amine functionalized pendant groups. Even though none of the polymers in the series synthesized demonstrated an optimal composition where there was high antimicrobial activity and low hemolysis, these polymers were shown to be degradable and lose their membrane active properties as a result, which could prove useful in certain applications. Polyethylene imines (PEIs) have also been investigated as possible antimicrobials.[102] Even though linear PEIs are well documented as being quite toxic to all cells, low molecular weight (~4400 Da) linear PEI has been shown to be somewhat more effective against bacteria. However, the introduction of branching points into the architecture of PEI results in a polymer that does not demonstrate significant hemolysis up to very high concentrations and is selective for *S. aureus* over *E. coli*. Branched polymers with a star architecture have additionally been synthesized and demonstrated very desirable properties.[103] The polymers were synthesized using either a generation three or four polyamidoamine dendrimer that served as a macroinitiator for the ring opening polymerization of valine and lysine *N*-carboxyanhydrides. These polymers demonstrated effectiveness against bacteria and very good compatibility with mammalian cells.

5.4 Conclusion

We have given an overview of synthetic cationic water-soluble antimicrobial polymers that have been developed over the past decades and described the rationale and characteristics of each class of antimicrobial polymers.

These polymers have distinct advantages over traditional antibiotics such as a lower probability of developing microbial resistance. A sustained effort in the design and optimization of such polymers can result in their deployment as an alternative to tackling emergent bacterial infections. However, the effectiveness of such antimicrobial polymers should to be improved and the toxicity to mammalian cells decreased to enable clinical use.

References

1. D. Gu, N. Dong, Z. Zheng, D. Lin, M. Huang, L. Wang, E. W.-C. Chan, L. Shu, J. Yu and R. Zhang, A fatal outbreak of ST11 carbapenem-resistant hypervirulent Klebsiella pneumoniae in a Chinese hospital: a molecular epidemiological study, *Lancet Infect. Dis.*, 2018, **18**(1), 37–46.
2. T. Tängdén and C. Giske, Global dissemination of extensively drug-resistant carbapenemase-producing Enterobacteriaceae: clinical perspectives on detection, treatment and infection control, *J. Intern. Med.*, 2015, **277**(5), 501–512.
3. D. R. Giacobbe, V. Del Bono, E. M. Trecarichi, F. De Rosa, M. Giannella, M. Bassetti, A. Bartoloni, A. Losito, S. Corcione and M. Bartoletti, Risk factors for bloodstream infections due to colistin-resistant KPC-producing Klebsiella pneumoniae: results from a multicenter case–control–control study, *Clin. Microbiol. Infect.*, 2015, **21**(12), 1106. e1–1106. e8.
4. Y. Wang, G.-B. Tian, R. Zhang, Y. Shen, J. M. Tyrrell, X. Huang, H. Zhou, L. Lei, H.-Y. Li and Y. Doi, Prevalence, risk factors, outcomes, and molecular epidemiology of mcr-1-positive Enterobacteriaceae in patients and healthy adults from China: an epidemiological and clinical study, *Lancet Infect. Dis.*, 2017, **17**(4), 390–399.
5. J. O'Neill, Tackling drug-resistant infections globally: final report and recommendations. in *The Review on Antimicrobial Resistance*, 2016.
6. J. M. Munita and C. A. Arias, Mechanisms of Antibiotic Resistance, *Microbiol. Spectrum*, 2016, **4**(2), VMBF-0016-2015.
7. K. Bush and G. A. Jacoby, Updated functional classification of β-lactamases, *Antimicrob. Agents Chemother.*, 2010, **54**(3), 969–976.
8. A. Dönhöfer, S. Franckenberg, S. Wickles, O. Berninghausen, R. Beckmann and D. N. Wilson, Structural basis for TetM-mediated tetracycline resistance, *Proc. Natl. Acad. Sci. U. S. A.*, 2012, **109**(42), 16900–16905.
9. Antibiotic Resistance Threats in the United States, *Centres for Disease Control and Prevention*, US Department of Health and Human Services, 2013.
10. F. Rose and G. Swain, Bisdiguanides having antibacterial activity, *J. Chem. Soc.*, 1956, **850**, 4422–4425.
11. H. Bratt and D. E. Hathway, Characterization of the urinary polymer-related material from rats given poly [biguanide-1, 5-diylhexamethylene hydrochloride], *Macromol. Chem. Phys.*, 1976, **177**(9), 2591–2605.

12. G. F. De Paula, G. I. Netto and L. H. C. Mattoso, Physical and chemical characterization of poly (hexamethylene biguanide) hydrochloride, *Polymers*, 2011, 3(2), 928–941.
13. L. P. O'Malley, K. Z. Hassan, H. Brittan, N. Johnson and A. N. Collins, Characterization of the biocide polyhexamethylene biguanide by matrix-assisted laser desorption ionization time-of-flight mass spectrometry, *J. Appl. Polym. Sci.*, 2006, **102**(5), 4928–4936.
14. K. Kaehn, Polihexanide: a safe and highly effective biocide, *Skin Pharmacol. Physiol.*, 2010, **23**(Suppl. 1), 7–16.
15. A. Welk, C. Splieth, G. Schmidt-Martens, C. Schwahn, T. Kocher, A. Kramer and M. Rosin, The effect of a polyhexamethylene biguanide mouthrinse compared with a triclosan rinse and a chlorhexidine rinse on bacterial counts and 4-day plaque re-growth, *J. Clin. Periodontol.*, 2005, **32**(5), 499–505.
16. M. Brecx, E. Decker, H. Freitag, G. Maier, C. Von Ohle, *The Effect of Polihexanide on Dental Biofilm Formation In Vivo*, Conf Pan Eur Fed Int Assoc Dent Res, London, UK, 2008.
17. A. Piatkowski, N. Drummer, A. Andriessen, D. Ulrich and N. Pallua, Randomized controlled single center study comparing a polyhexanide containing bio-cellulose dressing with silver sulfadiazine cream in partial-thickness dermal burns, *Burns*, 2011, **37**(5), 800–804.
18. N.-O. Hübner and A. Kramer, Review on the efficacy, safety and clinical applications of polihexanide, a modern wound antiseptic, *Skin Pharmacol. Physiol.*, 2010, **23**(Suppl. 1), 17–27.
19. G. Müller and A. Kramer, Biocompatibility index of antiseptic agents by parallel assessment of antimicrobial activity and cellular cytotoxicity, *J. Antimicrob. Chemother.*, 2008, **61**(6), 1281–1287.
20. A. Davies, M. Bentley and B. S. Field, Comparison of the action of vantocil, cetrimide and chlorhexidine on *Escherichia coli* and its spheroplasts and the protoplasts of gram positive bacteria, *J. Appl. Microbiol.*, 1968, **31**(4), 448–461.
21. T. Ikeda, S. Tazuke and M. Watanabe, Interaction of biologically active molecules with phospholipid membranes: I. Fluorescence depolarization studies on the effect of polymeric biocide bearing biguanide groups in the main chain, *Biochim. Biophys. Acta, Biomembr.*, 1983, **735**(3), 380–386.
22. N. F. Kamaruzzaman, R. Firdessa and L. Good, Bactericidal effects of polyhexamethylene biguanide against intracellular Staphylococcus aureus EMRSA-15 and USA 300, *J. Antimicrob. Chemother.*, 2016, **71**(5), 1252–1259.
23. K. Chindera, M. Mahato, A. K. Sharma, H. Horsley, K. Kloc-Muniak, N. F. Kamaruzzaman, S. Kumar, A. McFarlane, J. Stach, T. Bentin and L. Good, The antimicrobial polymer PHMB enters cells and selectively condenses bacterial chromosomes, *Sci. Rep.*, 2016, **6**, 23121.
24. W. A. Jacobs, The bactericidal properties of the quaternary salts of hexamethylenetetramine: I. The problem of the chemotherapy of experimental bacterial infections, *J. Exp. Med.*, 1916, **23**(5), 563–568.

25. W. Jacobs, M. Heidelberger and H. Amoss, The bactericidal properties of the quaternary salts of hexamethylenetetramine: II. The relation between constitution and bactericidal action in the substituted benzylhexamethylenetetraminium salts, *J. Exp. Med.*, 1916, **23**(5), 569–576.
26. W. A. Jacobs, M. Heidelberger and C. G. Bull, The bactericidal properties of the quaternary salts of hexamethylenetetramine: III. The relation between constitution and bacterricidal action in the quaterrnary salts obtained from halogenacetyl compounds, *J. Exp. Med.*, 1916, **23**(5), 577–599.
27. G. Domagk, Eine neue klasse von desinfektionsmitteln, *DMW - Dtsch. Med. Wochenschr.*, 1935, **61**(21), 829–832.
28. Y. Xue, H. Xiao and Y. Zhang, Antimicrobial polymeric materials with quaternary ammonium and phosphonium salts, *Int. J. Mol. Sci.*, 2015, **16**(2), 3626–3655.
29. J. S. Parent, A. Penciu, S. A. Guillén-Castellanos, A. Liskova and R. A. Whitney, Synthesis and characterization of isobutylene-based ammonium and phosphonium bromide ionomers, *Macromolecules*, 2004, **37**(20), 7477–7483.
30. N. Kawabata, Capture of micro-organisms and viruses by pyridinium-type polymers and application to biotechnology and water purification, *Prog. Polym. Sci.*, 1992, **17**(1), 1–34.
31. B. Dizman, M. O. Elasri and L. J. Mathias, Synthesis and antimicrobial activities of new water-soluble bis-quaternary ammonium methacrylate polymers, *J. Appl. Polym. Sci.*, 2004, **94**(2), 635–642.
32. L. Timofeeva and N. Kleshcheva, Antimicrobial polymers: mechanism of action, factors of activity, and applications, *Appl. Microbiol. Biotechnol.*, 2011, **89**(3), 475–492.
33. E.-R. Kenawy, S. Worley and R. Broughton, The chemistry and applications of antimicrobial polymers: a state-of-the-art review, *Biomacromolecules*, 2007, **8**(5), 1359–1384.
34. T. Ikeda, H. Hirayama, H. Yamaguchi, S. Tazuke and M. Watanabe, Polycationic biocides with pendant active groups: molecular weight dependence of antibacterial activity, *Antimicrob. Agents Chemother.*, 1986, **30**(1), 132–136.
35. T. Ikeda, S. Tazuke and Y. Suzuki, Biologically active polycations, 4. Synthesis and antimicrobial activity of poly(trialkylvinylbenzylammonium chloride)s, *Macromol. Chem. Phys.*, 1984, **185**(5), 869–876.
36. R. Tejero, D. López, F. López-Fabal, J. L. Gómez-Garcés and M. Fernández-García, Antimicrobial polymethacrylates based on quaternized 1, 3-thiazole and 1, 2, 3-triazole side-chain groups, *Polym. Chem.*, 2015, **6**(18), 3449–3459.
37. N. Kawabata, T. Hayashi and T. Matsumoto, Removal of bacteria from water by adhesion to cross-linked poly (vinylpyridinium halide), *Appl. Environ. Microbiol.*, 1983, **46**(1), 203–210.
38. N. Kawabata and M. Nishiguchi, Antibacterial activity of soluble pyridinium-type polymers, *Appl. Environ. Microbiol.*, 1988, **54**(10), 2532–2535.

39. G. Li, J. Shen and Y. Zhu, Study of pyridinium-type functional polymers. II. Antibacterial activity of soluble pyridinium-type polymers, *J. Appl. Polym. Sci.*, 1998, **67**(10), 1761–1768.
40. T. Eren, A. Som, J. R. Rennie, C. F. Nelson, Y. Urgina, K. Nüsslein, E. B. Coughlin and G. N. Tew, Antibacterial and hemolytic activities of quaternary pyridinium functionalized polynorbornenes, *Macromol. Chem. Phys.*, 2008, **209**(5), 516–524.
41. E. Littmann and C. Marvel, , Cyclic quaternary ammonium salts from halogenated aliphatic tertiary amines, *J. Am. Chem. Soc.*, 1930, **52**(1), 287–294.
42. C. Gibbs, E. Littmann and C. Marvel, Quaternary ammonium salts from halogenated alkyl dimethylamines. II. The polymerization of gamma-halogenopropyldimethylamines, *J. Am. Chem. Soc.*, 1933, **55**(2), 753–757.
43. M. Lehman, C. Thompson and C. Marvel, Quaternary ammonium salts from halogenated alkyl dimethylamines. III. omega-bromo-heptyl-,-octyl-,-nonyl-and-decyl-dimethylamines, *J. Am. Chem. Soc.*, 1933, **55**(5), 1977–1981.
44. A. Rembaum, W. Baumgartner and A. Eisenberg, Aliphatic ionenes, *J. Polym. Sci., Part C: Polym. Lett.*, 1968, **6**(3), 159–171.
45. T. Tsutsui, Ionene polymers: preparation, properties and applications, in *Developments in Ionic Polymers—2*, Springer, 1986, pp. 163–189.
46. A. Strassburg, F. Kracke, J. Wenners, A. Jemeljanova, J. Kuepper, H. Petersen and J. C. Tiller, Nontoxic, hydrophilic cationic polymers—identified as class of antimicrobial polymers, *Macromol. Biosci.*, 2015, **15**(12), 1710–1723.
47. S. Liu, R. J. Ono, H. Wu, J. Y. Teo, Z. C. Liang, K. Xu, M. Zhang, G. Zhong, J. P. Tan and M. Ng, Highly potent antimicrobial polyionenes with rapid killing kinetics, skin biocompatibility and in vivo bactericidal activity, *Biomaterials*, 2017, **127**, 36–48.
48. Z. Geng and M. Finn, Thiabicyclononane-Based Antimicrobial Polycations, *J. Am. Chem. Soc.*, 2017, **139**(43), 15401–15406.
49. A. Rembaum, H. Rile and R. Somoano, V. Kinetics of formation of high charge density ionene polymers, *J. Polym. Sci., Part C: Polym. Lett.*, 1970, **8**(7), 457–466.
50. T. Ikeda, H. Yamaguchi and S. Tazuke, Phase separation in phospholipid bilayers induced by biologically active polycations, *Biochim. Biophys. Acta, Biomembr.*, 1990, **1026**(1), 105–112.
51. T. Narita, R. Ohtakeyama, M. Nishino, J. Gong and Y. Osada, Effects of charge density and hydrophobicity of ionene polymer on cell binding and viability, *Colloid Polym. Sci.*, 2000, **278**(9), 884–887.
52. S.-W. Wang, W. Liu and R. H. Colby, Counterion dynamics in polyurethane-carboxylate ionomers with ionic liquid counterions, *Chem. Mater.*, 2011, **23**(7), 1862–1873.
53. B. Bauer, H. Strathmann and F. Effenberger, Anion-exchange membranes with improved alkaline stability, *Desalination*, 1990, **79**(2–3), 125–144.

54. K. J. Noonan, K. M. Hugar, H. A. Kostalik IV, E. B. Lobkovsky, H. C. D. Abruña and G. W. Coates, Phosphonium-functionalized polyethylene: a new class of base-stable alkaline anion exchange membranes, *J. Am. Chem. Soc.*, 2012, **134**(44), 18161–18164.
55. S. T. Hemp, M. Zhang, M. H. Allen, S. Cheng, R. B. Moore and T. E. Long, Comparing ammonium and phosphonium polymerized ionic liquids: thermal analysis, conductivity, and morphology, *Macromol. Chem. Phys.*, 2013, **214**(18), 2099–2107.
56. A. Kanazawa, T. Ikeda and T. Endo, Novel polycationic biocides: synthesis and antibacterial activity of polymeric phosphonium salts, *J. Polym. Sci., Part A: Polym. Chem.*, 1993, **31**(2), 335–343.
57. A. Kanazawa, T. Ikeda and T. Endo, Polymeric phosphonium salts as a novel class of cationic biocides. II. Effects of counter anion and molecular weight on antibacterial activity of polymeric phosphonium salts, *J. Polym. Sci., Part A: Polym. Chem.*, 1993, **31**(6), 1441–1447.
58. A. Kanazawa, T. Ikeda and T. Endo, Polymeric phosphonium salts as a novel class of cationic biocides. IV. Synthesis and antibacterial activity of polymers with phosphonium salts in the main chain, *J. Polym. Sci., Part A: Polym. Chem.*, 1993, **31**(12), 3031–3038.
59. A. Kanazawa, T. Ikeda and T. Endo, Polymeric phosphonium salts as a novel class of cationic biocides. VII. Synthesis and antibacterial activity of polymeric phosphonium salts and their model compounds containing long alkyl chains, *J. Appl. Polym. Sci.*, 1994, **53**(9), 1237–1244.
60. A. Kanazawa, T. Ikeda and T. Endo, Polymeric phosphonium salts as a novel class of cationic biocides. IX. Effect of side-chain length between main chain and active group on antibacterial activity, *J. Polym. Sci., Part A: Polym. Chem.*, 1994, **32**(10), 1997–2001.
61. A. Kanazawa, T. Ikeda and T. Endo, A novel approach to mode of action of cationic biocides morphological effect on antibacterial activity, *J. Appl. Microbiol.*, 1995, **78**(1), 55–60.
62. E.-R. Kenawy, F. I. Abdel-Hay, A. E.-R. R. El-Shanshoury and M. H. El-Newehy, Biologically active polymers: synthesis and antimicrobial activity of modified glycidyl methacrylate polymers having a quaternary ammonium and phosphonium groups, *J. Controlled Release*, 1998, **50**(1), 145–152.
63. B. Gao, Q. Liu and Y. Li, Preparation of water-insoluble antibacterial materials with surface-grafted material PSt/SiO2 and their antibacterial activity, *J. Polym. Environ.*, 2010, **18**(4), 474–483.
64. T. Qiu, Q. Zeng and N. Ao, Preparation and characterization of chlorinated nature rubber (CNR) based polymeric quaternary phosphonium salt bactericide, *Mater. Lett.*, 2014, **122**, 13–16.
65. N. K. Brogden and K. A. Brogden, Will new generations of modified antimicrobial peptides improve their potential as pharmaceuticals?, *Int. J. Antimicrob. Agents*, 2011, **38**(3), 217–225.

66. S. Rotem and A. Mor, Antimicrobial peptide mimics for improved therapeutic properties, *Biochim. Biophys. Acta, Biomembr.*, 2009, **1788**(8), 1582–1592.
67. Y. Hamuro, J. P. Schneider and W. F. DeGrado, De novo design of antibacterial β-peptides, *J. Am. Chem. Soc.*, 1999, **121**(51), 12200–12201.
68. D. Liu and W. F. DeGrado, De novo design, synthesis, and characterization of antimicrobial β-peptides, *J. Am. Chem. Soc.*, 2001, **123**(31), 7553–7559.
69. E. A. Porter, B. Weisblum and S. H. Gellman, Mimicry of host-defense peptides by unnatural oligomers: antimicrobial β-peptides, *J. Am. Chem. Soc.*, 2002, **124**(25), 7324–7330.
70. E. A. Porter, B. Weisblum and S. H. Gellman, Use of parallel synthesis to probe structure-activity relationships among 12-helical β-peptides: Evidence of a limit on antimicrobial activity, *J. Am. Chem. Soc.*, 2005, **127**(32), 11516–11529.
71. E. A. Porter, X. Wang, H.-S. Lee, B. Weisblum and S. H. Gellman, Antibiotics: Non-haemolytic β-amino-acid oligomers, *Nature*, 2000, **404**(6778), 565.
72. K. Kirshenbaum, A. E. Barron, R. A. Goldsmith, P. Armand, E. K. Bradley, K. T. Truong, K. A. Dill, F. E. Cohen and R. N. Zuckermann, Sequence-specific polypeptoids: a diverse family of heteropolymers with stable secondary structure, *Proc. Natl. Acad. Sci. U. S. A.*, 1998, **95**(8), 4303–4308.
73. J. A. Patch and A. E. Barron, Helical peptoid mimics of magainin-2 amide, *J. Am. Chem. Soc.*, 2003, **125**(40), 12092–12093.
74. N. P. Chongsiriwatana, J. A. Patch, A. M. Czyzewski, M. T. Dohm, A. Ivankin, D. Gidalevitz, R. N. Zuckermann and A. E. Barron, Peptoids that mimic the structure, function, and mechanism of helical antimicrobial peptides, *Proc. Natl. Acad. Sci. U. S. A.*, 2008, **105**(8), 2794–2799.
75. R. Kapoor, M. W. Wadman, M. T. Dohm, A. M. Czyzewski, A. M. Spormann and A. E. Barron, Antimicrobial peptoids are effective against Pseudomonas aeruginosa biofilms, *Antimicrob. Agents Chemother.*, 2011, **55**(6), 3054–3057.
76. I. S. Radzishevsky, S. Rotem, D. Bourdetsky, S. Navon-Venezia, Y. Carmeli and A. Mor, Improved antimicrobial peptides based on acyl-lysine oligomers, *Nat. Biotechnol.*, 2007, **25**(6), 657–659.
77. B. P. Mowery, S. E. Lee, D. A. Kissounko, R. F. Epand, R. M. Epand, B. Weisblum, S. S. Stahl and S. H. Gellman, Mimicry of antimicrobial host-defense peptides by random copolymers, *J. Am. Chem. Soc.*, 2007, **129**(50), 15474–15476.
78. B. P. Mowery, A. H. Lindner, B. Weisblum, S. S. Stahl and S. H. Gellman, Structure–activity relationships among random nylon-3 copolymers that mimic antibacterial host-defense peptides, *J. Am. Chem. Soc.*, 2009, **131**(28), 9735–9745.
79. S. Chakraborty, R. Liu, J. J. Lemke, Z. Hayouka, R. A. Welch, B. Weisblum, K. S. Masters and S. H. Gellman, Effects of cyclic vs

acyclic hydrophobic subunits on the chemical structure and biological properties of nylon-3 copolymers, *ACS Macro Lett.*, 2013, **2**(8), 753–756.
80. S. Chakraborty, R. Liu, Z. Hayouka, X. Chen, J. Ehrhardt, Q. Lu, E. Burke, Y. Yang, B. Weisblum and G. C. Wong, Ternary nylon-3 copolymers as host-defense peptide mimics: beyond hydrophobic and cationic subunits, *J. Am. Chem. Soc.*, 2014, **136**(41), 14530–14535.
81. R. Liu, J. M. Suárez, B. Weisblum, S. H. Gellman and S. M. McBride, Synthetic Polymers Active against Clostridium difficile Vegetative Cell Growth and Spore Outgrowth, *J. Am. Chem. Soc.*, 2014, **136**(41), 14498–14504.
82. R. F. Epand, B. P. Mowery, S. E. Lee, S. S. Stahl, R. I. Lehrer, S. H. Gellman and R. M. Epand, Dual mechanism of bacterial lethality for a cationic sequence-random copolymer that mimics host-defense antimicrobial peptides, *J. Mol. Biol.*, 2008, **379**(1), 38–50.
83. S. G. Hovakeemian, R. Liu, S. H. Gellman and H. Heerklotz, Correlating antimicrobial activity and model membrane leakage induced by nylon-3 polymers and detergents, *Soft Matter*, 2015, **11**(34), 6840–6851.
84. H. Choi, S. Chakraborty, R. Liu, S. H. Gellman and J. C. Weisshaar, Single-Cell, Time-Resolved Antimicrobial Effects of a Highly Cationic, Random Nylon-3 Copolymer on Live *Escherichia coli*, *ACS Chem. Biol.*, 2015, **11**(1), 113–120.
85. S. Mankoci, R. L. Kaiser, N. Sahai, H. A. Barton and A. Joy, Bactericidal Peptidomimetic Polyurethanes with Remarkable Selectivity against *Escherichia coli*, *ACS Biomater. Sci. Eng.*, 2017, **3**(10), 2588–2597.
86. K. Kuroda and W. F. DeGrado, Amphiphilic Polymethacrylate Derivatives as Antimicrobial Agents, *J. Am. Chem. Soc.*, 2005, **127**(12), 4128–4129.
87. K. Kuroda, G. A. Caputo and W. F. DeGrado, The role of hydrophobicity in the antimicrobial and hemolytic activities of polymethacrylate derivatives, *Chem. – Eur. J.*, 2009, **15**(5), 1123–1133.
88. E. F. Palermo, I. Sovadinova and K. Kuroda, Structural determinants of antimicrobial activity and biocompatibility in membrane-disrupting methacrylamide random copolymers, *Biomacromolecules*, 2009, **10**(11), 3098–3107.
89. E. F. Palermo and K. Kuroda, Chemical structure of cationic groups in amphiphilic polymethacrylates modulates the antimicrobial and hemolytic activities, *Biomacromolecules*, 2009, **10**(6), 1416–1428.
90. K. E. Locock, T. D. Michl, J. D. Valentin, K. Vasilev, J. D. Hayball, Y. Qu, A. Traven, H. J. Griesser, L. Meagher and M. Haeussler, Guanylated polymethacrylates: a class of potent antimicrobial polymers with low hemolytic activity, *Biomacromolecules*, 2013, **14**(11), 4021–4031.
91. A. Punia, A. Mancuso, P. Banerjee and N.-L. Yang, Nonhemolytic and antibacterial acrylic copolymers with hexamethyleneamine and poly (ethylene glycol) side chains, *ACS Macro Lett.*, 2015, **4**(4), 426–430.

92. L. M. Thoma, B. R. Boles and K. Kuroda, Cationic methacrylate polymers as topical antimicrobial agents against Staphylococcus aureus nasal colonization, *Biomacromolecules*, 2014, **15**(8), 2933–2943.
93. D. J. Phillips, J. Harrison, S.-J. Richards, D. E. Mitchell, E. Tichauer, A. T. Hubbard, C. Guy, I. Hands-Portman, E. Fullam and M. I. Gibson, Evaluation of the antimicrobial activity of cationic polymers against mycobacteria: toward antitubercular macromolecules, *Biomacromolecules*, 2017, **18**(5), 1592–1599.
94. K. Hu, N. W. Schmidt, R. Zhu, Y. Jiang, G. H. Lai, G. Wei, E. F. Palermo, K. Kuroda, G. C. Wong and L. Yang, A critical evaluation of random copolymer mimesis of homogeneous antimicrobial peptides, *Macromolecules*, 2013, **46**(5), 1908–1915.
95. U. Baul, K. Kuroda and S. Vemparala, Interaction of multiple biomimetic antimicrobial polymers with model bacterial membranes, *J. Chem. Phys.*, 2014, **141**(8), 084902.
96. I. Sovadinova, E. F. Palermo, R. Huang, L. M. Thoma and K. Kuroda, Mechanism of polymer-induced hemolysis: nanosized pore formation and osmotic lysis, *Biomacromolecules*, 2010, **12**(1), 260–268.
97. M. F. Ilker, K. Nüsslein, G. N. Tew and E. B. Coughlin, Tuning the hemolytic and antibacterial activities of amphiphilic polynorbornene derivatives, *J. Am. Chem. Soc.*, 2004, **126**(48), 15870–15875.
98. K. Lienkamp, A. E. Madkour, A. Musante, C. F. Nelson, K. Nusslein and G. N. Tew, Antimicrobial polymers prepared by ROMP with unprecedented selectivity: a molecular construction kit approach, *J. Am. Chem. Soc.*, 2008, **130**(30), 9836–9843.
99. G. J. Gabriel, A. E. Madkour, J. M. Dabkowski, C. F. Nelson, K. Nüsslein and G. N. Tew, Synthetic mimic of antimicrobial peptide with nonmembrane-disrupting antibacterial properties, *Biomacromolecules*, 2008, **9**(11), 2980–2983.
100. K. Lienkamp, K. N. Kumar, A. Som, K. Nüsslein and G. N. Tew, ''Doubly selective'' antimicrobial polymers: How do they differentiate between bacteria?, *Chem. – Eur. J.*, 2009, **15**(43), 11710–11714.
101. M. Mizutani, E. F. Palermo, L. M. Thoma, K. Satoh, M. Kamigaito and K. Kuroda, Design and synthesis of self-degradable antibacterial polymers by simultaneous chain-and step-growth radical copolymerization, *Biomacromolecules*, 2012, **13**(5), 1554–1563.
102. K. A. Gibney, I. Sovadinova, A. I. Lopez, M. Urban, Z. Ridgway, G. A. Caputo and K. Kuroda, Poly (ethylene imine) s as antimicrobial agents with selective activity, *Macromol. Biosci.*, 2012, **12**(9), 1279–1289.
103. S. J. Lam, N. M. O'Brien-Simpson, N. Pantarat, A. Sulistio, E. H. Wong, Y.-Y. Chen, J. C. Lenzo, J. A. Holden, A. Blencowe and E. C. Reynolds, Combating multidrug-resistant Gram-negative bacteria with structurally nanoengineered antimicrobial peptide polymers, *Nat. Microbiol.*, 2016, **1**, 16162.

CHAPTER 6

Focal Drug Delivery for Management of Oral Infections

DAVID POLAK,* RAWI ASSAD, DANIEL MOREINOS, YOAV PIETROKOVSKI AND NURIT BEYTH

Faculty of Dentistry, The Hebrew University-Hadassah, Jerusalem, Israel
*Email: polak@mail.huji.ac.il; rawidmd@gmail.com; dmoreinos@gmail.com; yoavpi@gmail.com; nuritb@ekmd.huji.ac.il

6.1 Introduction

The term "biofilm" is used to describe communities of microorganisms attached to a surface. The organisms are usually spatially organized in a three-dimensional structure and are enclosed in a matrix of extracellular material derived both from the cells themselves and from the environment.[1] Research in recent years has revealed that cells growing as biofilms have unique properties, some of which are of clinical significance. Biofilm bacteria are distinguished from planktonic bacteria by a number of features, one of which is hyper-resistance to antimicrobial agents.[2] In part, this can be attributed to alterations in the expression of a large number of genes in response to the proximity of a specific surface.[3] Furthermore, biofilm bacteria, as opposed to bacteria in the planktonic phase, often survive under extreme conditions, including an anoxic microenvironment, extreme pH and ionic strength.

A wide range of microorganisms, including bacteria, yeast and viruses, constitute the microflora found in the oral cavity. All of these groups, principally bacteria, may be related to oral infections. The bacterial diversity in the oral cavity is estimated to consist of more than 1300 different species,

Biomaterials Science Series No. 5
Antimicrobial Materials for Biomedical Applications
Edited by Abraham J. Domb, Konda Reddy Kunduru and Shady Farah

Published by the Royal Society of Chemistry, www.rsc.org

including Spirochaetes, Fusobacteria, Actinobacteria, Firmicutes, Bacteroidetes and Proteobacteria. The wide species diversity can be attributed to the variety of nutrients, the different forms of environment for colonization, including surfaces on the teeth, mucosa and tongue, and the possibility to survive as a biofilm. The biofilm offers three main advantages for the microorganism: (1) it provides a physical barrier to external insults; (2) it protects from the host immune system; and (3) it enables nutrient availability.

Biofilm formation may have crucial implications in medicine and in dentistry. One example is the serious complications encountered in implant surgery caused by implant-associated infection. In the oral cavity, biofilm accumulation can lead to the development of the most common diseases, *i.e.*, dental caries and periodontitis.[4]

6.2 Biofilms and Oral Infections

Biofilm is the main cause of localized diseases in the oral cavity, including caries, gingivitis, periodontitis, candidiasis, endodontic infections, orthodontic infections and peri-implantitis.[3] In the oral cavity, as in most natural environments, microorganisms exist predominantly in multispecies biofilms. Consequently, bacterial infections within the oral cavity mainly involve multispecies, making it difficult to specify the role of each one in such infections. Most oral infections originate in microorganisms normally found in the mouth; seldom is the origin an exogenous source. Regardless of whether the bacterial accumulation and biofilm formation occur on natural surfaces or on artificial materials, dental caries or periodontal disease may evolve.

6.2.1 Formation and Characteristics of Oral Biofilms

Formation of the biofilm depends on different variables, such as the species of bacteria, the physical and chemical properties of the underlying surface, environmental factors and essential gene products, and may be regulated by the quorum sensing system.[5] In the oral cavity, organisms are not distributed randomly, but selectively attach and grow on certain surfaces, forming dental plaque. The latter is an example of a microbial biofilm, sharing most of the features of other currently known biofilms, with antimicrobial resistance being of special relevance.[6] Potential bacterial attachment sites include the tooth and epithelial surfaces. Conditions on theses surfaces change with respect to oxygen levels, nutrient availability, saliva and gingival crevicular fluid secretion, masticatory forces and other variables such as oral hygiene. Generally, adhesion is facilitated in the initial stage by non-specific interactions, followed in the next stage by the production of specific molecular interactions. Primary bacterial adhesion entails a random encounter between an acquired pellicle or a conditioning film composed of proteins drawn from the saliva and planktonic bacteria. The initial colonizers of the oral biofilm are gram-positive species, primarily from the streptococcal group. The bacteria include facultative anaerobes, which are part of the resident microbial flora of the oral cavity and upper respiratory tract, but are also opportunistic pathogens in human diseases. These initial colonizers can influence the subsequent sequence of microbial colonization. Members of this group can stimulate host cells, and the biofilm itself generates a habitat for additional species, some of which are closely associated with the initiation and progression of dental caries and periodontal disease. The formation of oral biofilm has a pattern of colonization called autogenic succession, *i.e.*, the microorganisms themselves generate or induce local physicochemical changes that modify the microbial composition of the biofilm. The initial stage of biofilm formation depends on the strength of the bacterial attachment, as this step is reversible. Once attachment is established, major mass growth occurs mainly by bacterial cell division within the biofilm rather than by co-aggregation at the surface of the developing biofilm.[7] The first colonizers in the supra-gingival biofilm are of a relatively low diversity compared with those present in the mature biofilm. The initial colonizers include *Streptococcus sanguinis*, *Streptococcus mitis* and *Streptococcus oralis*, the pioneers in the formation of dental biofilm. Subsequently, other bacteria emerge and co-aggregate with these bacteria, including predominantly Gram-negative species, such as *Eikenella corrodens*, *Veillonella atypica* and *Prevotella loescheii*. After multiplication of the primary colonizers, additional bacteria, known as secondary or late colonizers, are incorporated. *Fusobacterium*, a late colonizer also recognized as a bridging microorganism, binds by co-aggregating with the pioneer species. Late colonizers include *Aggregatibacter actinomycetemcomitans*, *Prevotella intermedia*, *Treponema denticola* and *Porphyromonas gingivalis*.[8] Depending largely on the effectiveness of dental hygiene, other organisms join the biofilm. These late-arriving species have little ability to co-aggregate with the pioneers, but do bind with intermediates such as *Fusobacterium*. Biofilm development and maturation include interactions between bacteria such as physical contact, metabolic exchange, molecular communication and exchange of genetic material. Biofilm may accumulate on both the hard and soft oral tissues, growth and maturation continuing until the biofilm reaches a critical mass.

6.2.2 Biofilms and Oral Disease

6.2.2.1 Dental Caries

Dental caries, also known as tooth decay, is a destructive condition of the dental hard tissues (enamel, dentin and cementum). It originates in bacterial infection causing demineralization of the tissues due to fermentation of food debris on the tooth surface. As the hard tissues progressively break down, inflammation and death of vital pulp tissue may occur. Moreover, if untreated this may ultimately lead to periapical infection. The disease process involves mostly acidogenic plaque bacteria, including *Streptococcus mutans*, *Streptococcus sobrinus* and *Lactobacillus* spp.[9] These bacteria produce acid in the presence of fermentable carbohydrates such as sucrose, fructose and glucose. The acidity of the mineral content of the tooth surface tends to increase when lactic acid is produced, resulting in demineralization. Today, caries remains one of the most common diseases in the world.

6.2.2.2 *Periodontal and Peri-implant Diseases*

Periodontal diseases can involve one or more of the periodontal tissues, including alveolar bone, periodontal ligament, cementum and gingiva. The vast majority of the periodontal diseases are initiated by plaque biofilm that develops on the hard root surface adjacent to the soft tissues of the supporting periodontium. Periodontal diseases range from simple gingival inflammation to severe disease that results in major damage to the soft tissue and bone supporting the teeth, leading in some cases to tooth loss. When the deeper supporting structures, including the periodontal ligament and the alveolar bone, are harmed, pocket formation, bone loss and tooth loosening occurs. In periodontal disease bacterial infection is considered the initiating and maintaining factor for the destructive inflammatory response. Host response, nutrient availability and environmental changes in the gingival sulcus and periodontal pocket can affect disease progression. Moreover, compromised tissue metabolism and repair may enhance microbial virulence factors. *Porphyromonas gingivalis*, *Tanerella forsythia* and *Treponema denticola* are regarded as the major pathogens.

Implant systems are routinely used in the reconstruction of fully or partially edentulous individuals. Like natural teeth, the implants may be susceptible to the contaminated oral cavity. Biofilms formed on implant surfaces may result in a peri-implant tissue response very similar to that of periodontal tissues in a prone host, *i.e.*, an inflammatory reaction, with the loss of supporting bone in the tissues surrounding a functioning implant. Anaerobic Gram-negative organisms play a major role in peri-implantitis,[10] inflammatory alterations in tissues adjacent to the implant leading to bone destruction and progressive loss of support and terminating in implant failure.

6.2.2.3 *Endodontic Infection*

Endodontic infections are polymicrobial, with predominantly anaerobic bacteria and some facultative bacteria. A tooth with an infected non-vital pulp is a reservoir of infection that is isolated from the patient's immune response and will eventually produce a periradicular inflammatory response. When the microbes invade the periradicular tissues, abscesses and cellulitis may develop. The severity of the infection depends on the pathogenicity of the microbes and the resistance of the host. These may not only give rise to an immunopathogenic and protective response, but also can be destructive to the surrounding tissues and contribute to the adverse signs and symptoms.

6.2.3 The Challenge of Controlling Oral Biofilm

Biofilm bacteria exhibit significantly reduced susceptibility to antimicrobials compared with cultures grown in suspension.[11] Various factors such as biofilm impermeability,[12,13] low uptake of the antimicrobial molecules,[14,15] formation of an antagonizing microenvironment against antimicrobial activity and horizontal gene transfer of antibiotic resistance genes[16] have been related to increased resistance of the biofilm bacteria. Frequently, prevention of dental caries and periodontal diseases is accomplished by mechanical or non-specific control of the plaque biofilm. Nevertheless, the use of antimicrobial agents may serve as a beneficial complement to mechanical plaque control.[17] New approaches aim at rapid bacterial abolition, minimizing the risk of resistance evolvement, enhancing the safety of the adjacent host tissue and maintaining the biological equilibrium of the microflora. Moreover, the real-time of exposure to antimicrobial agents during tooth brushing and mouth rinsing can be extremely brief, and may amount to about 30 seconds, rather than the recommended 2 minutes.[18] Thus, adjunctive antimicrobial treatment may be beneficial.

Innovative technologies and drug delivery systems for the prevention of colonization and biofilm formation have been proposed. Anti-plaque agents are generally classified as preventive, removing or disrupting agents. However, they do not necessarily eradicate the biofilm microorganisms. Antimicrobial agents can be bacteriostatic when acting as bacterial growth inhibitors or bactericidal when causing bacterial death. Particularly in the oral cavity, such therapeutic strategies are expected to reach less accessible sites, where plaque mostly accumulates. The uptake and penetration of antimicrobial agents into biofilm are key considerations in the administration of therapeutics.[19] As conventional plaque control basically relies on the patient, it is strongly affected by non-compliance. Thus, many new anti-plaque strategies requiring minimal patient compliance and minimal professional healthcare intervention are advantageous. Proposed strategies include surface modification of tissues and materials to reduce adhesion and biofilm formation; antimicrobial-impregnated matrices; enhancement of drug penetration into biofilm; and using carrier systems to target antimicrobials to the biofilm surface.[20] The anti-plaque potential of various antimicrobials, including chlorhexidine, hexetidine, delmopinol, amine fluoride/stannous fluoride, triclosan and phenolic compounds, has been well characterized and is frequently found in mouthwashes and toothpastes.[17] Additionally, the use of nanoparticles as components incorporated into topical agents or into various dental materials has been gaining interest over the past decade.[21–23] As it is possible to modify and adjust the physical and chemical characteristics of the nanoparticles by altering their surface charge, hydrophobicity, *etc.*, they are potentially useful.[24]

6.3 Focal Delivery Systems Against Periodontal and Peri-implant Infection

Periodontal diseases are a group of infectious diseases manifesting as a continuous destruction of tooth-supporting tissues (*i.e.*, cementum, periodontal ligament and alveolar bone). The estimated prevalence of periodontal disease in the U.S. is 47% of the general population, of which 8.5% suffer from the severe form of the disease.[25] The aetiology of the disease is not fully understood, although its association with biofilm and bacterial infection is well established.[26] Most treatment modalities for the disease aim to eradicate the infection by means of mechanical or chemical treatment.

Peri-implant diseases are relatively new pathologies that are emerging along with the growing popularity of dental implants. The clinical characteristics of the disease are similar to those seen in periodontal diseases,[27] but its aetiology is mostly unclear. Therapeutic modalities for peri-implantitis are drawn mainly from the field of periodontics, but often these modes achieve limited success.

6.3.1 Traditional Periodontal and Peri-implant Therapy

Periodontal treatment consists of an anti-infective phase followed by corrective and maintenance phases. The anti-infective phase aims to eliminate the periodontal infection, and to re-establish non-inflamed periodontal tissues. The classic treatment modality consists of mechanical plaque debridement, along with strict patient oral hygiene practice. This treatment modality allows physical clearance of biofilm located beneath the gingival margin. The use of systemic antibiotics in periodontitis has been, and still is, a controversial issue. In the chronic type of the disease, the use of systemic antibiotics is not commonly recommended, although recent insights show a beneficial effect of this treatment.[28] In the aggressive type of the disease, the use of amoxicillin with metronidazole treatment is recommended and commonly used, mainly because of resistance of *Aggregatibacter actinomycetemcomitans* (the main pathogen associated with the disease) to treatment with either antibiotic alone.[29]

In periodontal treatment, all chemical agents that are used in the oral cavity without systemic exposure fall under the definition of focal delivery systems. Classic and commonly used focal delivery systems are dentifrices and mouth washes. Dentifrices serve as a detergent during tooth brushing, providing ease of use during biofilm removal by mechanical action. Additional active materials in the dentifrice have a role in maintaining and preventing pathological processes around the teeth (see Section 6.5.3).

6.3.2 Focal Controlled Agents in Periodontitis

6.3.2.1 *Chlorhexidine Gluconate*

Although systemically applied antibiotics have proven to be effective, they may promote the development of bacterial resistance.[30] Thus, attempts have been made to develop local delivery devices based on antiseptic agents for subgingival application. PerioChip® is the most widely tested delivery device of this category, which contains 2.5 mg chlorhexidine gluconate in a biodegradable matrix of hydrolyzed gelatin (CHX chip).

Soskolne *et al.* evaluated the safety and efficacy of CHX chip in a randomized, blinded, multi-center study of 118 patients with moderate periodontitis in a split-mouth design.[31] A control group included scaling and root planing (SRP) alone, while the test group included SRP combined with application of CHX chip in periodontal pockets larger than 5–8 mm. The average probing depth reduction in the CHX chip-treated sites was significantly greater than those receiving SRP alone at both 3 and 6 months post-treatment, with a mean difference of 0.42 mm at 6 months. Furthermore, clinical attachment level (CAL) was more improved at the chip-treated sites than the SRP sites, with a statistically significant difference 6 months post-treatment. An analysis of patients with 7–8 mm initial probing depths showed greater reduction in probing depth and CAL in the chip group than those treated with SRP at both 3 and 6 months, with mean differences in probing depth and CAL at 6 months of 0.71 mm and 0.56 mm, respectively. Jeffcoat *et al.* evaluated the efficacy of CHX chip when used as an adjunct to SRP to reduce probing depth and improve CAL in adult periodontitis.[32] Two double-blind, randomized, placebo-controlled multi-center clinical trials (with five centers in each study) were conducted. Results show a significant reduction for CHX chip compared with control treatments observed 9 months post-treatment in probing depth (CHX chip plus SRP 0.95 ± 0.05 mm; SRP alone 0.65 ± 0.05 mm, $P < 0.001$; placebo chip plus SRP 0.69 ± 0.05 mm, $P < 0.001$) and CAL (CHX chip plus SRP 0.75 ± 0.06 mm; SRP alone 0.58 ± 0.06 mm, $P < 0.05$; placebo chip plus SRP 0.55 ± 0.06 mm, $P < 0.05$). In the study, adverse effects were minor and included transient toothache, pain, tenderness, aching, throbbing, soreness, discomfort or sensitivity. All were observed in the chlorhexidine group *versus* placebo (P 0.042). Another multicenter study by Paolantonio *et al.* also showed probing depth reduction and gain in clinical attachment in CHX chip treatment *versus* SRP alone.[33]

Mizrak *et al.* showed significant positive effects on the percentage of subgingival spirochetes when CHX chips were used after SRP compared to SRP alone.[34] In contrast, Grisi *et al.* showed in a randomized single-blind study with parallel design that the CHX chip did not provide any clinical or microbiological benefit beyond that achieved with conventional SRP, after a 9 month period.[35]

Still, the majority of studies advocate the beneficial clinical effect of PerioChip® over standard periodontal care.

6.3.2.2 *Minocycline Ointment and Microspheres (Arestin)*

van Steenberghe *et al.* conducted a double-blind, randomized, parallel-arms study to evaluate the long-term safety and efficacy of subgingivally administered minocycline ointment *versus* vehicle control.[36] For pockets ≥5 mm, a mean reduction of 1.9 mm probing depth was observed in test sites, *versus* 1.2 mm in control sites. Sites with a baseline probing depth of ≥7 mm and bleeding index of >2 showed an average of 2.5 mm reduction with minocycline *versus* 1.5 mm in the control. Attachment gains of 0.9 mm and 1.1 mm were observed in minocycline-treated sites, with baseline probing depth of ≥5 mm and ≥7 mm, respectively, compared with 0.5 mm and 0.7 mm values, respectively, in control sites. Furthermore, subgingival administration of minocycline ointment was well tolerated.

Moreover, Williams *et al.* reported on the efficacy and safety of locally administered microencapsulated minocycline.[37] The results indicate that minocycline microspheres plus SRP provided substantial reduction in probing depth over SRP alone or SRP plus vehicle. The differences were statistically significant at 1 month post-treatment and maintained their effect throughout the trial (9 months). Van Dyke *et al.* evaluated the efficacy of the minocycline microspheres in patients with moderate-to-severe periodontitis in a four-arm parallel design, randomized, open-label evaluator-blinded study.[38] Patients with at least two teeth with pocket depth ≥6 mm and gingival crevicular fluid prostaglandin E2 levels of above 66.2 $ng\,mL^{-1}$ were set as the study population. For 6 months, one group received SRP followed by one dose per pocket of minocycline microspheres (MPTS), another group received SRP alone, a third group received MPTS alone and a fourth group provided sham no treatment control. Substantially greater reduction in probing depth and clinical attachment gain was observed for the SRP + MPTS group compared with all other groups. Probing depth reduction and attachment gain at month 3 in the SRP + MPTS group *versus* SRP alone was statistically significant.

6.3.2.3 Doxycycline Hyclate (Atridox)

Garrett *et al.* evaluated clinical parameters following local delivery of doxycycline hyclate or traditional SRP in a group of patients undergoing supportive periodontal therapy.[39] Results show that both doxycycline hyclate without mechanical instrumentation or SRP alone were equally effective as supportive periodontal therapy over the 9 month study period. Wennstrom *et al.* conducted a 6 month multicenter trial examining two different approaches to non-surgical treatment in chronic periodontitis patients with the use of locally delivered controlled-release doxycycline.[40] Patients were randomly assigned either SRP followed by a subgingival application of an 8.5% w/w doxycycline polymer at sites with a remaining probing depth of ≥5 mm (SRP group), or to a group which received application of doxycycline in sites with a probing depth of ≥5 mm followed by SRP (debridement group). Results show that the proportion of sites showing residual pockets of ≤4 mm was significantly higher in the "debridement" group than in the "SRP" group at 3 months, while no difference was observed at 6 months. Overall, the results indicate that simplified subgingival instrumentation combined with local application of doxycycline in deep periodontal sites can be considered as a justified approach for non-surgical treatment of chronic periodontitis.

6.3.2.4 Metronidazole Gel (Elyzol)

Griffiths *et al.* compared the clinical effect of SRP with the same treatment adjunct with subgingival application of 25% metronidazole gel in patients with chronic periodontitis.[41] At the end of the study, the mean reductions for probing pocket depth were 1.0 mm (SRP) compared to 1.5 mm (SRP + gel), and for CAL they were 0.4 mm (SRP) compared to 0.8 mm (SRP + gel). The combination therapy of SRP + gel was superior to the conventional treatment of SRP alone, and these differences were maintained for 9 months. Stelzel *et al.* conducted a randomized single-blind study with split-mouth design,[42] and showed minor advantages following application of a metronidazole 25% dental gel as adjunctive therapy to subgingival scaling. Similarly, Ainamo *et al.* found that the differences between SRP alone and SRP plus gel were clinically insignificant.[43]

6.3.2.5 Tetracycline (Actisite)

Newman *et al.* conducted a multicenter study comparing the efficacy of SRP alone *versus* tetracycline fiber therapy as an adjunctive modality to the SRP in the treatment of localized recurrent periodontitis sites in a supportive maintenance program.[44] At 1, 3 and 6 months, adjunctive fiber therapy showed significantly better reduction of probing depth and bleeding on probing (BOP) compared with SRP alone. At 6 months, fiber therapy was found to be significantly better in promoting clinical attachment gain than SRP alone. Tonetti *et al.* evaluated the effectiveness of adjunctive tetracycline fibers in the control of bleeding on probing in mandibular class II furcation during maintenance care, in a randomized single-blind multicenter clinical trial.[45] Results indicate that both test and control subjects show significant improvements of BOP and probing depth at 3 and 6 months. The tetracycline fiber treatment showed robust improvements of BOP and probing depth at 3 months compared with control, without a difference between groups at 6 months. No differences were observed in changes in CAL.

6.3.3 Focal Controlled Agents in Peri-implantitits

Peri-implantitis is an inflammatory process caused by microorganisms affecting the tissues around osseointegrated implant in function, resulting in a loss of supporting bone. Due to the lack of knowledge on its pathogenesis, the most current treatment modalities aim to clear bacterial deposits of the implant surface.

Salvi *et al.*, evaluated the clinical and radiographic changes occurring after adjunctive local delivery of minocycline microspheres for the treatment of peri-implantitis.[46] Non-surgical mechanical treatment of peri-implantitis lesions with adjunctive local delivery of microencapsulated minocycline led to a beneficial clinical effect, sustained for 12 months. Renvert *et al.* evaluated the clinical and microbiological outcome of repeated local administration of minocycline microspheres in cases of peri-implantitis.[47]

32 subjects with at least one implant with a probing depth ≥4 mm combined with bleeding and/or exudate on probing and the presence of putative pathogenic bacteria were included in the study. At baseline, subjects were randomly assigned to receive local minocycline microspheres or chlorhexidine gel following debridement at baseline, and at 30, 90 and 180 days. The use of minocycline resulted in significant improvements in probing depths compared to chlorhexidine at all tested times. Mean probing depth reduction was 0.6 mm at 12 months in the deepest sites following minocycline treatment. Also, significant differences between groups were found in BOP levels throughout the experiment. Both treatments resulted in a marked reduction in the indicator bacteria.

6.4 Focal Delivery Systems Against Endodontic Infection

The dental pulp is enclosed in a compartment of hard tooth structure, and even a small increase in tissue pressure due to inflammation may result in great pain.[48] If left untreated, the pulp tissue will undergo necrosis and become a rich breeding ground for bacterial infection. Bacteria and their by-products inside the pulp canal can seep outside the canal system into the surrounding tissues and cause destructive inflammatory response at the tooth-supporting bone.

Endodontic treatment aims to clear infection from the tooth's canal system and seal the system with a special filling material. The treatment modalities can be non-surgical or surgical endodontic treatment. Non-surgical endodontic treatment is usually a localized procedure that involves cleaning, shaping and obturation of the canal system. The most important factor for a successful outcome is debridement of the root canal and surrounding tissues. When systemic involvement is diagnosed, including malaise fever and lymphatic involvement, an adjacent antibiotics regimen is included in the treatment.

During endodontic treatment, the cleaning procedure has into two main stages: (1) mechanical debridement and shaping of the canal; and (2) chemical disinfection. Mechanical cleaning and shaping is achieved by rotary files in an effort to scrape the infected canal walls. Most rotary files are today are based on NiTi alloy in different shapes and mechanical properties. Despite advancements in file fabrication and improvement of metallurgic and file structure, this modality reaches up to 60% of the canal wall.[49,50] Untouched canal wall surface may harbor pulp tissue and bacteria which may lead to treatment failure.[51] Most of the untouched canal walls are cleaned by antibacterial irrigation. The most common irrigation is sodium hypochlorite solution, which can dissolve organic matter. In general, the aim of disinfection irrigation is to reduce bacterial load to such a level that the patient's immune response will allow complete healing of the infected site.

Since endodontic irrigation is applied to the root canal system only, this modality is, in fact, a focal drug delivery.

6.4.1 Canal Irrigation

The greatest challenge of any root canal irrigation material is to reach all the sites not reached by mechanical files. The efficacy of various irrigation materials is measured by their ability to clear biological substance in general and bacteria specifically.

Many irrigation materials are available. We focus on popular and available materials used in modern endodontics.

6.4.1.1 Sodium Hypochlorite

Sodium hypochlorite (NaOCl) is one of the most commonly used irrigation solutions in endodontics. Sodium hypochlorite is a proteolytic agent with great antibacterial and tissue-dissolving capabilities.[52] Different concentrations are used in the clinic, ranging from 0.5% to 6%. Studies have shown that a higher concentration of sodium hypochlorite improves the dissolving capability of tissue and biofilm, whereas the antibacterial properties are unrelated to concentration.[53] Several modifications of this application have been tested in an effort to increase its efficiency, including warming the solution before use, ultrasonic activating, pH buffering, changing its concentration and the volume of the solution used.[52]

6.4.1.2 Chlorhexidine

Chlorhexidine is a strong antiseptic.[54] In endodontics, it is usually used in a 2% solution both as an irrigant and as intracanal medication between appointments. Unlike sodium hypochlorite, chlorhexidine lacks tissue-dissolving capabilities, thus making it unfit as a single irrigant.[55] Chlorhexidine has been advocated as a useful addition to other irrigants in cases of persistent infections caused by Gram-positive bacteria such as *Enterococcus faecalis*.[56]

6.4.1.3 Iodine Potassium Iodide

Iodine potassium iodide (IKI) was first used as an irrigant in endodontic treatment nearly a century ago, due to its highly potent antibacterial properties.[52] However, several drawbacks, such as inability to disrupt biofilm, discoloration of the tooth and allergic reactions to iodine made such agents less favorable.[57]

6.4.1.4 Ethylenediaminetetra-acetic Acid

Ethylenediaminetetra-acetic acid (EDTA) is a chelating agent. EDTA lacks antibacterial properties (susceptibilities of *E. faecalis*)[58] and is used as an addition to antibacterial agents such as sodium hypochlorite. The role of EDTA is to remove the inorganic material, therefore allowing the antibacterial agents better access to the bacteria in the canal.[59]

6.4.1.5 *Qmix*

Qmix is a solution made of an antibacterial agent (CHX) and a calcium chelator for smear layer removal (EDTA), dissolved in a detergent (disodium dehydrate).[60] This mixture was proven effective against several types of bacteria, including *E. faecalis*.[61]

6.4.1.6 *MTAD*

MTAD is a solution made of 3% doxycycline, 4.25% citric acid and 0.5% Tween 80.[62] It has been recommended as a final irrigation after conventional disinfection protocol. This mixture has been shown effective against several types of bacteria in biofilm, but not against *E. faecalis*.[63]

6.4.2 Intracanal Medication

Elimination of bacteria from the root canal is one of the main factors leading to treatment success.[64] There is much controversy in the literature as to whether a one- or two-visit treatment protocol should be undertaken.[65] If root canal treatment cannot be completed in one visit the remaining bacteria in the canal can grow and multiply, thus jeopardizing the treatment. In any case, where a root canal treatment cannot be finished in a single appointment an antibacterial intracanal medicament should be applied.

6.4.2.1 *Calcium Hydroxide*

Calcium hydroxide is the most popular intracanal medication in use today.[66] It has been used in endodontics for nearly a century. Hammarstrom showed that intracanal use of calcium hydroxide in infected root canals was effective in eliminating the infection, but also delayed healing.[67] Calcium hydroxide must come into direct contact in order to kill bacteria.[68] In addition to its antibacterial activities, calcum hydroxide can neutralize bacterial lipopolysaccharides, thus further reducing bacterial damage to the host.[69] In addition, the use of calcium hydroxide can help dissolve necrotic tissues from the canal.[70] Several studies advocate the use of calcium hydroxide as a long-term canal dressing;[71] however, Andreasen *et al.* showed that leaving calcium hydroxide for longer than 2 months can weaken the dentinal walls.[72]

6.4.2.2 *Chlorhexidine*

The properties of this medication are listed above as an irrigant. Chlorhexidine can be used as a medication between appointments when applied in the form of a 2% gel. Its concentration and antibacterial activity diminishes over time, but it remains active for at least 12 weeks inside the root canal.[73]

6.4.2.3 Iodine

IKI is a powerful disinfectant. It produces a potent vapor which effectively reaches and disinfects irregular areas of the canal.[74] Iodine can be used alone or as an addition to calcium hydroxide paste.

6.4.2.4 Formocresol

Formocresol is a formaldehyde solution, which despite being mutagenic, carcinogenic and highly toxic, has been used widely in endodontics. It is still used in pediatric dentistry to seal deciduous teeth.[75]

6.4.2.5 Antibiotic–Steroid Paste

Ledermix is antibiotic–steroid paste used since the late 1960s. It is composed of triamcinolone, for the relief of symptoms related to endodontic inflammation, and demeclocycline, a wide-spectrum antibiotic.[76]

6.4.2.6 Bioactive Glass

A new type of interappointment medication based on bioactive glass, SiO_2–Na_2O–CaO–P_2O_5, has been shown to eliminate bacteria from root canals after 5 days in an *in vitro* study.[77] The mechanism by which the bioactive glass affects bacteria is not clear and may be due to an alkaline modification at the canal environment. However, currently, bioactive glass is not available for clinic use.

6.5 Focal Drug Agents Against Caries Lesions

Dental caries is the localized destruction of calcified tissue by enzymatic lysis. If left unchecked, this leads to cavity formation reaching tooth-structured layers (enamel, dentin and dental pulp). The aetiology of caries is generally attributed to acid-producing bacteria, *via* fermentable carbohydrate metabolism. Two main groups of bacteria are associated with dental caries: Mutans *Streptococci* (mainly *Streptococcus mutans* and *Streptococcus sobrinus*) and *Lactobacilli* species.[9]

The residual intact enamel and dentin tissues are clinically crucial for the success of the prosthetic treatment, which include plastic restorations for small lesions and casted or computer-aided design/computer-aided manufacturing (CAD/CAM) fabricated restorations for larger lesions.[78,79] Moreover, time is a key factor in the progress of the carious lesion,[80] and a rapid resolution of bacterial infection prevents extensive tissue loss and enables conservative solution.

Treatment strategies for caries include prevention, remineralization and restoration. In addition, antimicrobial agents may be delivered in various forms including toothpaste, mouthwash, varnish, restorative material, cement, *etc.* Caries prevention can be achieved by keeping a constant intra-oral concentration of antimicrobial agents with a controlled rate. These agents varies in their active ingredients and their carriers.

6.5.1 Fluoride

Fluoride is one of the most investigated anti-carious agents.[81] The ability of fluoride to prevent the development of dental caries involves several mechanisms, including reduction in the acid solubility of enamel, promotion of enamel remineralization, inhibition of glucose uptake and utilization by acidogenic bacteria, and possibly bacteriostatic or bactericidal properties.[82,83]

Slow-release fluoride devices are a potentially cost-effective method of reducing dental caries in caries high-risk patients. Such devices currently include a copolymer membrane, glass beads and a mixture of NaF and hydroxyapatite. The copolymer membrane device consists of a small pellet with an inner core of hydroxyethyl methacrylate (HEMA)/methyl methacrylate (MMA) copolymer mixture, containing NaF. This device is a membrane-controlled reservoir of fluoride owing to the HEMA/MMA copolymer membrane surrounding the core.[84] The fluoride glass bead system is based on its ability to gradually release fluoride upon exposure to saliva. The latest device has the form of a disk placed within a plastic bracket, allowing its replacement and installation. Both copolymer membrane and glass bead systems are usually attached to the buccal surface (inner part of the cheek) adjacent to the first permanent molar. Studies in animals and humans demonstrated that the use of fluoride slow-release devices in copolymer membrane and glass bead as carriers efficiently protect masticatory surfaces, which are normally fluoride-devoid areas.[84] However, retention rates are still a grave problem for such devices and require further improvement. An additional diffusion control device is the Hydroxyapatite-Eudragit RS100 F system, based on a mixture of hydroxyapatite, NaF, and Eudragit RS100.

Furthermore, fluoride may be immersed in restorative materials such as dental cements and filling materials. The most investigated material in this context is glass ionomer, by which aluminum fluorosilicate glass is incorporated in to glass ionomer cement. Not only do these cements release fluoride for extended periods, they are also capable of absorbing fluoride from toothpastes and releasing it later, thereby serving as a fluoride reservoir. Despite the numerous advantages of fluoride-impregnated dental materials, disadvantages include difficulties in controlling the rate of fluoride release. However, the fluoride slow-release devices have proved to be beneficial and cost-effective.[81]

6.5.2 Chlorhexidine

CHX is the most common substance used as an oral antiseptic, with low toxicity and broad antibacterial activity. It is a strong base that binds bacteria and negatively charged molecules, leading to a bactericidal and bacteriostatic effect. CHX has a wide range of action against Gram-positive, Gram-negative, aerobic, anaerobic, yeast and fungus organisms.[85,86]

CHX was found to be effective against *Streptococcus mutans* in the plaque biofilm in human mouths,[87] and to a lesser extent against *Lactobacilli*. Elimination of *Streptococcus mutans* can be achieved by intensive, highly concentrated and repeated application of CHX.

CHX varnish may be used as a controlled-release agent. A Cochrane systematic review assessed the effects of chlorhexidine containing oral products on the prevention of dental caries in children and adolescents.[88] Eight randomized controlled trials were included that evaluated the effects of CHX varnishes (1%, 10% or 40% concentration) and CHX gel (0.12%) on primary or permanent teeth. The results did not provide evidence that CHX varnish or gel reduces tooth decay or reduces the bacteria that encourage tooth decay.[88] A study by Forgie *et al.*[89] assessed the efficacy of CHX varnish in reducing caries increase in adolescents at high risk ofcaries, and failed to show any caries-reducing beneficial effect. Incorporating CHX into glass ionomer cements (GICs) can affect antibacterial properties against cariogenic bacteria.[90] Reports show that CHX interacts with fluoride synergistically as a bacterial inhibitor. However, a high concentration of CHX embedded in the GIC results in changes in physical and mechanical properties, *e.g.* decreased bonding potential and increased setting time.

6.5.3 Triclosan

Triclosan is a broad-spectrum antibacterial agent with low toxicity. When combined with a copolymer as a substantive agent, this formula can be added to toothpaste and reduces plaque accumulation as well as gingival inflammation. While CHX may have a greater antimicrobial effect, triclosan is more compatible with other toothpaste ingredients. In addition, triclosan does not stain teeth or have an unpleasant taste, two common disadvantages of CHX.[91]

Disadvantages of triclosan are its solubility and short-term half-life clearance. Solutions to these drawbacks are its emulsion, using dentifrice compositions that contain cyclodextrin, or a composition containing sodium lauryl sulfate as a surfactant.[92,93] Reports show that triclosan-loaded micelles composed of polyvinylmethyl ether/maleic acid copolymer prolong triclosan retention in the oral cavity.[94,95] Dendrimers are also used as triclosan carriers, and indeed, Zhou *et al.* showed in an *in vitro* study that triclosan-loaded dendrimers released this agent for a long period.[96]

Similarly to CHX, triclosan may be incorporated in GICs. It has been proposed that a material containing triclosan can immobilize bacteria, which is retained in its carrier material and provides long-term anti-cariogenic activity.[98] The non-releasing bactericide of triclosan gives it advantageous properties over chlorhexidine as an antibacterial additive in GICs. In addition, 2.5% triclosan added to GIC showed more antimicrobial activity compared with 2.5% CHX incorporated GIC against *Lactobacillus acidophilus* and *Streptococcus mutans*, with similar composite mechanical strength.[90]

6.5.4 Calcium Phosphate

Calcium phosphate-based systems increase the availability of calcium and phosphate ions in oral fluids. The presence of these ions in the environment surrounding the tooth is crucial to remineralization. Agents providing calcium phosphate release include casein phosphopeptides (CPP) alone or as CPP–ACP (CPP with amorphous calcium phosphate). CPP is also believed to have an antibacterial and buffering effect and the ability to interfere in the growth and adherence of *Streptococcus mutans* and *Streptococcus sobrinus*.[97] Controlled release of CPP–ACP is achieved by carriers such as toothpaste, dental cream and chewing gum.[98–100]

6.6 Conclusions

In the oral cavity focal drug delivery may offer many advantages as a therapeutic means for preventing or treating dental diseases. Moreover, slow release and focal treatment may overcome disadvantages such as low dosage and limited penetration into the surrounding tissues, while minimizing the risk of undesired side effects.

References

1. P. Watnick and R. Kolter, Biofilm, city of microbes, *J. Bacteriol.*, 2000, **182**(10), 2675–2679.
2. A. Amano, I. Nakagawa and S. Hamada, Studying initial phase of biofilm formation: molecular interaction of host proteins and bacterial surface components, *Methods Enzymol.*, 1999, **310**, 501–513.
3. P. D. Marsh and D. J. Bradshaw, Dental plaque as a biofilm, *J. Ind. Microbiol.*, 1995, **15**(3), 169–175.
4. P. D. Marsh, Microbial ecology of dental plaque and its significance in health and disease, *Adv. Dent. Res.*, 1994, **8**(2), 263–271.
5. C. Hannig and M. Hannig, The oral cavity–a key system to understand substratum-dependent bioadhesion on solid surfaces in man, *Clin. Oral. Investig.*, 2009, **13**(2), 123–139.
6. D. Herrera, *et al.*, Antimicrobial therapy in periodontitis: the use of systemic antimicrobials against the subgingival biofilm, *J. Clin. Periodontol.*, 2008, **35**(8 Suppl), 45–66.
7. P. E. Kolenbrander, Oral microbial communities: biofilms, interactions, and genetic systems, *Annu. Rev. Microbiol.*, 2000, **54**, 413–437.
8. P. E. Kolenbrander, *et al.*, Bacterial interactions and successions during plaque development, *Periodontol. 2000*, 2006, **42**, 47–79.
9. J. M. Hardie, Oral microbiology: current concepts in the microbiology of dental caries and periodontal disease, *Br. Dent. J.*, 1992, **172**(7), 271–278.
10. N. U. Zitzmann and T. Berglundh, Definition and prevalence of peri-implant diseases, *J. Clin. Periodontol.*, 2008, **35**(8 Suppl), 286–291.

11. R. M. Donlan, Role of biofilms in antimicrobial resistance, *ASAIO J.*, 2000, **46**(6), S47–S52.
12. K. Lewis, Riddle of biofilm resistance, *Antimicrob. Agents Chemother.*, 2001, **45**(4), 999–1007.
13. T. F. Mah and G. A. O'Toole, Mechanisms of biofilm resistance to antimicrobial agents, *Trends Microbiol.*, 2001, **9**(1), 34–39.
14. K. D. Xu, G. A. McFeters and P. S. Stewart, Biofilm resistance to antimicrobial agents, *Microbiology*, 2000, **146**(Pt 3), 547–549.
15. M. C. Walters, 3rd, *et al.*, Contributions of antibiotic penetration, oxygen limitation, and low metabolic activity to tolerance of Pseudomonas aeruginosa biofilms to ciprofloxacin and tobramycin, *Antimicrob. Agents Chemother.*, 2003, **47**(1), 317–323.
16. A. P. Roberts and P. Mullany, Genetic basis of horizontal gene transfer among oral bacteria, *Periodontol 2000*, 2006, **42**, 36–46.
17. P. C. Baehni and Y. Takeuchi, Anti-plaque agents in the prevention of biofilm-associated oral diseases, *Oral Dis.*, 2003, **9**(Suppl 1), 23–29.
18. F. J. van der Ouderaa, Anti-plaque agents. Rationale and prospects for prevention of gingivitis and periodontal disease, *J. Clin. Periodontol.*, 1991, **18**(6), 447–454.
19. P. S. Stewart, Diffusion in biofilms, *J. Bacteriol.*, 2003, **185**(5), 1485–1491.
20. A. W. Smith, Biofilms and antibiotic therapy: is there a role for combating bacterial resistance by the use of novel drug delivery systems? *Adv. Drug Delivery Rev.*, 2005, **57**(10), 1539–1550.
21. N. Beyth, *et al.*, Surface antimicrobial activity and biocompatibility of incorporated polyethylenimine nanoparticles, *Biomaterials*, 2008, **29**(31), 4157–4163.
22. N. Beyth, *et al.*, Antibacterial activity of dental composites containing quaternary ammonium polyethylenimine nanoparticles against Streptococcus mutans, *Biomaterials*, 2006, **27**(21), 3995–4002.
23. N. Beyth, *et al.*, Polyethyleneimine nanoparticles incorporated into resin composite cause cell death and trigger biofilm stress in vivo, *Proc. Natl. Acad. Sci. U. S. A.*, 2010, **107**(51), 22038–22043.
24. A. E. Nel, *et al.*, Understanding biophysicochemical interactions at the nano-bio interface, *Nat. Mater.*, 2009, **8**(7), 543–557.
25. P. I. Eke, *et al.*, Prevalence of Periodontitis in Adults in the United States: 2009 and 2010, *J. Dent. Res.*, 2012, **91**(10), 914–920.
26. Informational Paper: The Pathogenesis of Periodontal Diseases, *J. Periodontol.*, 1999, **70**(4), 457–470.
27. N. P. Lang and T. Berglundh, Periimplant diseases: where are we now?–Consensus of the Seventh European Workshop on Periodontology, *J. Clin. Periodontol.*, 2011, **38**(Suppl 11), 178–181.
28. N. Cionca, *et al.*, Microbiologic testing and outcomes of full-mouth scaling and root planing with or without amoxicillin/metronidazole in chronic periodontitis, *J. Periodontol.*, 2010, **81**(1), 15–23.
29. M. J. Pavicic, *et al.*, Microbiological and clinical effects of metronidazole and amoxicillin in Actinobacillus actinomycetemcomitans-associated

periodontitis. A 2-year evaluation, *J. Clin. Periodontol.*, 1994, **21**(2), 107–112.
30. C. L. Ventola, The antibiotic resistance crisis: part 1: causes and threats, *P T*, 2015, **40**(4), 277–283.
31. W. A. Soskolne, *et al.*, Sustained local delivery of chlorhexidine in the treatment of periodontitis: a multi-center study, *J. Periodontol.*, 1997, **68**(1), 32–38.
32. M. K. Jeffcoat, *et al.*, Adjunctive use of a subgingival controlled-release chlorhexidine chip reduces probing depth and improves attachment level compared with scaling and root planing alone, *J. Periodontol.*, 1998, **69**(9), 989–997.
33. M. Paolantonio, *et al.*, Clinical and microbiologic effects of subgingival controlled-release delivery of chlorhexidine chip in the treatment of periodontitis: a multicenter study, *J. Periodontol.*, 2008, **79**(2), 271–282.
34. T. Mizrak, *et al.*, Effect of a controlled-release chlorhexidine chip on clinical and microbiological parameters and prostaglandin E2 levels in gingival crevicular fluid, *J. Periodontol.*, 2006, 77(3), 437–443.
35. D. C. Grisi, *et al.*, Effect of a controlled-release chlorhexidine chip on clinical and microbiological parameters of periodontal syndrome, *J. Clin. Periodontol.*, 2002, **29**(10), 875–881.
36. D. van Steenberghe, *et al.*, A 15-month evaluation of the effects of repeated subgingival minocycline in chronic adult periodontitis, *J. Periodontol.*, 1999, **70**(6), 657–667.
37. R. C. Williams, *et al.*, Treatment of periodontitis by local administration of minocycline microspheres: a controlled trial, *J. Periodontol.*, 2001, **72**(11), 1535–1544.
38. T. E. Van Dyke, *et al.*, Enhancing the value of scaling and root-planing: Arestin clinical trial results, *J. Int. Acad. Periodontol.*, 2002, **4**(3), 72–76.
39. S. Garrett, *et al.*, The effect of locally delivered controlled-release doxycycline or scaling and root planing on periodontal maintenance patients over 9 months, *J. Periodontol.*, 2000, **71**(1), 22–30.
40. J. L. Wennstrom, *et al.*, Utilisation of locally delivered doxycycline in non-surgical treatment of chronic periodontitis. A comparative multi-centre trial of 2 treatment approaches, *J. Clin. Periodontol.*, 2001, **28**(8), 753–761.
41. G. S. Griffiths, *et al.*, Comparison of clinical outcomes following treatment of chronic adult periodontitis with subgingival scaling or subgingival scaling plus metronidazole gel, *J. Clin. Periodontol.*, 2000, **27**(12), 910–917.
42. M. Stelzel and L. Flores-de-Jacoby, Topical metronidazole application as an adjunct to scaling and root planing, *J. Clin. Periodontol.*, 2000, **27**(6), 447–452.
43. J. Ainamo, *et al.*, Clinical responses to subgingival application of a metronidazole 25% gel compared to the effect of subgingival scaling in adult periodontitis, *J. Clin. Periodontol.*, 1992, **19**(9 Pt 2), 723–729.

44. M. G. Newman, K. S. Kornman and F. M. Doherty, A 6-month multi-center evaluation of adjunctive tetracycline fiber therapy used in conjunction with scaling and root planing in maintenance patients: clinical results, *J. Periodontol.*, 1994, **65**(7), 685–691.
45. M. S. Tonetti, *et al.*, A controlled multicenter study of adjunctive use of tetracycline periodontal fibers in mandibular class II furcations with persistent bleeding, *J. Clin. Periodontol.*, 1998, **25**(9), 728–736.
46. G. E. Salvi, *et al.*, Adjunctive local antibiotic therapy in the treatment of peri-implantitis II: clinical and radiographic outcomes, *Clin. Oral. Implants Res.*, 2007, **18**(3), 281–285.
47. S. Renvert, *et al.*, Mechanical and repeated antimicrobial therapy using a local drug delivery system in the treatment of peri-implantitis: a randomized clinical trial, *J. Periodontol.*, 2008, **79**(5), 836–844.
48. H. J. Van Hassel, Physiology of the human dental pulp, *Oral Surg., Oral Med., Oral Pathol.*, 1971, **32**(1), 126–134.
49. F. Paque, D. Ganahl and O. A. Peters, Effects of root canal preparation on apical geometry assessed by micro-computed tomography, *J. Endod.*, 2009, **35**(7), 1056–1059.
50. O. A. Peters, K. Schonenberger and A. Laib, Effects of four Ni-Ti preparation techniques on root canal geometry assessed by micro computed tomography, *Int. Endod. J.*, 2001, **34**(3), 221–230.
51. S. J. Card, *et al.*, The effectiveness of increased apical enlargement in reducing intracanal bacteria, *J. Endod.*, 2002, **28**(11), 779–783.
52. M. Zehnder, Root canal irrigants, *J. Endod.*, 2006, **32**(5), 389–398.
53. J. F. Siqueira, Jr., *et al.*, Efficacy of instrumentation techniques and irrigation regimens in reducing the bacterial population within root canals, *J. Endod.*, 2002, **28**(3), 181–184.
54. D. G. Maki, M. Ringer and C. J. Alvarado, Prospective randomised trial of povidone-iodine, alcohol, and chlorhexidine for prevention of infection associated with central venous and arterial catheters, *Lancet*, 1991, **338**(8763), 339–343.
55. E. Ercan, *et al.*, Antibacterial activity of 2% chlorhexidine gluconate and 5.25% sodium hypochlorite in infected root canal: in vivo study, *J. Endod.*, 2004, **30**(2), 84–87.
56. M. E. Vianna, *et al.*, In vitro evaluation of the antimicrobial activity of chlorhexidine and sodium hypochlorite, *Oral Surg., Oral Med., Oral Pathol., Oral Radiol. Endod.*, 2004, **97**(1), 79–84.
57. L. Spangberg, B. Engstrom and K. Langeland, Biologic effects of dental materials. 3. Toxicity and antimicrobial effect of endodontic antiseptics in vitro, *Oral Surg., Oral Med., Oral Pathol.*, 1973, **36**(6), 856–871.
58. A. Kishen, *et al.*, An investigation on the antibacterial and antibiofilm efficacy of cationic nanoparticulates for root canal disinfection, *J. Endod.*, 2008, **34**(12), 1515–1520.
59. M. Torabinejad, *et al.*, Clinical implications of the smear layer in endodontics: a review, *Oral Surg., Oral Med., Oral Pathol., Oral Radiol. Endod.*, 2002, **94**(6), 658–666.

60. L. Dai, *et al.*, The effect of QMix, an experimental antibacterial root canal irrigant, on removal of canal wall smear layer and debris, *J. Endod.*, 2011, **37**(1), 80–84.
61. S. Devaraj, N. Jagannathan and P. Neelakantan, Antibiofilm efficacy of photoactivated curcumin, triple and double antibiotic paste, 2% chlorhexidine and calcium hydroxide against Enterococcus fecalis in vitro, *Sci. Rep.*, 2016, **6**, 24797.
62. M. Torabinejad, *et al.*, The antimicrobial effect of MTAD: an in vitro investigation, *J. Endod.*, 2003, **29**(6), 400–403.
63. B. M. Newberry, *et al.*, The antimicrobial effect of biopure MTAD on eight strains of Enterococcus faecalis: an in vitro investigation, *J. Endod.*, 2007, **33**(11), 1352–1354.
64. C. H. Stuart, *et al.*, Enterococcus faecalis: its role in root canal treatment failure and current concepts in retreatment, *J. Endod.*, 2006, **32**(2), 93–98.
65. C. Sathorn, P. Parashos and H. H. Messer, Effectiveness of single-versus multiple-visit endodontic treatment of teeth with apical periodontitis: a systematic review and meta-analysis, *Int. Endod. J.*, 2005, **38**(6), 347–355.
66. A. Almyroudi, *et al.*, The effectiveness of various disinfectants used as endodontic intracanal medications: An in vitro study, *J. Endod.*, 2002, **28**(3), 163–167.
67. L. E. Massarstrom, *et al.*, Effect of calcium hydroxide treatment on periodontal repair and root resorption, *Endod. Dent. Traumatol.*, 1986, **2**(5), 184–189.
68. A. Farhad and Z. Mohammadi, Calcium hydroxide: a review, *Int. Dent. J.*, 2005, **55**(5), 293–301.
69. K. E. Safavi and F. C. Nichols, Alteration of biological properties of bacterial lipopolysaccharide by calcium hydroxide treatment, *J. Endod.*, 1994, **20**(3), 127–129.
70. G. Hasselgren, B. Olsson and M. Cvek, Effects of calcium hydroxide and sodium hypochlorite on the dissolution of necrotic porcine muscle tissue, *J. Endod.*, 1988, **14**(3), 125–127.
71. U. Sjogren, *et al.*, The Antimicrobial Effect of Calcium Hydroxide as a Short-Term Intracanal Dressing, *Int. Endod. J.*, 1991, **24**(3), 119–125.
72. J. O. Andreasen, B. Farik and C. Munksgaard, Long-term calcium hydroxide as a root canal dressing may increase risk of root fracture, *Dental Traumatol.*, 2002, **18**(3), 134–137.
73. A. Podbielski, A. Spahr and B. Haller, Additive antimicrobial activity of calcium hydroxide and chlorhexidine on common endodontic bacterial pathogens, *J. Endod.*, 2003, **29**(5), 340–345.
74. L. Spangberg, M. Rutberg and E. Rydinge, Biologic effects of endodontic antimicrobial agents, *J. Endod.*, 1979, **5**(6), 166–175.
75. N. S. Farooq, *et al.*, Success rates of formocresol pulpotomy and indirect pulp therapy in the treatment of deep dentinal caries in primary teeth, *Pediatr. Dent.*, 2000, **22**(4), 278–286.

76. I. Heling and M. Pecht, Efficacy of Ledermix Paste in Eliminating Staphylococcus-Aureus from Infected Dentinal Tubules Invitro, *Endod. Dental Traumatol.*, 1991, 7(6), 251–254.
77. T. Waltimo, *et al.*, Antimicrobial effect of nanometric bioactive glass 45S5, *J. Dent. Res.*, 2007, **86**(8), 754–757.
78. D. N. Ricketts and N. B. Pitts, Traditional operative treatment options, *Monogr. Oral Sci.*, 2009, **21**, 164–173.
79. P. Ahlholm, *et al.*, Digital Versus Conventional Impressions in Fixed Prosthodontics: A Review, *J. Prosthodont.*, 2016, 35–41.
80. N. B. Pitts, *et al.*, Dental caries, *Nat. Rev. Dis. Primers*, 2017, **3**, 17030.
81. J. D. Featherstone, Delivery challenges for fluoride, chlorhexidine and xylitol, *BMC Oral Health*, 2006, **6**(Suppl 1), S8.
82. I. R. Hamilton, Biochemical effects of fluoride on oral bacteria, *J. Dent. Res.*, 1990, **69**, 660–667, discussion 682–683.
83. Y. Iijima and T. Koulourides, Fluoride incorporation into and retention in remineralized enamel, *J. Dent. Res.*, 1989, **68**(8), 1289–1292.
84. J. P. Pessan, *et al.*, Slow-release fluoride devices: a literature review, *J. Appl. Oral Sci.*, 2008, **16**(4), 238–246.
85. H. Hajizadeh, M. Ghavamnasiri and S. Majidinia, Randomized clinical evaluation of the effect of chlorhexidine on postoperative sensitivity of posterior composite resin restorations, *Quintessence Int.*, 2013, **44**(10), 793–798.
86. O. Tulunoglu, *et al.*, The effect of cavity disinfectants on microleakage in dentin bonding systems, *J. Clin. Pediatr. Dent.*, 1998, **22**(4), 299–305.
87. M. H. Anderson, A review of the efficacy of chlorhexidine on dental caries and the caries infection, *J. Calif. Dent. Assoc.*, 2003, **31**(3), 211–214.
88. T. Walsh, J. M. Oliveira-Neto and D. Moore, Chlorhexidine treatment for the prevention of dental caries in children and adolescents, *Cochrane Database Syst. Rev.*, 2015, (4), Cd008457.
89. A. H. Forgie, *et al.*, A randomised controlled trial of the caries-preventive efficacy of a chlorhexidine-containing varnish in high-caries-risk adolescents, *Caries Res.*, 2000, **34**(5), 432–439.
90. T. M. Hafshejani, *et al.*, Antibacterial glass-ionomer cement restorative materials: A critical review on the current status of extended release formulations, *J Controlled Release*, 2017, **262**, 317–328.
91. P. Riley and T. Lamont, Triclosan/copolymer containing toothpastes for oral health, *Cochrane Database Syst. Rev.*, 2013, (12), CD010514.
92. F. S. Panagakos, *et al.*, Advanced oral antibacterial/anti-inflammatory technology: A comprehensive review of the clinical benefits of a triclosan/copolymer/fluoride dentifrice, *J. Clin. Dent.*, 2005, **16**(Suppl), S1–S19.
93. A. Gaffar, *et al.*, Antiplaque effects of dentifrices containing triclosan/copolymer/NaF system versus triclosan dentifrices without the copolymer, *Am. J. Dent.*, 1990, **3**, S7–S14.
94. F. Chen, *et al.*, Tooth-binding micelles for dental caries prevention, *Antimicrob. Agents Chemother.*, 2009, **53**(11), 4898–4902.

95. F. Chen, *et al.*, Triclosan-loaded tooth-binding micelles for prevention and treatment of dental biofilm, *Pharm. Res.*, 2010, **27**(11), 2356–2364.
96. Y. Zhou, *et al.*, Triclosan-loaded poly(amido amine) dendrimer for simultaneous treatment and remineralization of human dentine, *Colloids Surf., B*, 2014, **115**, 237–243.
97. A. Rao and N. Malhotra, The role of remineralizing agents in dentistry: a review, *Compend. Contin. Educ. Dent.*, 2011, **32**(6), 26–33, quiz 34, 36.
98. M. V. Morgan, *et al.*, The anticariogenic effect of sugar-free gum containing CPP-ACP nanocomplexes on approximal caries determined using digital bitewing radiography, *Caries Res.*, 2008, **42**(3), 171–184.
99. D. L. Bailey, *et al.*, Regression of post-orthodontic lesions by a remineralizing cream, *J. Dent. Res.*, 2009, **88**(12), 1148–1153.
100. S. K. Rao, *et al.*, Study of the efficacy of toothpaste containing casein phosphopeptide in the prevention of dental caries: a randomized controlled trial in 12- to 15-year-old high caries risk children in Bangalore, India, *Caries Res.*, 2009, **43**(6), 430–435.

CHAPTER 7

Photodynamic Antimicrobial Polymers

COLIN P. MCCOY,* JESSICA V. MOORE AND MATTHEW P. WYLIE

School of Pharmacy, Queen's University Belfast, 97 Lisburn Road, Belfast BT9 7BL, Northern Ireland, UK
*Email: c.mccoy@qub.ac.uk

7.1 Introduction

A healthcare-associated infection (HAI) is defined as an infection acquired by a patient during the course of their treatment and developed either as a direct result of medical treatment or through contact with a healthcare environment. HAIs pose a considerable risk to patient health and treatment outcomes, with a recent UK-based study reporting a HAI incidence of 4.5%. HAIs affect not only the patient, but also the economy, with these infections estimated to cost Europe more than €7 billion annually.[1,2] At least half of all reported HAIs are associated with the use of indwelling medical devices, such as central venous and urinary catheters.[3] However, contamination of surfaces within healthcare facilities represents another common pathway for HAI spread and can facilitate both intra- and inter-ward transmission of hospital pathogens.[4–6]

An extensive range of antimicrobial strategies to reduce the risk of infection associated with biomedical devices and healthcare fomites has been investigated in the past half century, such as antibiotic-eluting coatings, anti-adherent surfaces and high-level disinfectants. Of the many antimicrobial strategies developed to address these issues, triggered

Biomaterials Science Series No. 5
Antimicrobial Materials for Biomedical Applications
Edited by Abraham J. Domb, Konda Reddy Kunduru and Shady Farah

Published by the Royal Society of Chemistry, www.rsc.org

antimicrobial systems offer a distinct advantage by providing antimicrobial action only when triggered by an external stimulus (*e.g.* light, temperature, pH or electric), thus avoiding propagation of antibiotic resistance commonly caused by exposure of microorganisms to sub-lethal concentrations of antimicrobial agents.

Light-triggered systems are among the best-researched stimulius-responsive platforms. For example, they have been used to produce self-healing polymers, through photocrosslinking reactions, and polymeric drug-delivery systems capable of releasing a drug upon light irradiation by breakdown of photolabile linkages or reversible photoinduced structural changes.[7,8] In addition, light-sensitive compounds, known as photosensitisers, can generate cytotoxic reactive oxygen species (ROS) upon irradiation with visible light in an oxygenated environment. The ability of photosensitisers to produce ROS has been investigated and exploited in a range of areas. The first, most established application is their use in photodynamic therapy (PDT) to treat a range of medical conditions caused by the hyperproliferation of abnormal or unwanted cells, including ophthalmic diseases (*e.g.* age-related macular degeneration[9]), dermatological conditions and various forms of cancer.[10,11] PDT involves the delivery of photosensitisers specifically to a target site, and upon accumulation, the area is illuminated with an appropriate light source to initiate photosensitisation of the compound, leading to the generation of ROS which destroy neighbouring unwanted cells and microvasculature. PDT offers a minimally invasive procedure to specifically target malignant cells without the adverse effects commonly associated with chemo- or radio-therapeutic procedures. While PDT for the treatment of cancer and other medical conditions is beyond the remit of this chapter, extensive literature reviews on the development and progress of this area are available.[12,13]

More recently, with antimicrobial resistance being a global concern, PDT has been exploited to treat microbial infections, especially in drug-resistant bacterial infections.[14] This light-triggered antimicrobial therapy is known as 'photodynamic antimicrobial chemotherapy' (PACT) and often involves the irradiation of organic photosensitisers, which generate cytotoxic species capable of eradicating a wide range of microorganisms.[15] A vast array of photosensitiser compounds exist, including naturally occurring organic molecules, such as hypocrellin A and B pigments isolated from the parasitic fungi *Hypocrella bambuase*, or those produced *via* synthetic or semi-synthetic means.[16] While the majority of photosensitisers are organic, some inorganic metal oxides, such as titanium dioxide (TiO_2), have also displayed photosensitive cytotoxic properties.[17] Despite the wide variety of organic photosensitisers, they each produce cytotoxic ROS *via* the same photochemical processes. These processes are summarised in the Jablonski model, which is presented in Figure 7.1.

As illustrated in Figure 7.1, when a photosensitiser molecule is irradiated with a certain wavelength of light the energy supplied (E) causes the molecule to be excited to an elevated energy level or excited singlet state.

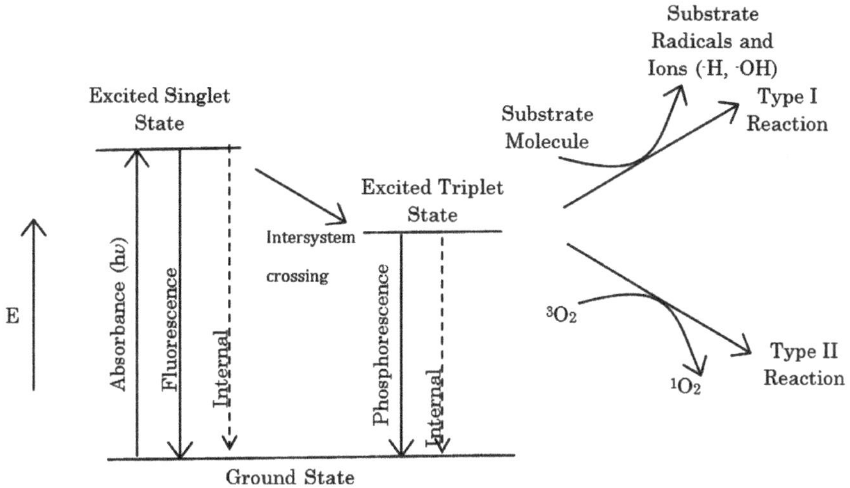

Figure 7.1 Jablonski model detailing the two types of mechanisms that can occur during the photosensitisation reaction.

However, the excited photosensitiser is unstable in this form, and may decay by fluorescent emission or undergo intersystem crossing to an excited triplet state, which has a longer lifespan. The excited triplet state can then partake in one of two photochemical pathways: type I (electron or hydrogen transfer) or type II (energy transfer), each leading to the formation of cytotoxic ROS. The type I pathway involves the triplet state photosensitisers interacting with substrates, such as biomolecules and water, to produce radical ions that can then react with oxygen to form cytotoxic species. Type II photochemical reactions involve the transfer of energy to molecular oxygen ($^3O_2$) to form singlet oxygen ($^1O_2$).[18–23]

Photosensitisers can be described as catalysts due to the fact the processes described can be repeated once the molecule returns to its ground state.[22] However, this efficiency displayed by photosensitisers is limited by photobleaching; the photosensitiser molecule's ability to perform further photosensitisation reactions is decreased when the generated ROS reacts with the photosensitiser itself. The rate of photobleaching, or photobleaching quantum yield, varies greatly between photosensitisers and depends on several factors.[18] Large aromatic photosensitiser molecules tend to aggregate in aqueous solutions which can reduce their $^1O_2$ yield, but this also makes them more resistant to photobleaching, while structures with lower oxidation potential and those containing electron-donating substituents tend to bleach at an enhanced rate.[24]

PACT has been studied extensively against a range of microbial targets and potent lethal effects against bacteria, fungi, viruses, parasites, protozoa and multi-species biofilms have been reported.[15,25–27] As such, PACT has been established as an innovative and effective alternative to conventional antimicrobial therapies with minimal risk of the development of bacterial resistance.[28] At present, there are no documented reports of any microorganisms truly resistant to this therapy.[28–30] This is likely due to two main reasons. Firstly, unlike antibiotics which target a specific bacterial cell structure or function, PACT-generated ROS exert an indiscriminate, multisite attack against the cell. Secondly, the transient nature of the ROS, especially $^1O_2$, prevents microorganisms developing protective mechanisms to negate the cytotoxic effects of ROS and makes microbial resistance to PACT techniques unlikely.[28,29,31]

The application of PACT in clinical treatments has been extensively researched and has been trialled in the treatment of soft tissue infections, such as acne, and treatment of superficial dental infections, while methylene blue (MB) is now licensed for oral photodisinfection and photodecontamination of blood plasma.[29,32,33] However, more recently, the incorporation of photosensitisers into polymers has been investigated as a method to broaden the clinical areas in which PACT can be utilised for infection control. This has facilitated the use of PACT in applications more commonly associated with conventional antimicrobials, such as incorporation into indwelling medical devices to prevent medical device-associated infection.[22,34] In addition, the use of polymeric platforms has allowed the manufacture of PACT-enhanced paints and hard surfaces to address the risk of cross-transmission of infection within hospitals.[35,36] This chapter briefly discusses the main factors that affect the efficacy of photodynamic antimicrobial polymers, followed by a detailed discussion on the current development and application of photodynamic antimicrobial polymers in biomedical devices and their implementation into other areas of the healthcare environment to prevent transmission of HAIs.

7.2 Photodynamic Antimicrobial Polymers—Important Factors for Optimal Antimicrobial Efficacy

Photodynamic antimicrobial polymers are polymers which exhibit potent antimicrobial properties when irradiated with a light source, although in some cases antimicrobial activity is also maintained in non-irradiated conditions, but often at attenuated or sub-optimal levels. This triggered or tuneable activity is often viewed as advantageous due to its potential to reduce unnecessary antimicrobial exposure, thus helping to prevent propagation of antimicrobial resistance or causing human toxicity, as is commonly seen with overexposure to conventional antibiotics.

Photodynamic antimicrobial polymers consist of a polymeric system with a photoactive agent physically or chemically incorporated. These photoactive agents mainly consist of organic photosensitisers, such as phenothiazines or porphyrins, or metallic photocatalysts, such as TiO_2 or zinc oxide. However, to develop an effective photodynamic antimicrobial polymer several factors must be considered to fully maximise antimicrobial activity. The main factors to consider are briefly discussed here.

7.2.1 Photosensitiser Class, Structure and Concentration

Organic photosensitisers are the most commonly used compounds to functionalise materials with photoactive properties. These photosensitisers can be classified into several classes according to their base chemical structure, with porphyrins, phenothiazines and phthalocyanines the most commonly researched. The physicochemical properties of photosensitisers, such as lipophilicity, water solubility and net charge, are important aspects to consider.[37] These properties can affect the ability of the photosensitisers to bond or associate with the polymer during production.

Light absorption properties of the photosensitisers, such as the wavelength of maximum absorbance (λ_{max}) and its associated extinction coefficient, are important to consider.[15] Photosensitisers with strong absorption in the red region of the UV-visible spectrum, such as porphyrins and phenothiazines, are often considered ideal for clinical applications; longer-wavelength red light can penetrate human tissue to a depth of almost 5 mm, making it more advantageous for PDT of deeper tissue targets. In contrast, scattering and absorption of low-wavelength violet or blue light by human tissue is high, restricting the use of photosensitisers with maximum absorption in this region of the spectrum to superficial treatment of infections and tumours.[38] Additionally, use of UV/blue light-absorbing photosensitisers, such as psoralen, can make the skin and eyes sensitive to sunlight.[39]

Singlet oxygen quantum yield efficiency (Φ_Δ) is a measure of the efficiency with which a photosensitiser can use light energy to convert oxygen from its ground state to reactive singlet oxygen ($^1O_2$). $^1O_2$ is regarded as the most cytotoxic ROS for antimicrobial activity and thus by the definition of Φ_Δ it is logical that an effective photosensitiser for PACT will possess a high Φ_Δ to ensure generation of sufficient $^1O_2$ quantities to kill nearby pathogens.[40] However, as reported by Phoenix and Harris, some photosensitisers, such as Pyronin Y, may exhibit potent antimicrobial activity despite possessing a low Φ_Δ, suggesting it is a photosensitiser with efficient type I photosensitisation ROS production.[40]

Furthermore, the concentration of a photosensitiser incorporated into a polymer can impact the structural properties of the polymer carrier. Loss of optical transparency, reduced elasticity or mechanical strength may occur, therefore this must be considered when specific properties, such as appearance, flexibility or strength, are important for the intended application.[41] In addition, increased concentrations of photosensitisers do not equate to improved antimicrobial performance. High concentrations of photosensitisers in solution can lead to photosensitiser aggregation or dimer formation, which reduces the $^1O_2$ yield efficiency of the photosensitisers. Additionally, increased pigmentation associated with high photosensitiser concentrations can reduce light transmission throughout, isolating $^1O_2$ generation to the illuminated surface only.[42,43]

7.2.2 Light Source

As described, the antimicrobial efficacy of a photosensitiser is directly related to its ability to undergo photosensitisation which leads to the type I or type II generation of cytotoxic ROS. Type II reactions produce extremely reactive $^1O_2$, and as previously explained the Φ_Δ of a photosensitiser plays an important role in how efficiently this process occurs. However, the irradiation source also plays a principal role in the efficacy of PDT; the emission wavelength(s) of a light source should be complimentary to the λ_{max} of a photosensitiser in order to maximise ROS production. However, this is not always the case, as a typical UV-visible spectrum of a porphyrin shows a large absorption peak (Soret peak) between 400 and 430 nm, followed by a series of smaller absorption peaks (Q bands) beyond 600 nm, with activation of even the smaller absorption peaks resulting in significant $^1O_2$ generation.[44,45] Incandescent and halogen lamp sources produce a full spectrum of wavelengths with emission intensity increasing to the maximum intensity in the near-infrared end of the spectrum. Conversely, daylight possesses high emissions throughout the visible spectrum with a peak in the violet-blue end of the spectrum.[46] However, a more common approach in recent years has been the use of light-emitting diodes or laser light sources with a specific wavelength that closely matches the λ_{max} of a photosensitiser to improve excitation efficiency.

Light intensity can also affect the antimicrobial efficacy of a photosensitiser-incorporated polymer. The light intensity can be easily altered by changing the distance between the light source and the irradiation target, or by increasing the power output of a light source. Typically, a higher light intensity will deliver a greater light dose to the target sample, subsequently leading to improved photosensitiser excitation and antimicrobial performance.[47] However, increased light intensity or duration can lead to undesirable hyperthermic conditions at the target site and may also increase the rate of photobleaching.[48]

7.2.3 Application Environment

An important consideration when developing a photodynamic polymer for antimicrobial applications is to understand the environment and conditions it will be exposed to. When an indwelling medical device is modified to produce a photoactive polymer or when a photodynamic polymer is used as an antimicrobial surface in a clinical setting, the effect of bodily fluids (*e.g.* mucus, blood or other organic material) on the material's antimicrobial activity should be considered. The presence of bodily fluids could affect light transmission to the polymer surface, may sequester the photosensitiser or prevent diffusion of $^1O_2$, and in the case of an eluting coating, may adversely affect photosensitiser diffusion or cause photosensitiser inactivation.[49,50]

The presence of oxygen at the target site should also be considered. The majority of the most potent photosensitisers exert their antibacterial properties primarily by type II photosensitisation reactions, in which environmental oxygen is converted into ROS. Thus, in applications where there are low oxygen levels, such as the gastrointestinal tract or deep periodontal tissue, PACT efficacy may be severely compromised unless a photosensitiser possesses efficient generation of type I cytotoxic species.[51]

7.3 Biomedical Device Applications

Biomaterials can be described as any material implanted or inserted into the body to replace a physiological part or function.[52,53] Despite their benefits and ability to improve human health, they are responsible for at least half of all HAIs, including central venous catheter-related infections, catheter-associated urinary tract infections (UTIs) and ventilator-associated pneumonia (VAP).[3] The detrimental effects of device-related infections on patient morbidity and mortality, and the accompanying economic costs, has led to an increase in research efforts to develop biomaterials that will prevent or impede bacterial colonisation.[3] The development of photosensitiser-incorporated biomaterials, whereby photoactive compounds are physically or chemically incorporated into or onto medical polymers are increasingly being used as a bulk material to create anti-infective medical devices or used as a surface coating on pre-fabricated articles to combat against medical device-associated infections.

7.3.1 Catheters

A research group from University College London, led by Michael Wilson and Ivan Parkin, have published a number of articles relating to the use of PACT to address medical device-related infections, mainly catheter-associated infections. This group recognised that UTIs are the most common type of HAI in the UK, with 80% of these UTIs associated with urinary catheter use.[54] Early work from this group reported the ability to incorporate photosensitiser compounds into medically relevant polymers *via* a simplistic yet effective swell–encapsulation–shrink (SES) technique, whereby a polymer is immersed in an appropriate solvent system, containing the dissolved photosensitiser for a defined duration to allow the polymer to swell and take in the photosensitiser. After this swelling stage the solvent is allowed to evaporate, causing the polymer to return to its original size, trapping the photosensitiser within the polymer.[34] Using this method, Parkin and co-workers have incorporated MB, toluidine blue O (TBO), crystal violet (CV) and indocyanine green (ICG) into silicone, polyurethane and poly(vinyl chloride), and reported significant antimicrobial activity against a range of planktonic bacteria, including methicillin-resistant *Staphylococcus aureus*

(MRSA), *Staphylococcus epidermidis*, *Pseudomonas aeruginosa* and *Escherichia coli*, when irradiated with a laser light for up to 15 minutes.[34,54–62] Several of these studies incorporated a mixture of photosensitisers with gold nanoparticles (AuNPs) and found their combination enhanced activity compared to photosensitisers alone; bactericidal action of the resultant materials increased as AuNP size was reduced from 20 nm to 2 nm.[34,63] Although the exact mechanism of how this combination enhances antimicrobial activity is unclear, Parkin and co-workers observed a small increase in triplet state intensity in the presence of AuNPs, while electron-transfer between the AuNPs and a photosensitiser molecule, inducing a type I reaction, has also been postulated.[59] Narband *et al.* reported that cationic dyes, such as MB and TBO, coordinate around AuNPs and display significantly enhanced UV absorption maxima, which may explain the enhanced production of the cytotoxic ROS responsible for bacterial kill.[64]

Parkin and co-workers have also shown that the antimicrobial efficacy instilled by the incorporation of a MB–AuNP mixture into silicone can delay biofilm development and propagation.[54] MB–AuNP silicone surfaces were able to reduce the surface coverage of a *S. epidermidis* biofilm by >50%, compared to an unmodified silicone surface, after 6 hours of irradiation.[54] This study examined several irradiation protocols, including exposure to laser light (660 nm light generated from a 230 mW laser) for 5 minutes every 30 minutes, 10 minutes every 60 minutes and 20 minutes every 120 minutes. It was found that longer exposure times with short intervals between irradiation was optimal and highlighted that treatment of biofilms, or in cases of chronic infection, optimisation of the irradiation exposure schedule is crucial for successful treatment of more complex infections.[54]

Unlike previous materials produced by this group, involving incorporation of cationic photosensitisers, the anionic ICG photosensitiser was incorporated into polyurethane *via* the same SES technique.[57] With regards to microbiological studies, the materials were inoculated with two Gram-positive (MRSA and *S. epidermidis*) and two Gram-negative (*E. coli* and *P. aeruginosa*) bacteria and then illuminated with 808 nm light generated by a 250 mW laser for up to 15 minutes, equating to a total energy density of 31.83 $J\,cm^{-2}$. The Gram-negative microorganisms were less susceptible to photosensitisation than the Gram-positive bacteria, with a 0.5 $\log_{10}$ reduction compared to a 2 $\log_{10}$ reduction achieved, respectively. The reduced antimicrobial activity of this ICG-loaded polyurethane, compared to other MB- and TBO-containing materials, may be partly due to the fact that ICG is an anionic photosensitiser compared to the cationic MB and TBO. Anionic photosensitisers have been shown to be less effective than their cationic counterparts because the cell walls of bacteria have a net negative charge, therefore an attractive force between cationic photosensitisers and microbes exists. The authors suggested the potential of using the ICG-embedded polyurethane to produce intravenous (IV) catheters, rather than urinary catheters, as IV catheters are typically colonised by Gram-positive staphylococcal species, while Gram-negative bacteria are more commonly implicated with urinary catheter colonisation.[57]

While these photosensitiser-incorporated polymers have shown potential as anti-infective medical device surfaces, a drawback is the risk of photosensitisers leaching from the material. While Parkin and co-workers have shown leaching to be minimal and isolated mostly to the first 24 hours of immersion in solution, leaching of photosensitisers can be undesirable; it can cause loss of efficacy of the material over time, while their dye properties could lead to staining of tissue or other surfaces upon repeated or prolonged contact. To combat this, Parkin and co-workers reported an improved incorporation technique that involves the covalent attachment of TBO and MB to a functionalised silicone surface.[65] It was suggested these materials could not only be used in the production of photobactericidal catheters, but other photoactive antimicrobial surfaces. Microbial studies displayed significant activity against both *S. epidermidis* and *E. coli* when exposed to a 634 nm laser light source, reducing viability by 5 $\log_{10}$ after 4 minutes of exposure. This study showed that covalently bound photosensitiser silicone surfaces required significantly less photosensitiser than the materials prepared using the SES method in order to achieve comparable antibacterial results. This was due to the confinement of the covalently bound photosensitisers to the polymer surface, where $^1O_2$ generation can directly attack surface-attached microorganisms. The reduced concentration of photosensitiser required for the covalent attachment method also reduced the pigmentation of the surface, which could make these surfaces more attractive for applications where optical clarity or aesthetic appearance are important.[65]

Spagnul *et al.* have reported the incorporation of a phenothiazinium photosensitiser into an acrylamide hydrogel.[66] In this study, a novel phenothiazinium compound was synthesised and subsequently covalently immobilised into a polyacrylamide hydrogel to form a photoantimicrobial hydrogel. No photosensitiser leaching was detected from the hydrogel when immersed in an aqueous solution for over 1 week, although $^1O_2$ generation studies did indicate that the physical entrapment of the photosensitiser did diminish $^1O_2$ production compared to photosensitiser in solution when irradiated with white light.[66] Antibacterial activity of the photosensitiser–hydrogels was assessed against 10^6 CFU mL^{-1} *S. aureus* and *E. coli*; bacterial reductions of 3.3 $\log_{10}$ and 2.3 $\log_{10}$, respectively, were achieved after 25 minutes of irradiation with visible white light (fluence rate of 14.5 mW cm^{-2}). Interestingly, the photosensitiser–hydrogel retained its antimicrobial activity, albeit at a reduced level, when challenged with four cycles of inoculation with bacteria and illumination with light for 20 minutes. While the hydrogel in this study was intended for water sterilisation, the use of hydrogel coatings on medical devices, such as urinary catheters, is common and thus similar hydrogel photoantimicrobial systems could be applied to a range of devices or surfaces for infection control purposes.[66]

7.3.2 Endotracheal Tubes

Within intensive care units many critically ill patients will be intubated with an endotracheal tube (ETT) during their recovery. Unfortunately, ETTs provide a large surface for opportunistic pathogens to colonise and provide an unobstructed pathway into deep regions of the lung. ETTs are considered a major contributor towards the development of VAP, an infection associated with a considerable mortality rate and associated healthcare costs.[67]

To combat this, Biel *et al.* reported a significant reduction of polymicrobial biofilm viability on ETT surfaces coated with MB.[67] More specifically, Biel *et al.* developed a PDT treatment system which involved spraying a MB solution onto the inner lumen of the ETT using a 'solution application catheter', followed by exposing the surface to a 664 nm light source using a small-diameter optical fibre that is fed into the lumen *via* an ETT access port. Using a simulated *in vitro* ETT model setup, the ETTs were irradiated for two 12 minute durations, with a 5 minute dark interval between each irradiation, using a light delivery catheter with a fluence rate of 150 mW cm^{-1} of catheter length. MB-coated ETTs reduced the viability of a mixed biofilm of MRSA and a multi-drug resistant clinical isolate of *P. aeruginosa* by >3 $\log_{10}$ after a single treatment period.[67] Investigating the effect of the presence of sputum in the ETT on antimicrobial activity would be beneficial, with this viscous material potentially attenuating activity by reducing light transmission to the photosensitisers or by impeding $^1O_2$ or photosensitiser diffusion through the matrix. A similar approach has also been reported by Berra *et al.*, who coated the inner lumen of an ETT with rose bengal and introduced a UV light source *via* a connector port; however, no microbiological assessment of this coating was reported.[68]

7.3.3 Intraocular Lenses

A series of published papers by McCoy and co-workers focussed on the incorporation of porphyrin photosensitisers to the surface of hydrogel materials, with the overall aim of developing a photoactive intraocular lens (IOL) biomaterial.[22,29,69,70] These novel IOLs were intended to replace current IOLs, implanted during cataract surgery, which are associated with post-operative infectious endophthalmitis. This work used a convenient method of electrostatic attachment of ionic porphyrins to the surface of hydrogels containing oppositely charged monomers. Surface localisation of the photosensitisers was desired due to the benefits associated. Firstly, it allowed concentration of this agent's antimicrobial activity at the point where it is needed to prevent bacterial colonisation. Upon irradiation with visible light, which passes into the eye, the photosensitiser would catalytically generate $^1O_2$, which would have a cytotoxic effect on adherent microorganisms. Secondly, the transient nature of $^1O_2$ limits the distance in which it can exert its cidal effect to a few micrometres, thus allowing control of the phototoxic effects to the target pathogenic cells and limiting toxicity to host cells.

A tetrakis(4-*N*-methylpyridyl)porphyrin (TMPyP)-incorporated poly(2-hydroxyethyl methacrylate-*co*-methacrylic acid) copolymer was one of the most promising candidates, achieving a 2 $\log_{10}$ reduction of *S. epidermidis* relative to the control sample, when irradiated with halogen light (4300 lux) for 1 hour.[22,29,69,70]

7.3.4 Oral and Dental Applications

One of the most common biofilms found in the human body are dental plaques on the surface of teeth and epithelial cells within the oral cavity. If undisturbed, these multi-species biofilms can lead to serious periodontal diseases, such as gingivitis or periodontitis. While mechanical removal or treatment with traditional antimicrobial agents are frequently used treatments for periodontal disease, PACT has been investigated as an alternative. Although initial studies using photosensitiser solutions for the treatment of periodontal infections have shown promise, single treatments of PACT as an independent or adjunct treatment are often insufficient to eradicate multi-species biofilms.[71] This has been attributed to several factors, such as in-activation of the photosensitisers, poor photosensitiser penetration into the biofilm and the ability of some bacteria to expel photosensitisers, such as MB, from the cell using efflux pumps.[72,73]

In an attempt to overcome these issues, Klepac-Ceraj *et al.* developed cationic and anionic poly(lactide-*co*-glycolic acid) (PLGA) nanoparticles loaded with MB, comparing their activity with MB solution.[71] PLGA is a biodegradable and biocompatible polymer that quenches the photodynamic properties of MB when encapsulated within the PLGA nanoparticle, but when in an aqueous environment the nanoparticle undergoes hydrolytic degradation, slowly releasing MB. Due to the small diameter ($<$220 nm) and the positive charge of the cationic PLGA nanoparticles, they facilitate bacterial intracellular uptake of MB and protect MB from binding to extracellular materials. Following PLGA degradation, MB is released and regains its photodynamic antimicrobial properties when illuminated with an appropriate light source. Klepac-Ceraj *et al.* found that the cationic PLGA-MB nanoparticles performed the best, reducing mixed planktonic bacteria isolated from supra- and sub-gingival dental plaque samples when irradiated with 665 nm light, with a power density of 100 mW cm^{-2} and light fluence of 30 J cm^{-2}. Biofilms from these dental plaque samples were also cultivated for 48 hours and upon exposure to cationic PLGA–MB nanoparticles under red light conditions biofilm viability was reduced by 48%. These results indicate that these PDT–nanoparticle formulations could be effective in eradicating both planktonic bacteria and biofilm infections provided repeated irradiation protocols are developed and optimised.[71]

A similar approach has been described by Nagahara *et al.*, who manufactured PLGA nanoparticles loaded with ICG to treat *Porphyromonas gingivalis* infection.[74] However, in this study the surface of the ICG-loaded PLGA nanoparticles were coated with chitosan to form mucoadhesive nanoparticles to aid uptake by bacteria. Fluorescent microscopy confirmed that ICG-nanoparticles without the chitosan coating, which possessed a negative charge, were unable to adhere to *P. gingivalis*, while those coated with chitosan, which provided the surface with a net positive charge, were capable of adhering to the negatively charged bacteria. Moreover, the chitosan-coated particles caused *P. gingivalis* to agglutinate, which facilitated improved bactericidal effects by ensuring intimate contact between bacteria and $^1O_2$-generating molecules.[74] When irradiated with an 805 nm diode laser for 1 minute (0.5 W, repeated pulse every 100 ms), the viability of a solution of 10^8 CFU mL^{-1} *P. gingivalis* mixed with chitosan-coated ICG nanoparticles was significantly reduced by approximately 2 log_{10} compared to a solution without the presence of nanoparticles. Irradiation cycles >1 minute led to further reductions in viability up to a 4 log_{10} reduction achieved if treatment continued for 5 minutes. However, treatment cycles extending beyond 1 minute were associated with a hyperthermal effect, which could provide bactericidal effects not solely attributable to PDT. Furthermore, this hyperthermal effect was associated with a temperature rise of 5–15 °C, which could be associated with periodontal tissue damage.[74] Several studies have shown a threshold temperature increase of 7 °C to avoid major tissue damage, while it is recommended temperature increases resulting from treatment procedures be kept below 5 °C.[75]

To address this issue, further work on this ICG-nanoparticle technology has been conducted by Sasaki *et al.* and optimised for treatment of deep-lying subgingival infections.[76] A simulated gingival model, made from fresh beef, was developed to assess the delivery of light through tissue. The 810 nm laser diode used in this study required a peak power output of 0.7 W to achieve a 2 log_{10} reduction in *P. gingivalis* viability during direct irradiation of the inoculum. At this power output, a total transmittable energy of 0.4 W was detected by a power meter at the surface of the bacterial solution. To achieve a similar transmitted energy level through the simulated gingival model, a peak power of 2 W was required. However, this increased power led to considerable temperature elevation in tissue when using a similar protocol to that described by Nagahara *et al.* To combat this, intermittent irradiation periods coupled with air cooling reduced temperature elevation to <3 °C at 5 minutes. Using these refined conditions, namely 2 W power output (100 ms pulse) with an intermittent break of 10 s every 60 s coupled with air cooling, *P. gingivalis* viability was reduced by 78%, 97% and 99.99% after 1, 3 and 5 minutes of irradiation, respectively. This study highlighted that polymeric photosensitiser-loaded nanoparticles can be effectively used to treat both external- and *trans*-gingival infections when coupled with an appropriate laser irradiation protocol, which can remove the need for invasive debridement procedures and decrease the risk of antibiotic resistance development.[76]

Shrestha *et al.* reported conjugation of rose bengal, *via* amide bonds, with the natural polymer chitosan for the treatment of infected dentin.[77] It was postulated that the rose bengal–chitosan conjugate (CSRB) could provide a dual effect on infected dentin by cross-linking dentin collagen to provide structural support, while also being capable of generating $^1O_2$ to treat infection. Antibacterial efficacy of the CSRB was conducted against *Enterococcus faecalis*, a highly prevalent bacterium found in persistent infections related to root canal surgery, in its planktonic and biofilm form. Samples were irradiated with a 540 nm fibre optic with a light dose of 5 or 10 J cm^{-2} for planktonic bacteria and 20, 40 or 60 J cm^{-2} for biofilm studies. The combination of the antibacterial properties of chitosan and $^1O_2$ generation by CSRB provided an extremely potent antimicrobial polymer surface, achieving >7 $\log_{10}$ reductions of planktonic *E. faecalis* at 0.1 mg mL^{-1} and complete eradication at 0.3 mg mL^{-1} of CSRB in dark conditions and complete eradication at both concentrations when irradiated with light. When challenged with an *E. faecalis* biofilm, 0.3 mg mL^{-1} CSRB was less effective in dark conditions, achieving an ~0.5 $\log_{10}$ reduction after 15 minutes, but showed potent activity in light conditions with ~2, 2.5 and 3 $\log_{10}$ reductions in biofilm viability when irradiated with a light dose of 20, 40 and 60 J cm^{-2}, respectively.[77] Chitosan has also been used to create a mucoadhesive gel capable of delivering TBO into the oral mucosa.[78,79]

Oral candidiasis is an oral infection commonly associated with infants and elderly patients but has seen a dramatic rise in incidence in recent decades due to the increased spread of HIV/AIDS in young adults, of whom 84% suffer from oral candidiasis.[80] The condition can be difficult to manage with many antifungal agents possessing only fungistatic properties, while *Candida albicans* has shown increasing resistance to conventional antifungal treatments. Donnelly *et al.* investigated the use of PACT as an alternative treatment for oral candidiasis by formulating a mucoadhesive patch containing TBO.[80] The patch was formed from an aqueous blend of poly(methylvinylether-*co*-maleic anhydride) (PMVE/MA) and the plasticiser tripropylene glycol methyl ether, followed by dissolution of TBO in the aqueous polymer solution. The film produced was then attached to a thin PVC backing film. Release studies found patches containing 50 or 100 mg cm^{-2} TBO produced a TBO concentration of 6.54 mg mL^{-1} and 11.3 mg mL^{-1} in the release medium, respectively, after 6 hours. Antifungal assessment of TBO solutions showed that concentrations of 2 mg mL^{-1} or higher, when irradiated with a light dose of 200 J cm^{-2} over 30 minutes, were able to eradicate 10^5 CFU mL^{-1} of planktonic *C. albicans*. While a 5 mg mL^{-1} TBO solution could eradicate a *C. albicans* biofilm after 3 hours irradiation with a 200 J cm^{-2} light dose, release studies through a Cuprophan® membrane, used to mimic photosensitiser diffusion through a biofilm, were only capable of releasing a maximum of 0.5 mg mL^{-1} TBO after 6 hours. Therefore, the patches would require longer pre-irradiation application times to allow sufficient release of the TBO to be effective in chronic oral candidiasis.[80] The patch described could also be potentially used to treat vulvovaginal candidiasis.

7.3.5 Wound Dressings and Superficial Infection Management

Many initial delivery systems of photosensitisers to the skin or mucosa involved direct application of photosensitiser solutions, which suffer from poor retention at the application site. This can have a significant impact on the success of PDT treatment, as it often requires irradiation periods varying from several minutes to hours to be successful. Typically, topical formulations produced to deliver active agents across the skin or mucosal membranes require sufficient contact with the target tissue to provide a high concentration of the agent to facilitate a diffusion gradient. Therefore, to achieve this, considerable research has been conducted on thermo-responsive and bioadhesive polymers to increase contact with the skin/mucosa, allowing for greater retention of the active at the target site when compared to application of a solution.[81]

Several groups have reported the incorporation of photosensitisers into bioadhesive polymers to improve PDT to the skin for both cancer and antimicrobial therapies. Junqueira *et al.* reported the development of a poloxamer 407-Carbopol 934P polymer with bioadhesive and thermoresponsive properties, which was subsequently loaded with MB.[81] The three polymer formulations displayed optimal sol–gel transition temperatures and rheological properties for application and retention on skin. A formulation consisting of 20% w/w poloxamer, 0.15% w/w Carbopol and 0.25% w/w MB was found to provide the fastest release of MB in aqueous conditions, with 50% of MB released in 18 hours.[81] While antimicrobial tests were not conducted, this concept of a photosensitiser-loaded bioadhesive polymer could be exploited for treatment of skin and mucosal infections.

Similarly, Donnelly *et al.* described the incorporation of meso-tetra (*N*-methyl-4-pyridyl) porphine (TMP) and MB into a cross-linked poly(vinyl alcohol) hydrogel to treat infected wounds.[82] A loading of 1 mg mL^{-1} of either photosensitiser did not adversely affect hydrogel properties and using a simulated wound model, each hydrogel released 16–17% and 26–27% of the photosensitiser payload after 6 hours at 25 °C and 37 °C, respectively. In addition, they investigated the extent to which light transmission to the wound site is affected by the presence of both biological fluids, using calf serum of a defined thickness to mimic wound exudates, and increasing concentrations of the photosensitisers released. Light fluence decreased almost linearly with increasing serum thickness, although this only equated to a 15% decrease at the maximum depth examined. Higher concentrations of photosensitiser also caused a reduction in light transmission and indicated that a greater delivery of photosensitiser may not result in a proportional improvement in treatment efficacy.[82] Hydrogel samples loaded with 10 μg mL^{-1} MB or TMP displayed potent antibacterial activity when illuminated with a light dose of 100 J cm^{-2} at 635 nm with a >4 $\log_{10}$ reduction in planktonic MRSA cell viability, while irradiation of gels with 50 μg mL^{-1} of TMP eradicated 4 $\log_{10}$ of a MRSA biofilm under the same irradiation conditions. Moreover, unlike MB, TMP antibacterial activity was not affected by the presence of calf serum, used to mimic the presence of extracellular material within a wound infection. This difference was likely due to the different protein binding affinities of the photosensitisers with the serum, with MB possibly more prone to protein binding, which prevents intracellular uptake by bacteria and therefore decreases efficacy.[82]

Donnelly and co-workers have also utilised microneedles to aid transdermal delivery of photosensitisers, with MB-loaded dissolving microneedle arrays, made with the water-soluble PMVE/MA copolymer, capable of releasing MB upon dissolution of the needles in aqueous intradermal tissue.[83] An *in vitro* assessment of MB permeation using a modified Franz-cell showed that 5% MB-loaded microneedles delivered 2.36 $mg\,mL^{-1}$ MB after 30 minutes of insertion, while microbiological studies, using a similar MB concentration of 2.5 $mg\,mL^{-1}$, were shown to reduce viability of *S. aureus*, *E. coli* and *C. albicans* by 99.5, 100 and 99.9%, respectively, upon 5 minutes irradiation with a Paterson lamp (fluence rate of 100 $mW\,cm^{-2}$). Caffarel-Salvador *et al.* postulated that this microneedle-facilitated PACT could be used as a rapid method to treat both acute and chronic wound infections, as its repeated application is unlikely to encourage resistance due to the $^1O_2$-mediated antimicrobial effects.[83]

Donnelly *et al.* also reported the preparation of PLGA nanoparticles loaded with Nile red (NR), which were subsequently mixed into an aqueous blend of PMVE/MA and used to form microneedle arrays.[84] *In vitro* depth penetration studies were performed; the NR-PLGA nanoparticle-loaded microneedle array was applied to neonatal porcine skin using a spring-loaded applicator (11 N force), with the NR distribution in the skin analysed following an array application time of 6 hours, using a jacketed Franz-type diffusion cell. After 6 hours, 383.63 $ng\,cm^{-3}$ NR was detected at a depth of 1.125 mm below the skin surface. This was significantly better than skin samples treated with a traditional transdermal control patch, which achieved no permeation beyond 1 mm. Importantly, NR was not detected in the receiver compartment of the diffusion cell, indicating a skin-specific delivery system with no inadvertent transdermal delivery, ensuring maximum drug concentrations in the target area. While this study by Donnelly *et al.* was directed towards treatment of cancerous deep-skin lesions, such as basal cell carcinoma, this delivery method of polymeric photosensitiser-loaded nanoparticles into deeper areas of the skin could be useful to treat a range of soft-tissue infections.[84]

7.3.6 Gastrointestinal Infections

Recent advances in the ability to illuminate internal organs of the body, such as the gastrointestinal tract, using fibre optics has facilitated the treatment of cancerous lesions not previously directly targetable with PDT, and has also led to the ability to treat infections commonly associated with the gastrointestinal tract. Initial research on gastrointestinal tract PDT primarily involved the systemic delivery of photosensitisers, causing adverse effects such as skin sensitisation. Cassidy *et al.* reported the first example of a PACT treatment that can be specifically delivered to the colon to minimise unwanted side effects associated with systemic delivery.[85] Cassidy *et al.* formulated a pH-sensitive Eudragit® (copolymer of methyl methacrylate and methacrylic acid (2 : 1)) formulation containing the photosensitisers 5-aminolevulinic acid (ALA), MB and TMP, using a hot-melt extrusion technique to form a colon-specific oral delivery system. 5% w/w photosensitiser in Eudragit® S100 was mixed thoroughly with a plasticiser until a homogenous mixture was formed and then manually fed into a twin-screw extruder. The study examined photosensitiser release from the formulations over 6 hours in a dissolution chamber set up to simulate *in vivo* gastrointestinal conditions, with formulations exposed to an acidic pH for 2 hours, followed by exposure to pH 7.4 conditions for 4 hours. The results showed that the formulated tablets containing MB or TMP displayed no drug release during the first 2 hours of exposure to acidic pH 1 conditions; however 31% of MB and 50% of TMP were released after 4 hours in the neutral pH 7.4 conditions. Formulations containing ALA showed a small amount of release (<5%) in acidic conditions, with a total of 21% of ALA released after 6 hours. Microbiological tests involved 250 $\mu g\,mL^{-1}$ of each photosensitiser being challenged with 10^7 CFU mL^{-1} *Bacteroides fragilis* and *E. faecalis*, with samples irradiated with 630 nm light emitted from a Paterson lamp, with a 600 mW power and delivering a 100 $J\,cm^{-2}$ light dose. In anaerobic conditions, MB, TMP and ALA reduced *B. fragilis* viability by 1.89, 3.66 and 0.63 $\log_{10}$, respectively, and reduced *E. faecalis* viability by 3.27, 4.05 and 1.06 $\log_{10}$, respectively.[85] The potent results for MB and, especially, TMP were obtained in anaerobic conditions, suggesting that both photosensitisers were able to exert antibacterial properties *via* type I photosensitisation mechanisms due to the lack of oxygen making type II $^1O_2$ generation negligible. In contrast, ALA activity appeared dependent on type II photosensitisation. In addition to these initial anaerobic antibacterial studies, Cassidy *et al.* investigated the addition of an oxygen generator, known as tetrachlorodecaoxide (TCDO), at a concentration of 1% w/v, on photosensitiser activity in anaerobic conditions. In the presence of TCDO, the viability of *B. fragilis* was reduced by 2.82, 3.86 and 7.73 $\log_{10}$ for MB, TMP and ALA, respectively. This indicated that the presence of oxygen, a requirement for $^1O_2$ generation, significantly enhanced the antibacterial properties of MB and ALA, while TMP activity increased only slightly. Conversely, the enhancing effect of TCDO on photosensitiser activity was less pronounced against *E. faecalis*, with MB, TMP and ALA achieving 3.34, 4.67 and 0.96 $\log_{10}$ reductions in viability, respectively, which did not vary greatly from results achieved in the absence of TCDO. Thus, MB and TMP-loaded Eudragit® formulations could offer an attractive delivery system to target colon infections, while the inclusion of an oxygen-generating compound, such as TCDO, could provide the ability to use PACT in anaerobic or hypoxic environments, previously seen as impractical for effective PDT.[85]

7.4 Photoactive Antimicrobial Surfaces for Infection Control in Clinical Environments

Surfaces within healthcare facilities, such as walls, door handles and floors, provide uncontested substrates for microbial colonisation. Many different bacterial species have been shown to maintain viability for long periods of time upon various hospital surfaces, with MRSA, vancomycin-resistant *Enterococci* and *Clostridium difficile* able to survive for up to 6 weeks under ambient conditions.[5,86–88] While hand hygiene and stringent disinfection protocols remain the primary method to reduce and prevent microbial contamination of hospital surfaces, the introduction of antimicrobial surfaces could provide additional protection against HAIs. In particular, due to their light-activated properties, photosensitisers are considered ideal agents to provide antimicrobial properties to clinical surfaces without encouraging bacterial resistance, as has been linked with some previous antibacterial-loaded materials, such as triclosan.[89,90]

7.4.1 Polymer Coatings and Films

Several photosensitiser-incorporated polymer coatings have been developed to provide a versatile antimicrobial technology to protect surfaces within clinical settings. Wilson and co-workers reported the development of cellulose acetate coatings loaded with TBO and rose bengal.[91–94] These specific photosensitisers were chosen because they strongly absorb light at wavelengths typically emitted by fluorescent lights already commonplace in hospitals, thus exerting a continuous antimicrobial effect with ambient illumination. The production of these coatings involved dissolving 50 mg mL^{-1} of cellulose acetate and 100 μg mL^{-1} of each photosensitiser in acetone (final concentration of each photosensitiser being 25 μM), which formed a thin coating following evaporation of the acetone. A series of published papers by this group provide a comprehensive assessment of the coating's antimicrobial activity in both the laboratory and clinical environment.[91–94]

Firstly, under laboratory conditions, the materials prepared were challenged with 10^6 CFU mL^{-1} bacteria (*S. aureus*, MRSA, *C. difficile* and *E. coli*), 10^5 CFU mL^{-1} *C. albicans* and 10^6 PFU mL^{-1} bacteriophage. Viable cells were enumerated after 2, 4, 6 and 16 hours of illumination with a 25 W fluorescent lamp (3700 ± 20 lux). The coating caused 100% reduction in viability, relative to control, of *C. difficile*, *S. aureus*, MRSA and *E. coli* after 4, 6, 6, and 16 hours of light irradiation, respectively.[91] This result was particularly significant, as both *C. difficile* and MRSA are major causative organisms of HAIs and are associated with considerable patient mortality,[95,96] while 88% and 91% reductions in *C. albicans* and bacteriophage, respectively, were achieved after 16 hours of illumination. Additional work found that photobleaching did not appear to be an issue in the short term, with similar *S. aureus* reductions achieved after 6 hours, even after the coating was exposed to seven cycles of 16 hour light and 8 hour dark periods. The authors noted the requirement for further studies on the longevity of the coating, with the possibility of the coating solution being sprayed onto the surfaces regularly, with the evaporation of the solvent to leave the renewed coating.[91]

Wilson and co-workers also examined the efficacy of these surface coatings in the presence of saliva and horse serum to mimic bacterial transmission through aerosolised droplets from the oral cavity or respiratory tract, which act as a major mode of transmission of nosocomial infections.[92] In this study, nebulised solutions of phosphate-buffered saline (PBS), human saliva and horse serum, inoculated with *S. aureus*, were deposited onto cellulose acetate coatings with and without TBO or rose bengal, and placed under a fluorescent lamp (3700 ± 20 lux) for 6 hours. After 6 hours, 99.8%, 97.6% and 78.9% reductions in cell viability were reported for PBS, saliva and horse serum, respectively, indicating that the photosensitiser-incorporated surfaces were still capable of achieving bacterial reductions in the presence of organic soiling. However, appreciable kills were also obtained for the photosensitiser-free cellulose acetate coating controls after 6 hours of illumination: 72.4%, 88.1% and 12.3%, respectively. The authors attributed this to activation of an endogenous photosensitiser within *S. aureus* with illumination of the control samples with a fluorescent light. This self-induced cidal effect of *S. aureus* was not reported in previous studies by this research group.[92]

The antimicrobial efficacy of these coatings was further assessed in a dental clinic to establish their clinical benefit.[93] Prepared cellulose acetate coatings containing TBO and rose bengal, and no photosensitiser (control coating) were placed under a fluorescent lamp (3700 ± 90 lux) in a dental clinic for 24 hours. A total of 42 different microorganisms were cultured from the photosensitiser-incorporated coatings, with the main contaminating microbes typically originating from the oral cavity, respiratory tract, skin or were common environmental bacteria. The photosensitiser-incorporated coatings were found to cause statistically significant reductions of both aerobic and anaerobic microorganism survival compared to the control coatings. More specifically, a 64% median reduction in aerobes was achieved, while an 82% median reduction of anaerobes occurred; however, the range over which the median values were calculated were 0–96% and 0–100%, respectively. Overall, the authors appreciated that their coatings would not be the sole solution to reduce contamination in clinical areas, but their use in these environments would still prove beneficial in reducing the bioburden and help decrease the spread of HAIs.[93]

The SES technique described in section 7.3.1 was further employed by Wilson and co-workers to incorporate MB and MB + 2 nm AuNPs into silicone to create a photoactive antimicrobial coating for clinical surfaces. The activity of these silicone samples was also assessed both in a dental clinic and a laboratory.[97] The experiments performed in the dental clinic involved assessing both aerobic and anaerobic microbial growth on settle plates, control polymer samples, MB- and MB + AuNP-loaded samples following 48 hours in the clinic. The coating surfaces were exposed to an average light intensity of 2305 lux in this setting. Modest reductions were obtained for the MB and MB + AuNP samples, relative to the control silicone. The MB + AuNP materials exhibited greater activity than the MB samples, with 54.8% and 71.2% mean reductions of aerobic and anaerobic microbes, respectively. In contrast, antimicrobial assessments performed in the laboratory involved inoculation of the samples with 10^8 CFU mL^{-1} MRSA and illumination with light, of similar intensity as that in the clinic. When the MB and MB + AuNP coatings were irradiated with light for 24 hours, >2 log_{10} and 4 log_{10} reductions of MRSA, compared to the control, were achieved for the respective samples. Additionally, protecting the samples from light for 24 hours resulted in an increase in viable MRSA numbers colonising the materials, confirming that bacterial reductions obtained by the illuminated photosensitiser-incorporated samples were solely due to the cidal action brought about by the photosensitisers when activated with light. The authors suggested that the greater antimicrobial kills achieved by the materials in the laboratory setting, in comparison to the dental clinic, were due to a range of factors that could have affected microbial fluctuation in the clinical environment, which were not an issue in a controlled laboratory setting. Also, the samples in the clinic could have been colonised by a range of different microorganisms that may be less susceptible to the antimicrobial action of the polymers, or could have been suspended in an organic media, such as saliva, which could also impede photosensitiser activity. This highlights the need to assess these photodynamic surfaces in simulated or real clinical environments to fully understand their clinical benefit.[97]

Wilson, Parkin and co-workers have also reported the potent antimicrobial efficacy of silicone and polyurethane polymers functionalised with photosensitisers, with further studies also incorporating MB, CV, ICG, acridine orange and boron-dipyrromethene (BODIPY).[57,58,63,98–100] Of these studies, possibly the most significant advancement has been the combination of photosensitiser dye with metal nanoparticles, such as AuNPs, which causes a synergistic effect on the photosensitiser's antimicrobial properties. This combination was able to induce a 38% increase in MB-triplet state generation, resulting in improved ROS production.[59,101] These surfaces have also been used to combat infection transmission from electronic devices by incorporating MB and CV into commercial mobile phone screen protectors.[102] In this study, an aqueous solution containing 200 mL of 700 ppm MB solution, 200 mL of 0.001 M CV solution and 5 mL of a 2 nm AuNP solution was heated to 90 °C and commercial screen protectors dipped into the solution for up to 60 minutes, cooled and rinsed with water to obtain antimicrobial screen protectors. The modified polymer films were resistant to photosensitiser leaching during both finger contact or cleaning with a wet medical wipe, and while a slight purple discolouration was noticeable it did not adversely affect device usability. The modified screen protectors were capable of reducing MRSA viability by at least 3 log_{10} after 3 hours and *E. coli* viability by 1.6 log_{10} after 6 hours, when irradiated with a fluorescent lamp. These screen protectors could easily be applied to numerous devices within clinical environments to prevent microbial contamination and cross-transmission of high-touch surfaces.[102]

McCoy *et al.* employed a hot-melt extrusion technique to physically incorporate photosensitisers into high-density poly(ethylene) polymers, with the successful production of a range of photosensitiser-containing materials displaying photoactive antimicrobial and/or anti-adherent characteristics, that could be easily applied to common hospital fomites, such as bed rails, keypads and medical equipment.[35] The photosensitisers incorporated were the cationic TMPyP, TBO, MB and the neutral 5,10,15,20-tetraphenyl-21H,23H-porphine (TPP), with a bilayer material produced: an antimicrobial photosensitiser-incorporated layer and a backing layer to provide mechanical support. A range of techniques were employed to characterise the polymers produced and assess how they would perform for their intended end applications. Results revealed that incorporation of the porphyrins, TMPyP and TPP, had little effect on the mechanical properties of the films; however, addition of the phenothiazine dyes, TBO and MB, had a slightly negative impact on the mechanical strength of the photosensitiser-incorporated layer. Touch and cleaning tests found that only the TPP-incorporated polymers exhibited minimal leaching of the photosensitiser after 10 000 touches, equating 0.0045% of the total TPP loading. The second test, wiping the produced films up to 500 times (×50 increments) with a medical wipe moistened with 1% Tween 20, a model for typical cleaning agents used on hospital surfaces, showed slight staining of the wipes used to clean the TPP, TMPyP and MB materials. There was no staining of the wipe used to clean the TBO-containing films. Assessment of bacterial adherence to the photosensitiser-incorporated materials was also performed by inoculating the samples with MRSA and *E. coli* ($\sim 1.6\times 10^4$ CFU cm^{-2}) and illuminating them for 2 hours with a halogen lamp (intensity 1.34 mW cm^{-2}). TMPyP-containing polymers (0.40% w/w) produced the most pronounced antimicrobial effect; a 3.61 $\log_{10}$ reduction of viable MRSA and a 1.51 $\log_{10}$ reduction of *E. coli*. These findings support the potential for this material to be applied, as a covering or coating, to surfaces and inanimate objects found in the clinical area.[35]

While the polymer coatings and films discussed here have demonstrated pronounced antimicrobial activity, there are often disadvantages associated with photosensitiser-loaded materials produced through physical incorporation techniques. For example, leaching of the photosensitiser from the polymer coating can lead to loss of efficacy and undesirable staining. Therefore, a number of groups, such as those discussed later, have employed techniques to chemically bind the photosensitisers to polymer supports, thus allowing the photoactive material to better withstand cleaning and prolonging its longevity.

Felgenträger *et al.* reported the successful polymerisation of a porphyrin photosensitiser with polyurethane to obtain a non-leaching photodynamic antimicrobial coating that could be easily applied to surfaces in clinical settings.[103] 5-(4-Hydroxyphenyl)-10,15,20-triphenylporphyrin (TPP-OH) was synthesised through the addition of a hydroxyl group to the neutral TPP, which then facilitated the polymerisation of the functionalised porphyrin with the isocyanate (–CNO) groups in polyurethane. The polyurethane/photosensitiser mixture was subsequently sprayed onto a poly(methyl methacrylate) (p(MMA)) base polymer using an airbrush, which was then heated at 20 °C for 1 hour to allow crosslinking of the polyurethane and TPP-OH. This resulted in a ~30 μm thick polyurethane/photosensitiser coating on top of the 2 mm thick p(MMA). The thin coating was considered necessary to allow faster oxygen diffusion into the coating and to ensure the diffusion of generated ROS out to the surface to have its desired microbicidal effect. Leaching studies showed no release of the hydrophobic TPP-OH into water; however, photosensitiser was detected when experiments were carried out with ethanol, which was likely due to the release of photosensitisers that did not polymerise with the polyurethane. Microbiological studies challenged polyurethane coatings containing 0, 1×10^{-4} and 2×10^{-4} M TPP-OH against *S. aureus*. An inoculum of ~6.3×10^{6} to 6.3×10^{7} CFU cm^{-2} was placed on sample surfaces and then irradiated with 50 mW cm^{-2} light for up to 30 minutes. Polyurethane coatings containing the highest concentration of TPP-OH achieved the greatest bacterial reduction of 3.2 log_{10} after 30 minutes.[103]

Extensive research on porphyrin immobilisation has been conducted by a French research group led by Krausz and Sol.[104–112] Krausz and co-workers initially reported a 'one-pot, two-step' synthesis in which porphyrin molecules are initially covalently attached to cellulose, followed by an esterification reaction with lauric acid to form a self-sterilising biodegradable porphyrinated polymer film.[104,106] Initial films containing 0.52–1.1% protoporphyrin IX (PpIX) showed bactericidal activity against *S. aureus* and *E. coli*, with no leaching of the porphyrin, when assessed using a modified Kirby–Bauer test and irradiated with four 150 W tungsten bulbs with a total fluence rate of 1.7 mW cm^{-2} for 24 hours. One drawback associated with covalently grafting photosensitisers to a surface can be the impact of steric hindrance between neighbouring photosensitiser molecules, which can limit the amount of photosensitiser that can be decorated on a polymer surface. However, Krausz and co-workers investigated the effect of adding 4- or 11-carbon chain spacers when attaching porphyrins to cellulose, and found the addition of the longer 11-carbon chain improved grafting percentage on the polymer surface. However, the 4-carbon spacer proved to have the greatest antimicrobial action, as it displayed the greatest activity with the lowest photosensitiser content.[104,106]

Bacterial spores are a major protective mechanism of bacteria that permit bacterial strains, such as *C. difficile*, to survive in unfavourable conditions and to resist the lethal effects of most chemical disinfectants for long periods of time.[113] These spores can then undergo rapid conversion to their vegetative state when the unfavourable conditions are removed. This protective mechanism can allow bacteria to colonise surfaces despite regular cleaning and thus remain one of the biggest challenges in preventing transmission of infection within healthcare facilities. For example, the resilient nature of *C. difficile* spores allows them to continue contaminating surfaces for up to 5 months in a vacant hospital room.[21,113] Due to the indiscriminate killing pathways provided by PACT-generated $^1O_2$, several studies have demonstrated the sporicidal properties of photosensitisers against *Bacillus* and *Clostridium* spores; thus, photoactive antimicrobial polymers could provide an antimicrobial surface capable of neutralising both the vegetative and spore forms of bacteria.[114,115] Zerdin *et al.* have incorporated an anthraquinone derivative (AQ) into an ethylene–acrylic copolymer using a covalent attachment technique to produce a photoactive polymer film.[116] The AQ-films were shown to enhance inactivation of *B. cereus* spores when irradiated with UV-A light compared to UV-A light alone. Zerdin *et al.* determined that this enhanced effect was due to a $^1O_2$-driven process, despite the low Φ_Δ of AQ. This finding indicates that similar polymers functionalised with more efficient photosensitisers, such as a phenothiazinium dye, which have been shown to be effective against *Bacillus* and *Clostridium* spores, could provide further improved activity against spores and may also allow this effect to be achieved with light sources more commonly found in healthcare environments.[116]

7.4.2 Antimicrobial Textiles

Textiles within the hospital environment, such as bedding, patients' gowns, healthcare personnel clothing and ward privacy curtains can harbour bacteria and viruses, which can then colonise individuals upon contact.[117] In a recent study on MRSA contamination of nurse and care assistant staff clothing in long-term care facilities, >500 samples were taken from uniforms and their pockets, with MRSA contamination found on up to 80% of uniforms and 60% of trouser pockets.[118] Similar studies have also found that antibiotic-resistant *Staphylococci* and *Enterococci* strains can survive for several months on common fabric materials, such as cotton, polypropylene and polyester.[119,120] Thus, PACT has been seen as one approach to overcome this issue, with the incorporation of photosensitisers into fabrics to produce photoactivated textiles.

With many fabrics used within healthcare facilities commonly manufactured completely or partly from cotton, which consists of cellulose fibres, much of PACT research related to antimicrobial fabrics has concentrated on the incorporation of photosensitisers to cellulose materials. Krausz and Sol and co-workers have reported covalent attachment of natural and synthetic porphyrins to cotton and cellulose fabrics to obtain antimicrobial textiles which could be applied in various applications, such as clothing, fabrics and paper within food and healthcare settings.[109,111,112] This group recognised the requirement of these textiles to withstand repeated exposure to harsh detergents and washing. They therefore preferred to investigate irreversible chemical modifications to fabrics, such as the covalent bonding of biocidal porphyrinic moieties, rather than incorporation through physical methods, such as impregnating, blending or coating the materials with the biocides, which may result in short-term antimicrobial activity only. Cotton was seen as an ideal starting material, as it is a natural fibre composed of the organic cellulose polymer which, unlike nylon fibres, has a plethora of attachment sites to covalently bond photosensitisers to its surface through substitution of a number of its many available hydroxyl groups.[107–109]

Initial studies on porphyrin-grafted cellulose fibres involved the synthesis of cationic, anionic and neutral amino porphyrins, which were modified to allow the covalent attachment to cotton. Of the anionic, neutral and cationic porphyrins investigated, the latter showed the greatest antimicrobial activity; the percentages of *S. aureus* growth inhibition were 37%, 93.7% and 100%, respectively, when irradiated with white light for 24 hours. The anionic porphyrin was least effective, likely due to electrostatic repulsion with the negatively charged bacterial cell wall. Interestingly, cationic porphyrin-grafted cellulose displayed an 80% reduction in *S. aureus* viability in dark conditions, with the authors attributing this to the charge of the quaternary ammonium groups present on this porphyrin, which have been shown in literature to cause disorganisation of bacterial cell membranes. In contrast, all the modified fabrics had no photoactive action against the model Gram-negative bacterium, *E. coli*.[108,109] Unfortunately, this is a common drawback of porphyrin-mediated PACT with few porphyrins capable of effective broad-spectrum activity under low-intensity irradiation.[121,122]

One issue with incorporating biocidal agents onto non-cellulose fabrics, such as nylon, is the lack of available grafting sites on the material's surface, limiting the quantity of active immobilised on the surface. To overcome this limitation, Tobiesen and Michielsen grafted poly(acrylic acid) (PAA) onto nylon to act as a scaffold, thus increasing possible grafting sites by almost 1000-fold.[123,124] Bozja and Sherrill and co-workers covalently bound protoporphyrin IX (PPIX) and zinc PPIX (ZnPPIX) firstly to PAA-nylon films, and following optimisation, these photosensitisers were incorporated onto PAA-nylon fibres; the fibres provide larger surface areas for increased photosensitiser attachment and contact with bacteria, thus potentially improving antimicrobial activity.[117,124] The antimicrobial efficacy of these PPIX- and ZnPPIX-modified fibres against *S. aureus* and *E. coli* was investigated, with the effect of light intensity (10 000, 40 000 and 60 000 lux) and exposure time (5, 15 and 30 minutes) determined. As expected, both increasing the light intensity supplied and length of time they were exposed to this light elicited greater microbial reductions. Overall, the ZnPPIX-grafted fibres exhibited greater antimicrobial activity at all light intensities and exposure times than the PPIX-grafted counterpart. The greatest microbial reduction was seen with the ZnPPIX samples illuminated with a light fluence of 40 000 lux, where there was a 94% reduction in *S. aureus* viability, relative to the control. Only the ZnPPIX materials had a cidal effect on the Gram-negative *E. coli* when exposed to the highest intensity of light; a 30% reduction of *E. coli*, relative to the dark control, was achieved. The authors concluded, that with room light typically 30 000 lux, these porphyrin-bound nylon fibres could be used in the production of textiles commonly found in the clinical setting, including laboratory coats for healthcare personnel and ward privacy curtains, reducing colonisation and spread of *S. aureus*.[117,124]

7.4.3 Antimicrobial Polymeric Paints

High touch frequency surfaces, such as door handles, furniture, switches and taps are known to harbour microbes and can be a key contributor to the transmission of infection throughout hospitals. To combat this, the introduction of antimicrobial metals, particularly copper, has become popular due to its contact-kill antimicrobial properties.[125] However, the significant initial cost and logistics of replacing hospital fixtures and fittings with a copper alternative has so far limited the widespread introduction of these surfaces.

A cheaper and more logistically friendly alternative has been reported by Parkin and co-workers, who have described the manufacture of a light-activated antimicrobial paint that can be applied to high touch surfaces as an alternative to copper fixture installation.[36] The combination of MB, CV, Safranin O and AuNPs, into a vinyl acrylic copolymer emulsion paint was developed using a similar SES technique.[34,36] A glass slide was first coated in a 9:1 paint:water mixture using a dip coater (30 second immersion and withdrawal rate of 120 cm min^{-1}) and left to dry for 24 hours. Following this, the slides were soaked for 2 hours in either a solution containing (1) 5×10^{-4} mol dm^{-3} CV, MB, Safranin O and 10% v/v 2 nm AuNPs; (2) 5×10^{-4} mol dm^{-3} CV, MB and 10% v/v 2 nm AuNPs or (3) 5×10^{-4} mol dm^{-3} Safranin O and 10% v/v 2 nm AuNPs. Samples were then dried for a further 24 hours, washed with deionised water and dried. Characterisation studies demonstrated that the materials exhibited negligible leaching when wiped strongly with Clinell® wipes. However, leaching of the photosensitisers into PBS was reported, which plateaued within 24 hours, and was attributed to the release of weakly attached photosensitisers on the surface of the materials. Painted surfaces were inoculated with 10^6 CFU mL^{-1} *S. aureus* and *E. coli* and irradiated with light from a 28 W fluorescent lamp (3500 ± 250 lux). All the photosensitiser-incorporated samples irradiated with light caused a complete kill, below the limit of detection, of both microorganisms within 4 hours. Interestingly, there was also a decrease in viable *S. aureus*, below the detectable limit, for the materials containing CV, MB and Safranin O and the materials containing CV and MB when in darkness. The paint containing Safranin O also brought about a significant reduction in *S. aureus* numbers when in dark conditions, but did not cause complete kill. However, none of the materials exhibited toxicity to *E. coli* when in darkness. Overall, the authors noted the potential benefit this antimicrobial paint will have on inhibiting the spread of nosocomial infections if applied to high-touch surfaces in the clinical environment. While some applications would find the bright colour of the paint undesirable, the distinct blue colour of the paint would meet UK National Health Service regulations, as they require door finger plates, handles, and light switches to be a contrasting colour to the door or walls to help aid visually impaired individuals.[36]

In a further study, an aqueous solution of CV was mixed with an acrylic latex paint to produce paint mixtures containing 250, 500, 750 and 1000 ppm CV concentrations, which were painted onto glass substrates.[126] The bright violet paint showed a small amount (<0.03%) of CV leaching over 5 days when immersed in PBS. Antimicrobial testing against *E. coli* showed a 1.21 $\log_{10}$ and >3 $\log_{10}$ reduction in viability in dark and white light conditions, respectively, after 6 hours.[126]

More recently, Hwang *et al.* reported similar results with a paint containing either CV or TBO that also boasted a superhydrophobic surface.[127] Hwang *et al.* reported a method in which 1 g of a silicone polymer, 1H,1H,2H,2H-perfluorooctyltriethoxysilane (PFOTES), was dissolved in 99 mL ethanol to form a stock polymer solution. 4 g of TiO_2 nanoparticles was added to 40 mL of the polymer solution to make an initial white paint which could then be converted to a blue or violet paint upon the addition of 40 mg of TBO or CV, respectively. Glass slides painted with the white, blue and violet paints reduced *S. aureus* and *E. coli* adherence by >3 $\log_{10}$ and 2 $\log_{10}$, respectively. Upon irradiation with white light (3900–5300 lux) for 4 hours, the TBO-containing blue paint reduced *S. aureus* and *E. coli* viability by 3.2 and 2.6 $\log_{10}$ reductions, respectively, while the CV paint reduced viability of both bacteria below the limit of detection (<10 CFU), indicating >4.5 $\log_{10}$ reductions in each case. In addition to the potent antimicrobial properties of the paints reported by Hwang *et al.*, the durability of the coatings was demonstrated with a contact angle >158° and hysteresis <3.4° maintained after 10 cycles of abrasion with sandpaper, with no loss in coating colour or thickness.[127]

Polyurethane is commonly used to coat leather products to aid cleaning of the material and also to make them more durable. With many chairs and examination beds within healthcare and care home facilities made with leather, the production and application of a photosensitiser-containing polyurethane coating would create a photoactive antimicrobial surface to help prevent the spread of pathogens. Hong and Sun developed a polyurethane coating containing up to 1.43% w/w rose bengal or benzophenone that could be painted onto leather.[128] After application, the painted leather was placed in an oven at 35 °C for 3 minutes and then left to cure at room temperature for 2 days. Painted leather containing 1.43% w/w benzophenone reduced *E. coli* and *S. aureus* cell viability by >2 $\log_{10}$ and >6 $\log_{10}$, respectively, after irradiation with a UVA lamp for 1.5 hours, while 1.43% w/w rose bengal paint under the same conditions achieved >6 $\log_{10}$ reductions for both bacteria. Under fluorescent light irradiation both 1.43% w/w benzophenone and rose bengal paints achieved a 6 $\log_{10}$ reduction of *E. coli*. This study also subjected painted-leather samples to an abrasion test using a crockmeter, with both benzophenone and rose bengal paints able to withstand up to 500 and 1000 cycles, respectively, without significant loss in antimicrobial activity against *E. coli*.[128]

7.5 Conclusions

The development of photodynamic antimicrobial polymers for a plethora of both biomedical and environmental applications over the past two decades has provided a novel and potent method to control and prevent HAIs, while also minimising the risk of bacterial resistance common with traditional antimicrobials. However, in order to be successfully implemented into everyday clinical use it is crucial to address several factors, such as costs associated with designing bespoke light-emitting medical devices to facilitate PACT in indwelling device applications, as well as improved clinical assessment of efficacy when exposed to mature or more complex chronic infections. Furthermore, the longevity and durability of photoactive polymers incorporated into hospital fomites and surroundings must be assessed in real clinical situations to allow proper assessment of their clinical benefit and cost effectiveness before full clinical translation can occur. However, the major advantages of photodynamic antimicrobial polymers over conventional antimicrobial surfaces, namely a lack of bacterial resistance development and catalytic nature, warrants continued research and development of these promising surfaces towards effective prevention of HAIs.

References

1. European Centre for Disease Prevention and Control. Annual epidemiological report on communicable diseases in Europe 2008: Report on the state of communicable diseases in the EU and EEA/EFTA countries; 2008.
2. Health Protection Scotland. National Point Prevalence Survey of Healthcare Associated Infection and Antimicrobial Prescribing 2016. Health Protection Scotland, NHS National Services Scotland, Meridian Court, 5 Cadogan Street, Glasgow G2 6QE; 2017.
3. J. P. Guggenbichler, O. Assadian, M. Boeswald and A. Kramer, Incidence and clinical implication of nosocomial infections associated with implantable biomaterials – catheters, ventilator-associated pneumonia, urinary tract infections, *GMS Krankenhhyg Interdiszip*, 2011, **6**(1). Doc18.
4. J. A. Otter, S. Yezli, J. A. G. Salkeld and G. L. French, Evidence that contaminated surfaces contribute to the transmission of hospital pathogens and an overview of strategies to address contaminated surfaces in hospital settings, *Am. J. Infect. Control.*, 2013, **41**(5), S6–S11.
5. B. Hota, Contamination, disinfection, and cross-colonization: are hospital surfaces reservoirs for nosocomial infection? *Clin. Infect. Dis.*, 2004, **39**(8), 1182–1189.
6. V. Russotto, A. Cortegiani, S. M. Raineri and A. Giarratano, Bacterial contamination of inanimate surfaces and equipment in the intensive care unit, *J. Intensive Care.*, 2015, **3**, 54.

7. D. Habault, H. Zhang and Y. Zhao, Light-triggered self-healing and shape-memory polymers, *Chem. Soc. Rev.*, 2013, **42**(17), 7244–7256.
8. C. Alvarez-Lorenzo, L. Bromberg and A. Concheiro, Light-sensitive intelligent drug delivery systems, *Photochem. Photobiol.*, 2009, **85**(4), 848–860.
9. A. Kawczyk-Krupka, A. M. Bugaj, M. Potempa, K. Wasilewska, W. Latos and A. Sieron, Vascular-targeted photodynamic therapy in the treatment of neovascular age-related macular degeneration: Clinical perspectives, *Photodiagn. Photodyn. Ther.*, 2015, **12**(2), 161–175.
10. Z. Huang, A review of progress in clinical photodynamic therapy, *Technol. Cancer Res. Treat.*, 2005, **4**(3), 283–293.
11. G. Jori, M. Camerin, M. Soncin, L. Guidolin and O. Coppellotti, Chapter 1. Antimicrobial Photodynamic Therapy: Basic Principles. Photodynamic Inactivation of Microbial Pathogens, *Compr. Ser. Photochem. Photobiol. Sci.*, 2011, 1–18.
12. D. E. Dolmans, D. Fukumura and R. K. Jain, Photodynamic therapy for cancer, *Nat. Rev. Cancer*, 2003, **3**(5), 380–387.
13. P. Agostinis, K. Berg, K. A. Cengel, T. H. Foster, A. W. Girotti and S. O. Gollnick, *et al.*, Photodynamic Therapy of Cancer: An Update, *CA Cancer J. Clin.*, 2011, **61**(4), 250–281.
14. WHO. Antimicrobial Resistance, Global Report on Survelliance World Health Organisation. 2014.
15. M. Wainwright, Photodynamic antimicrobial chemotherapy (PACT), *J. Antimicrob. Chemother.*, 1998, **42**(1), 13–28.
16. Z. J. Diwu and J. W. Lown, Photosensitization with Anticancer Agents 9. Hypocrellins and Their Use in Photosensitization, *Photochem. Photobiol.*, 1990, **52**(3), 609–616.
17. J. W. Liou and H. H. Chang, Bactericidal Effects and Mechanisms of Visible Light-Responsive Titanium Dioxide Photocatalysts on Pathogenic Bacteria, *Arch. Immunol. Ther. Ex.*, 2012, **60**(4), 267–275.
18. A. P. Castano, T. N. Demidova and M. R. Hamblin, Mechanisms in photodynamic therapy: part one-photosensitizers, photochemistry and cellular localization, *Photodiagn. Photodyn. Ther.*, 2004, **1**(4), 279–293.
19. L. B. Josefsen and R. W. Boyle, Photodynamic therapy and the development of metal-based photosensitisers, *Met.- Based Drugs.*, 2008, **2008**, 276109.
20. I. J. MacDonald and T. J. Dougherty, Basic principles of photodynamic therapy, *J. Porphyrins phthalocyanines*, 2001, **5**(2), 105–129.
21. S. Noimark, C. W. Dunnill and I. P. Parkin, Shining light on materials – A self-sterilising revolution, *Adv. Drug Delivery Rev.*, 2013, **65**(4), 570–580.
22. C. P. Mccoy, R. A. Craig, S. M. McGlinchey, L. Carson, D. S. Jones and S. P. Gorman, Surface localisation of photosensitisers on intraocular lens biomaterials for prevention of infectious endophthalmitis and retinal protection, *Biomaterials*, 2012, **33**(32), 7952–7958.
23. Z. Huang, A review of progress in clinical photodynamic therapy, *Technol. Cancer Res. Treat.*, 2005, **4**(3), 283–293.

24. R. Bonnett and G. Martinez, Photobleaching of sensitisers used in photodynamic therapy, *Tetrahedron*, 2001, **57**(47), 9513–9547.
25. R. F. Donnelly, P. A. McCarron and M. M. Tunney, Antifungal photodynamic therapy, *Microbiol. Res.*, 2008, **163**(1), 1–12.
26. L. Ryskova, V. Buchta and R. Slezak, Photodynamic antimicrobial therapy, *Cent. Eur. J. Biol.*, 2010, **5**(4), 400–406.
27. T. Dai, Y. Y. Huang and M. R. Hamblin, Photodynamic therapy for localized infections-State of the art, *Photodiagn. Photodyn.*, 2009, **6**(3–4), 170–188.
28. T. Maisch, Resistance in antimicrobial photodynamic inactivation of bacteria, *Photochem. Photobiol. Sci.*, 2015, **14**(8), 1518–1526.
29. R. A. Craig, C. P. McCoy, S. P. Gorman and D. S. Jones, Photosensitisers – the progression from photodynamic therapy to anti-infective surfaces, *Expert Opin. Drug Delivery*, 2015, **12**(1), 85–101.
30. G. B. Kharkwal, S. K. Sharma, Y. Y. Huang, T. Dai and M. R. Hamblin, Photodynamic therapy for infections: clinical applications, *Lasers Surg. Med.*, 2011, **43**(7), 755–767.
31. M. Wainwright and K. B. Crossley, Photosensitising agents – circumventing resistance and breaking down biofilms: a review, *Int. Biodeterior. Biodegrad.*, 2004, **53**(2), 119–126.
32. M. Wainwright, 6 – Dyes for the medical industry, in *Handbook of Textile and Industrial Dyeing*, ed. M. Clark, Woodhead Publishing, 2011, vol. 2, pp. 204–230.
33. M. Wainwright, Photodynamic medicine and infection control, *J. Antimicrob. Chemother.*, 2012, **67**(4), 787–788.
34. S. Perni, C. Piccirillo, J. Pratten, P. Prokopovich, W. Chrzanowski and I. P. Parkin, *et al.*, The antimicrobial properties of light-activated polymers containing methylene blue and gold nanoparticles, *Biomaterials*, 2009, **30**(1), 89–93.
35. C. P. McCoy, E. J. O'Neil, J. F. Cowley, L. Carson, A. T. De Baroid and G. T. Gdowski, *et al.*, Photodynamic Antimicrobial Polymers for Infection Control, *PLoS One*, 2014, **9**(9), e108500.
36. S. J. Beech, S. Noimark, K. Page, N. Noor, E. Allan and I. P. Parkin, Incorporation of crystal violet, methylene blue and safranin O into a copolymer emulsion; the development of a novel antimicrobial paint, *RSC Adv.*, 2015, **5**(33), 26364–26375.
37. E. Alves, M. A. Faustino, M. G. Neves, A. Cunha, J. Tome and A. Almeida, An insight on bacterial cellular targets of photodynamic inactivation, *Future Med. Chem.*, 2014, **6**(2), 141–164.
38. C. Ash, M. Dubec, K. Donne and T. Bashford, Effect of wavelength and beam width on penetration in light-tissue interaction using computational methods, *Laser Med Sci.*, 2017, **32**(8), 1909–1918.
39. W. L. Morison, S. Marwaha and L. Beck, PUVA-induced phototoxicity: incidence and causes, *J. Am. Acad. Dermatol.*, 1997, **36**(2 Pt 1), 183–185.
40. D. A. Phoenix and F. Harris, Phenothiazinium-based photosensitizers: antibacterials of the future? *Trends Mol. Med.*, 2003, **9**(7), 283–285.

41. C. P. McCoy, E. J. O'Neil, J. F. Cowley, L. Carson, Á. T. De Baróid and G. T. Gdowski, *et al.*, Photodynamic Antimicrobial Polymers for Infection Control, *PLoS One*, 2014, **9**(9), e108500.
42. M. N. Usacheva, M. C. Teichert and M. A. Biel, The role of the methylene blue and toluidine blue monomers and dimers in the photoinactivation of bacteria, *J. Photochem. Photobiol., B*, 2003, **71**(1–3), 87–98.
43. M. Tanaka, M. Kinoshita, Y. Yoshihara, N. Shinomiya, S. Seki and K. Nemoto, *et al.*, Optimal Photosensitizers for Photodynamic Therapy of Infections Should Kill Bacteria but Spare Neutrophils, *Photochem. Photobiol.*, 2012, **88**(1), 227–232.
44. A. T. De Baroid, C. P. McCoy, R. A. Craig, L. Carson, G. P. Andrews and D. S. Jones, *et al.*, Optimization of singlet oxygen production from photosensitizer-incorporated, medically relevant hydrogels, *J. Biomed. Mater. Res., Part B*, 2017, **105**(2), 320–326.
45. Z. Luksiene and A. Zukauskas, Prospects of photosensitization in control of pathogenic and harmful micro-organisms, *J. Appl. Microbiol.*, 2009, **107**(5), 1415–1424.
46. C. D. Elvidge, D. M. Keith, B. T. Tuttle and K. E. Baugh, Spectral identification of lighting type and character, *Sensors*, 2010, **10**(4), 3961–3988.
47. G. P. Tegos and M. R. Hamblin, Phenothiazinium antimicrobial photosensitizers are substrates of bacterial multidrug resistance pumps, *Antimicrob. Agents Chemother.*, 2006, **50**(1), 196–203.
48. F. Cieplik, D. M. Deng, W. Crielaard, W. Buchalla, E. Hellwig and A. Al-Ahmad, *et al.*, Antimicrobial photodynamic therapy – what we know and what we don't, *Crit. Rev. Microbiol.*, 2018, **44**(5), 571–589.
49. R. F. Donnelly, P. A. McCarron, C. M. Cassidy, J. S. Elborn and M. M. Tunney, Delivery of photosensitisers and light through mucus: Investigations into the potential use of photodynamic therapy for treatment of Pseudomonas aeruginosa cystic fibrosis pulmonary infection, *J. Controlled Release*, 2007, **117**(2), 217–226.
50. M. Wilson and J. Pratten, Lethal photosensitisation of Staphylococcus aureus in vitro: effect of growth phase, serum, and pre-irradiation time, *Lasers Surg. Med.*, 1995, **16**(3), 272–276.
51. T. Maisch, J. Baier, B. Franz, M. Maier, M. Landthaler and R. M. Szeimies, *et al.*, The role of singlet oxygen and oxygen concentration in photodynamic inactivation of bacteria, *Proc. Natl. Acad. Sci. U. S. A.*, 2007, **104**(17), 7223–7228.
52. D. F. Williams, On the nature of biomaterials, *Biomaterials*, 2009, **30**(30), 5897–5909.
53. J. Park, R. S. Lakes, *Biomaterials: An Introduction*, Springer, 2007.
54. S. Perni, P. Prokopovich, I. P. Parkin, M. Wilson and J. Pratten, Prevention of biofilm accumulation on a light-activated antimicrobial catheter material, *J. Mater. Chem.*, 2010, **20**(39), 8668–8673.
55. S. Perni, P. Prokopovich, C. Piccirillo, J. Pratten, I. P. Parkin and M. Wilson, Toluidine blue-containing polymers exhibit potent

bactericidal activity when irradiated with red laser light, *J. Mater. Chem.*, 2009, **19**(18), 2715–2723.
56. P. Prokopovich, S. Perni, C. Piccirillo, J. Pratten, I. P. Parkin and M. Wilson, Frictional properties of light-activated antimicrobial polymers in blood vessels, *J. Mater Sci.:Mater. Med.*, 2010, **21**(2), 815–821.
57. S. Perni, J. Pratten, M. Wilson, C. Piccirillo, I. P. Parkin and P. Prokopovich, Antimicrobial Properties of Light-activated Polyurethane Containing Indocyanine Green, *J. Biomater. Appl.*, 2011, **25**(5), 387–400.
58. A. J. T. Naik, S. Ismail, C. Kay, M. Wilson and I. P. Parkin, Antimicrobial activity of polyurethane embedded with methylene blue, toluidene blue and gold nanoparticles against Staphylococcus aureus; illuminated with white light, *Mater. Chem. Phys.*, 2011, **129**(1–2), 446–450.
59. S. Noimark, C. W. Dunnill, C. W. M. Kay, S. Perni, P. Prokopovich and S. Ismail, *et al.*, Incorporation of methylene blue and nanogold into polyvinyl chloride catheters; a new approach for light-activated disinfection of surfaces, *J. Mater. Chem.*, 2012, **22**(30), 15388–15396.
60. S. Noimark, M. Bovis, A. J. MacRobert, A. Correia, E. Allan and M. Wilson, *et al.*, Photobactericidal polymers; the incorporation of crystal violet and nanogold into medical grade silicone, *RSC Adv.*, 2013, **3**(40), 18383–18394.
61. M. J. Bovis, S. Noimark, J. H. Woodhams, C. W. M. Kay, J. Weiner and W. J. Peveler, *et al.*, Photosensitisation studies of silicone polymer doped with methylene blue and nanogold for antimicrobial applications, *RSC Adv.*, 2015, **5**(68), 54830–54842.
62. A. Patir, G. B. Hwang, S. P. Nair, E. Allan and I. P. Parkin, Photobactericidal Activity of Dual Dyes Encapsulated in Silicone Enhanced by Silver Nanoparticles, *ACS Omega*, 2018, **3**(6), 6779–6786.
63. S. Perni, C. Piccirillo, A. Kafizas, M. Uppal, J. Pratten and M. Wilson, *et al.*, Antibacterial Activity of Light-Activated Silicone Containing Methylene Blue and Gold Nanoparticles of Different Sizes, *J. Cluster Sci.*, 2010, **21**(3), 427–438.
64. N. Narband, M. Uppal, C. W. Dunnill, G. Hyett, M. Wilson and I. P. Parkin, The interaction between gold nanoparticles and cationic and anionic dyes: enhanced UV-visible absorption, *Phys. Chem. Chem. Phys.*, 2009, **11**(44), 10513–10518.
65. C. Piccirillo, S. Perni, J. Gil-Thomas, P. Prokopovich, M. Wilson and J. Pratten, *et al.*, Antimicrobial activity of methylene blue and toluidine blue O covalently bound to a modified silicone polymer surface, *J. Mater. Chem.*, 2009, **19**(34), 6167–6171.
66. C. Spagnul, J. Greenman, M. Wainwright, Z. Kamil and R. W. Boyle, Synthesis, characterization and biological evaluation of a new photoactive hydrogel against Gram-positive and Gram-negative bacteria, *J. Mater. Chem. B*, 2016, **4**(8), 1499–1509.
67. M. A. Biel, C. Sievert, M. Usacheva, M. Teichert, E. Wedell and N. Loebel, *et al.*, Reduction of Endotracheal Tube Biofilms Using

Antimicrobial Photodynamic Therapy, *Lasers Surg. Med.*, 2011, **43**(7), 586–590.

68. L. Berra, F. Curto, G. Li Bassi, P. Laquerriere, B. Pitts and A. Baccarelli, *et al.*, Antimicrobial-coated endotracheal tubes: an experimental study, *Intensive Care Med.*, 2008, **34**(6), 1020–1029.
69. C. Parsons, C. P. McCoy, S. P. Gorman, D. S. Jones, S. E. Bell and C. Brady, *et al.*, Anti-infective photodynamic biomaterials for the prevention of intraocular lens-associated infectious endophthalmitis, *Biomaterials*, 2009, **30**(4), 597–602.
70. C. Brady, S. E. Bell, C. Parsons, S. P. Gorman, D. S. Jones and C. P. McCoy, Novel porphyrin-incorporated hydrogels for photoactive intraocular lens biomaterials, *J. Phys. Chem. B*, 2007, **111**(3), 527–534.
71. V. Klepac-Ceraj, N. Patel, X. Song, C. Holewa, C. Patel and R. Kent, *et al.*, Photodynamic effects of methylene blue-loaded polymeric nanoparticles on dental plaque bacteria, *Lasers Surg. Med.*, 2011, **43**(7), 600–606.
72. G. P. Tegos, K. Masago, F. Aziz, A. Higginbotham, F. R. Stermitz and M. R. Hamblin, Inhibitors of Bacterial Multidrug Efflux Pumps Potentiate Antimicrobial Photoinactivation, *Antimicrob. Agents Chemother.*, 2008, **52**(9), 3202–3209.
73. A. Kishen, M. Upadya, G. P. Tegos and M. R. Hamblin, Efflux Pump Inhibitor Potentiates Antimicrobial Photodynamic Inactivation of *Enterococcus faecalis* Biofilm, *Photochem. Photobiol.*, 2010, **86**(6), 1343–1349.
74. A. Nagahara, A. Mitani, M. Fukuda, H. Yamamoto, K. Tahara and I. Morita, *et al.*, Antimicrobial photodynamic therapy using a diode laser with a potential new photosensitizer, indocyanine green-loaded nanospheres, may be effective for the clearance of Porphyromonas gingivalis, *J. Periodontal Res.*, 2013, **48**(5), 591–599.
75. S. Elavarasu, D. Naveen and A. Thangavelu, Lasers in periodontics, *J. Pharm. BioAllied Sci.*, 2012, **4**(Suppl 2), S260–S263.
76. Y. Sasaki, J. Hayashi, T. Fujimura, Y. Iwamura, G. Yamamoto and E. Nishida, *et al.*, New Irradiation Method with Indocyanine Green-Loaded Nanospheres for Inactivating Periodontal Pathogens, *Int. J. Mol. Sci.*, 2017, **18**(1), 154.
77. A. Shrestha, M. R. Hamblin and A. Kishen, Characterization of a Conjugate between Rose Bengal and Chitosan for Targeted Antibiofilm and Tissue Stabilization Effects as a Potential Treatment of Infected Dentin, *Antimicrob. Agents Chemother.*, 2012, **56**(9), 4876–4884.
78. T. B. Graciano, T. S. Coutinho, C. B. Cressoni, C. D. Freitas, M. B. R. Pierre and S. A. D. Pereira, *et al.*, Using chitosan gels as a toluidine blue O delivery system for photodynamic therapy of buccal cancer: In vitro and in vivo studies, *Photodiagn Photodyn.*, 2015, **12**(1), 98–107.
79. M. T. J. Garcia, C. D. Freitas, T. B. Graciano, T. S. Coutinho, C. B. Cressoni and S. A. D. Pereira, *et al.*, Chitosan-based mucoadhesive

gel for oral mucosal toluidine blue O delivery: The influence of a non-ionic surfactant, *Photodiagn. Photodyn.*, 2017, **20**, 48–54.
80. R. F. Donnelly, P. A. McCarron, M. M. Tunney and A. David Woolfson, Potential of photodynamic therapy in treatment of fungal infections of the mouth. Design and characterisation of a mucoadhesive patch containing toluidine blue O, *J. Photochem. Photobiol., B*, 2007, **86**(1), 59–69.
81. M. V. Junqueira, F. B. Borghi-Pangoni, S. B. S. Ferreira, B. R. Rabello, N. Hioka and M. L. Bruschi, Functional Polymeric Systems as Delivery Vehicles for Methylene Blue in Photodynamic Therapy, *Langmuir*, 2016, **32**(1), 19–27.
82. R. F. Donnelly, C. M. Cassidy, R. G. Loughlin, A. Brown, M. M. Tunney and M. G. Jenkins, *et al.*, Delivery of Methylene Blue and meso-tetra (N-methyl-4-pyridyl) porphine tetra tosylate from cross-linked poly(vinyl alcohol) hydrogels: A potential means of photodynamic therapy of infected wounds, *J. Photochem. Photobiol., B*, 2009, **96**(3), 223–231.
83. E. Caffarel-Salvador, M. C. Kearney, R. Mairs, L. Gallo, S. A. Stewart and A. J. Brady, *et al.*, Methylene Blue-Loaded Dissolving Microneedles: Potential Use in Photodynamic Antimicrobial Chemotherapy of Infected Wounds, *Pharmaceutics*, 2015, 7(4), 397–412.
84. R. F. Donnelly, D. I. J. Morrow, F. Fay, C. J. Scott, S. Abdelghany and R. R. T. Singh, *et al.*, Microneedle-mediated intradermal nanoparticle delivery: Potential for enhanced local administration of hydrophobic pre-formed photosensitisers, *Photodiagn. Photodyn.*, 2010, 7(4), 222–231.
85. C. M. Cassidy, M. M. Tunney, D. L. Caldwell, G. P. Andrews and R. F. Donnelly, Development of Novel Oral Formulations Prepared via Hot Melt Extrusion for Targeted Delivery of Photosensitizer to the Colon, *Photochem. Photobiol.*, 2011, **87**(4), 867–876.
86. J. A. Otter and G. L. French, Survival of nosocomial bacteria and spores on surfaces and inactivation by hydrogen peroxide vapor, *J. Clin. Microbiol.*, 2009, **47**(1), 205–207.
87. P. Espinal, S. Marti and J. Vila, Effect of biofilm formation on the survival of Acinetobacter baumannii on dry surfaces, *J. Hosp. Infect.*, 2012, **80**(1), 56–60.
88. D. C. Esteves, V. C. Pereira, J. M. Souza, R. Keller, R. D. Simoes and L. K. W. Eller, *et al.*, Influence of biological fluids in bacterial viability on different hospital surfaces and fomites, *Am. J. Infect. Control.*, 2016, **44**(3), 311–314.
89. B. D. Kalyon and U. Olgun, Antibacterial efficacy of triclosan-incorporated polymers, *Am. J. Infect. Control.*, 2001, **29**(2), 124–125.
90. L. M. Junker and A. G. Hay, Effects of triclosan incorporation into ABS plastic on biofilm communities, *J. Antimicrob. Chemother.*, 2004, **53**(6), 989–996.
91. V. Decraene, J. Pratten and M. Wilson, Cellulose acetate containing toluidine blue and rose bengal is an effective antimicrobial coating

when exposed to white light, *Appl. Environ. Microb.*, 2006, 72(6), 4436–4439.

92. V. Decraene, J. Pratten and M. Wilson, Novel light-activated antimicrobial coatings are effective against surface-deposited Staphylococcus aureus, *Curr. Microbiol.*, 2008, **57**(4), 269–273.
93. V. Decraene, J. Pratten and M. Wilson, Assessment of the Activity of a Novel Light-Activated Antimicrobial Coating in a Clinical Environment, *Infect. Control Hosp. Epidemiol.*, 2008, **29**(12), 1181–1184.
94. M. Wilson, Light-activated antimicrobial coating for the continuous disinfection of surfaces, *Infect. Control Hosp. Epidemiol.*, 2003, **24**(10), 782–784.
95. Public Health England. *C. difficile* infections: quarterly counts by acute trust and CCG, and financial year counts and rates by acute trust and CCG, up to financial year 2015 to 2016. 2016. p. Table 7a.
96. Public Health England. Thirty-day all-cause fatality subsequent to MRSA, MSSA and *E. coli* bacteraemia and *C. difficile* infection, 2016/17. 2017.
97. S. Ismail, S. Perni, J. Pratten, I. Parkin and M. Wilson, Efficacy of a Novel Light-Activated Antimicrobial Coating for Disinfecting Hospital Surfaces, *Infect. Control Hosp. Epidemiol.*, 2011, **32**(11), 1130–1132.
98. S. Noimark, J. Weiner, N. Noor, E. Allan, C. K. Williams and M. S. P. Shaffer, *et al.*, Dual-Mechanism Antimicrobial Polymer-ZnO Nanoparticle and Crystal Violet-Encapsulated Silicone, *Adv. Funct. Mater.*, 2015, **25**(9), 1367–1373.
99. W. J. Peveler, S. Noimark, H. Al-Azawi, G. B. Hwang, C. R. Crick and E. Allan, *et al.*, Covalently Attached Antimicrobial Surfaces Using BODIPY: Improving Efficiency and Effectiveness, *ACS Appl. Mater. Interfaces*, 2018, **10**(1), 98–104.
100. S. Noimark, E. Salvadori, R. Gomez-Bombarelli, A. J. MacRobert, I. P. Parkin and C. W. M. Kay, Comparative study of singlet oxygen production by photosensitiser dyes encapsulated in silicone: towards rational design of anti-microbial surfaces, *Phys. Chem. Chem. Phys.*, 2016, **18**(40), 28101–28109.
101. S. Noimark, E. Allan and I. P. Parkin, Light-activated antimicrobial surfaces with enhanced efficacy induced by a dark-activated mechanism, *Chem. Sci.*, 2014, **5**(6), 2216–2223.
102. K. Page, A. Correia, M. Wilson, E. Allan and I. P. Parkin, Light-activated antibacterial screen protectors for mobile telephones and tablet computers, *J. Photochem. Photobiol., A*, 2015, **296**, 19–24.
103. A. Felgenträger, T. Maisch, A. Spath, J. A. Schroder and W. Baumler, Singlet oxygen generation in porphyrin-doped polymeric surface coating enables antimicrobial effects on Staphylococcus aureus, *Phys. Chem. Chem. Phys.*, 2014, **16**(38), 20598–20607.
104. M. Krouit, R. Granet, P. Branland, B. Verneuil and P. Krausz, New photoantimicrobial films composed of porphyrinated lipophilic cellulose esters, *Bioorg. Med. Chem. Lett.*, 2006, **16**(6), 1651–1655.

105. M. Krouit, R. Granet and P. Krausz, Photobactericidal plastic films based on cellulose esterified by chloroacetate and a cationic porphyrin, *Bioorg. Med. Chem.*, 2008, **16**(23), 10091–10097.
106. M. Krouit, R. Granet and P. Krausz, Photobactericidal films from porphyrins grafted to alkylated cellulose – synthesis and bactericidal properties, *Eur. Polym. J.*, 2009, **45**(4), 1250–1259.
107. C. Ringot, V. Sol, R. Granet and P. Krausz, Porphyrin-grafted cellulose fabric: New photobactericidal material obtained by "Click-Chemistry" reaction, *Mater. Lett.*, 2009, **63**(21), 1889–1891.
108. C. Ringot, N. Saad, R. Granet, P. Bressollier, V. Sol and P. Krausz, Meso-functionalized aminoporphyrins as efficient agents for photo-antibacterial surfaces, *J. Porphyr. Phthalocyanines*, 2010, **14**(11), 925–931.
109. C. Ringot, V. Sol, M. Barriere, N. Saad, P. Bressollier and R. Granet, *et al.*, Triazinyl Porphyrin-Based Photoactive Cotton Fabrics: Preparation, Characterization, and Antibacterial Activity, *Biomacromolecules*, 2011, **12**(5), 1716–1723.
110. A. Memmi, R. Granet, M. Aouni, A. Bakhrouf and P. Krausz, Synthesis of new photobactericidal polymers by "click chemistry" and a study of their biological activity, *E-Polymers*, 2012, **12**(1), 040.
111. J. P. Mbakidi, K. Herke, S. Alves, V. Chaleix, R. Granet and P. Krausz, *et al.*, Synthesis and photobiocidal properties of cationic porphyrin-grafted paper, *Carbohydr. Polym.*, 2013, **91**(1), 333–338.
112. J. K. Nzambe Ta keki, T. S. Ouk, R. Zerrouki, P. A. Faugeras, V. Sol and F. Brouillette, Synthesis and photobactericidal properties of a neutral porphyrin grafted onto lignocellulosic fibers, *Mater. Sci. Eng. C-Mater. Biol. Appl.*, 2016, **62**, 61–67.
113. A. D. Russell, Bacterial resistance to disinfectants: present knowledge and future problems, *J. Hosp. Infect.*, 1999, **43**(Suppl), S57–S68.
114. L. De Sordi, M. A. Butt, H. Pye, D. Kohoutova, C. A. Mosse and G. Yahioglu, *et al.*, Development of Photodynamic Antimicrobial Chemotherapy (PACT) for Clostridium difficile, *PLoS One*, 2015, **10**(8), e0135039.
115. T. N. Demidova and M. R. Hamblin, Photodynamic inactivation of Bacillus spores, mediated by phenothiazinium dyes, *Appl. Environ. Microbiol.*, 2005, **71**(11), 6918–6925.
116. K. Zerdin, M. A. Horsham, R. Durham, P. Wormell and A. D. Scully, Photodynamic inactivation of bacterial spores on the surface of a photoactive polymer, *React. Funct. Polym.*, 2009, **69**(11), 821–827.
117. J. Bozja, J. Sherrill, S. Michielsen and I. Stojiljkovic, Porphyrin-based, light-activated antimicrobial materials, *J. Polym. Sci., Part A: Polym. Chem.*, 2003, **41**(15), 2297–2303.
118. P. Gaspard, E. Eschbach, D. Gunther, S. Gayet, X. Bertrand and D. Talon, Meticillin-resistant Staphylococcus aureus contamination of healthcare workers' uniforms in long-term care facilities, *J. Hosp. Infect.*, 2009, **71**(2), 170–175.

119. A. N. Neely and M. P. Maley, Survival of enterococci and staphylococci on hospital fabrics and plastic, *J. Clin. Microbiol.*, 2000, **38**(2), 724–726.
120. S. P. Watson, M. O. Clements and S. J. Foster, Characterization of the starvation-survival response of Staphylococcus aureus, *J. Bacteriol.*, 1998, **180**(7), 1750–1758.
121. E. Alves, L. Costa, C. M. Carvalho, J. P. Tome, M. A. Faustino and M. G. Neves, *et al.*, Charge effect on the photoinactivation of Gram-negative and Gram-positive bacteria by cationic meso-substituted porphyrins, *BMC Microbiol.*, 2009, **9**, 70.
122. S. Banfi, E. Caruso, L. Buccafurni, V. Battini, S. Zazzaron and P. Barbieri, *et al.*, Antibacterial activity of tetraaryl-porphyrin photosensitizers: an in vitro study on Gram negative and Gram positive bacteria, *J. Photochem. Photobiol., B*, 2006, **85**(1), 28–38.
123. F. A. Tobiesen and S. Michielsen, Method for grafting poly(acrylic acid) onto nylon 6,6 using amine end groups on nylon surface, *J. Polym. Sci., Part A: Polym. Chem.*, 2002, **40**(5), 719–728.
124. J. Sherrill, S. Michielsen and I. Stojiljkovic, Grafting of light-activated antimicrobial materials to nylon films, *J. Polym. Sci., Part A: Polym. Chem.*, 2003, **41**(1), 41–47.
125. C. D. Salgado, K. A. Sepkowitz, J. F. John, J. R. Cantey, H. H. Attaway and K. D. Freeman, *et al.*, Copper Surfaces Reduce the Rate of Healthcare-Acquired Infections in the Intensive Care Unit, *Infect. Control Hosp. Epidemiol.*, 2013, **34**(5), 479–486.
126. G. B. Hwang, E. Allan and I. P. Parkin, White Light-Activated Antimicrobial Paint using Crystal Violet, *ACS Appl. Mater. Interfaces*, 2016, **8**(24), 15033–15039.
127. G. B. Hwang, A. Patir, E. Allan, S. P. Nair and I. P. Parkin, Superhydrophobic and White Light-Activated Bactericidal Surface through a Simple Coating, *ACS Appl. Mater. Interfaces*, 2017, **9**(34), 29002–29009.
128. K. H. Hong and G. Sun, Photoactive Antimicrobial Agents/Polyurethane Finished Leather, *J. Appl. Polym. Sci.*, 2010, **115**(2), 1138–1144.

CHAPTER 8

Antimicrobial Biomaterials in Ophthalmology

DEBARUN DUTTA,*[a] RENXUN CHEN,[b] NARESH KUMAR[b] AND MARK D. P. WILLCOX*[a]

[a] School of Optometry and Vision Science, University of New South Wales, Australia; [b] School of Chemistry, University of New South Wales, Australia
*Email: debarun.dutta@unsw.edu.au; m.willcox@unsw.edu.au

8.1 Introduction

Ophthalmology is a discipline of medicine that deals with the anatomy, physiology, diagnosis, and management of the diseases affecting the eyeball and orbit. Similar to other body parts, increasing numbers of biomaterials and implants are used in this field for various reasons ranging from vision restoration to sustained drug delivery. Worldwide, there are more than 140 million contact lens users.[1] Intraocular lenses (IOLs) are one of the most common implantable biomaterials that are used for restoration of clear vision following cataractous crystalline lens removal. It is estimated that more than 20 million cataract surgeries are performed every year.[2] The global market for IOLs is estimated at more than US$4.0 billion in 2018 and expected to reach US$5.49 billion by 2022.[3] The global contact lens market is predicted to grow to US$10 billion in 2022.[4] There are a variety of other extraocular and intraocular biomaterials that are used in the area of ophthalmology; punctal plugs, silicone sponge implants, prosthetic eyes, glaucoma drainage devices, keratoprosthesis, intra-stromal corneal ring, and scleral buckles are few of them.

Biomaterials Science Series No. 5
Antimicrobial Materials for Biomedical Applications
Edited by Abraham J. Domb, Konda Reddy Kunduru and Shady Farah

Published by the Royal Society of Chemistry, www.rsc.org

Microbial contamination of biomaterials remains a complex and troublesome issue for healthcare services. Development of post-surgical intraocular infection (endophthalmitis) is associated with poor prognosis and loss of vision.[5] Contact lens wear is one of the biggest risk factors for the development of corneal infection (keratitis) associated with ulceration and vision loss.[6] Figure 8.1 shows contamination of a contact lens surface by initial adhesion of planktonic *Pseudomonas aeruginosa*, a leading opportunistic pathogenic bacterium responsible for contact lens-induced keratitis. Punctal plugs are known to be associated with canaliculitis and dacryocystitis;[7] prosthetic eyes with chronic eye socket infection. Scleral buckle infection is one of the contributing factors for post-operative development of endophthalmitis and panophthalmitis, and the incidence rate can vary from 0.5% to 5.6%.[8–12] These infections are associated with the development of microbial biofilms, which are heterogenous mixtures of bacteria species covered with extracellular polymeric substances (EPS). These biofilms can be as much as 1000 times more resistant to antimicrobial treatment.[13]

This chapter examines strategies for the development of antimicrobial biomaterials in the field of Ophthalmology.

8.2 Antiadhesive Biomaterials

The adhesion process of a microorganism to a biomaterial is primarily driven by the interaction between the surface characteristics of the biomaterial and the microorganism. Antiadhesive biomaterials aim to discourage microbial adhesion and subsequent colonisation of the surface by interfering with the normal microbial adhesion procedure by modifying the surface characteristics of the biomaterial. Alterations can be achieved by changing

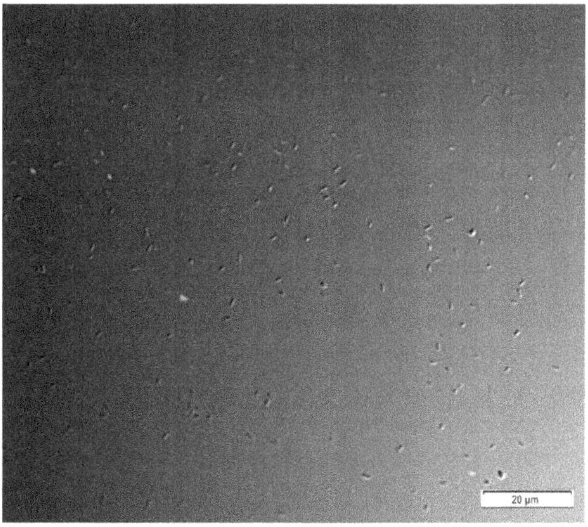

Figure 8.1 Examples of initial (within 30 minutes) *Pseudomonas aeruginosa* adhesion to a contact lens surface.

surface chemical characteristics, topography, and hydrophilicity. Typically, new molecules have been incorporated onto biomaterials *via* surface grafting, plasma treatment, surface interpenetrating network, or by simple soaking. Although these methods do not have an active "killing" mechanism and thus may not be termed as "antimicrobial", they are being widely studied as a mechanism that can significantly reduce bacterial colonisation and have the potential to reduce biomaterial-related infections.

Increasing the surface hydrophilicity of a biomaterial can reduce the adhesion of opportunistic microorganisms,[14] and a variety of synthetic and natural polymers have been investigated to enhance the surface hydrophilicity of ophthalmic biomaterials. Poly(ethylene glycol) (PEG) is one of the most popular polymer surface coatings employed due to its hydrophilicity and *in vivo* biocompatibility. Covalently grafted PEG on to poly(methyl methacrylate) (PMMA)-based ocular prostheses enhances their surface hydrophilicity.[15] Two distinct methods have been used to bind the PEG to the ocular prostheses; "grafting from", where PEG was polymerised from the PMMA surface, and "grafting to", involving covalent binding of PEG polymers to a amine-functionalised PMMA surface. These techniques can significantly reduce the adhesion of *P. aeruginosa* and *Staphylococcus epidermidis.*[16] *P. aeruginosa* is a leading pathogen responsible for the development of a variety of acute eye infections, including microbial keratitis.[17] *S. epidermidis* is an abundant opportunistic bacterium that is believed to reside in the ocular surface and is often associated with acute and chronic eye inflammation and infection.

8.3 Antimicrobial Biomaterials

8.3.1 Metallic Antimicrobials

Metals, in particular silver (Ag), copper (Cu), zinc (Zn), gold (Au), and platinum (Pt), have been used alone or in combination, often as nanoparticles, to impart antimicrobial activity to contact lenses,[17–34] contact lens cases,[35–40] corneal replacements,[41] IOLs,[42] and in nanoparticle form as additives to contact lens disinfecting solutions.[43,44] The use of metals as antimicrobial agents has a long history, with transition metals of the d-block (V, Ti, Cr, Co, Ni, Cu, Zn, Tb, W, Ag, Cd, Au, and Hg) and metals and metalloids from groups 13–16 reported as having antimicrobial activity.[45] Often the metals are used alone, but combinations they have been reported to show synergistic activities. For example, the combination of Au/Pt, Ag/Cu or Ag/Pt shows some synergism against planktonic and biofilm forms of *Enterococcus faecium* and *Acinetobacter baumannii.*[46]

The mechanisms of toxicity of metals to microbes are not always fully understood, but several potential mechanisms have been reported for some metals, and this chapter outlines what is known of the mechanism of activity of those metals used to date in ophthalmic devices. Copper, in the form Cu(II), can increase the intracellular concentration of reactive oxygen species.[47] Overproduction of reactive oxygen species can lead to DNA damage and inhibition of the activities of certain enzymes.[47] Bacterial ferric-ion reductases can be susceptible to site-specific inactivation by copper, silver, and zinc,[47] and silver can become incorporated into the Cu(I)-binding site of superoxide dismutase, abolishing its activity.[47] Superoxide dismutase is an important antioxidant enzyme in the removal of the superoxide radical which can cause many types of cellular damage. Another mechanism of action of metals is to impair the function of bacterial membranes. Silver can compromise the integrity of the cytoplasmic membrane of *S. aureus* and *Escherichia coli*, and can disrupt the activity of the bacterial electron transport chain that is membrane localised.[47] Copper has been reported to cause peroxidation of lipids in membranes.[47]

8.3.1.1 Silver

Silver has been used for many years in ophthalmology as a prophylaxis for neonatal gonococcal conjunctivitis.[48] The only commercially available silver-containing ophthalmic devices are contact lens cases. Silver-containing lens cases were probably first patented in 1994.[49] There are several versions of these cases, with apparent differences in the amount of silver that is released from them.[37] *In vitro* studies have shown that these cases can prevent the colonisation of a range of Gram-positive and Gram-negative bacteria, including *S. aureus* and *P. aeruginosa*[35,37] as well as having a smaller effect ($\leq$0.6 $\log_{10}$ reduction) on the mould *Fusarium solani*.[37] These *in vitro* evaluations showed that the maximum effect was observed after incubation with the microbes for $\geq$10 hours, which is more than the minimum manufacturers' recommended disinfection time for the cases with their disinfecting solutions ($\sim$6 hours). Other *in vitro* studies using these cases have shown that there was a relatively higher likelihood of transmission of bacteria from lens cases to contact lenses in the case due to the low adhesion force of bacteria to the silver-impregnated cases.[40] However, in the presence of the appropriate disinfecting solution, even though bacteria may be transferred to lenses they were likely to be dead and transmission was in any case lower in the presence of the disinfectants.[36,40] Another study demonstrated that, even though bacterial colonisation on silver-impregnated lens cases was lower than on other cases, the addition of tissue wiping of the lens cases after disinfection resulted in the lowest number of residual microbes on the cases.[38] Studies have examined whether silver cases could reduce microbial colonisation *in vivo*. In one report, 26–38% of silver lens cases were colonised by microbes compared to 63–67% of normal cases,[35] while in another report using the same lens cases the difference was less pronounced, with 71% of silver cases being colonised compared to 82% of normal cases.[39] The first trial to be reported demonstrated that silver cases were colonised by a less diverse range of microbes,[35] and the second publication showed that silver cases were colonised by lower numbers of bacteria (particularly Gram-positive bacteria): 1.7 $\log_{10}$ compared to 4.1 $\log_{10}$ colony-forming units per case well.[39] In overall, is it clear that silver could be developed as a potent antimicrobial agent for reducing infections and inflammations for various ophthalmic applications. Further *in vivo* and clinical studies are required for development of silver antimicrobial clinical applications.

There has been an increase in publications on the use of silver in contact lenses, and several patent studies,[50–53] especially in the 2000s from Johnson & Johnson Vision Care,[54–65] but as of 2018 no silver-impregnated contact lenses are commercially available. Silver has been incorporated into lenses in a number of ways, by impregnation with silver nitrate and reduction to silver ions forming nanoparticles,[19] by drying lenses and soaking in silver nanoparticles,[24] by simply soaking in colloidal silver nanoparticles,[28] by depositing silver nanoparticles on lens surfaces using matrix-assisted pulsed laser evaporation,[20] by incorporation of silver nitrate or nanoparticles into lens monomers and subsequent curing,[18,29] and by activating lens surfaces by immersing in dopamine prior to immersing in solutions of silver nanoparticles, either in one step or for several steps to produce a multilayered surface coating.[34]

Silver lenses have been made using either silicone hydrogel[19,20] or hydrogel materials.[18,22,24,25,29,31] Nissen and Furkert reported that a silver coating on a hydrogel lens was able to reduce colonisation of *P. aeruginosa* by more than 4.0 $\log_{10}$, but that of *S. aureus* by less than 2.0 $\log_{10}$ colony-forming units.[22] Silver nanoparticles in lenses have better antimicrobial activity against *S. aureus* and *P. aeruginosa* compared with *E. coli* and *Bacillus subtilis*,[19] and good activity against *S. aureus* and *P. aeruginosa* after imbibition with 10 or 20 ppm nanoparticles.[24,28] After deposition of silver nanoparticles on the surface of lenses, the lenses had good activity against *S. aureus* and *E. coli.*[20] After incorporation of nanoparticles into lenses and subsequent curing, the lenses (hydrogel disks) had good activity against *P. aeruginosa*, but less activity against *S. aureus*.[18] Lenses soaked in 20 ppm silver nanoparticles were also able to inhibit the viability of *Acanthamoeba* trophozoites.[28]

No *et al.* produced a hydrogel contact lens containing both silver and titanium oxide nanoparticles, and showed that these lenses could block ultraviolet light transmission, but they did not examine their antimicrobial efficacy.[23] Hydrogel lenses have been produced using silver and platinum nanoparticles,[21,31] and silver and gold nanoparticles.[30] The lenses were able to block ultraviolet light transmission,[21] but no antimicrobial activities were reported for the silver/platinum lenses. This same group has also produced a lens containing zinc oxide nanoparticles,[26] but have yet to report on their antimicrobial activity. Tuby *et al.*[27] produced a zinc-doped copper oxide coating on the surface of a silicone hydrogel lens by soaking the lens in a zinc–copper solution while applying high-intensity ultrasonication. These zinc–copper oxide-coated lenses reduced adhesion of *P. aeruginosa* by more than 4 $\log_{10}$ and that of *S. aureus* by more than 2.5 $\log_{10}$ colony-forming units.[27] Silver nanoparticles in a hydrogel were able to inhibit the growth of *S. aureus* and *E. coli*, but a combination of silver nanoparticles and graphene oxide was needed to inhibit the growth of moulds.[33]

Using silver-based antimicrobial lenses for therapeutic purposes has been examined. A silver nanoparticle coating on a hydrogel lenses, produced using dopamine, reduced the severity of keratitis produced by either *P. aeruginosa* or *Aspergillus fumigatus.*[34] A combination of silver nanoparticle, graphene oxide, and voriconazole in lenses was also able to reduce the severity of keratitis caused by *A. fumigatus* in a mouse model, but it appeared that it was the voriconazole that had the most beneficial effect.[33] The only paper that has been published to date on silver lenses in humans concluded that wearing silver-containing lenses did not alter the normal ocular microbiota.[66]

Several studies have reported that metal nanoparticles in lenses do not affect key aspects of contact lenses, including visible light transmission, at least to a degree that would affect vision,[19] nor obstruct oxygen transmission[20] (although the combination of silver and platinum nanoparticles did slightly reduced oxygen permeability).[21] An additional benefit is that the silver may also reduce protein deposition on lenses.[20] One of the issues with metal-containing lenses is that they often have an undesirable pigment, typically a yellow-brown colouration (Figure 8.2).[21,26,34] Alarcon *et al.* produced a slight blue tint with silver nanoparticles in collagen by incorporating thiol-modified LL-37 peptide.[41] Another potential problem with their use *in vivo* is complexation and precipitation of silver by host macromolecules such as proteins and DNA.[67] Silver can also cause the condition argyrosis in eyes. Argyrosis is a deposition of silver as a blue-grey discolouration in tissues.[68] This has been shown to occur when a person wore a silver-nitrate coated contact lens.[68]

Silver nanoparticles have been used in a few other ophthalmic devices, including being incorporated into collagen for corneal replacement. This was facilitated by the use of thiol-modified LL-37 peptide, and while results showed good efficacy of the silver/LL-37 collagen against *P. aeruginosa*, no controls were incorporated for the effect of the LL-37 peptide, which has known antimicrobial activity.[41] Therefore it is uncertain whether the antimicrobial activity was produced by silver or the peptide. Silver has been incorporated into IOLs by diffusion of silver nitrate into a readymade IOL[42] or silver being incorporated during the synthesis of the IOL.[69]

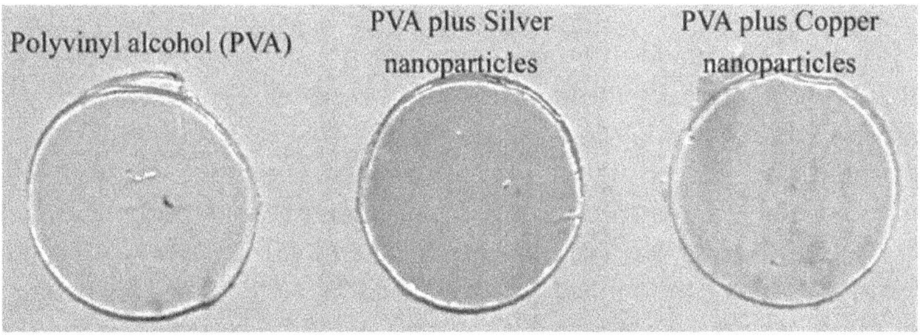

Figure 8.2 Examples of metal-containing contact lenses. Polymers of polyvinyl alcohol containing silver or copper nanoparticles.

Unfortunately, the authors[42] did not test the IOL for antimicrobial activity. Finally, silver or gold nanoparticles have been incorporated into solutions used to disinfect contact lenses. Silver nanoparticles added to a series of commercially available contact lens disinfecting solutions or to chlorhexidine did not produce any added anti-*Acanthamoeba* effect,[43,44] but the combination of gold nanoparticles and chlorhexidine reduced the viability of *Acanthamoeba*, and also reduced the toxicity of *Acanthamoeba* to human endothelial cells.[43]

There is the possibility of metal-containing devices selecting for resistant microbes. Microbial mechanisms that could be used to reduce the effectiveness of metal compounds include reduced uptake of the metal, efflux pumps to remove metals, sequestration in the inner or outer layers of microbes, and metabolic bypass.[47] Genes involved in resistance to silver have been isolated from Enterobacteriacceae, oral bacteria and *S. aureus*.[70] Furthermore, these silver resistance genes, *sil*, have been isolated from silver-containing creams used to treat burns.[70] Another set of genes that are more commonly associated with resistance to copper, *cus*, can also confer low-level resistance to silver.[70] These genes are often carried on mobile genetic elements such as plasmids, that also confer resistance to antibiotics and other metals (*e.g.* mercury).[70] Their presence on mobile genetic elements means that they could be transferred rapidly, and even not as the result of exposure to metals, but as a consequence of exposure to antibiotics. There have been no reports examining the presence of resistance to silver and the associated genes from isolates from silver-containing lens cases, but this would be useful in future studies.

8.3.2 Selenium

Selenium (Se), in its natural form as organoselenium, is an essential nutrient with a World Health Organization (WHO) recommended daily intake of 40 μg per day. Certain organoseleniums have the ability to form superoxide and hydrogen peroxide *in vitro* and *in vivo*. Selenium in the −2 oxidation state such as selenide (RSe^-) reacts with thiols, such as glutathione, cysteine (CysSG), or dithiothreitol D $(SH)_2$, to produce superoxide ($O^{\bullet -}$) and hydrogen peroxide (H_2O_2).[71] This catalytic mechanism has been used to produce short-lived superoxides from organoselenium compounds covalently attached to material surfaces, without release or depletion of the compounds, which is ideal for ophthalmological and biomedical applications in general. For the controlled production of superoxide by surface-attached selenium compounds, the presence of oxygen and reducing agents such as thiol groups or other electron-donating groups (such as NADPH-dependent reductase, membrane proteins) is required.

Selenium-coated contact lenses and contact lens cases have been produced.[72–77] They are active *in vitro* against *P. aeruginosa* and show good *in vivo* safety in a rabbit contact lens model, which examined the effects on the cornea over 2 months of lens wear.[74] In another study, monoselenide has been found to be more active compared to selenocysteine (diselenide) after covalent attachment to contact lenses *via* amide bonds.[77] A more comprehensive investigation of the *in vitro* antimicrobial effects of organoselenium compounds coated on contact lenses showed complete inhibition of the adhesion of *P. aeruginosa*, *Serratia marcescens* and *S. aureus* on the lens surface.[73] In addition, selenium incorporated into contact lens cases (polypropylene) could prevent biofilm formation on the case surface by *S. aureus.*[72]

Selenium coating coupled to a silyation reagent on scleral buckles have been evaluated for their safety in respective rabbit models and demonstrated to have no adverse effects.[78,79] Consistent with the *in vivo* studies that demonstrate safety of selenium in various applications, a selenium-containing dental sealant has received Food and Drug Administration 510(k) clearance and has been demonstrated to be safe and effective in *in vitro* and clinical studies.[80] The future application of selenium-coated devices in ophthalmology is promising. However, further *in vivo* antimicrobial testing in animal models as well as clinical human trials would be required to bring selenium-treated antimicrobial ophthalmic devices to market.

8.3.3 Antibiotics

There have been various attempts to use a spectrum of antibiotics incorporated into ophthalmic biomaterials, such as IOLs, contact lenses, punctum plugs, and scleral implants, for the reduction of bacterial contamination and colonisation. Several approaches have been used to incorporate antibiotics onto IOLs, including soaking, impregnation, covalent immobilisation, and coating with drug and polymer layers.

In 1952, Klein and Millwood made one of the earliest investigations with chloramphenicol-releasing calcium alginate caps for the control of *P. aeruginosa* corneal infections.[81] A calcium alginate mould *in vitro* was produced and trimmed to fit at the ocular surface, making it into a contact lens-like device. Recently, moxifloxacin has been soaked into commercially available hydrophobic acrylic AcrySof™ (Alcon Inc.) and hydrophilic collamer Affinity™ (Staar Inc.) IOLs could deliver clinically significant antibiotic levels that exceeded the inhibitory concentration for most of the common pathogens responsible for post-operative endophthalmitis.[82] The release of chloramphenicol and dexamethasone by hydrogel IOLs implanted at the anterior chamber of rabbit eyes has been investigated,[83] and showed that *in vivo* release of the antibiotics could be a viable option for reducing post-cataract surgery endophthalmitis.

Combining IOLs with one or more drug-loaded inserts is possibly the most direct method used to achieve the intraocular release of antibiotics following cataract surgery.[84,85]

Similar intraocular release could be achieved by immersion of IOLs with concentrated aqueous or hydroalcoholic solutions of antibiotics. This is one of the most popular methods of intraocular delivery of antibiotics.

Commercially available hydrophilic acrylic IOLs have affinity for fourth-generation fluoroquinolones such as gatifloxacin and moxifloxacin.[86] Rabbit studies have shown that these antibiotic-soaked IOLs can progressively release antibiotics over 6–24 hours in the anterior chamber of the eye to achieve concentrations above the minimum inhibitory concentrations needed to affect common ocular pathogens.[86,87]

Applying layer-by-layer deposition of oppositely charged polymers can be used to coat rigid IOLs with antibiotics. PMMA IOLs coated with successive layers of poly(sodium 4-styrenesulfonate) containing ampicillin and poly-(ethylenimine) provided sustained release of ampicillin over 7 days.[88] The fluoroquinolone antibiotic norfloxacin has been combined with IOL monomers and then co-polymerisation into poly(2-hydroxyethyl methacrylate) (pHEMA) IOLs. This gave a sustained norfloxacin release *in vitro*,[89] and the released antibiotics could inhibit *S. epidermidis* colonisation of the anterior chamber of the eye.[89]

Repairing a retinal detachment with scleral implants or scleral buckles is becoming a common ophthalmological procedure. A scleral buckle made of co-poly(methylacrylate-2-hydroxyethyl acrylate) was one of the first hydrophilic implants designed in 1980.[90] This buckle could absorb and release six (gentamycin, polymyxin B, lincomycin, staphcillin, chloramphenicol, and bacitracin) water-soluble antibiotics that were active against *E. coli* and *Sarcina lutea* (*Micrococcus luteus*).[91] The antibiotics were gradually released from the scleral buckle *in vitro* and achieved sustained antibacterial activity for 30 days. After implantation of lincomycin- or methicillin-soaked buckles in rabbit eyes the aqueous humour showed antimicrobial activity against *S. lutea* for 5 and 14 days, respectively.

Contact lenses can be used as a therapeutic device for delivery of ocular drugs. Controlled delivery of antibiotics has been achieved using several approaches, such as incorporation of a drug reservoir consisting of a poly-[lactic-*co*-glycolic acid] (PLGA) film, incorporation of anionic or cationic groups into lens monomer mixtures during synthesis, additional drug–diffusion barriers such as vitamin E, and the use of drug nanoparticles.[92–99] Similar to IOLs, the most common procedure to adsorb and release antibiotics has been to soak the contact lenses in antibiotics. One of the first investigation for antibiotic delivery with contact lenses used soaking of commercial lenses in gentamicin, chloramphenicol and carbenicillin.[100] Ciprofloxacin and norfloxacin (second-generation fluoroquinolones) have been extensively investigated for ways of obtaining their sustained release from commercially available or custom-made lenses.[94,101–103] Uptake (by soaking) and release of the antifungal agent natamycin has been investigated using commercially available silicone hydrogel and hydrogel contact lenses. Although all the contact lenses released clinically relevant concentrations of natamycin, the release lasted for only 30 minutes, indicating that further improvement was required for a sustained drug-releasing technology.[104] Acrylic acid has been used in order to improve the release of ciprofloxacin[103] and norfloxacin[105] using molecular imprinting. Alternatively, liposomes loaded with levofloxacin can slow drug release for up to 120 hours.[106] Using ciprofloxacin and PLGA at a 1 : 1 ratio between two layers of contact lenses can enhance drug release kinetics for up to 3 days.[107] Using a similar method to incorporate the antifungal drug econazole, release of effective fungicidal concentrations for up to 21 days was achieved.[108] Vitamin E barriers have also been used to slow release of another antifungal, fluconazole from commercial contact lenses.[109]

Although use of antibiotics has revolutionised global disease control, frequent and unrestricted use of antibiotics has been associated with the development of antibiotic resistance by microorganisms.[110] The WHO has defined antibiotic resistance as a major global health issue for this generation. The rate of antibiotic resistance in microbial isolates from ocular infections is increasing.[111,112] A randomised clinical trial (RCT) that included 500 participants with corneal infection showed that the most commonly isolated microorganisms such as *P. aeruginosa* and *Streptococcus pneumoniae* demonstrated increased resistance against fourth generation fluoroquinolone antibiotic moxifloxacin.[113] More importantly, increased antibiotic resistance was associated with reduced visual outcome, larger corneal infiltrates, and longer healing period.[113] Recently, Subedi *et al.*[114] reviewed the antibiotic resistance of *P. aeruginosa* isolated from eye infections, and found that corneal isolates have the highest antibiotic resistance, which may reach up to 33% against ciprofloxacin. This indicates that development of antibiotic-releasing biomaterials may not be a long-term solution as a prophylactic device in alleviating ophthalmic infections. In fact, these devices may increase the risk of developing antibiotic resistance, thus should be used with caution. However, antibiotic-releasing devices will continue to be suitable as therapeutic devices for treatment of active ocular infection.

8.3.4 Antimicrobial Peptides

Antimicrobial peptides (AMPs) are part of the immune system of all multicellular organisms including plants and animals.[115–118] AMPs are generally short in length and consist of 15–100 amino acids. Features of AMPs thought to be important in their antimicrobial activity include amphipathicity and typically cationicity (sometimes anionicity). They are mostly positively charged due to the high number of positively charged (net charge higher than +2) amino acids such as arginine and lysine. However, negatively charged peptides (net charge from −1 to −7) consisting of five residues to approximately 70 residues have been recently reported to have high antimicrobial activity.[119] AMPs have broad-spectrum antimicrobial activity, including activity against bacteria (including antibiotic-resistant strains), viruses, fungi, protozoa, and parasites.[117,120–127] AMPs are also known to have healing and immunomodulatory activities.[116,123,128,129] Microbial membrane interactions appear to be the primary mode of antimicrobial function of all AMPs where the AMP interacts with the negatively charged microbial membranes, leading to membrane destabilisation, increased cell permeability, and eventually cell death.[130,131] There are a variety of models proposed for interaction between AMPs and bacterial cell membranes, such as producing toroidal pores, barrel staves, electroporation, and charged lipid clustering.[131] Due to their variety of modes of action and low inhibitory concentration, bacteria find it difficult to gain resistance against these molecules.[132]

Protamine, melittin, melimine, magainin, esculentin, lactoferricin, cecropins, bovine lactoferrampin, and cathelicidins have been investigated in attempts to produced antimicrobial biomaterials.[128,133–136] These peptides are either soaked or immobilised on the biomaterial surface. Soaking with AMPs could be an effective option with the initial burst of high antimicrobial dose, although antimicrobial-releasing systems which result in release of sub-inhibitory concentrations can potentially contribute to microbial resistance.[137]

The β-defensin family of peptides and the cathelicidin LL-37 have been tested with contact lens materials to produce antimicrobial activity.[135,138,139] The benefit with these peptides is that they are physiologically present at the ocular surface and likely to be biocompatible with ocular tissue at their inhibitory concentrations. Both of these peptides have high antibacterial, antifungal, and antiviral activity, and incorporation of these AMPs onto ophthalmic biomaterials resulted in high antimicrobial effects *in vitro*.[140] Esculentins (Esc) are a family of α-helical membrane-active AMPs derived from amphibian skin.[140] These natural peptides have broad-spectrum antimicrobial activity against a variety of Gram-negative and Gram-positive bacteria. Esc(1-21) and Esc(1-21)-1c are novel chimeric peptides derived from esculentin, and are showned to have high bactericidal activity with negligible cytotoxicity and maintain high biostability.[141] Incorporation of these peptides onto contact lens surfaces produced high bactericidal activity against *P. aeruginosa*.[141] Importantly, these antimicrobial coatings were not cytotoxic to mammalian cells and did not interfere with the contact lens physical parameters.[141] Table 8.1 details the AMPs that have been investigated with the development of antimicrobial biomaterials in ophthalmology. To date, only the peptides melimine and Mel4 have been tested for safety and efficacy in human eyes.

Table 8.1 Antimicrobial peptides used in ophthalmic biomaterials.

Antimicrobial peptides	Primary structure of the peptides
LL-37	L-L-G-D-F-F-R-K-S-K-E-K-I-G-K-E-F-K-R-I-V-Q-R-I-K-D-F-L-R-N-L-V-P-R-T-E-S
Melimine	T-L-I-S-W-I-K-N-K-R-K-Q-R-P-R-V-S-R-R-R-R-R-R-G-G-R-R-R-R
Mel4	K-N-K-R-K-R-R-R-R-R-R-G-G-R-R-R-R
Esculentin (1-21)[a]	G-I-F-S-K-L-A-G-K-K-I-K-N-***L***-L-I-***S***-G-L-K-G
Esculentin (1-21)-1c	GI-F-S-K-L-A-G-K–K-I-K-N-L-L-I–S-G-L-K-G

[a]D-amino acids are presented in bold italics.

8.3.4.1 Melimine

Melimine and its derivative Mel4 are chimeric peptides consisting of 29 and 17 amino acid residues, respectively. These are derived from the naturally occurring peptides melittin and protamine. Protamine has high activity against Gram-negative bacteria, but is less active against Gram-positive bacteria.[142,143] Melittin has good activity against Gram-positive bacteria but it is relatively less active against Gram-negative bacteria.[144] However, the chimeric peptides melamine and Mel4 have broad-spectrum antimicrobial activity *in vitro*.[145] Repeated exposure of bacteria tomelimine at sub-minimal inhibitory concentrations does not induce bacterial resistance.[144] Both peptides have been immobilised onto contact lenses and have been shown to retain their broad-spectrum antimicrobial activity against a variety of ocular pathogens, including drug-resistant *P. aeruginosa* and *S. aureus* bacteria, *F. solani*, *Candida albicans* and *Acanthamoeba*.[134]

Melimine immobilised onto contact lenses has been extensively investigated *in vivo* in animal models and human clinical trials.[130,146,147] Melimine immobilised onto contact lenses can reduce ocular inflammation associated with bacterial contamination of contact lenses in animals model studies.[148] Melimine-coated lenses can also reduce the incidence and severity of infections associated with contact lenses in a rabbit model study.[149] Figure 8.3 shows the appearance of a healthy rabbit eye (Figure 8.3A) following wear of a *P. aeruginosa*-contaminated melimine antimicrobial contact lens. This figure also shows the development of microbial keratitis (corneal infection) in rabbit eyes following the use of *P. aeruginosa*-contaminated control contact lenses (Figure 8.3B–E).

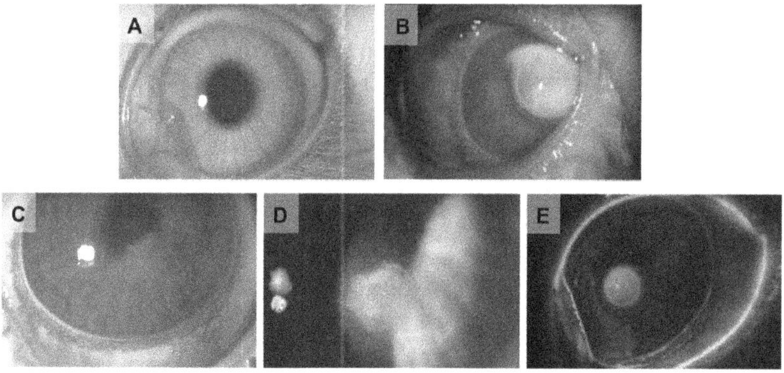

Figure 8.3 Representative images of slit lamp findings of rabbit eyes. (A) Healthy rabbit eye mostly seen with antimicrobial contact lens wear (7.5× magnification); (B) microbial keratitis in a rabbit eye following *P. aeruginosa*-contaminated control contact lens wear; (C) conjunctival hyperaemia during microbial keratitis (12× magnification); (D) accumulation of corneal infiltratives at the ulcer (32× magnification); (E) epithelial break overlying the corneal ulcer; imaged with fluorescein staining and Wratten 12 filter (7.5× magnification).

The safety and efficacy of these lenses have also been investigated in non-dispensing and dispensing human clinical trials. The first non-dispensing trial showed that melimine-coated contact lenses were biocompatible with human eyes.[146] Mel4 peptide-coated contact lenses were investigated in a similar fashion; they showed broad-spectrum antimicrobial activity against a variety of ophthalmic pathogens and were not cytotoxic *in vitro*.[150] This was followed by further investigations that demonstrated that these lenses were biocompatible *in vivo* with rabbit animal model studies.[150] This was followed by 1 week of daily wear of the Mel4-coated contact lenses, which confirmed their biocompatibility with human eyes over 1 week.[150] A recent phase II/III RCT has investigated the efficacy of Mel4-coated antimicrobial contact lenses in reducing corneal infiltrative events. This RCT showed that these antimicrobial contact lenses were biocompatible with humans over many months of wear. Figure 8.4 displays a representative photograph of a contact lens-related acute red eye (CLARE) event observed with control lens wear during the human clinical trial. This RCT showed that the Mel4-coated contact lenses have the capacity to reduce corneal infiltrative (keratitis) events by 50%.[151] When worn on an extended-wear schedule (*i.e.* 24 hours of lens wear each day for up to 14 days), Mel4-coated contact lenses retained their antimicrobial activity for up to 1 week.[152] This is a significant finding for the development of antimicrobial biomaterials in ophthalmology in general, and antimicrobial contact lenses in particular.

Bacterial resistance against AMPs is relatively rare. This is mainly because AMPs have multiple mechanism of bactericidal action, making it difficult for the bacteria to gain resistance.[153] Repeated exposure of AMPs to ophthalmic bacterial isolates at the sub-inhibitory concentration showed no rapid development of resistance.[145] However, a few recent reports have emphasised the emergence of AMP resistance by selective bacterial species. Resistance mechanisms of many Gram-negative bacteria involve alteration of the lipopolysaccharide molecule of the outer membrane, which could reduce the interaction of AMPs to the inner membrane of a bacteria.[154] Species of the

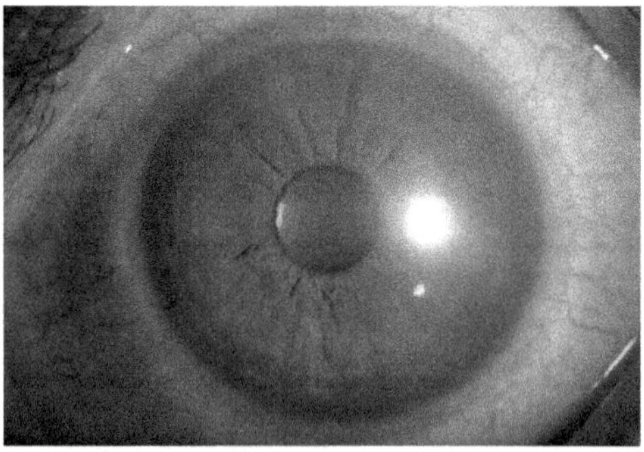

Figure 8.4 Representative photograph of contact lens-induced acute red eye (CLARE) with continuous control contact lens wear during human clinical trial.

genus *Burkholderia* have extreme AMP resistance, and IspH, a $[4Fe-4S]^{2+}$ enzyme involved in the synthesis of isoprenoids could play a role in the development of resistance by modifying the lipid composition of the inner layer of this bacterium.[155] Furthermore, RpoE (putative antibiotic resistance regulator) and BCAL2831 regulators may modulate genes such as *mucD*, leading to an increase in AMP resistance. Development of resistance of *P. aeruginosa* to colistin (polymixin E) is a multistep procedure, and mutations in the transcriptional regulators are necessary for functionalising and increasing the resistance.[156] AMPs are some of the last resort of our antibiotic arsenal for development of safer antimicrobial biomaterials and should be used carefully at the clinical environment.

8.3.5 Quorum-sensing Inhibitors—Fimbrolides and Dihydropyrrolones

A critical role in the formation of mature and differentiated biofilms is played by a bacterial cell-to-cell communication system known as quorum sensing. Quorum sensing is a mechanism by which bacteria regulate the expression of specific genes in response to the density of their cell populations.[157] Quorum sensing is produced by the release of diffusible signal molecules called autoinducers that increase in concentration as a function of cell density; when the concentration of autoinducer reaches a certain threshold density the accumulated signalling compounds interact with cellular receptors, which controls the expression of a set of specific target genes. Quorum sensing-controlled genes encode for proteins that play a crucial role in biofilm development, for instance they are involved in the building of the EPS matrix or the irreversible adhesion of the bacteria onto surfaces.[158]

Beside biofilm maturation, a large number of other specialised processes are reported to be regulated by density-dependent signalling molecules, including antibiotic production, bioluminescence, genetic competence, sporulation, swarming motility, and virulence.[157,159] Quorum sensing enables bacteria to make collective decisions with respect to the expression of specific set of genes.

P. aeruginosa strains isolated from contact lens-associated keratitis produce quorum sensing molecules.[160] *P. aeruginosa* mutants deficient in the production of quorum sensing-signalling molecules form abnormal biofilms,[161] and cannot infect mouse corneas.[162] This indicates that disruption of this system may be useful in controlling the colonisation of ophthalmic devices by *P. aeruginosa* and other bacteria. Nature provides fascinating examples of quorum sensing disruption for the prevention of biofouling or biofilm formation. *Delisea pulchra*, a red marine alga found on the south-east coast of Australia, secretes a range of halogenated compounds called fimbrolides which can inhibit bacterial colonisation.[163] These fimbrolides consist in general of a furan ring structure with a substituted acyl chain at the C-3 position and a bromine substitution at the C-4 position. The major difference between these compounds is in the number of halogen substituents and the presence or absence of oxygen functionality in the butyl side chain (Figure 8.5).[164]

Due to the structural similarities of fimbrolides to acyl homoserine lactone (AHL), the quorum sensing autoinducers used by Gram-negative bacteria, fimbrolides affect the interaction between AHLs and the regulatory proteins (LuxR or LuxR homologues) by competitively binding to the receptor site, thereby displacing the cognate AHL autoinducer.[165,166] In addition to the inhibitory effect of fimbrolides towards the AHL-regulated quorum sensing systems, fimbrolides have also been found to disrupt AI-2-mediated quorum sensing systems of *Vibrio harveyi*, *E. coli*, *B. subtilis* and oral streptococci.[167–169] This AI-2 quorum sensing system is used by both Gram-negative and Gram-positive bacteria, indicating that this class of quorum sensing inhibitors is not limited to Gram-negative species. However, natural fimbrolides are often cytotoxic. To overcome such limitations, a range of analogues with varying side-chain length, substitution and attachment point to the fimbrolide ring, as well as different substituents in the ring have been synthesised.[169–172] In the process of imitating structures of antibacterial fimbrolides, a related structure, 1,5-dihydropyrrol-2-one (DHP), was also found to possess antibacterial properties. The main difference between the two structures is the replacement of the lactone oxygen of the heterocyclic ring with a nitrogen atom, thus yielding a pyrrol-2-one (lactam) (Figure 8.6). This lactam ring system is known to be hydrolytically more stable compared to an equivalent lactone, and is therefore less susceptible to lactonolysis (ring opening) in physiological conditions.[173,174]

Fimbrolides can be covalently bound to polymers using plasma-functionalised surfaces.[175–177] A fimbrolide using this strategy has been successfully covalently bound to commercially available contact lenses.[178] These fimbrolide-coated lenses reduced the adhesion of *P. aeruginosa*, *S. aureus*, *S. marcescens*, and the protozoan *Acanthamoeba*, with a reduction of cells on surfaces between 67% and 92%. The fimbrolide-coated lenses were safe in rat and guinea-pig models and a human clinical trial.[179] In the guinea-pig model, the animals wore the fimbrolide-coated contact lenses for

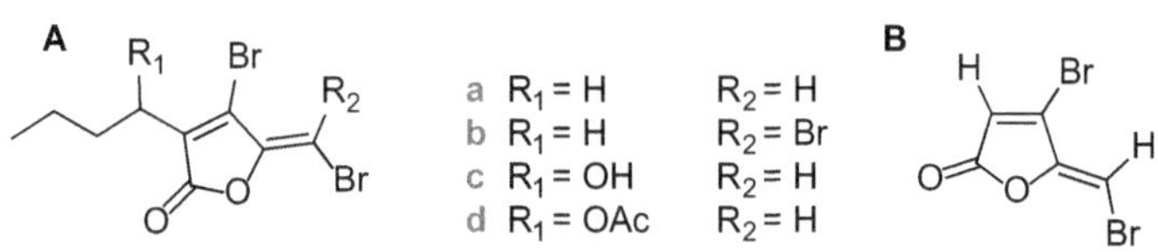

Figure 8.5 (A) Halogenated fimbrolide isolated from *Delisea pulchra* and (B) a synthetic fimbrolide, 4-bromo-5-(bromomethylene)-2(5*H*)-fimbrolide.

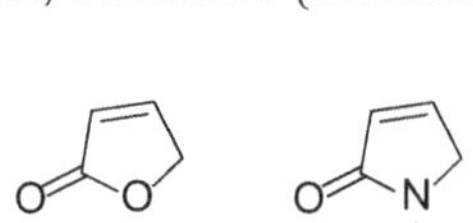

Figure 8.6 Chemical structure of lactone (left) and pyrrol-2-one (right) core "nucleus".

1 month. The results showed no differences in their ocular responses; neither conjunctival redness nor corneal staining were found to be different between test and control lenses. Human volunteers wore fimbrolide-coated lenses for 1 day and 1 night, and no difference in the ocular response was shown between the control and coated lenses.[178]

A potential advantage of fimbrolide and DHP coatings is that these may not induce development of resistance in microbes. This has been postulated due to the fact that they do not (always) kill the microbes on which they act. They merely turn off the production of toxins or ability to adhere, meaning there would be less selection pressure for resistance to develop. However, this has yet to be tested.

8.3.6 Other Antimicrobial Strategies

Capsular polysaccharides play an important role in bacterial coaggregation and production of biofilms. Nonsteroidal anti-inflammatory drugs (NSAIDs) can reduce production of bacterial polysaccharides. Salicylic acid, diclofenac, and ketorolac inhibit the colonisation of *P. aeruginosa*, *S. epidermidis*, *S. pneumoniae*, and *Haemophilus influenzae* to contact lenses.[180] Teichoic acid is an essential component of the Gram-positive bacterial cell wall and a crucial component of slime produced by Gram-positive bacteria for the formation of biofilm. Salicylic acid-soaked contact lenses can reduce production of the teichoic acid by *S. epidermidis*,[181] and inhibit bacterial biofilm formation.[182–184] Salicylic acid-releasing contact lenses can inhibit *Acanthamoeba*[185] and the attachment of *A. castellanii* trophozoites to contact lenses.[182] The release kinetics of NSAIDs from these contact lenses have not been reported. Initial burst release of high concentrations of NSAID might be associated with wearer toxicity.[186]

Covalent immobilisation of *N*,*N*-hexylmethyl-polyethylenimine, a long-chain hydrophobic polycation that has broad antibacterial, antifungal, and antiviral properties,[187] on to Boston Keratoprosthesis material reduced biofilm formation of *S. aureus*.[188] Furthermore, the antimicrobial surface was biocompatible *in vitro* and when tested with rabbit animal models. The Boston Keratoprosthesis is the most widely used artificial cornea and one of the very few treatment options for multiple corneal graft failure, Stevens–Johnson syndrome, ocular cicatricial pemphigoid and other autoimmune diseases, ocular burns, and other conditions with poor prognosis with traditional penetrating keratoplasty. Given this, the development of antimicrobial surface by *N*,*N*-hexyl,methyl-polyethylenimine on Boston Keratoprosthesis is an important step towards the development of ophthalmic antimicrobial biomaterials.

8.4 Conclusion

A variety of microorganisms are implicated in the development of eye infections associated with ophthalmic biomaterials. Controlled antimicrobial activity or targeted antimicrobial drug delivery in the ophthalmic biomaterial field has immense potential to reduce the burden of these ophthalmic infections and inflammations. Despite investigations with various promising technologies, only a handful of them are available in the clinic. Recently, many promising strategies are being tested in phase II/III human clinical trials, and soon may enter industrial production lines as an antimicrobial biomaterial in the area of ophthalmology.

References

1. R. M. Klevens, J. R. Edwards and C. L. Richards Jr., *et al.*, *Public Health Rep.*, 2007, **122**, 160–166.
2. J. C. Erie, *Ophthalmology*, 2014, **121**, 2–4.
3. Rohan. Dallas, USA, PR Newswire, 2013, http://www.prnewswire.com/news-releases/biomaterials-market-worth-884-billion—2017-199824911.html, Accessed on 13th August 2018.
4. R. Tiwari. Pune, India, MarketWatch, 2017, https://www.marketwatch.com/press-release/global-contact-lenses-market-worth-10000-million-usd-by-2022-2017-02-14-11203121, Accessed on 13th October 2018.
5. A. R. Shirodkar, A. Pathengay and H. W. Flynn Jr., *et al.*, *Am. J. Ophthalmol.*, 2012, **153**, 391–398 e392.
6. F. Stapleton, K. Edwards and L. Keay, *et al.*, *Ophthalmology*, 2012, **119**, 1516–1521.
7. N. Jehangir, G. Bever, S. M. Mahmood and M. Moshirfar, *J. Ophthalmol.*, 2016, **2016**, 9312340.
8. W. E. Smiddy, D. Miller and H. W. Flynn Jr., *Ophthalmic. Surg.*, 1993, **24**, 440–445.
9. J. Deutsch, R. K. Aggarwal and E. M. Eagling, *Eye*, 1992, **6**(Pt 6), 570–573.
10. J. C. Pastor, I. Fernandez and E. Rodriguezde la Rua, *et al.*, *Br. J. Ophthalmol.*, 2008, **92**, 378–382.
11. Q. Sun, T. Sun and Y. Xu, *et al.*, *Curr. Eye Res.*, 2012, **37**, 492–499.
12. J. Joseph, A. Pathengay, V. Michael, B. Raju, S. Sharma and T. Das, *Clin. Exp. Ophthalmol.*, 2006, **34**, 390–391.
13. J. J. Harrison, H. Ceri, C. A. Stremick and R. J. Turner, *Environ. Microbiol.*, 2004, **6**, 1220–1227.
14. D. Dutta, N. Cole and M. Willcox, *Mol. Vis.*, 2012, **18**, 14–21.
15. J. Ko, K. Cho and S. W. Han, *et al.*, *Colloids Surf., B*, 2017, **158**, 287–294.
16. P. Kingshott, H. Thissen and H. J. Griesser, *Biomaterials*, 2002, **23**, 2043–2056.
17. D. Dutta and M. D. Willcox, *Eye Contact Lens*, 2014, **40**, 312–324.
18. B. S. Fazly Bazzaz, B. Khameneh, M. M. Jalili-Behabadi, B. Malaekeh-Nikouei and S. A. Mohajeri, *Cont. Lens Anterior Eye*, 2014, **37**, 149–152.
19. F. M. Helaly, S. M. El-Sawy, A. I. Hashem, A. A. Khattab and R. M. Mourad, *Cont. Lens Anterior Eye*, 2017, **40**, 59–66.
20. G. Huang, W. H. Tse and J. Zhang, *RSC Adv.*, 2016, **6**, 67166–67172.
21. T. H. Kim and A. Y. Sung, *J. Nanosci. Nanotechnol.*, 2013, **13**, 5966–5975.

22. S. Nissen and F. H. Furkert, *Der Ophthalmologe: Zeitschrift der Deutschen Ophthalmologischen Gesellschaft*, 2000, **97**, 640–643.
23. J. W. No, D. H. Kim, M. J. Lee, D. H. Kim, T. H. Kim and A. Y. Sung, *J. Nanosci. Nanotechnol.*, 2015, **15**, 8016–8022.
24. M. Shayani Rad, B. Khameneh, Z. Sabeti, S. A. Mohajeri and B. S. Fazly Bazzaz, *Curr. Eye Res.*, 2016, **41**, 1286–1293.
25. A. Y. Sung and T. H. Kim, *Korean J. Chem. Eng.*, 2012, **29**, 686–691.
26. A. Y. Sung and T. H. Kim, *J. Biomater. Nanobiotechnol.*, 2013, **4**, 309–315.
27. R. Tuby, S. Gutfreund and I. Perelshtein, *et al.*, *ChemNanoMat*, 2016, **2**, 547–551.
28. M. D. P. Willcox, E. B. H. Hume, A. K. Vijay and R. Petcavich, *J. Opt.*, 2010, **3**, 143–148.
29. K.-H. Ye, S.-H. Cho and A.-Y. Sung, *J. Korean Chem. Soc.*, 2009, **53**, 542–546.
30. K.-H. Ye, T.-H. Kim and A.-Y. Sung, *J. Korean Chem. Soc.*, 2010, **54**, 228–233.
31. K.-H. Ye and A.-Y. Sung, *J. Korean Chem. Soc.*, 2010, **54**, 310–316.
32. A. Xiao, C. Dhand, C. M. Leung, R. W. Beuerman, S. Ramakrishna and R. Lakshminarayanan, *J. Mater. Chem. B*, 2018, **6**, 2171–2186.
33. J. F. Huang, J. Zhong and G. P. Chen, *et al.*, *ACS Nano*, 2016, **10**, 6464–6473.
34. X. Liu, J. Chen and C. Qu, *et al.*, *ACS Biomater. Sci. Eng.*, 2018, **4**, 1568–1579.
35. C. F. Amos and M. D. George, *Cont. Lens Anterior Eye*, 2006, **29**, 247–255.
36. P. B. Vermeltfoort, J. M. Hooymans, H. J. Busscher and H. C. van der Mei, *J. Biomed. Mater. Res. B Appl. Biomater.*, 2008, **87**, 237–243.
37. J. Dantam, H. Zhu and F. Stapleton, *Invest. Ophthalmol. Vis. Sci.*, 2011, **52**, 51–57.
38. Y. T. Wu, H. Zhu, M. Willcox and F. Stapleton, *Eye Contact Lens*, 2011, **37**, 365–369.
39. J. Dantam, H. Zhu and M. Willcox, *et al.*, *Invest. Ophthalmol. Vis. Sci.*, 2012, **53**, 1641–1648.
40. W. Qu, H. J. Busscher, H. C. van der Mei and J. M. Hooymans, *Cornea*, 2013, **32**, 326–331.
41. E. I. Alarcon, B. Vulesevic and A. Argawal, *et al.*, *Nanoscale*, 2016, **8**, 6484–6489.
42. T. Badur, H.-C. Kim and N. Hampp, *Proceeding of SPIE*, 2017, **10078**, 1007809-1007801–1007809-1007810.
43. Y. Aqeel, R. Siddiqui, A. Anwar, M. R. Shah and N. A. Khan, *Antimicrob. Agents Chemother.*, 2016, **60**, 1283–1288.
44. M. Niyyati, R. Sasani and M. Mohebali, *et al.*, *Iran J. Parasitol.*, 2018, **13**, 180–185.
45. R. J. Turner, *Microb. Biotechnol.*, 2017, **10**, 1062–1065.
46. M. Y. Vaidya, A. J. McBain, J. A. Butler, C. E. Banks and K. A. Whitehead, *Sci. Rep.*, 2017, 7, 5911.
47. J. A. Lemire, J. J. Harrison and R. J. Turner, *Nat. Rev. Microbiol.*, 2013, **11**, 371–384.

48. D. L. Moore and N. E. MacDonald, I. D. Canadian Paediatric Society, C. Immunization, *Can. J. Infect. Dis. Med. Microbiol.*, 2015, **26**, 122–125.
49. A. J. Dziabo, A. A. Holmes, C. B. Anger, J. C. Baker and L. Peng, Antimicrobial lenses and lens care systems, *US Pat.*, US5515117A, 1994.
50. S. Sakuma, K. Atsumi and A. Inoue, Antibacterial contact lens, *JP Pat.*, JPH0476518A, 1993.
51. R. Chromecek, J. Vodnansky and J. Manuch, Apparatus for preserving hydrophilic gels, more particularly ocular contact lenses, *US Pat.*, US3591329A, 1971.
52. Y. Qiu and X. Qian, Antimicrobial medical devices including silver nanoparticles, *US Pat.*, US 2008/0181931 A1, 2008.
53. S. P. Sawan, S. Subramanyam and A. Yurkovetskly, Contact-killing antimicrobial devices, *US Pat.*, US5817325A, 2000.
54. O. Rathore, D. Zanini, F. Neely, D. E. Riederer, D. Vanderlaan and A.-M. Anderson, Antimicrobial lenses displaying extended efficacy, processes to prepare them and methods of their use, *US Pat.*, US 2004/0151755 A1, 2004.
55. O. Rathore and A. Alli, Antimicrobial lenses, processes to prepare them and methods of their use, *US Pat.*, US 2008/0273168 A1, 2008.
56. B. M. Karandikar, B. I. Gibbins and K. A. Cornell, Antimicrobial silver compositions, *US Pat.*, US 2007/0003603 A1, 2007.
57. J. B. Enns, A.-M. Meyers and F. L. Neely, Antimicrobial contact lenses and methods for their production, *US Pat.*, US 2004/0213827 A1, 2004.
58. S. Mahadevan, A. Knanolkar, O. Rathore, Y. Li, G. W. Walker and T. R. Rooney, Antimicrobial polymeric articles, processes to prepare them and methods of their use, *US Pat.*, US 2008/0102122 A1, 2008.
59. G. A. Hill, K. Young, H. Chaouk and O. Rathore, Basic processes to prepare antimicrobial contact lenses, *US Pat.*, US 2008/0241225 A1, 2008.
60. F. L. Neely and A. Alli, Antimicrobial contact lenses and methods for their production, *US Pat.*, US 2005/0260249 A1, 2005.
61. D. C. Turner, A. Alli and J. D. Ford, *et al.*, Antimicrobial lenses and methods of their use, *US Pat.*, US 2003/0043341 A1, 2003.
62. A.-M. Anderson, O. Rathore and A.-M. Wong Meyers, Antimicrobial lenses, processes to prepare them and methods of their use, *US Pat.*, US 2004/0150788 A1, 2004.
63. A.-C. Nayiby, S. Brown-Skrobot, A.-M. Wong Meyers, F. Neely, B. Pall and O. Rathore, Antimicrobial lenses, processes to prepare them and methods of their use, *US Pat.*, US 2005/0117112 A1, 2005.
64. D. Zanini, A. Alli, J. Ford, R. B. Steffen, D. Vanderlaan and J. Petisce, Antimicrobial contact lenses and methods for their production, *US Pat.*, US 2003/0044447 A1, 2000.
65. D. G. Vanderlaan, A. Meyers and S. Brown-Skrobot, Antimicrobial contact lenses containing activated silver and methods for their production, *WO Pat.*, WO2002062402 A1, 2000.
66. C. Lakkis, F. Anastasopoulos, J. Slater and L. May, *Assoc. Res. Vis. Ophthalmol.*, 2011, 6477.

67. B. S. Atiyeh, M. Costagliola, S. Hayek and S. A. Dibo, *Burns*, 2007, **33**, 139–148.
68. S. C. Hau and S. J. Tuft, *Cornea*, 2009, **28**, 703–705.
69. F. R. Christ, in *Intraocular lens with antimicrobial activity*, ed. U. States, Implemed Inc, United States, 1998.
70. J. L. Hobman and L. C. Crossman, *J. Med. Microbiol.*, 2014, **64**, 471–497.
71. J. E. Spallholz, *Free Radical Biol. Med.*, 1994, **17**, 45–64.
72. T. W. Reid, P. Tran and C. Jarvis, *et al.*, *Invest. Ophthalmol. Visual Sci.*, 2012, **53**, 6081.
73. P. Tran, A. Hamood and C. Jarvis, *et al.*, *Invest. Ophthalmol. Vis. Sci.*, 2012, **53**, 6086.
74. S. M. Mathews, J. E. Spallholz, M. J. Grimson, R. R. Dubielzig, T. Gray and T. W. Reid, *Cornea*, 2006, **25**, 806–814.
75. P. Pham, P. Tran, E. Huynh, A. Hamood, R. Hanes and T. W. Reid, *Invest. Ophthalmol. Vis. Sci.*, 2014, **55**, 4641.
76. P. L. Tran, E. Huynh and P. Pham, *et al.*, *Eye Contact Lens*, 2017, **43**, 110–115.
77. R. C. Mercer, J. Cortez, P. Tran, T. Mosley, A. Hamood and T. W. Reid, *Invest. Ophthalmol. Vis. Sci.*, 2010, **51**, 1520.
78. M. E. Jansen, D. Good, S. Sidhu, J. Kaufman, T. Mosley and T. W. Reid, *Invest. Ophthalmol. Vis. Sci.*, 2012, **53**, 3742.
79. P. Tran, A. Arnett and C. Jarvis, *et al.*, *Transl. Vis. Sci. Technol.*, 2017, **6**, 1.
80. P. Tran, A. Hamood and T. Mosley, *et al.*, *J. Dent. Res.*, 2013, **92**, 461–466.
81. M. Klein and E. G. Millwood, *Br. J. Ophthalmol.*, 1952, **36**, 679–682.
82. J. Shaw, E. F. Smith, R. U. Desai, B. Enriquez and A. Schrier, *J. Ocul. Pharmacol. Ther.*, 2010, **26**, 587–589.
83. T. P. Heyrman, M. L. McDermott, J. L. Ubels and H. F. Edelhauser, *J. Cataract Refract Surg.*, 1989, **15**, 169–175.
84. S. Eperon, L. Bossy-Nobs, I. K. Petropoulos, R. Gurny and Y. Guex-Crosier, *Int. J. Pharm.*, 2008, **352**, 240–247.
85. R. C. Siqueira, E. R. Filho and S. L. Fialho, *et al.*, *Ophthalmologica*, 2006, **220**, 338–342.
86. G. Kleinmann, D. J. Apple and J. Chew, *et al.*, *J Cataract Refract Surg.*, 2006, **32**, 1717–1721.
87. L. T. Davis, N. Kumar and L. M. Nijm, *et al.*, *J. Chromatogr. B Analyt. Technol. Biomed. Life Sci.*, 2010, **878**, 2421–2426.
88. S. Manju and S. Kunnatheeri, *Pharm. Dev. Technol.*, 2010, **15**, 379–385.
89. N. N. Kruk, B. M. Dzhagarov, V. A. Galievsky, V. S. Chirvony and P. Y. Turpin, *J. Photochem. Photobiol., B*, 1998, **42**, 181–190.
90. C. L. Schepens and F. Acosta, *Surv. Ophthalmol.*, 1991, **35**, 447–453.
91. M. F. Refojo, F. L. Leong, I. M. Chan and F. I. Tolentino, *Retina*, 1983, **3**, 45–49.
92. F. M. V. T. Thakkar, T. G. Soni, M. C. Gohel and T. R. Gandhi, *Int. J. Pharm. Sci. Res.*, 2009, **1**, 1–14.

93. K. Ranch, H. Patel, L. Chavda, A. Koli, F. Maulvi and R. K. Parikh, *J. Appl. Pharm. Sci.*, 2017, 7, 101–108.
94. A. Hui, M. Willcox and L. Jones, *Invest. ophthalmol. Vis. Sci.*, 2014, **55**, 4896–4904.
95. D. Nguyen, A. Hui and A. Weeks, *et al.*, *Materials*, 2012, **5**, 684–698.
96. J. B. Ciolino, C. H. Dohlman and D. S. Kohane, *Semin. Ophthalmol.*, 2009, **24**, 156–160.
97. C. J. White, A. Tieppo and M. E. Byrne, *J. Drug Delivery Sci. Technol.*, 2011, **21**, 369–384.
98. L. Xinming, C. Yingde and A. W. Lloyd, *et al.*, *Contact Lens Anterior Eye*, 2008, **31**, 57–64.
99. L. C. Bengani, K. H. Hsu, S. Gause and A. Chauhan, *Expert Opin. Drug Delivery*, 2013, **10**, 1483–1496.
100. M. R. Jain and S. Lal, *Indian J. Ophthalmol.*, 1983, **31**, 645–647.
101. A. Hui, A. Boone and L. Jones, *Eye Contact Lens*, 2008, **34**, 266–271.
102. M. Bajgrowicz, C. M. Phan, L. N. Subbaraman and L. Jones, *Invest. Ophthalmol. Vis. Sci.*, 2015, **56**, 2234–2242.
103. A. Hui, H. Sheardown and L. Jones, *Materials*, 2012, **5**, 85–107.
104. C. M. Phan, L. N. Subbaraman and L. Jones, *Eye Contact Lens*, 2013, **39**, 162–168.
105. C. Alvarez-Lorenzo, F. Yanez, R. Barreiro-Iglesias and A. Concheiro, *J. Controlled Release*, 2006, **113**, 236–244.
106. A. Danion, I. Arsenault and P. Vermette, *J. Pharm. Sci.*, 2007, **96**, 2350–2363.
107. J. B. Ciolino, T. R. Hoare and N. G. Iwata, *et al.*, *Invest. Ophthalmol. Vis. Sci.*, 2009, **50**, 3346–3352.
108. J. B. Ciolino, S. P. Hudson and A. N. Mobbs, *et al.*, *Invest. Ophthalmol. Vis. Sci.*, 2011, **52**, 6286–6291.
109. C. C. Peng, J. Kim and A. Chauhan, *Biomaterials*, 2010, **31**, 4032–4047.
110. P. Nordmann, T. Naas and L. Poirel, *Emerg. Infect. Dis.*, 2011, **17**, 1791–1798.
111. M. D. Willcox, *Clin. Experimental Optom.*, 2011, **94**, 161–168.
112. M. D. Willcox, *Clin. Ophthalmol.*, 2012, **6**, 919–924.
113. P. Lalitha, M. Srinivasan and P. Manikandan, *et al.*, *Clin. Infect. Dis.*, 2012, **54**, 1381–1387.
114. D. Subedi, A. K. Vijay and M. Willcox, *Clin. Exp. Optom.*, 2018, **101**, 162–171.
115. T. Ganz, *Integr. Comp. Biol.*, 2003, **43**, 300–304.
116. S. S. Kolar and A. M. McDermott, *Cell. Mol. Life Sci.*, 2011, **68**, 2201–2213.
117. A. M. McDermott, *Ophthalmol. Res.*, 2009, **41**, 60–75.
118. A. M. McDermott, *Ocul. Surf.*, 2004, **2**, 229–247.
119. F. Harris, S. R. Dennison and D. A. Phoenix, *Curr. Protein Pept. Sci.*, 2009, **10**, 585–606.
120. M. Zasloff, *Nature*, 2002, **415**, 389–395.
121. M. G. Scott and R. E. Hancock, *Crit. Rev. Immunol.*, 2000, **20**, 407–431.

122. J. A. Hoffmann, F. C. Kafatos, C. A. Janeway and R. A. Ezekowitz, *Science*, 1999, **284**, 1313–1318.
123. R. E. Hancock, *Lancet Infect. Dis.*, 2001, **1**, 156–164.
124. K. Hamamoto, Y. Kida, Y. Zhang, T. Shimizu and K. Kuwano, *Microbiol. Immunol.*, 2002, **46**, 741–749.
125. T. Murakami, M. Niwa, F. Tokunaga, T. Miyata and S. Iwanaga, *Chemotherapy*, 1991, **37**, 327–334.
126. M. Masuda, H. Nakashima and T. Ueda, *et al.*, *Biochem. Biophys. Res. Commun.*, 1992, **189**, 845–850.
127. D. Dutta, N. Cole, N. Kumar and M. D. Willcox, *Invest. Ophthalmol. Vis. Sci.*, 2013, **54**, 175–182.
128. F. Costa, I. F. Carvalho, R. C. Montelaro, P. Gomes and M. C. Martins, *Acta Biomater.*, 2011, **7**, 1431–1440.
129. R. E. Hancock and H. G. Sahl, *Nat. Biotechnol.*, 2006, **24**, 1551–1557.
130. R. Rasul, N. Cole, D. Balasubramanian, R. Chen, N. Kumar and M. D. Willcox, *Int. J. Antimicrob. Agents*, 2010, **35**, 566–572.
131. L. T. Nguyen, E. F. Haney and H. J. Vogel, *Trends Biotechnol.*, 2011, **29**, 464–472.
132. L. T. Nguyen, E. F. Haney and H. J. Vogel, *Trends Biotechnol.*, 2011, **29**, 464–472.
133. R. Rasul, *School of Optometry and Vision Science*, The University of New South Wales, Sydney, Australia, 2010.
134. D. Dutta, N. Cole, N. Kumar and M. D. Willcox, *Invest. Ophthalmol. Vis. Sci.*, 2013, **54**, 175–182.
135. D. Dutta, N. Kumar and M. Willcox, *Biofouling*, 2016, **32**, 429–438.
136. R. Chen, N. Cole, D. Dutta, N. Kumar and M. D. Willcox, *J. Biomed. Mater. Res. B Appl. Biomater.*, 2016, **105**(8), 2612–2617.
137. D. I. Andersson and D. Hughes, *FEMS Microbiol. Rev.*, 2011, **35**, 901–911.
138. A. McDermott, C. M. Santos, S. S. Kolar, A. Kumar and C. Cai, *ARVO. Fort Lauderdale: Invest. Ophthalmol. Vis. Sci.*, 2010, **51**, 3433.
139. S. S. Kolar, A. Kumar, C. Cai and A. McDermott, *ARVO. Fort Lauderdale: Invest. Ophthalmol. Vis. Sci.*, 2010, **53**, 3333.
140. M. L. Mangoni, V. Luca and A. M. McDermott, *Peptides*, 2015, **71**, 286–295.
141. B. Casciaro, D. Dutta and M. R. Loffredo, *et al.*, *Biopolymers*, 2017, DOI: 10.1002/bip.23074.
142. A. Aspedon and E. A. Groisman, *Microbiology*, 1996, **142**(Pt 12), 3389–3397.
143. C. Johansen, T. Gill and L. Gram, *J. Appl. Bacteriol.*, 1995, **78**, 297–303.
144. M. D. Willcox, E. Hume, Y. Aliwarga, N. Kumar and N. Cole, *J. Appl. Microbiol.*, 2008, **105**, 1817–1825.
145. Y. Aliwarga, E. B. H. Hume, J. Lan and M. D. Willcox, *Clin. Exp. Ophthalmol.*, 2001, **29**, 157–160.
146. D. Dutta, J. Ozkan and M. D. Willcox, *Optom. Vis. Sci.*, 2014, **91**, 570–581.
147. D. Dutta, B. Kamphuis and B. Ozcelik, *et al.*, *Optom. Vis. Sci.*, 2018.

148. N. Cole, E. B. Hume, A. K. Vijay, P. Sankaridurg, N. Kumar and M. D. Willcox, *Invest. Ophthalmol. Vis. Sci.*, 2010, **51**, 390–395.
149. D. Dutta, A. K. Vijay, N. Kumar and M. D. Willcox, *Invest. Ophthalmol. Vis. Sci.*, 2016, **57**, 5616–5624.
150. D. Dutta, T. Zhao, K. B. Cheah, L. Holmlund and M. D. P. Willcox, *Contact Lens Anterior Eye*, 2017, **40**, 175–183.
151. D. Dutta, P. Kalaiselvan, N. Konda, *et al.*, *American Academy of Optometry*, San Antonio, USA, 2018.
152. P. Kalaiselvan, D. Dutta, S. Sharma and M. D. Willcox, *Invest. Ophthalmol. Vis. Sci.*, 2018, **59**, 1766.
153. L. O. Brandenburg, J. Merres, L. J. Albrecht, D. Varoga and T. Pufe, *Polymers-Basel*, 2012, **4**, 539–560.
154. V. I. Band and D. S. Weiss, *Antibiotics*, 2015, **4**, 18–41.
155. S. A. Loutet and M. A. Valvano, *Front. Microbiol.*, 2011, **2**, 159.
156. N. Jochumsen, R. L. Marvig and S. Damkiaer, *et al.*, *Nat. Commun.*, 2016, **7**, 13002.
157. M. B. Miller and B. L. Bassler, *Annu. Rev. Microbiol.*, 2001, **55**, 165–199.
158. R. M. Donlan, *Emerg. Infect. Dis.*, 2002, **8**, 881–890.
159. M. P. DeLisa and W. E. Bentley, *Microb. Cell Fact.*, 2002, **1**, 5.
160. H. Zhu, S. J. Thuruthyil and M. D. Willcox, *J. Med. Microbiol.*, 2002, **51**, 1063–1070.
161. D. G. Davies, M. R. Parsek, J. P. Pearson, B. H. Iglewski, J. W. Costerton and E. P. Greenberg, *Science*, 1998, **280**, 295–298.
162. H. Zhu, R. Bandara and T. C. Conibear, *et al.*, *Invest. Ophthalmol. Vis. Sci.*, 2004, **45**, 1897–1903.
163. R. Maximilien, R. d. Nys and C. Holmstrom, *et al.*, *Aquat. Microbial Ecol.*, 1998, **15**, 233–246.
164. M. Hentzer and M. Givskov, *J. Clin. Invest.*, 2003, **112**, 1300–1307.
165. M. Givskov, R. de Nys and M. Manefield, *et al.*, *J. Bacteriol.*, 1996, **178**, 6618–6622.
166. T. B. Rasmussen, M. Manefield and J. B. Andersen, *et al.*, *Microbiology*, 2000, **146**, 3237–3244.
167. D. Ren, J. J. Sims and T. K. Wood, *Environ. Microbiol.*, 2001, **3**, 731–736.
168. D. Ren, J. J. Sims and T. K. Wood, *Lett. Appl. Microbiol.*, 2002, **34**, 293–299.
169. J. Lonn-Stensrud, F. C. Petersen, T. Benneche and A. A. Scheie, *Oral Microbiol. Immunol.*, 2007, **22**, 340–346.
170. M. Manefield, T. B. Rasmussen and M. Henzter, *et al.*, *Microbiology*, 2002, **148**, 1119–1127.
171. H. Wu, Z. Song and M. Hentzer, *et al.*, *J. Antimicrob. Chemother.*, 2004, **53**, 1054–1061.
172. S. Reverchon, B. Chantegrel, C. Deshayes, A. Doutheau and N. Cotte-Pattat, *Bioorg. Med. Chem. Lett.*, 2002, **12**, 1153–1157.
173. E. A. Yates, B. Philipp and C. Buckley, *et al.*, *Infect. Immun.*, 2002, **70**, 5635–5646.
174. J. T. Byers, C. Lucas, G. P. C. Salmond and M. Welch, *J. Bacteriol.*, 2002, **184**, 1163–1171.

175. E. B. H. Hume, J. Baveja and B. Muir, *et al.*, *Biomaterials*, 2004, **25**, 5023–5030.
176. S. A. Al-Bataineh, L. G. Britcher and H. J. Griesser, *Surf. Sci.*, 2006, **600**, 952–962.
177. S. A. Al-Bataineh, R. Luginbuehl, M. Textor and M. Yan, *Langmuir*, 2009, **25**, 7432–7437.
178. H. Zhu, A. Kumar and J. Ozkan, *et al.*, *Optom. Vis. Sci.*, 2008, **85**, 292–300.
179. H. Zhu, A. Kumar and J. Ozkan, *et al.*, *Optom. Vis. Sci.*, 2008, **85**, 292–300.
180. B. M. K. Bandara, P. R. Sankaridurg and M. D. P. Willcox, *Curr. Eye Res.*, 2004, **29**, 245–251.
181. E. Muller, S. Takeda, H. Shiro, D. Goldmann and G. B. Pier, *J. Infect. Dis.*, 1993, **168**, 1211–1218.
182. T. K. Beattie, A. Tomlinson, D. V. Seal and A. K. McFadyen, *Optom. Vis. Sci.*, 2011, **88**, 1422–1432.
183. B. F. Farber, H. C. Hsieh, E. D. Donnenfeld, H. D. Perry, A. Epstein and A. Wolff, *Ophthalmology*, 1995, **102**, 831–836.
184. A. Tomlinson, P. A. Simmons, D. V. Seal and A. K. McFadyen, *Ophthalmology*, 2000, **107**, 112–117.
185. T. K. Beattie and A. Tomlinson, *Eye Contact Lens*, 2009, **35**, 316–319.
186. J. C. Venter, *Pharmacol. Rev.*, 1982, **34**, 153–187.
187. N. M. Milovic, J. Wang, K. Lewis and A. M. Klibanov, *Biotechnol. Bioeng.*, 2005, **90**, 715–722.
188. I. Behlau, K. Mukherjee and A. Todani, *et al.*, *Biomaterials*, 2011, **32**, 8783–8796.

CHAPTER 9

Metal-based Antimicrobials

N. K. MONYCH, N. GUGALA AND R. J. TURNER*

Department of Biological Sciences, University of Calgary, 2500 University Dr. NW, Calgary, Alberta T2N 1N4, Canada
*Email: turnerr@ucalgary.ca

9.1 Background and History of Metal-based Antimicrobials

9.1.1 Antibiotic Resistance Era

The introduction of antibiotics originated from the discovery of penicillin by Sir Alexander Fleming in 1928, thus fueling a new medical era.[1] Antibiotics were once regarded as miracle drugs, capable of curing illnesses and infections that would have previously been considered death sentences for many.[2] Despite this, by 1945, Fleming himself addressed concern regarding the overuse and abuse of antibiotics. However, his apprehensions were largely ignored and antibiotics were prescribed extensively, a trend that continues to this day. It is now widely accepted that a strong correlation exists between the overuse of antibiotics in agriculture and medicine and the development of antibiotic resistance in microorganisms.[3–6] Antibiotic-resistant bacteria were identified as early as the 1940s, only a few years after the initial discovery of penicillin.[7] At the time, this was not seen as a major cause for concern, due to the further modifications being made to beta-lactam antibiotics.[8] These new developments briefly slowed the antibiotic resistance mechanisms that were acquired or evolved by microbes, such as the production of lactamases.[1,3] Despite this, a continuous cycle

Biomaterials Science Series No. 5
Antimicrobial Materials for Biomedical Applications
Edited by Abraham J. Domb, Konda Reddy Kunduru and Shady Farah

Published by the Royal Society of Chemistry, www.rsc.org

developed in which the discovery of a novel antibiotic was followed by the progression of resistance. This was a recurring theme, particularly during the 1940s to 1970s,[6] a period marked by the identification of numerous novel classes of antibiotics. After this time period, no new classes were discovered, until recently. Instead, as a means of combating resistance, we relied on chemical modifications to existing classes.[6,9] Nonetheless, this strategy is now in jeopardy as resistance now exists for all classes of antibiotics currently available, and new novel antibiotics are simply not being developed by the pharmaceutical industry, or are in short supply.[3]

There is no doubt that new approaches are required to address this concern. Increased awareness regarding the proper use of antibiotics is one of the first steps. Another vital aspect is the development of antimicrobials that are effective against a large variety of pathogenic bacteria, considered broad-spectrum, displaying multiple cellular targets.[10] Different approaches have been taken to solve this issue, such as searching within bodies of water and soil sediments for microbes that produce novel antibiotics.[11] Other groups have begun developing methods to slow the progression of resistance by coupling existing antibiotics with other antimicrobial agents.[12]

In this chapter, we present the use of metal antimicrobial agents as enhancements to the current antimicrobial regime, a practice that is very quickly gaining popularity. A variety of metals have shown antimicrobial efficacy against multiple strains of bacteria, even those that display multidrug resistance.[13,14] Furthermore, metals generally have multiple targets within a cell, a characteristic that may be ideal for opposing resistance and increasing the effectiveness of an antimicrobial agent.[10,15,16]

9.1.2 Metals and Their Biological Importance

Metals can be defined based on their physical properties. The majority of metals are malleable in addition to being efficient conductors of heat and electricity. Metals will generally donate their electrons to form cationic species in solutions. Metalloids are intermediate species that are physically more similar to non-metals, but under certain circumstances, metalloids can conduct electricity and are considered semi-conductors. Both metals and metalloids can be further distinguished by their essentiality to biotic life. Those that are termed essential are necessary for an organism's survival, as they participate intimately in the biochemical processes. Some essential metals include sodium (Na), potassium (K), magnesium (Mg), calcium (Ca), molybdenum (Mo), manganese (Mn), iron (Fe), cobalt (Co), copper (Cu), zinc (Zn)[17] and other rare exceptions. Essential metals are used in various cellular processes and often required for proper enzymatic function and cell signalling.[17] Non-essential metals are not essential for biotic life, but may still play a role in specialized cellular processes that are organism-dependent.[17] Many non-essential metals are toxic at very low concentrations, such as arsenic (As) and mercury (Hg).[17] In fact, while essential metals are necessary for biological processes, any metal is toxic at elevated concentrations and/or a specific speciation state. However, the opposite is also true: deficiencies in certain metals can also lead to impaired biological function.[17] Certain metals are able to mimic the coordination chemistry of essential metals, such as iron (Fe) or zinc (Zn), yet these metals are often unable to perform the same function. Lead (Pb) provides an example of this, as it is able to replace multiple divalent and monovalent cations, thus perturbing the cell's natural homeostasis.[18] Metals that have multiple oxidation states and are redox active are also capable of changing the oxidation state of the biomolecules within the cell, potentially resulting in the development of harmful reactive oxygen/nitrogen species (RONS).[18] RONS have been demonstrated, mostly *in vitro*, to cause DNA damage, lipid peroxidation and damage to cellular membranes and proteins.[19] As a result of the characteristics described, metal-based antimicrobials (MBAs) have been used for millennia as effective antimicrobial agents.[15,16]

9.1.3 A Brief History of Metal-based Antimicrobials

Although metals are being suggested as a 'new' means of combating the threat of antibiotic resistance, they have in fact been used for millennia as antimicrobial agents. Copper is thought to be one of the oldest metals used, first recorded in an ancient Egyptian medical text that dates back to 2600–2200 BCE.[20] This text describes the use of copper by the Egyptains to sterilize water and chest wounds.[20] Silver is also a metal with ancient claims, dating back to the period of Alexander the Great, in which water vessels were coated in silver as a means of maintaining water sterility during long military campaigns.[21] Over the centuries, various civilizations have taken advantage of ancestral and neighbouring knowledge regarding the use of metals as antimicrobials, including great civilizations such as the Phoenicians, Greeks, Romans and Egyptians.[21] In India, the use of metals was also practiced, particularly for the treatment of diseases and infections, as recorded in Ayurvedic medical texts from 400–200 BCE.[22] The use of gold (Au), silver (Ag), iron (Fe), copper (Cu), lead (Pb), tin (Sn) and mercury (Hg) is referenced for various medical ailments, including the treatment of eye disorders, respiratory disorders, abscesses and sexually transmitted diseases, among others.[22]

Over the past 200 years metals have been used broadly in medicine. During the 19th and 20th centuries, silver and copper were used extensively for the treatment of wounds and diseases such as syphilis and lupus.[20,21] Other 'heavy metals', which are defined as dense elements with elevated toxicity, such as mercury and arsenic, were also used to cure diseases like syphilis, malaria and skin disorders.[15] The use of metals for medicinal purposes saw a drastic decline in the early 20th century, coinciding with the discovery and use of antibiotics. Despite this, and as a result of the rise in antibiotic resistance, metals have regained popularity as antimicrobials. Silver is seeing a resurgence in use for wound dressings and eye drops,[21] among other examples. Copper is suggested for use as a non-porous surface in hospitals due to its ability to kill harmful bacteria upon contact.[20] Furthermore, there is a continuous increase in research geared towards developing new formulations of metal–metal, metal–nanomaterial and metal–antibiotic antimicrobials.[23–25] With the ever-expanding development of metal formulations for antimicrobial use, it is imperative that the underlying mechanism(s) of action are understood.

9.2 Mechanisms of Metal-based Antimicrobial (MBA) Toxicity to Bacteria

9.2.1 Metal Binding Affinity and Toxicity

All metals are capable of exerting toxic effects at elevated concentrations. Each metal typically displays distinct mechanisms of toxicity, which are generally reliant on speciation. However, general mechanisms of metal toxicity can also be attributed to a particular metal ion or group of metals. Unlike antibiotics, metal ions commonly have multiple targets and can affect multiple components of bacterial cells. These include proteins, membranes, DNA, pathways involved in nutrient uptake, the overall redox potential of the cell and the electron transport chain (Figure 9.1).

One of the most universal mechanisms of metal toxicity is centred upon the binding affinity of metals to various biological components of the cell. Metal ions are commonly surrounded by ligands in a specific arrangement—referred to as coordination chemistry.[26,27] Ligand field theory is able to predict the preferred coordination of a metal based on its electronic properties.[28] In biology, the most common ligands for metals can be found in amino acids of proteins. Amino acids containing sulfur, nitrogen and oxygen are common donors due to their Lewis base characteristics.[26] Still, other biological molecules, such as membrane lipids, DNA and metabolites, which contain these elements, may also participate in metal coordination. The hard–soft acid–base theory is another important factor when considering the binding capabilities of metals. The theory is based upon the Irving–Williams series ($Mn^{2+} < Fe^{2+} < Co^{2+} < Ni^{2+} < Cu^{2+} < Zn^{2+}$), which predicts the stability of divalent metals based on their ionic radius and ionization potential.[29] Compounds with both a higher oxidation state and electronegativity as well as a low polarizability and small size can generally be classified as 'hard'.[30] Compounds with a lower oxidation state and electronegativity coupled with a higher polarizability and size can generally be classified as 'soft'. Soft acids have strong interactions with soft bases, while hard bases will interact more strongly with hard acids.[30] Compounds with intermediate properties can be considered borderline and will preferentially bind with each other, although they are able to interact with hard or soft acids and bases as well. Soft acids include copper (Cu^{+}), gold (Au^{+}), silver (Ag^{+}), mercury (Hg^{+}/Hg^{2+}) and cadmium (Cd^{2+}), which have a strong binding affinity for soft bases such as phenyls, thiols, and thioethers.[30,31] Hard acids include sodium (Na^{+}), potassium (K^{+}), magnesium (Mg^{2+}), calcium (Ca^{2+}), chromium (Cr^{3+}), aluminum (Al^{3+}), gallium (Ga^{3+}), cobalt (Co^{3+}) and iron (Fe^{3+}), which bind hard bases such as carbonates, sulfates, carboxylates, nitrates, alcohols, amines, phosphates and ethers.[31] Borderline acids include copper (Cu^{2+}), zinc (Zn^{2+}), lead (Pb^{2+}), bismuth (Bi^{3+}), nickel (Ni^{2+}), cobalt (Co^{2+}) and iron (Fe^{2+}), which bind borderline bases such as anilines, imidazoles, pyridines, nitriles and azides.[31] Metals with a strong affinity for particular biomolecules, and/or in excess, can displace essential metals that are typically bound, disrupting the normal function of these components. The aforementioned theories provide the groundwork for understanding the mechanisms of metal toxicity in microbial organisms.

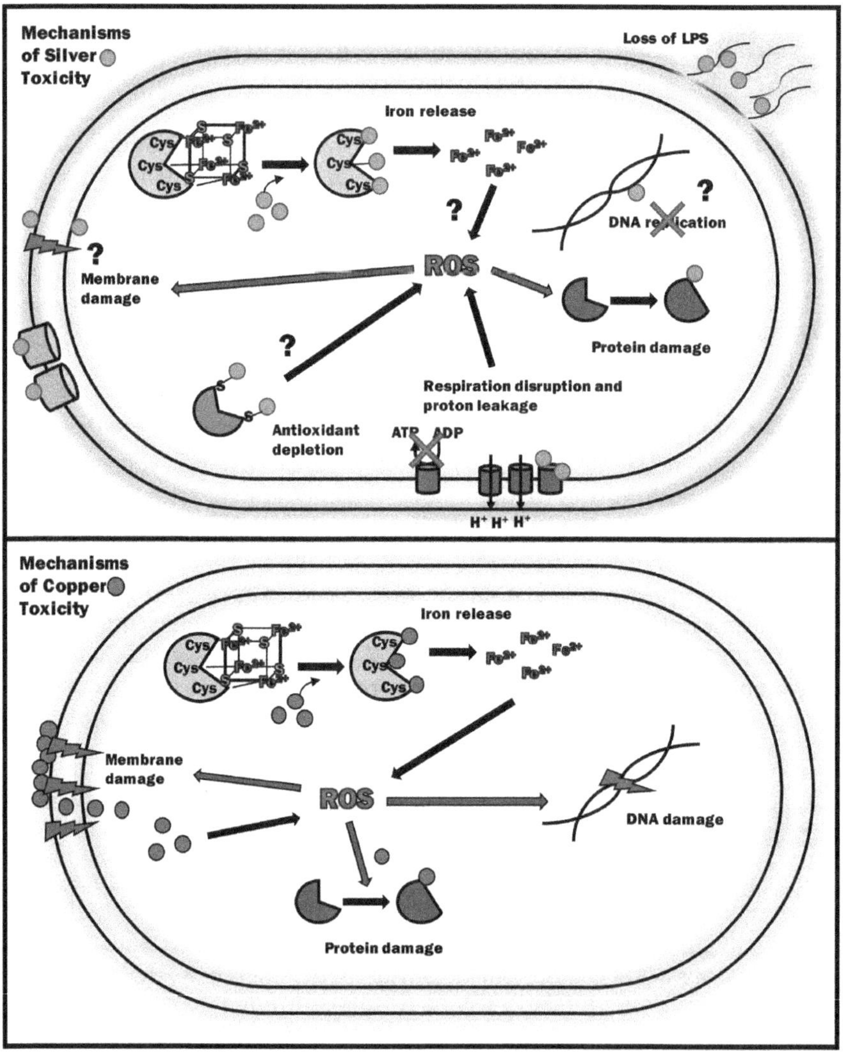

9.2.2 Reactive Oxygen Species and Oxidative Stress

Within the bacterial cell, numerous reactions are controlled by the cell's redox potential. Exclusive of a tightly controlled redox potential, certain reactions may no longer become favourable increasing the production of reactive oxygen/nitrogen species. Many metals are biologically redox active—having more than one oxidation state, and will readily participate in reduction–oxidation reactions. These metals are also capable of potentiating and propagating reactive oxygen species. The production of reactive oxygen/nitrogen/sulfur (RONS) species by metals, either directly or indirectly, is thought to be a key mechanism of metal toxicity.[15,16,32] Once RONS are produced, they are capable of causing cell-wide damage. For example, RONS are capable of disrupting iron–sulfur clusters within proteins, thus rendering them inactive. This may lead to the propagation of RONS throughout the cell, potentially causing DNA and membrane damage.

The production of RONS *via* metals commonly occurs through the catalysis of Fenton chemistry. Fenton chemistry involves the production of a hydroxyl radical $OH^{\bullet -}$ and hydroxide OH^- through the oxidation of Fe^{2+} by H_2O_2.[19] The resulting Fe^{3+} can be reduced by $O_2^{\bullet -}$ or other cellular reducing agents.[19] This net reaction, which results in $OH^{\bullet -}$, is favourable and can be propagated further.[19] While the lifespan of a radical is short-lived, the existence of these molecules, when present at sufficient concentrations, will result in damage to cellular biomolecules that are in close proximity to the source of radical production. In this way, the production of nitrogen-centered, carbon-centered, thiol and peroxyl radicals can cause damage to proteins, DNA, polysaccharides and lipids.[19] Iron is not the only metal believed to participate in Fenton reactions; however, it is one of the only

Figure 9.1 Mechanisms of silver and copper toxicity within bacteria. Silver and copper are two of the more readily available metal-based antimicrobials (MBAs) in medical settings; therefore, an understanding of how they work is imperative. One of the more common explanations for metal toxicity is due to the production of reactive oxygen/nitrogen species (RONS). Both silver and copper may bind cysteine residues within iron–sulfur (Fe–S) clusters, resulting in release of free iron into the cell leading to RONS production.[41,43] Production of RONS may then cause protein damage, lipid peroxidation and DNA damage.[19] Similarly, binding of cysteine residues (Cys) within proteins can lead to protein damage and dysfunction.[43] Both copper and silver have been observed to disrupt the membrane and lyse cells. DNA is another target for metals, although in the case of silver and copper neither have displayed mutagenicity or genotoxicity.[54,55] It was previously proposed that silver may induce a conformational change and halt DNA replication, although the data behind this have been questioned.[56,139] Therefore, the direct effect of these metals on DNA is still unclear. Silver also causes loss of lipopolysaccharides (LPS) in bacteria, although the mechanism underlying this is also unclear.[139] It is clear that both silver and copper have multiple targets within the cell; however, there are still many unanswered questions regarding their mechanism of action.

elements demonstrated to catalyze these reactions *in vivo*.[33] Copper, chromium, cobalt, vanadium and nickel have all been shown to catalyze Fenton-like reactions *in vitro*.[33] While these metals are able to generate RONS directly, others are suspected to be indirectly responsible for increased RONS.[34]

Additional mechanisms indirectly responsible for RONS production include the release of free iron, as well as interactions with antioxidant metabolites. The first involves binding of particular soft metals to thiol groups found on proteins or metabolites. The formation of iron–sulfur [Fe–S] clusters within a protein is a highly regulated, and vital, process.[35] These clusters take part in a diverse set of functions within a cell, some of which include electron transfer, transcriptional regulation, iron storage, enzyme activity, substrate binding and activation.[35] Iron present within [Fe–S] clusters can be in either an Fe^{2+} or Fe^{3+} oxidation state, which are borderline and hard acids, respectively. The presence of sulfur within the cluster allows for binding of soft metals such as copper and silver (Figure 9.1) as well as mercury, cadmium and zinc. Soft metals are able to displace iron within the cluster, resulting in its release. Liberated iron that is not quickly bound or coordinated by other biomolecules is able to participate in Fenton chemistry, as described earlier, resulting in oxidative damage.

Another indirect mechanism of oxidative stress occurs through metal interactions with metabolite antioxidants. One of the most common antioxidants found within bacteria is glutathione.[36] This antioxidant contains cysteine residues that are capable of being oxidized in order to maintain the redox state of the cell.[36] Glutathione is then regenerated through reduction by glutathione reductase.[36] Soft metals, including copper and silver, may bind the thiols on antioxidants such as glutathione[32,37] (Figure 9.1). This binding can result in the formation of an oxidized disulfide bond or coordinated thiols that are no longer able to react with radicals as a means of removing them from the cell.[32] The reduction in this antioxidant content results in greater abundance of naturally occurring RONS from cellular respiration going unchecked, leading to subsequent cell damage.[32,33] For example, silver has also been linked to disruption of the respiratory chain, although the mechanism remains unknown, which may increase RONS production[34] (Figure 9.1).

9.2.3 Proteins

Approximately half of all proteins require metals for their function, either as catalytic or structural components.[38] Binding of the proper metal to a protein is largely dependent on the placement and coordination of amino acids surrounding the binding site.[38,39] This creates a precise metal-binding pocket within the protein that preferentially binds a specific metal. The process of metal binding is accomplished *via* strict coordination chemistry that is coupled to metal chaperones and regulated by metal concentrations within the cell.[38] Nonetheless, a number of proteins have less-specific binding pockets; hence a broader range of metals can fit within these sites.

As a result, metal replacement in such binding sites may be anticipated to cause protein and enzyme dysfunction.

Specific amino acids within proteins are also vulnerable to oxidation, these include lysine, arginine, proline, threonine, leucine and histidine.[40] Not only can specific side chains be attacked by RONS, but RONS can cause protein cross-linking or the breakage of peptide bonds.[40] Redox active metals, as mentioned in Section 9.2.2, which can produce RONS, may be responsible for the aforementioned damage, either directly or indirectly. In the former case, the oxidation of amino acid residues that are in close proximity to metal binding sites has been observed.[40]

The exact number of [Fe–S] proteins is not fully known. This is mainly due to the difficulty of maintaining the integrity of these sites during the process of protein purification. However, it is well established that [Fe–S] clusters within proteins play an integral role in many cellular processes. As noted, metals with greater binding affinity to sulfur than iron may cause the release of iron into the cell, subsequently resulting in increased concentrations of RONS. Furthermore, this binding can also affect the function of [Fe–S]-containing enzymes (Figure 9.1). An example of this includes dehydratases.[41–43] The [Fe–S] cluster present in this class of enzymes is part of the catalytic centre, and disruption, *via* metal interaction, may lead to loss in activity.[43] Furthermore, the inactivation of these enzymes will likely result in reduced cell growth due to their presence in multiple metabolic pathways.[43] Of the soft metals, silver (Ag^{+}), mercury (Hg^{2+}), cadmium (Cd^{2+}), zinc (Zn^{2+}) and copper (Cu^{2+}) are all capable of targeting the [Fe–S] clusters found in dehydratases with activity both *in vitro* and *in vivo*.[41,43] The mechanism by which this occurs is independent of RONS production, since it takes place by direct interaction with the [Fe–S] cluster[43] (Figure 9.1). Preferential binding of the metal to cysteines or sulfur atoms involved in the cluster is likely the cause of this.[43]

9.2.4 Cell Membranes

The lipid cell membrane is a prime target for toxicity, as this is the initial point of interaction with the external environment.[44] Disruption of the membrane can result in cell lysis and consequently cell death. The membrane is the site of vital cellular processes, such as the electron transport chain. It is also an important barrier between the cell and its surrounding environment. This feature of the membrane ensures the correct redox potential, and electrochemical and nutrient gradients are maintained.[44] The membrane contains transporters responsible for the uptake of important nutrients and the release of excess materials in order to maintain homeostasis. The same transporters responsible for uptake of essential metals may also be used for the import of toxic metals due to their similar properties.[39]

Upon the exposure of the cell to silver (Ag^{+}), it has been shown that bacterial membranes detach from the cell wall.[45] This observation is solely based on electron microscopy images; however, these images alone are not enough to expose the underlying mechanisms that may cause detachment, if this is indeed a mechanism of action. One potential mechanism is based upon the ability of silver to induce proton leakage, subsequently disrupting the proton motive force and electron transport chain[34] (Figure 9.1). This is thought to be a result of silver's ability to bind specific cysteine-containing cellular respiratory enzymes, thereby disrupting their function and potentially increasing intracellular RONS.[34]

Copper (Cu) and cadmium (Cd) have been suggested to cause lipid peroxidation in bacteria and yeast. It has been shown that upon bacterial exposure to copper surfaces, the membrane becomes depolarized and the disruption of the outer membrane occurs[46] (Figure 9.1). This can be explained based on the ability of copper to induce RONS production, specifically hydroxyl radicals, thus causing lipid peroxidation[46] (Figure 9.1). Cadmium is thought to induce lipid peroxidation in yeast with similar consequences as copper and bacteria.[47] While lipid peroxidation is the suggested mechanism, the underlying technique to quantify lipid peroxidation has not been fully validated in bacteria.[16]

Furthermore, interactions between lipid head groups and metal ions may lead to changes in the fluidity of the membrane.[48] Due to the functional groups associated with the polar head of lipids, charged metal ions can interact and bind to the membrane surface. Metal ions of cobalt (Co), mercury (Hg) and cadmium (Cd) have been shown to influence cell membrane melting temperatures and lateral mobility of various lipid types in the membrane.[49]

9.2.5 Nutrient Uptake

Certain metals are capable of interfering with nutrient uptake and homeostasis within the cell. Gallium (Ga), often considered a 'Trojan horse' has an almost identical ionic radius to iron (Fe).[50] Hence, the potential for iron replacement by gallium is a foreseeable event. Gallium does not have the capability to be reduced like iron at biologically relevant reduction potentials.[50] This interferes with the function of proteins within the cell. Gallium has been observed to decrease the uptake of iron into the cell. It is thought to do this through the repression of *pvdS*, an important regulator activated in response to low iron.[50] The protein PvdS is a transcriptional regulator that controls genes involved in pyoverdine synthesis and other iron-response genes.[50] The mechanism by which gallium is able to disrupt *pvdS* activity is yet unknown; however, the resulting disruption of iron homeostasis is likely a component of gallium toxicity.[50]

9.2.6 DNA Damage and Mutation

As discussed previously, the production of RONS can result in DNA damage. This primarily occurs through Fenton chemistry involving iron (Fe).

Cadmium (Cd^{2+}), beryllium (Be^{2+}) and chromium (Cr^{6+}) have displayed mutagenic abilities.[51] Arsenic (As^{3+}), nickel (Ni^{2+}), cadmium (Cd^{2+}), cobalt (Co^{2+}) and lead (Pb^{2+}), when combined with DNA-damaging agents, such as UV light, also display increased mutagenicity.[52] The latter is an indication of the metal interference with DNA repair mechanisms, rather than direct contact. While there is *in vitro* evidence of DNA damage after metal exposure, the underlying mechanism of this damage remains largely unclear, with the exception of iron.[53]

Other metals with high bacterial toxicity such as copper (Cu) and silver (Ag) have not been shown to result in large mutagenic or genotoxic activities upon exposure.[54,55] However, based on the chemistry of copper and silver, it is possible that these metal ions bind to DNA in the absence of direct DNA damage (Figure 9.1). In the case of silver, it has also been suggested that DNA replication may be halted and a conformational change may occur[56] (Figure 9.1). This is highly suspect and more recent evidence indicates that although silver is capable of binding both DNA and RNA, it does not induce a significant conformational change.[57]

9.2.7 Metal Nanoparticles

The above mechanisms represent those primarily due to metal ion exposure. Recent literature has focused on generating new and improved metal formulations and compounds that display enhanced antimicrobial activities. In particular, the production and use of metal nanoparticles has gained considerable traction. Nanoparticles, in general, are structures composed of connected atoms and range from 10 to 100 nm in size. Nanomaterials exhibit unique features when compared to their bulk counterparts, mainly caused by their nanometer dimensions. Since metal nanoparticles have only recently been suggested for their antimicrobial properties, many of their mechanisms remain a mystery. Nanostructures feature a high surface area-to-volume ratio: the surface area is much larger in comparison to their size. Given this, the nature of the interaction between the cell and nanostructures is different than that of the free ion. It is hypothesized that interactions between nanosized formulations and bacterial cells lead to the activation of a broad range of antimicrobial mechanisms, therefore improving the antimicrobial efficacy of these nanomaterials. Furthermore, the antimicrobial activity of nanostructures can be influenced by other chemical–physical parameters, such as their shape, chemical modification (conjugation to organic antibiotics) or surface coatings.

In the past decade, metal-based nanoparticles, such as silver (Ag), gold (Au), titanium oxide (TiO_2), copper oxide (CuO) and zinc oxide (ZnO), have been investigated for their antimicrobial efficacy.[58] Other elemental formulations including selenium have also been recently explored.[59] Silver nanoparticles have seen the most traction.[60] One of the most probable mechanisms of toxicity is the production of RONS, which is linked to the presence of pro-oxidant groups and active redox cycling along the particle's plasmon surface, and particle–cell interactions.[61] Due to their small size and large surface area, large electrostatic interactions between positive groups on their surfaces and negative groups found on membrane lipids may take place, leading to membrane permeabilization.[62] It is through this interaction that nanoparticles are able to enter the cell, leading to the disruption or loss of the cell's chemiosmotic potential.[62]

Nanoparticles composed of transition metals, in the presence of light, are capable of photo-killing. Here, titanium nanomaterials are gaining attention. The light generates and releases electrons that are subsequently used by the metal oxide nanoparticles to generate the superoxide radicals.[62] In turn, this results in damage, such as changes in cell membrane properties, calcium permeability and protein and DNA damage.[62]

The aforementioned mechanisms primarily refer to mechanisms that are based on the structure rather than composition of the nanoparticle. However, an important aspect of nanoparticle toxicity is the release of metal ions. Dependent on metal composition, the mechanisms of toxicity may be similar to those mentioned for metal ions. This includes protein damage, genotoxicity and membrane damage due to metal ions. In addition, primary toxicity may be either completely or partially due to the metal ions it contains. In the case of zinc oxide (ZnO) nanoparticles, the primary cause of toxicity is likely due to the release of zinc (Zn^{2+}). However, in the case of copper oxide (CuO) nanoparticles, release of copper (Cu^{2+}) plays only a partial role in toxicity.[63] Interestingly, for iron oxide (Fe_2O_3), cobalt oxide (Co_3O_4), chromium oxide (Cr_2O_3) and nickel oxide (NiO), the metal ions themselves had no impact on toxicity.[63]

Research into MBA nanomaterials is presently undergoing a revolution. Many of these studies evaluate the physicochemical characteristics of the nanomaterial, while also presenting the antimicrobial activity. For application approval, some simple tissue culture and hemolysis data is presented evaluating acute exposure. However, very little has been done to investigate their actual mechanisms of antimicrobial action or their chronic effects in higher organisms.

9.3 Current Applications of Metal-based Antimicrobials

The progression of antimicrobial resistance has warranted the production of alternative antimicrobial agents,[64] such as metals. However, the accompanying increase in metal resistance displayed by microbial organisms has not hindered this movement.[65] In fact, the development and production of novel metal-based antimicrobials (MBAs) has seen an exponential increase in the past several decades, a trend that is displaying no signs of reduction. There is a great deal of documentation regarding the efficacy of MBAs, or lack thereof. Regardless of mixed data and the accompanying conclusions, many of these studies have led to the development of products for consumer use. In general, MBAs can be found in a wide range of products currently being investigated, with some now available for consumer use (Table 9.1). Whereas the use of MBAs is a practice that stretches back thousands of years, we are just now beginning to tap into the endless possibility of MBAs in the form of additives, formulations and devices.

The prevalence of MBAs is not limited to healthcare settings, in fact, consumers can purchase many products in-store and online. Examples include clothing (http://info.lululemon.com/design/fabrics-technology/silverescent), deodorant (www.niveamen.in/products/SILVER-PROTECT) and antibacterial glass (www.agc-glass.eu). Additionally, coating services for a range of products, from flooring to kitchen utensils and food storage containers are offered (www.biocote.com, www.silverclear.ca). Furthermore, medical devices can be coated in silver (http://coatings2go.com) and copper (www.antimicrobialcopper.org/uk).

9.4 Consequences of Using Metal-based Antimicrobials

9.4.1 Bacterial Resistance

Resistance to MBAs must be considered when exploring their potential for use. Bacteria are highly adaptive, and coupled with fast replication rates, this makes for the efficient development of resistance under pressure. Similar mechanisms of resistance deployed in regard to antibiotics are also used in the presence of metals. In fact, the natural presence of metals in the environment has resulted in innate immunity, at least at 'environmental concentrations', for microbes within the immediate area.[137] The most common type of metal resistance mechanism used by bacteria are metal efflux transporters (Figure 9.2). The genes for these are often part of an operon coding for metal chelator molecules or proteins. Some include multiple transporters to create a full efflux system that often stretches from the cytoplasm to the outside of the cell.[15] Efflux transporters have been found on plasmids, making them easily transferrable through horizontal gene transfer, and on their chromosomes.[15] In some cases, similar genes are found on both chromosomal and plasmid DNA.[15] For example, the efflux system CusCFBA confers resistance to copper as well as silver. It is a multicomponent system which includes CusCBA, a resistance-nodulation-cell division (RND)-type tripartite pump, along with CusF, a metallochaperone.[15] CusF is found within the periplasm of several Gram-negative bacteria, preventing the uptake of copper to the cytoplasm.[138] Whereas CusCFBA is encoded on bacterial chromosomal DNA, a secondary copper-resistance system, Pco, is found on plasmid DNA and contains proteins with similar functions to the Cus system. Another RND-type pump system, CzcCBA, has broader specificity and is capable of effluxing cobalt (Co^{2+}), zinc (Zn^{2+}), nickel (Ni^{2+}) and cadmium (Cd^{2+}).[138]

Table 9.1 Metal-based antimicrobials under investigation and currently available for consumer and healthcare use.

Application	Description	Examples
Additives and formulations	Metals are currently being dosed into standing products to prevent bacterial growth and deliver antimicrobial properties. Here, the entirety of the formulation commonly contains evenly distributed metal ions that may be released upon contact	Ag-doped hydroxyapatite[69] Bi-doped calcium phosphate for root canal filling[70] Ga-doped phosphate glass[71] and other phosphocalcic compounds[72] Wound dressings combined with Ag,[65,73–76] and Hydrofiber® [77] Cu,[78–80] Ag[81] and Ag–cellulose fiber[82] biocomposites for wound healing Ag,[83] Ag–lactoferrin/xylitol,[84] Ag–amino acid,[85] Cu–chitosan polyethylene glycol[86] hydrogels Ag[87] and Cu[88] polymers Textiles,[89] such as socks[90–92] and respiratory face masks[93] impregnated with Cu
	Additives and formulations used in healthcare settings are designed to be stable and compatible with the host.[66] Since metals such as Ag and Cu, are regarded as safe for consumption (in reasonably low doses) they are now more common than ever, being combined with existing antimicrobials and antibiotics to produce additive and even synergistic affects, enhancing existing antimicrobial properties	Sulfadiazine and Ag–chlorhexidine formulations[94–97] Beta-lactam antibiotic combined with Ag-nanoparticles[98] Bismuth–norfloxacin[99] and –tobramycin formulations[100] Bismuth subsalicylate/salts[99] Cu combined with quaternary ammonium cations[101] and other binder components[102] Gallium-maltolate[103] and -desferrioxamine[104] combinations Chitosan–Zn complexes[105,106]
	Metal-based nanostructures (loosely defined as having at least one dimension between 0.1 nm and 100 nm) can be produced as simple or composite assemblies for antimicrobial use.[67] Recent advancements in nanotechnology have permitted the production of novel combinations with purposeful properties that allow for particular applications[68]	Nanoparticles of Zn-oxide,[107] Ag,[68,107–114] Cu,[115,116] and Au[107] Antimicrobial nanofiber mats produced from Ag-nanoparticles[117] Ag-nanoparticles on textile fabrics[118] Antimicrobial polymers with various metal nanoparticles[119]

Table 9.1 (*Continued*)

Application	Description	Examples
Coatings and surfaces	Metals are commonly found as coatings on the surfaces of a variety of products. Here, metal ions are not evenly distributed throughout the entirety of a product. Additionally, the main mechanism of toxicity may be a result of contact killing, in which metals ions generally remain bound to the surface and/or may be locally released	Ti surfaces for medical[120] and dental implants,[121] as well as endotracheal tubes coated with TiO_2[122] Ag-nano coatings on human dentine,[123] plastic catheters[124] Cu alloys against multidrug-resistant nosocomial pathogens,[125] *Escherichia coli* O157[126] and for the prevention of osteomyelitis[127] Cu surfaces for the prevention of hospital contamination[128–130] Cu-oxide coatings on non-porous solid surfaces[131] Ag-treated catheters[132–135] and endotracheal tubes[122,136]

Certain metals are vulnerable to chemical modification, rendering them either insoluble or nontoxic (Figure 9.2). The most common of these is the reduction or oxidation of the metal to the less toxic species (Figure 9.2). The oxidation of Cu^+ to Cu^{2+}, the less toxic form of copper by CueO is an example of this mechanism.[15]

Certain bacteria have evolved mechanisms of targeting and sequestration of toxic metals (Figure 9.2). If the sequestration occurs within the cell or periplasm it is generally coupled to an efflux system. For example, silver resistance can occur *via* this mechanism, in which SilE binds silver (Ag) within the periplasm during the initial stage of exposure and may transfer silver to SilF, a silver chaperone, for efflux.[15] Extracellular sequestration can also occur within the lipopolysaccharides of Gram-negative bacteria or the exo-polysaccharides of biofilms, thus binding metals prior to their entrance into the cell.[16] This mechanism provides an explanation for the enhanced resistance of microbial biofilms to MBAs and other antimicrobials.[16]

9.4.2 Responsible Use of Metal-based Antimicrobials

While the mechanisms of metal toxicity have yet to be fully elucidated, it is likely that metals disrupt bacterial cells *via* an array of mechanisms and multifactorial biochemical targets. This is unlike antibiotics, which commonly target one biomolecule. In this way, bacterial cells are not able to devote ample effort toward protecting and/or rebuilding a single biomolecule. Instead, they must delegate resources to each site of attack, diminishing the potential for specific resistance mechanisms to arise.

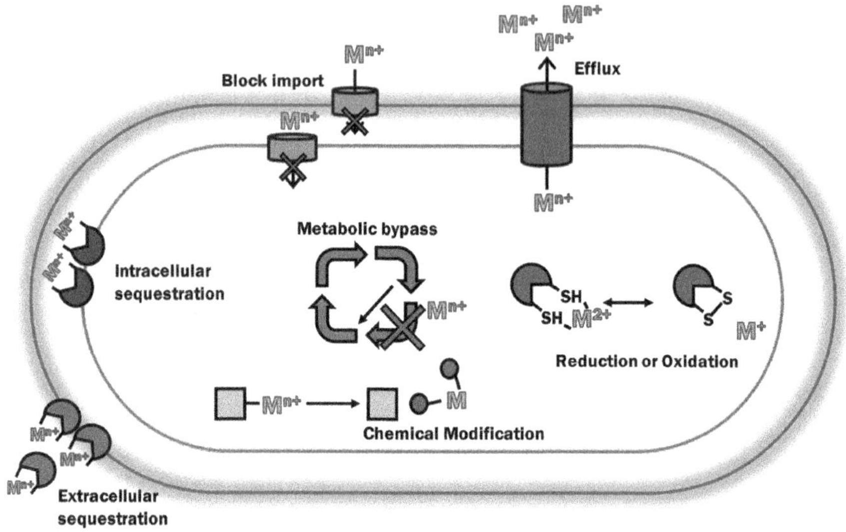

Figure 9.2 Resistance mechanisms identified for various metal-based antimicrobials (MBAs) within bacteria. Bacteria are capable of exporting various metals (M^{n+}) from the cytoplasm to outside of the cell.[15] This can occur in combination with sequestration of the metal prior to efflux.[15] Metals may also be chemically modified, reduced or oxidized into a less toxic form.[15] Preventing the metal from entrance into the cell is another common mechanism. This can be accomplished through extracellular sequestration of the metal or blocking the import mechanism by which the metal gains entrance to the cell.[15,140] Finally, metabolic processes are highly dynamic and if an individual step within a pathway is blocked due to the presence of a metal, other reactions may compensate for that step.[16,141] Not only can these resistance mechanisms occur individually, but in many cases they are coupled together as a whole resistance pathway.

Evolving metal resistance is a potential cause for concern. It is important that we do not abuse metals for their antimicrobial properties as we have done, and continue to do with antibiotics. It is highly imperative that their use is regulated, not only due to the potential for resistance, but also due to their potential eukaryotic cytotoxic effects. MBAs should be used with caution, at the correct dosage against the appropriate community of microbes. Metal or metal-coated medical implants or surfaces are likely the best deployment of MBAs. There is strong potential for use as topical treatments such as colloids and creams or impregnation in the fibres of wound bandages.

It is important to be wary of the aforementioned resistance mechanisms, most of which are carried on plasmids that may be transferrable between bacteria. In this sense, not only is it important to regulate metal usage in clinical settings, but also to determine the correct dosage of a metal to entirely kill and eradicate microbes without allowing resistant variants to arise. Although metals present a number of advantages (as well as disadvantages) over antibiotics, largely due to their broad spectrum of activity, there is still much left to be understood regarding the mechanisms of toxicity and resistance displayed by these antimicrobials. A lack in this understanding will lead to misuse, decreased effectiveness and ultimately the loss efficacy of these antimicrobials, as we are currently experiencing with antibiotics.

Acknowledgements

This work was supported by a bridge grant from Canadian Institutes of Health Research as well as a discovery grant from the Natural Sciences and Engineering Research Council of Canada to RJT.

References

1. J. Davies and D. Davies, Origins and Evolution of Antibiotic Resistance, *Microbiol. Mol. Biol. Rev.*, 2010, **74**, 417–433.
2. A. J. Alanis, Resistance to antibiotics: Are we in the post-antibiotic era?, *Arch. Med. Res.*, 2005, **36**, 697–705.
3. M. Onofrei, J. Hunt, J. Siemienczuk, D. R. Touchette and B. Middleton, A first step towards translating evidence into practice: Heart failure in a community practice-based research network, *Inform. Prim. Care*, 2004, **12**, 139–145.
4. L. Sun, E. Y. Klein and R. Laxminarayan, Seasonality and temporal correlation between community antibiotic use and resistance in the United States, *Clin. Infect. Dis.*, 2012, **55**, 687–694.
5. C. Llor and L. Bjerrum, Antimicrobial resistance: Risk associated with antibiotic overuse and initiatives to reduce the problem, *Ther. Adv. Drug Saf.*, 2014, **5**, 229–241.
6. R. I. Aminov, A brief history of the antibiotic era: Lessons learned and challenges for the future, *Front. Microbiol.*, 2010, **1**, 1–7.
7. H. P. Treffers, Studies on resistance to antibiotics; the action of penicillin on some Gram-positive and Gram-negative organisms and its potentiation by various inhibitors, *Yale J. Biol. Med.*, 1946, **18**, 609–623.
8. R. J. Stedman, J. R. E. Hoover, A. W. Chow, M. M. Dolan, N. M. Hall and R. J. Ferlauto, Semisynthetic penicillins. II. structure-activity studies on the 2-biphenylyl side chain 1, *J. Med. Chem.*, 1964, 7, 251–255.
9. J. P. Dirlam, B. H. Jaynes and M. R. Jefson, Antibacterial Agents, *Annu. Rep. Med. Chem.*, 1995, Geneva.
10. S. B. Singh, K. Young and L. L. Silver, What is an "ideal" antibiotic? Discovery challenges and path forward, *Biochem. Pharmacol.*, 2017, **133**, 63–73.
11. M. Kamjam, P. Sivalingam, Z. Deng and K. Hong, Deep sea *Actinomycetes* and their secondary metabolites, *Front. Microbiol.*, 2017, **8**, 760.
12. H. Chi and H. Holo, Synergistic antimicrobial activity between the broad spectrum bacteriocin Garvicin KS and Nisin, Farnesol and

Polymyxin B against Gram-positive and Gram-negative bacteria, *Curr. Microbiol.*, 2018, **75**, 272–277.

13. M. K. S. Ballo, S. Rtimi, S. Mancini, J. Kiwi, C. Pulgarin, J. M. Entenza and A. Bizzini, Bactericidal activity and mechanism of action of copper-sputtered flexible surfaces against multidrug-resistant pathogens, *Appl. Microbiol. Biotechnol.*, 2016, **100**, 5945–5953.
14. S. L. Percival, J. Thomas, S. Linton, T. Okel, L. Corum and W. Slone, The antimicrobial efficacy of silver on antibiotic-resistant bacteria isolated from burn wounds, *Int. Wound J.*, 2012, **9**, 488–493.
15. J. L. Hobman and L. C. Crossman, Bacterial antimicrobial metal ion resistance, *J. Med. Microbiol.*, 2015, **64**, 471–497.
16. J. A. Lemire, J. J. Harrison and R. J. Turner, Antimicrobial activity of metals: Mechanisms, molecular targets and applications, *Nat. Rev. Microbiol.*, 2013, **11**, 371–384.
17. W. Maret, The metals in the biological periodic system of the elements: Concepts and conjectures, *Int. J. Mol. Sci.*, 2016, **17**, 1–8.
18. M. Jaishankar, T. Tseten, N. Anbalagan, B. B. Mathew and K. N. Beeregowda, Toxicity, mechanism and health effects of some heavy metals, *Interdiscip. Toxicol.*, 2014, **7**, 60–72.
19. J. A. Imlay, Pathways of oxidative damage, *Annu. Rev. Microbiol.*, 2003, **57**, 395–418.
20. G. Grass, C. Rensing and M. Solioz, Metallic copper as an antimicrobial surface, *Appl. Environ. Microbiol.*, 2011, **77**, 1541–1547.
21. D. J. Barillo and D. E. Marx, Silver in medicine: A brief history BC 335 to present, *Burns.*, 2014, **40**, S3–S8.
22. Galib, M. Mashru, B. Patgiri, M. Barve, C. Jagtap and P. Prajapati, Therapeutic potentials of metals in ancient India: A review through Charaka Samhita, *J. Ayurveda Integr. Med.*, 2011, **2**, 55.
23. S. T. Khan, J. Musarrat and A. A. Al-Khedhairy, Countering drug resistance, infectious diseases, and sepsis using metal and metal oxides nanoparticles: Current status, *Colloids Surf., B*, 2016, **146**, 70–83.
24. M. Y. Vaidya, A. J. McBain, J. A. Butler, C. E. Banks and K. A. Whitehead, Antimicrobial efficacy and synergy of metal ions against *Enterococcus faecium*, *Klebsiella pneumoniae* and *Acinetobacter baumannii* in alanktonic and biofilm phenotypes, *Sci. Rep.*, 2017, **7**, 5911.
25. H. Deng, D. McShan, Y. Zhang, S. S. Sinha, Z. Arslan, P. C. Ray and H. Yu, Mechanistic study of the synergistic antibacterial activity of combined silver nanoparticles and common antibiotics, *Environ. Sci. Technol.*, 2016, **50**, 8840–8848.
26. L. Rulíšek and J. Vondrášek, Coordination geometries of selected transition metal ions (Co^{2+}, Ni^{2+}, Cu^{2+}, Zn^{2+}, Cd^{2+}, and Hg^{2+}) in metalloproteins, *J. Inorg. Biochem.*, 1998, **71**, 115–127.
27. K. L. Haas and K. J. Franz, Application of metal coordination chemistry to explore and manipulate cell biology, *Chem. Rev.*, 2009, **109**, 4921–4960.
28. J. S. Griffith and L. E. Orgel, Ligand-field theory, *Q. Rev., Chem. Soc.*, 1957, **11**, 381.

29. H. Irving and R. J. P. Williams, The stability of transition-metal complexes, *J. Chem. Soc.*, 1953, **637**, 3192–3210.
30. R. G. Pearson, Hard and soft acids and bases, *J. Chem. Educ.*, 1968, **45**, 581–587.
31. T. L. Ho, H. C. Ho and L. D. Hamilton, Biochemical significance of the hard and soft acids and bases principle, *Chem.-Biol. Interact.*, 1978, **23**, 65–84.
32. S. J. Stohs and D. Bagchi, Oxidative mechanisms in the toxicity of metal ions, *Free Radical Biol. Med.*, 1995, **18**, 321–336.
33. M. Valko, H. Morris and M. Cronin, Metals, toxicity and oxidative stress, *Curr. Med. Chem.*, 2005, **12**, 1161–1208.
34. K. B. Holt and A. J. Bard, Interaction of silver(I) ions with the respiratory chain of *Escherichia coli*: An electrochemical and scanning electrochemical microscopy study of the antimicrobial mechanism of micromolar Ag^+, *Biochemistry*, 2005, **44**, 13214–13223.
35. D. C. Johnson, D. R. Dean, A. D. Smith and M. K. Johnson, Structure, function, and formation of biological iron-sulfur clusters, *Annu. Rev. Biochem.*, 2005, **74**, 247–281.
36. H. Jay Forman, H. Zhang and A. Rinna Glutathione: Overview of its protective roles, measurement, and biosynthesis, *Mol. Aspects Med.*, 2009, **30**, 1–12.
37. M. M. Jones and W. K. Vaughn, HSAB theory and acute metal ion toxicity and detoxification processes, *J. Inorg. Nucl. Chem.*, 1978, **40**, 2081–2088.
38. K. J. Waldron, J. C. Rutherford, D. Ford and N. J. Robinson, Metalloproteins and metal sensing, *Nature*, 2009, **460**, 823–830.
39. Z. Ma, F. E. Jacobsen and D. P. Giedroc, Coordination chemistry of bacterial metal transport and sensing, *Chem. Rev.*, 2009, **109**, 4644–4681.
40. E. R. Stadtman and R. L. Levine, Free radical-mediated oxidation of free amino acids and amino acid residues in proteins, *Amino Acids*, 2003, **25**, 207–218.
41. L. Macomber and J. A. Imlay, The iron–sulfur clusters of dehydratases are primary intracellular targets of copper toxicity. *Proc. Natl. Acad. Sci. U. S. A.*, 2009, **106**, 8344–8349.
42. I. L. Calderón, A. O. Elías, E. L. Fuentes, G. A. Pradenas, M. E. Castro, F. A. Arenas, J. M. Pérez and C. C. Vásquez, Tellurite-mediated disabling of [4Fe-4S] clusters of *Escherichia coli* dehydratases, *Microbiology*, 2009, **155**, 1840–1846.
43. F. F. Xu and J. A. Imlay, Silver(I), mercury(II), cadmium(II), and zinc(II) target exposed enzymic iron-sulfur clusters when they toxify *Escherichia coli*, *Appl. Environ. Microbiol.*, 2012, **78**, 3614–3621.
44. R. M. Epand, C. Walker, R. F. Epand and N. A. Magarvey, Molecular mechanisms of membrane targeting antibiotics, *Biochim. Biophys. Acta., Biomembr.*, 2016, **1858**, 980–987.
45. W. K. Jung, H. C. Koo, K. W. Kim, S. Shin, S. H. Kim and Y. H. Park, Antibacterial activity and mechanism of action of the silver ion in

Staphylococcus aureus and *Escherichia coli*, *Appl. Environ. Microbiol.*, 2008, **74**, 2171–2178.
46. S. L. Warnes, V. Caves and C. W. Keevil, Mechanism of copper surface toxicity in *Escherichia coli* O157:H7 and *Salmonella* involves immediate membrane depolarization followed by slower rate of DNA destruction which differs from that observed for Gram-positive bacteria, *Environ. Microbiol.*, 2012, **14**, 1730–1743.
47. L. M. Costa-Moreira, B. A. A. Porto, F. Haddad-Ribeiro, F. S. Martins, M. A. B. C. Menezes, C. A. Rosa and M. J. Neves, Membrane damage by lipid peroxidation retains the cadmium constraint and is not the primary cause of K^+ extrusion in yeast, *Ann. Microbiol.*, 2016, **66**, 973–979.
48. J. Umbsaar, E. Kerek and E. J. Prenner, Cobalt and nickel affect the fluidity of negatively-charged biomimetic membranes, *Chem. Phys. Lipids*, 2018, **210**, 28–37.
49. E. Kerek, M. Hassanin, W. Zhang and E. J. Prenner, Preferential binding of inorganic mercury to specific lipid classes and its competition with cadmium, *Biochim. Biophys. Acta., Biomembr.*, 2017, **1859**, 1211–1221.
50. Y. Kaneko, M. Thoendel, O. Olakanmi, B. E. Britigan and P. K. Singh, The transition metal gallium disrupts *Pseudomonas aeruginosa* iron metabolism and has antimicrobial and antibiofilm activity, *J. Clin. Invest.*, 2007, **117**, 877–888.
51. A. Arlauskas, R. S. U. Baker, A. M. Bonin, R. K. Tandon, P. T. Crisp and J. Ellis, Mutagenicity of metal ions in bacteria, *Environ. Res.*, 1985, **36**, 379–388.
52. A. Hartwig, Current aspects in metal genotoxicity, *Biometals*, 1995, **8**, 3–11.
53. C. Angelé -Martí Nez, C. Goodman and J. Brumaghim, Metal-mediated DNA damage and cell death: mechanisms, detection methods, and cellular consequences, *Metallomics*, 2014, **6**, 1358.
54. K. Asakura, H. Satoh, M. Chiba, M. Okamoto, K. Serizawa, M. Nakano and K. Omae, Genotoxicity studies of heavy metals: Lead, bismuth, indium, silver and antimony, *J. Occup. Health*, 2009, **51**, 498–512.
55. L. Macomber, C. Rensing and J. A. Imlay, Intracellular copper does not catalyze the formation of oxidative DNA damage in *Escherichia coli*, *J. Bacteriol.*, 2007, **189**, 1616–1626.
56. Q. L. Feng, J. Wu, G. Q. Chen, F. Z. Cui, T. N. Kim and J. O. Kim, A mechanistic study of the antibacterial effect of silver ions on *Escherichia coli* and *Staphylococcus aureus*, *J. Biomed. Mater. Res.*, 2000, **52**, 662–668.
57. H. Arakawa, J. F. Neault and H. A. Tajmir-Riahi, Silver(I) complexes with DNA and RNA studied by fourier transform infrared spectroscopy and capillary electrophoresis, *Biophys. J.*, 2001, **81**, 1580–1587.
58. G. V. Vimbela, S. M. Ngo, C. Fraze, L. Yang and D. A. Stout, Antibacterial properties and toxicity from metallic nanomaterials, *Int. J. Nanomed.*, 2017, **12**, 3941–3965.
59. E. Piacenza, A. Presentato, E. Zonaro, J. A. Lemire, M. Demeter, G. Vallini, R. J. Turner and S. Lampis, Antimicrobial activity of biogenically produced spherical Se-nanomaterials embedded in organic material against

Pseudomonas aeruginosa and *Staphylococcus aureus* strains on hydroxyapatite-coated surfaces, *Microb. Biotechnol.*, 2017, **10**, 804–818.
60. S. Eckhardt, P. S. Brunetto, J. Gagnon, M. Priebe, B. Giese and K. M. Fromm, Nanobio silver: Its interactions with peptides and bacteria, and its uses in medicine, *Chem. Rev.*, 2013, **113**, 4708–4754.
61. A. Manke, L. Wang and Y. Rojanasakul, Mechanisms of nanoparticle-induced oxidative stress and toxicity, *BioMed Res. Int.*, 2013, **2013**, 1–15.
62. A. Raghunath and E. Perumal, Metal oxide nanoparticles as antimicrobial agents: a promise for the future, *Int. J. Antimicrob. Agents*, 2017, **49**, 137–152.
63. D. Wang, Z. Lin, T. Wang, Z. Yao, M. Qin, S. Zheng and W. Lu, Where does the toxicity of metal oxide nanoparticles come from: The nanoparticles, the ions, or a combination of both?, *J. Hazard. Mater.*, 2016, **308**, 328–334.
64. A. K. Thabit, J. L. Crandon and D. P. Nicolau, Antimicrobial resistance: impact on clinical and economic outcomes and the need for new antimicrobials, *Expert Opin. Pharmacother.*, 2015, **16**, 159–177.
65. S. Silver, L. T. Phung and G. Silver, Silver as biocides in burn and wound dressings and bacterial resistance to silver compounds, *J. Ind. Microbiol. Biotechnol.*, 2006, **33**, 627–634.
66. M. Gibson, Pharmaceutical preformulation and formulation: A practical guide from candidate drug selection to commercial dosage form, in *Org. Process Res. Dev.*, ed. M. Gibson, Interpharm/CRC, Boca Raton, FL, 2004, 2005, 596 + xi pp. {$}298.95. ISBN.
67. M. Guzman, J. Dille and S. Godet, Synthesis and antibacterial activity of silver nanoparticles against Gram-positive and Gram-negative bacteria, *Nanomedicine*, 2012, **8**, 37–45.
68. I. Sondi and B. Salopek-Sondi, Silver nanoparticles as antimicrobial agent: A case study on *E. coli* as a model for Gram-negative bacteria, *J. Colloid Interface Sci.*, 2004, **275**, 177–182.
69. C. Shi, J. Gao, M. Wang, J. Fu, D. Wang and Y. Zhu, Ultra-trace silver-doped hydroxyapatite with non-cytotoxicity and effective antibacterial activity, *Mater. Sci. Eng.*, 2015, **C55**, 497–505.
70. F. Chen, C. Liu and Y. Mao, Bismuth-doped injectable calcium phosphate cement with improved radiopacity and potent antimicrobial activity for root canal filling, *Acta Biomater.*, 2010, **6**, 3199–3207.
71. S. P. Valappil, D. Ready, E. A. Abou Neel, D. M. Pickup, W. Chrzanowski, L. A. O'Dell, R. J. Newport, M. E. Smith, M. Wilson and J. C. Knowles, Antimicrobial gallium-doped phosphate-based glasses, *Adv. Funct. Mater.*, 2008, **18**, 732–741.
72. S. Adzila, M. Murad and I. Sopyan, Doping Metal into Calcium Phosphate Phase for Better Performance of Bone Implant Materials, *Recent Patents Mater. Sci.*, 2012, **5**(1), 18–47.
73. C. Rigo, M. Roman, I. Munivrana, V. Vindigni, B. Azzena, C. Barbante and W. R. L. Cairns, Characterization and evaluation of silver release from four different dressings used in burns care, *Burns*, 2012, **38**, 1131–1142.

74. M. Ip, S. L. Lui, V. K. M. Poon, I. Lung and A. Burd, Antimicrobial activities of silver dressings: An *in vitro* comparison, *J. Med. Microbiol.*, 2006, **55**, 59–63.
75. V. Kostenko, J. Lyczak, K. Turner and R. J. Martinuzzi, Impact of silver-containing wound dressings on bacterial biofilm viability and susceptibility to antibiotics during prolonged treatment, *Antimicrob. Agents Chemother.*, 2010, **54**, 5120–5131.
76. S. Bergin and P. Wraight, Silver based wound dressings and topical agents for treating diabetic foot ulcers, *Cochrane Database Syst. Rev.*, 2006, CD005082.
77. P. G. Bowler, S. A. Jones, M. Walker and D. Parsons, Microbicidal properties of a silver-containing hydrofiber dressing against a variety of burn wound pathogens, *J. Burn Care Rehabil.*, 2004, **25**, 192–196.
78. S. Zhao, L. Li, H. Wang, Y. Zhang, X. Cheng, N. Zhou, M. N. Rahaman, Z. Liu, W. Huang and C. Zhang, Wound dressings composed of copper-doped borate bioactive glass microfibers stimulate angiogenesis and heal full-thickness skin defects in a rodent model, *Biomaterials*, 2015, **53**, 379–391.
79. J. Li, D. Zhai, F. Lv, Q. Yu, H. Ma, J. Yin, Z. Yi, M. Liu, J. Chang and C. Wu, Preparation of copper-containing bioactive glass/eggshell membrane nanocomposites for improving angiogenesis, antibacterial activity and wound healing, *Acta Biomater.*, 2016, **36**, 254–266.
80. X. Wang, F. Cheng, J. Liu, J. H. Smått, D. Gepperth, M. Lastusaari, C. Xu and L. Hupa, Biocomposites of copper-containing mesoporous bioactive glass and nanofibrillated cellulose: Biocompatibility and angiogenic promotion in chronic wound healing application, *Acta Biomater.*, 2016, **46**, 286–298.
81. P. Bober, J. Liu, K. S. Mikkonen, P. Ihalainen, M. Pesonen, C. Plumed-Ferrer, A. Von Wright, T. Lindfors, C. Xu and R. M. Latonen, Biocomposites of nanofibrillated cellulose, polypyrrole, and silver nanoparticles with electroconductive and antimicrobial properties, *Biomacromolecules*, 2014, **15**, 3655–3663.
82. E. Guibal, S. Cambe, S. Bayle, J. M. Taulemesse and T. Vincent, Silver/chitosan/cellulose fibers foam composites: From synthesis to antibacterial properties, *J. Colloid Interface Sci.*, 2013, **393**, 411–420.
83. C. Guo, L. Zhou and J. Lv, Effects of expandable graphite and modified ammonium polyphosphate on the flame-retardant and mechanical properties of wood flour-polypropylene composites, *Polym. Polym. Compos.*, 2013, **21**, 449–456.
84. M. C. B. Ammons, L. S. Ward and G. A. James, Anti-biofilm efficacy of a lactoferrin/xylitol wound hydrogel used in combination with silver wound dressings, *Int. Wound J.*, 2011, **8**, 268–273.
85. Z. Zhang, T. He, M. Yuan, R. Shen, L. Deng, L. Yi, Z. Sun and Y. Zhang, The in situ synthesis of Ag/amino acid biopolymer hydrogels as mouldable wound dressings, *Chem. Commun.*, 2015, **51**, 15862–15865.

86. S. K. Mishra, D. S. Mary and S. Kannan, Copper incorporated microporous chitosan-polyethylene glycol hydrogels loaded with naproxen for effective drug release and anti-infection wound dressing, *Int. J. Biol. Macromol.*, 2017, **95**, 928–937.
87. R. Kumar and H. Münstedt, Silver ion release from antimicrobial polyamide/silver composites, *Biomaterials*, 2005, **26**, 2081–2088.
88. R. S. Kumar and S. Arunachalam, DNA binding and antimicrobial studies of polymer-copper(II) complexes containing 1,10-phenanthroline and L-phenylalanine ligands, *Eur. J. Med. Chem.*, 2009, **44**, 1878–1883.
89. J. Gabbay, G. Borkow, J. Mishal, E. Magen, R. Zatcoff and Y. Shemer-Avni, Copper oxide impregnated textiles with potent biocidal activities, *J. Ind. Text.*, 2006, **35**, 323–335.
90. G. Borkow, R. C. Zatcoff and J. Gabbay, Reducing the risk of skin pathologies in diabetics by using copper impregnated socks, *Med. Hypotheses*, 2009, **73**, 883–886.
91. P. Dykes, Increase in skin surface elasticity in normal volunteer subjects following the use of copper oxide impregnated socks, *Ski. Res. Technol.*, 2015, **21**, 272–277.
92. R. C. Zatcoff, M. S. Smith and G. Borkow, Treatment of tinea pedis with socks containing copper-oxide impregnated fibers, *Foot*, 2008, **18**, 136–141.
93. G. Borkow, S. S. Zhou, T. Page and J. Gabbay, A novel anti-influenza copper oxide containing respiratory face mask, *PLoS One*, 2010, **5**, e11295.
94. W. Shao, H. Liu, X. Liu, S. Wang, J. Wu, R. Zhang, H. Min and M. Huang, Development of silver sulfadiazine loaded bacterial cellulose/sodium alginate composite films with enhanced antibacterial property, *Carbohydr. Polym.*, 2015, **132**, 351–358.
95. L. Lorente, M. Lecuona, A. Jiménez, R. Santacreu, L. Raja, O. Gonzalez and M. L. Mora, Chlorhexidine-silver sulfadiazine-impregnated venous catheters save costs, *Am. J. Infect. Control*, 2014, **42**, 321–324.
96. N. George, J. Faoagali and M. Muller, Silvazine® (silver sulfadiazine and chlorhexidine) activity against 200 clinical isolates, *Burns*, 1997, **23**, 493–495.
97. C. L. Fox, Silver Sulfadiazine—A new topical therapy for *Pseudomonas* in burns, *Arch. Surg.*, 1968, **96**, 184.
98. P. Li, J. Li, C. Wu, Q. Wu and J. Li, Synergistic antibacterial effects of beta-lactam antibiotic combined with silver nanoparticles, *Nanotechnology*, 2005, **16**, 1912–1917.
99. A. R. Shaikh, R. Giridhar and M. R. Yadav, Bismuth-norfloxacin complex: Synthesis, physicochemical and antimicrobial evaluation, *Int. J. Pharm.*, 2007, **332**, 24–30.
100. M. Halwani, S. Blomme, Z. E. Suntres, M. Alipour, A. O. Azghani, A. Kumar and A. Omri, Liposomal bismuth-ethanedithiol formulation enhances antimicrobial activity of tobramycin, *Int. J. Pharm.*, 2008, **358**, 278–284.

101. J. J. Harrison, R. J. Turner, D. A. Joo, M. A. Stan, C. S. Chan, N. D. Allan, H. A. Vrionis, M. E. Olson and H. Ceri, Copper and quaternary ammonium cations exert synergistic bactericidal and antibiofilm activity against *Pseudomonas aeruginosa*, *Antimicrob. Agents Chemother.*, 2008, **52**, 2870–2881.
102. V. A. Gant, M. W. D. Wren, M. S. M. Rollins, A. Jeanes, S. S. Hickok and T. J. Hall, Three novel highly charged copper-based biocides: Safety and efficacy against healthcare-associated organisms, *J. Antimicrob. Chemother.*, 2007, **60**, 294–299.
103. K. deleon, F. Balldin, C. Watters, A. Hamood, J. Griswold, S. Sreedharan and K. P. Rumbaugh, Gallium maltolate treatment eradicates *Pseudomonas aeruginosa* infection in thermally injured mice, *Antimicrob. Agents Chemother.*, 2009, **53**, 1331–1337.
104. E. Banin, A. Lozinski, K. M. Brady, E. Berenshtein, P. W. Butterfield, M. Moshe, M. Chevion, E. P. Greenberg and E. Banin, The potential of desferrioxamine-gallium as an anti-*Pseudomonas* therapeutic agent, *Proc. Natl. Acad. Sci. U. S. A.*, 2008, **105**, 16761–16766.
105. R. L. Patale and V. B. Patravale, O,N-carboxymethyl chitosan-zinc complex: A novel chitosan complex with enhanced antimicrobial activity, *Carbohydr. Polym.*, 2011, **85**, 105–110.
106. X. Wang, Y. Du and H. Liu, Preparation, characterization and antimicrobial activity of chitosan-Zn complex, *Carbohydr. Polym.*, 2004, **56**, 21–26.
107. J. F. Hernández-Sierra, F. Ruiz, D. C. Cruz Pena, F. Martínez-Gutiérrez, A. E. Martínez, A. de Jesús Pozos Guillén, H. Tapia-Pérez and G. Martínez Castañón, The antimicrobial sensitivity of *Streptococcus mutans* to nanoparticles of silver, zinc oxide, and gold, *Nanomedicine*, 2008, **4**, 237–240.
108. M. K. Rai, S. D. Deshmukh, A. P. Ingle and A. K. Gade, Silver nanoparticles: The powerful nanoweapon against multidrug-resistant bacteria, *J. Appl. Microbiol.*, 2012, **112**, 841–852.
109. D. R. Monteiro, L. F. Gorup, S. Silva, M. Negri, E. R. de Camargo, R. Oliveira, D. B. Barbosa and M. Henriques, Silver colloidal nanoparticles: antifungal effect against adhered cells and biofilms of *Candida albicans* and *Candida glabrata*, *Biofouling*, 2011, **27**, 711–719.
110. N. Durán, M. Durán, M. B. de Jesus, A. B. Seabra, W. J. Fávaro and G. Nakazato, Silver nanoparticles: A new view on mechanistic aspects on antimicrobial activity, *Nanomedicine.*, 2016, **12**, 789–799.
111. M. A. Radzig, V. A. Nadtochenko, O. A. Koksharova, J. Kiwi, V. A. Lipasova and I. A. Khmel, Antibacterial effects of silver nanoparticles on Gram-negative bacteria: Influence on the growth and biofilms formation, mechanisms of action, *Colloids Surf. B*, 2013, **102**, 300–306.
112. M. Rai, A. Yadav and A. Gade, Silver nanoparticles as a new generation of antimicrobials, *Biotechnol. Adv.*, 2009, **27**, 76–83.
113. J. S. Kim, E. Kuk, K. N. Yu, J. H. Kim, S. J. Park, H. J. Lee, S. H. Kim, Y. K. Park, Y. H. Park, C. Y. Hwang, Y. K. Kim, Y. S. Lee, D. H. Jeong and

M. H. Cho, Antimicrobial effects of silver nanoparticles, *Nanomedicine*, 2007, **3**, 95–101.
114. G. A. Sotiriou and S. E. Pratsinis, Antibacterial activity of nanosilver ions and particles, *Environ. Sci. Technol.*, 2010, **44**, 5649–5654.
115. C. Gunawan, W. Y. Teoh, C. P. Marquis and R. Amal, Cytotoxic origin of copper(II) oxide nanoparticles: Comparative studies with micron-sized particles, leachate, and metal salts, *ACS Nano*, 2011, **5**, 7214–7225.
116. G. Applerot, J. Lellouche, A. Lipovsky, Y. Nitzan, R. Lubart, A. Gedanken and E. Banin, Understanding the antibacterial mechanism of CuO nanoparticles: Revealing the route of induced oxidative stress, *Small*, 2012, **8**, 3326–3337.
117. A. M. Abdelgawad, S. M. Hudson and O. J. Rojas, Antimicrobial wound dressing nanofiber mats from multicomponent (chitosan/silver-nps/polyvinyl alcohol) systems, *Carbohydr. Polym.*, 2014, **100**, 166–178.
118. I. Perelshtein, G. Applerot, N. Perkas, G. Guibert, S. Mikhailov and A. Gedanken, Sonochemical coating of silver nanoparticles on textile fabrics (nylon, polyester and cotton) and their antibacterial activity, *Nanotechnology*, 2008, **19**, 245705.
119. H. Palza, Antimicrobial polymers with metal nanoparticles, *Int. J. Mol. Sci.*, 2015, **16**, 2099–2116.
120. S. Ferraris and S. Spriano, Antibacterial titanium surfaces for medical implants, *Mater. Sci. Eng.*, 2016, **C61**, 965–978.
121. M. R. Kaluđerović, J. P. Schreckenbach and H. L. Graf, Titanium dental implant surfaces obtained by anodic spark deposition – From the past to the future, *Mater. Sci. Eng.*, 2016, **C69**, 1429–1441.
122. K. M. Tarquinio, N. K. Kothurkar, D. Y. Goswami, R. C. Sanders, A. L. Zaritsky and A. M. Levine, Bactericidal effects of silver plus titanium dioxide-coated endotracheal tubes on *Pseudomonas aeruginosa* and Staphylococcus aureus, *Int. J. Nanomedicine*, 2010, **5**, 177–183.
123. A. Besinis, T. De Peralta and R. D. Handy, Inhibition of biofilm formation and antibacterial properties of a silver nano-coating on human dentine, *Nanotoxicology*, 2014, **8**, 745–754.
124. D. Roe, B. Karandikar, N. Bonn-Savage, B. Gibbins and J. B. Roullet, Antimicrobial surface functionalization of plastic catheters by silver nanoparticles, *J. Antimicrob. Chemother.*, 2008, **61**, 869–876.
125. O. Koseoglu Eser, A. Ergin and G. Hascelik, Antimicrobial activity of copper alloys against invasive multidrug-resistant nosocomial pathogens, *Curr. Microbiol.*, 2015, **71**, 291–295.
126. J. O. Noyce, H. Michels and C. W. Keevil, Use of copper cast alloys to control *Escherichia coli* O157 cross-contamination during food processing, *Appl. Environ. Microbiol.*, 2006, **72**, 4239–4244.
127. Y. Li, L. Liu, P. Wan, Z. Zhai, Z. Mao, Z. Ouyang, D. Yu, Q. Sun, L. Tan, L. Ren, Z. Zhu, Y. Hao, X. Qu, K. Yang and K. Dai, Biodegradable Mg-Cu alloy implants with antibacterial activity for the treatment of osteomyelitis: *In vitro* and *in vivo* evaluations, *Biomaterials*, 2016, **106**, 250–263.

128. A. L. Casey, D. Adams, T. J. Karpanen, P. A. Lambert, B. D. Cookson, P. Nightingale, L. Miruszenko, R. Shillam, P. Christian and T. S. J. Elliott, Role of copper in reducing hospital environment contamination, *J. Hosp. Infect.*, 2010, **74**, 72–77.
129. M. G. Schmidt, B. Von Dessauer, C. Benavente, D. Benadof, P. Cifuentes, A. Elgueta, C. Duran and M. S. Navarrete, Copper surfaces are associated with significantly lower concentrations of bacteria on selected surfaces within a pediatric intensive care unit, *Am. J. Infect. Control*, 2016, **44**, 203–209.
130. G. Faúndez, M. Troncoso, P. Navarrete and G. Figueroa, Antimicrobial activity of copper surfaces against suspensions of *Salmonella enterica* and *Campylobacter jejuni*, *BMC Microbiol.*, 2004, **4**, 1–7.
131. A. B. Monk, V. Kanmukhla, K. Trinder and G. Borkow, Potent bactericidal efficacy of copper oxide impregnated non-porous solid surfaces, *BMC Microbiol.*, 2014, **14**, 1–14.
132. F. Paladini, M. Pollini, A. Talà, P. Alifano and A. Sannino, Efficacy of silver treated catheters for haemodialysis in preventing bacterial adhesion, *J. Mater. Sci.: Mater. Med.*, 2012, **23**, 1983–1990.
133. A. Cochis, B. Azzimonti, C. Della Valle, R. Chiesa, C. R. Arciola and L. Rimondini, Biofilm formation on titanium implants counteracted by grafting gallium and silver ions, *J. Biomed. Mater. Res., Part A*, 2015, **103**, 1176–1187.
134. K. Davenport and F. X. Keeley, Evidence for the use of silver-alloy-coated urethral catheters, *J. Hosp. Infect.*, 2005, **60**, 298–303.
135. A. Ewald, S. K. Glückermann, R. Thull and U. Gbureck, Antimicrobial titanium/silver PVD coatings on titanium, *Biomed. Eng. Online*, 2006, **5**, 1–10.
136. M. H. Kollef, B. Afessa, A. Anzueto, C. Veremakis, K. M. Kerr, B. D. Margolis, D. E. Craven, P. R. Roberts, A. C. Arroliga, R. D. Hubmayr, M. I. Restrepo, W. R. Auger and R. Schinner, Silver-coated endotracheal tubes and incidence of ventilator-associated pneumonia: The NASCENT randomized trial, *JAMA*, 2008, **300**, 805–813.
137. M. R. Bruins, S. Kapil and F. W. Oehme, Microbial resistance to metals in the environment, *Ecotoxicol. Environ. Saf.*, 2000, **45**, 198–207.
138. D. H. Nies, Efflux-mediated heavy metal resistance in prokaryotes, *FEMS Microbiol. Rev.*, 2003, **27**, 313–339.
139. H. Li, Y. Gao, C. Li, G. Ma, Y. Shang and Y. Sun, A comparative study of the antibacterial mechanisms of silver ion and silver nanoparticles by Fourier transform infrared spectroscopy, *Vib. Spectrosc*, 2016, **85**, 112–121.
140. S. Langley and T. J. Beveridge, Effect of o-side-chain-lipopolysaccharide chemistry on metal binding, *Appl. Environ. Microbiol.*, 1999, **65**, 489–498.
141. J. Lemire, R. Mailloux, C. Auger, D. Whalen and V. D. Appanna, *Pseudomonas fluorescens* orchestrates a fine metabolic-balancing act to counter aluminium toxicity, *Environ. Microbiol.*, 2010, **12**, 1384–1390.

CHAPTER 10

Antimicrobial Quaternary Ammonium Polymers for Biomedical Applications

SHADY FARAH[a,b]

[a] The Wolfson Faculty of Chemical Engineering, Technion-Israel Institute of Technology, Haifa 3200003, Israel; [b] The Russell Berrie Nanotechnology Institute, Technion-Israel Institute of Technology, Haifa 3200003, Israel
Email: sfarah@technion.ac.il

10.1 Introduction

10.1.1 Biomedical Implants and the Problem of Infection

Biomedical implants are subject to infection and failure attributed to microbial accumulation and contamination on their surfaces. The surfaces act as reservoirs of microbes, which can in turn lead to the spread of infection and biofilm development. The process of bacterial contamination on surfaces can be widely divided into two major steps, starting with the initial adherence of few microorganisms to the surface, and the later step of biofilm development, usually in less than 24 h.[1–3] Infection and microbial contamination are among the most common causes of biomaterials and biomedical implants failure in modern medicine. Examples have been extensively reported in the literature, such as for dental composites, catheters and implanted devices.[4] Following the development of microorganism biofilms, the contamination issue is even more

Biomaterials Science Series No. 5
Antimicrobial Materials for Biomedical Applications
Edited by Abraham J. Domb, Konda Reddy Kunduru and Shady Farah

Published by the Royal Society of Chemistry, www.rsc.org

challenging. At this stage, microorganisms have been reported to exhibit extreme resistance to both antibiotics and the host's own immune system, resulting in the spread of more infections and several health issues, including mortality. For example, one of the major global healthcare issues is hospital-acquired infections, with more than 2 million cases and more than 100 000 deaths reported annually in the USA.[1,3] Accordingly, knowledge about microbe-contamination, including rapid spreading and resistance to treatment, and the growing public health awareness of pathogenic risk, have increased the importance of developing novel efficient antimicrobial materials that microbes are less resistant to, and applying them to surfaces and medical device coatings, among others. These materials are required to prevent complex epidemiological situations and infection risks, not only in a healthcare environment area including catheters, dental composites, surgical armaments, orthopaedics and other medical implants, but also in other areas such as food storage and packaging, water purification and decontamination systems, filters and general consumer markets.[1,5] Worldwide, several efforts have been invested in the development of different coating strategies and have been applied as treatment. Simple preventive approaches based on developing coatings capable of preventing microbial adhesion to the device or surface, include diamond-like carbon films (DLC), poly(ethylene glycol) coatings, *etc.*[5] Intensive research efforts in this field and the critical need for more potent antimicrobial coatings and modified surfaces have yielded the development of various families of antimicrobial materials.

Antimicrobial materials, particularly low molecular weight antimicrobial agents have been widely used as antimicrobial drugs, food preservatives and water and soil decontaminants. However, further use of these materials has been limited, due to their residual toxicity.[6] Alternatively, polymer-based antimicrobial agents, commonly named antimicrobial polymers have been found to be ideal to address this toxicity issue, given their non-volatile and chemically stable structure and design. Moreover, antimicrobial polymers can be easily utilized and bound to object surfaces, *i.e.* medical implants, using relatively simple modification chemistry and thus rendering them a long-term antimicrobial activity and adhesion properties. Antimicrobial polymers, given their high molecular weight and structure have been found to be minimally capable of permeating through biological membranes such as the skin, thus reducing their potential toxicity.[7] In addition, antimicrobial polymers giving their structure and design they have been identified with a high surface density of active functional groups, resulting in less microbial resistance and eventually an increased antimicrobial activity and decontamination efficiency.[8] Polycationic polymers are considered a leading category of the antimicrobial polymers, specifically quaternary ammonium (QA)-based polymers. In this chapter we overview QA polymer synthesis, recent advances and their biomedical application.

10.1.2 Quaternary Ammonium Compounds—Mechanism of Action

QA compounds are materials that usually contain four organic groups chemically bound to a nitrogen atom, acquiring it a permanent positive charge (Figure 10.1). Organic substituents can be either identical or varied in structure, composed of aryl, alkyl and heterocyclic groups[9] (see Figure 10.2 for representative structures). QA compounds have been found to exhibit a broad spectrum of antimicrobial activity and efficiency against both Gram-positive and Gram-negative bacteria, viruses, algae and fungi.[10]

Attaching QA functional groups to polymers was found to be a useful tool in the design of antimicrobial polymeric macromolecules, namely QA polymers. The hydrophobic/hydrophilic nature of the organic groups attached to the nitrogen atoms, density of nitrogen atoms that can be converted to be QA moieties, total charge and the counter ions have been found to significantly affect the biological activity of QA polymers.[8] It has been found that the preferred chemical structure of the repeating unit of QA polymers has at least one long alkyl chain among the four groups, to provide a hydrophobic segment. The alkyl chain should be long enough to be capable of disrupting the microbial cell wall (Figure 10.1b).[3,10,11] Although the detailed mechanism of action of the antibacterial effect of QA polymers is not fully clear, it was suggested that they cause lysis of the bacterial cells through non-specific interaction between the QA moieties by interacting with ions and negatively charged bacterial cell membranes.[7] Such interaction was found to destabilize the wall membrane, causing vital ion leakage and eventually leading to microbe death.[9] The most effective antimicrobial QA polymers were found to have at least one 4–8-carbon atoms alkyl chain-substituted QA moieties.[12] Conversely, very long alkyl chain-substituted QA polymers were found to be less effective, given their high hydrophobicity leading to less effective functional group organization and orientation with increased chances of alkyl chains folding and overlapping.

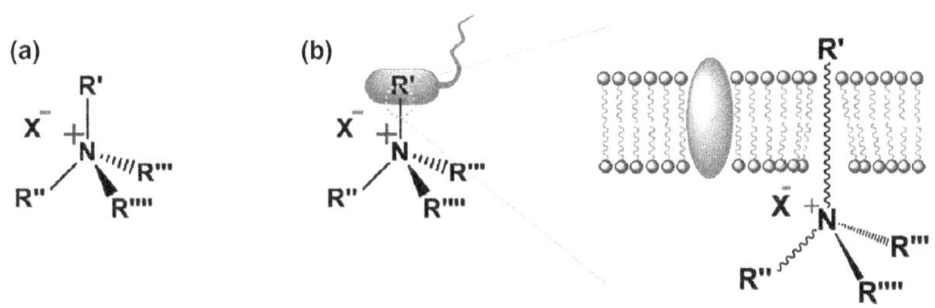

Figure 10.1 (a) Schematic structure of quaternary ammonium compound/moiety. R′-R′′′′ H-, alkyl-, aryl modifications; X is a negatively charged counter ion. (b) Interaction between quaternary ammonium moiety and bacterial cell wall, inducing destabilization and bacterial cell death.

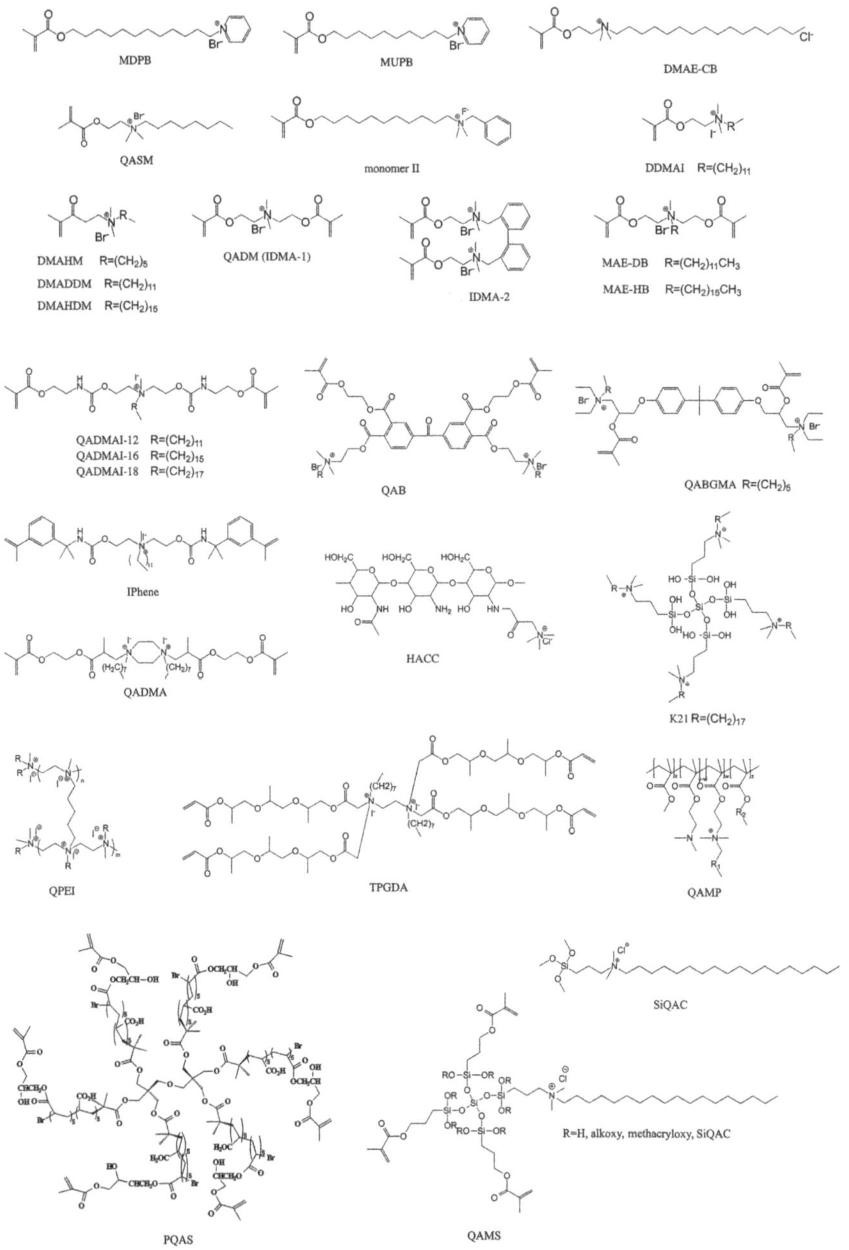

Figure 10.2 Chemical structures of representative quaternary ammonium compounds in antimicrobial biomedical materials.
Reprinted from *Prog. Polym. Sci,* **17**, Y. Jiao, L. Niu, S. Ma, J. Li, F. R. Tay, J.-H. Chen, Quaternary ammonium-based biomedical materials: State-of-the-art, toxicological aspects and antimicrobial resistance, 53–90, Copyright 2017, with permission from Elsevier.[28]

10.2 Antimicrobial Surface Strategies

The incorporation of antimicrobial materials into various surface coatings have been reported to result in surfaces with limited or without bacterial growth capability. The mode of action of these coatings can be widely divided into two main strategies: releasing and non-releasing antimicrobial coatings (Figure 10.3). The decision of which mode of action to follow is usually affected by the desired application and coating feasibility.[11,13,14] For example, in order to achieve antibacterial dental resin composites, initial efforts were focused on the simple and conventional approach where composites were soaked for various periods of time in the agent's solution or mixed with low molecular weight antibacterial agents, such as silver ions,[15] iodine, antibiotics[8,16] and QA compounds, *e.g.* benzalkonium chloride.[17] It was found that these agents can be released gradually over time, while maintaining a surface free of microbes as long as active agents being released. However, the limited space for the incorporation of potential active antimicrobial agents and the reported toxicity of some of the released antibiotics have minimized the use of such a technique for achieving antimicrobial dental composites.[8,16] Alternatively, long-lasting antimicrobial surfaces and composites were achieved by either physical or chemical incorporation of non-leachable antimicrobial active agents that can damage bacteria upon contact. These non-releasing long-acting surfaces were based on surface modification with high molecular weight QA compounds or the incorporation of crosslinked particulates of QA compounds.[3,18–22]

10.2.1 Non-releasing Antimicrobial Polymeric Surfaces

Hydrophobic polycationic polymers covalently attached or deposited upon the surface have been widely used to fabricate non-releasing and long-lasting antimicrobial surfaces. These surfaces have been shown to be highly efficient and to maintain long-term antimicrobial efficiency through inducing microbial death by interaction and contact with QA moieties, causing physical damage and destabilization to the cell wall of the microbe.[23]

Given that the nature of the organic substituents of QA moieties plays a key role in their bioactivity,[24] continuous efforts have been invested worldwide to

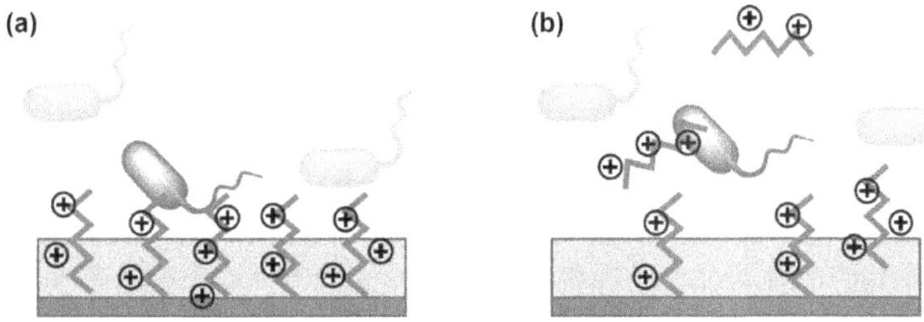

Figure 10.3 Antimicrobial polymeric surface strategies: (a) non-releasing and (b) releasing surfaces for biomedical applications.

synthesize active QA polymers with various QA substituents varied in their chemical composition, nature and alkyl length.[23–28] Several examples of QA moieties to be attached to polymeric backbones or to be polymerized leading to the formation of QA polymers have been reported (Figure 10.2). QA-based polymers have received much attention as antibacterial agents, due to their higher density of QA moieties and the capability to provide effective and long-term protection against bacterial colonization. Accordingly, recent and ongoing efforts have been focusing on the covalent attachment of poly(QA) materials to a different surfaces including metal, plastics and glass for long-term antimicrobial activity for various biomedical applications.[29–31] Following surface modification or embedding of QA polymers, it was found that the high density of these positively charged QA moieties help to keep the hydrophobic chains more separated and erected from the surface, thus increasing their efficiency compared to small QA molecules. Moreover, a higher density of the positively charged QA moietyu-based polymers was found to attract the partially negative microbes to the surfaces more effectively.[11,13] In biomedical applications, QA-diethylaminoethyl methacrylate (DEAEM) and its copolymerized acrylate-based derivatives, linear and cross-linked QA-poly(vinyl-*N*-pyridinium) (PVP) and QA-polyethyleneimine (PEI) are the most studied non-releasing surfaces based QA-positive coatings. These QA-based polymers have gained great interest for dental composites,[12,19–22] orthopaedic implants[28] and also for water disinfection,[32–35] due to their low cytotoxicity and safe profile.

Large portions of the non-releasing surfaces are based on the incorporation of QA polymer nanoparticles as a part of the therapeutic coating. Accordingly, in this chapter special attention is given to the synthesis of antimicrobial QA crosslinked nanoparticles of the aforementioned leading QA-polymer materials and their bioactivity following incorporation into coatings and composites (Section 10.3.4).

10.2.2 Releasing Antimicrobial Polymeric Surfaces

Antimicrobial agent-releasing surfaces are required for several biomedical applications, especially where there is limited direct contact between the surface and the microbe, or a short time is available for such interaction, for example, antimicrobial-releasing surfaces as part of water purification filters. In this case, in order to detoxify large volumes of running water, a different mode of action other than contact killing is required. The released agents may vary ranging from inorganic agents such as Ag and Cu released from surfaces containing silver or copper or their alloys,[3,36] to organic agents such as triclosan[37] and QA compound derivatives, including their QA polymeric formulations, up to biological agents such as bacteriophage-based surfaces.[3,5]

Out of the various biocidal polymeric groups under this strategy, *N*-halamine-based polymers have been found to exhibit unique biocidal properties.[38–40] An *N*-halamine can be defined as a chemical structure or moiety that contains one or more nitrogen–halogen covalent bonds.[41] Due to the fast oxidative properties of cyclic halogenated halamine compound moieties (N–Cl or N–Br),[42–44] they have been found to exhibit excellent biocidal activity leading to the inactivation of a wide spectrum of microorganisms, including Gram-positive and Gram-negative bacteria, fungi and viruses.[42,45–47] Their mechanism of action is attributed to targeting the biological receptor of the bacteria (*i.e.* thiol groups) resulting in cell inactivation.[41] Moreover, the rate of inactivation of the bacteria can be manipulated by the stability of the *N*-halogen bond and the rate of release.[48]

Following covalent incorporation of *N*-halamine moieties to polymeric materials as well as surface coatings, they have been found to exhibit controlled release of active agents from the surfaces. Various commercial polymers have been functionalized with *N*-halamine moieties and found to exhibit biocidal activity upon surface contact with pathogens, including nylon,[49] Kraton rubber,[50] PET,[51] cellulose[41,52] and other various surface coatings.[53] Accordingly, *N*-halamine derivatives have found their way to a wide variety of detoxification applications, such as in paints, biomedical surfaces, textiles and in water disinfection systems, thus improving world health.[3,13,38–40,54–59] However, for this strategy there is a need for new biocidal compounds that do not possess the limitations inherent in the disinfectants currently employed, *e.g.* bacterial resistance, loss of regeneration ability, long contact time and toxicity.[3,60]

Recently, throughout the world, different research groups have focused on improving antimicrobial efficiency and performance by combination with QA-based polymers. Given the wide range of possible chemical modifications and potential surface manipulation, such an approach has received considerable attention. One recent example has been reported by Yingfeng *et al.*,[61] where they presented the synthesis and antimicrobial properties of a tailored polymer-brush-grafted mesoporous silica with *N*-halamine and QA groups. First, a poly((3-acrylamidopropyl)trimethylammonium chloride) (PAPTMAC)-modified mesoporous SiO_2 was prepared using a polymer-brush-grafted method, treated with a 4,4′-azobis(4-cyanovaleric acid) (ACVA) as an initiator for polymerizing (3-acrylamidopropyl)trimethylammonium chloride (APTMAC) on the surface forming covalent bonding. Then, the mesoporous SiO_2 and the *N*-halamine precursor were subject to a chlorination reaction, resulting in successful transformation of N–H to N–Cl bonds. Analysis of the antimicrobial properties indicated that the combination of QA and *N*-halamine moieties in the synthesized polymers resulted in excellent antimicrobial properties, and within the short time of 10 min against both *Escherichia coli* Gram-negative bacteria O157 : H7 (ATCC 43895) (7.52 log) and 100% *Staphylococcus aureus* Gram-positive bacteria (ATCC 6538) (7.63 log). Such modified mesoporous SiO_2 with the added effective antimicrobial properties value could be a useful tool for water treatment and other filtration applications. Another study combining *N*-halamine and QA salt moieties was reported by Zhiqiang *et al.*[62] A regenerable antimicrobial polymeric resin poly(*p*-methylstyrene)-3-(5,5-dimethylhydantoin)-*co*-trimethyl ammonium chloride (PSHTMA) was synthesized in this study. First, commercially available crosslinked macroporous chloromethylated polystyrene (CMPS) resin reacted with both 5,5-dimethylhydantoin (DMH) salt and trimethylamine (TMA) in water to produce a polymeric resin conjugated hydantoinyl and QA salt moieties, *i.e.* PSHTMA, then PSHTMA resin was converted to an antimicrobial *N*-halamine structure, *i.e.* (Cl-PSHTMA) by chlorination reaction using sodium hypochlorite solution (Figure 10.4). This resin exhibited excellent antimicrobial activity within a short contact time of 1 min, 8 log inactivation of *E. coli* and 7 log inactivation of *S. aureus*. Another advantage of this QA-halamine combined moieties strategy is excellent regenerability. The released Cl-*N*-halamine moieties of the resin can be reloaded for several cycles of slow release following a simple re-chlorination reaction for regeneration of Cl-PSHTMA. Different preparation of active antimicrobial polymers having both functional halamine and QA moieties was reported by Xuehong *et al.*[63] In this study the synthesized polymer has the cyclic halamines structure substituted with QA moieties. First a precursor of 5,5-dimethyl-3-(chloroalkyl)hydantoin was synthesized and then was reacted with a high molecular weight poly(dimethylaminoethyl methacrylate), resulting in attachment of the *N*-halamine compound to the acrylate-based polymer, forming a quaternized nitrogen connected to the alkyl-*N*-halamine. The last step was halogenation either with NaOBr (sodium hypobromite) or NaOCl (sodium hypochlorite) to form N–Br or N–Cl, respectively, to the unsubstituted N on the cyclic halamine structure. These polymers exhibited superior antibacterial properties, non-toxicity and high chemical stability. An additional advantage of this synthetic process is that it can be performed under mild reaction conditions and using low-cost raw materials.

10.2.2.1 Microbial Resistance to Antimicrobial Agents

Microbial resistance to antimicrobial agents is a major concern, limiting the success of antimicrobial treatments, and thus more efforts should be made to

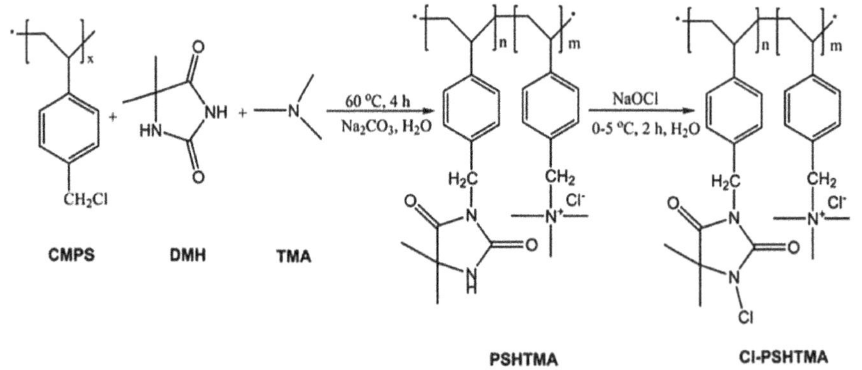

Figure 10.4 Schematic illustration for synthesis of Cl-PSHTMA. Reproduced from ref. 62 with permission from the Royal Society of Chemistry.

overcome it. According to a recent report released by World Health Organization (WHO), worldwide there is an increase in antimicrobial resistance-induced mortality, with an expectation that it will increase to an annual 10 million deaths by 2050.[64] While there are various parameters that could contribute to antimicrobial resistance, it was found that more reported cases had releasing-based antimicrobial strategies, especially with repeated antimicrobial courses, for example, implantable medical devices with antimicrobial-releasing coatings, in which bacteria are more likely to exhibit an increased resistance.[65] In such cases medical devices are associated with foreign body related infections starting with bacterial adhesion on the device's surface and progressing to later biofilm formation.[66] Insufficient release of antimicrobial agents or continuous release at low concentrations were reported to become sub-inhibitory and thus were among reasons to promote resistance evolution. However, antimicrobial resistance is mainly caused by bacterial biofilm development, where the bacteria are more protected from the antimicrobial agent's penetration, diffusion and antimicrobial impact/effect. Bacterial biofilms can be formed either by one or multiple bacterial species. On top of that, recent evidence clearly indicates that inhibited or deactivated bacteria can absorb large amount of antimicrobial agents, thus protecting other bacterial cells and increasing their survival and colonization.[67]

The WHO's report ranked the 12 drug-resistant bacterial species identified with the most threat to human health, for which new antimicrobial agents and antimicrobial coatings are urgently needed. Species were ranked in the following order:[64] *Acinetobacter baumannii*, *Pseudomonas aeruginosa*, Enterobacteriaceae, *Enterococcus faecium*, *Staphylococcus aureus*, *Helicobacter pylori*, *Campylobacter* species, *Salmonellae*, *Neisseria gonorrhoeae*, *Streptococcus pneumoniae*, *Haemophilus influenzae* and *Shigella* species. Accordingly, for addressing challenging bacterial resistance, better strategies of antimicrobial coatings should be developed and currently several approaches are subject to intensive study. One approach is designing antimicrobial coating with synergistic strategies of action where antimicrobial agents can be released from anti-adhesion surface coatings, thus reducing the chances of bacterial colonization and biofilm formation, and resultant bacterial resistance.[68–70] Another approach is by releasing varied antimicrobial agents capable of targeting microbes *via* different modes of action, thus reducing the microbe's resistance capability to antimicrobial treatments and coatings.

10.3 Antimicrobial Polymers Synthesis and Modifications

10.3.1 Quaternary Ammonium-based Polymers

During the past four decades, extensive efforts have been made worldwide to synthesize effective QA polymers with stable and active QA moieties.[7–9,28,71–73] It was found that embedding QA moieties, either as pendant groups or in the polymeric backbone, have resulted in enhanced efficacy over the corresponding small-molecule QA compounds. Nevertheless, such QA polymers were identified with prolonged shelf-life, increased selectivity and efficiency and significantly reduced residual toxicity.[8,73,74] Incorporation of these QA antimicrobial polymers into different coating surfaces was found to result in surfaces without bacterial growth capability, or at least limited growth compared to blank coatings. These coatings were subject to many potential chemical modifications to enhance their suitability for various medical needs. A comprehensive review by Ferreira and Zumbuehl highlights QA antimicrobial surface modification trends for various applications.[11]

For biomedical applications a wide range of polycations either homo-, co- or random polymers with various architecture structures including dendritic, linear or branched structures have been synthesized and tested. The following polyamines are the most commonly used and studied for biomedical applications: PEI, poly(2-dimethylaminoethyl methacrylate) (PDEAEM) and its acrylate versions, polyguanidines, poly(L-lysine) (PLL), polyamidoamine (PAMAM) dendrimers and (PVP) salts. Several extensive fundamental studies reported and summarized in several reviews and books discuss the therapeutic and antimicrobial functionality and biochemical and physiochemical properties of these polymers.[10] The following examples highlight the resultant antimicrobial activity of these polyamine-based polycationic polymers.

QA-PEI: both linear and crosslinked chemical structures, *i.e.* nanoparticles, have been found to significantly inhibit bacterial growth.[12,19–22,33,35,71,74–79] PEI has a branched chemical structure, with the ratio of 1 : 2 : 1 for the primary, secondary and tertiary amine groups, respectively. The abundance of amino groups along its backbone makes it easily accessible for chemical modification as well for quaternization, resulting in a high density of active antimicrobial groups and increased antimicrobial activity. Lin *et al.* proposed a novel, non-releasing strategy for creating bactericidal surfaces. The proposed strategy involves covalent coating with long hydrophobic polycationic chain-modified PEI. Such an approach was found to be effective against a variety of Gram-positive and Gram-negative bacteria.[71] The antibacterial activity was found to be dependent on the molecular weight of PEI and *N*-alkylation. It was found that *N*-alkylated PEI of low molecular weights, *i.e.* 0.8 and 2 kDa has a weak bactericidal activity due to short chain length, which probably does not penetrate bacterial cells. Alternatively, high molecular weight *N*-alkylated PEI exhibited excellent antimicrobial activity. In a study reported by De Prijck *et al.*,[80] PEI and dimethylaminoethyl methacrylate were covalently bound to the surface of polydimethylsiloxane (PDMS) or polymethylmethacrylate (PMMA) and quaternized before antimicrobial testing. These coatings were tested for prevention biofilm formation of *Candida albicans*, a major challenge for the safety of prosthesis deterioration in laryngectomized patients. These QA-based coatings were found to be promising for medical devices such as prostheses and catheters, with up to 92% reduction in bacterial growth.[80,81] Nanoparticle-based QA-PEI were found to be highly effective, even at low addition percentages such as 1% w/w to dental resin composites, and have been tested against variety of microbes. It was found to exhibit excellent biocide activity against challenging bacteria streptomycin-resistant *Enterococcus faecalis*, while moderate activity was found at lower loading percentages (Section 10.3.4).

Pyridinium-based polymers are compounds with a heterocyclic ring containing nitrogen atom that were found to possess a wide range of antibacterial activity following *N*-alkylation of the monomer units, gaining both hydrophobic nature and positive charge.[82] Alternative approaches for designing effective antimicrobial polymers were also reported.[83] First, insoluble poly(4-vinylpyridine-*co*-divinylbenzene) beads of block or random copolymer structures were synthesized on polystyrene and then the poly-(4-vinyl pyridine)s were quaternized with octyl iodide. It was found that quaternized polystyrene-block-poly(4-vinyl pyridine) demonstrates higher potency in the reduction of *Pseudomonas aeruginosa* and *Staphylococcus aureus* growth than quaternized polystyrene-random-poly(4-vinyl pyridine). This could be attributed to the different concentration of 4-vinyl pyridine groups on the surface of the tested specimens prepared from the two copolymers. In another study reported by Imazato and coworkers, they described the potential use of pyridinium-based polymers as antibacterial additives in restorative composite resins.[74,84] In a study with bonding resin render and dentin primer they showed that composite resins can acquire antibacterial properties following the addition of antibacterial monomer such as 12-methacryloyloxydodecyl pyridinium bromide (MDPB).[74,84] Also, these resins were found to significantly inhibit the bacterial growth on the surface following MDPB immobilization *via* resin polymerization. Nevertheless, the addition of MDPB to the bonding resin didn't induce any change in its bonding characteristics.[85,86]

Polyguanidine-based polymers are water soluble antimicrobial polymers with excellent antimicrobial activity against a wide spectrum of microbes and high safety profile. Their synthesis is considered relatively easy and straightforward *via* polycondensation of diamines with guanidinium salts and the fast removal of the resultant byproduct ammonia gas (NH_3). Polyguanidines and polybiguanide varied in their diamine length or acrylate monomers with pendant guandine and biguanide groups, respectively, have been reported and tested against both Gram-positive and Gram-negative bacteria. They were found to be effective at concentrations less than 200 $\mu g\,mL^{-1}$ as minimum inhibitory concentration (MIC).[87]

Overall, the antimicrobial bioactivity of QA-based polymers is affected by several factors, which should be studied for each polymer. The main parameters that affect activity are alkyl chain length and location, charge density and distribution, chemical structure and branching, molecular weight and QA moiety/moieties per repeating unit, hydrophobicity and hydrophilicity ratio and total alkylation and quaternization degrees. For maximum antimicrobial efficiency, each one of these parameters should be studied and optimized. Other parameters should be taken into account for synthesizing effective QA polymers including the physicomechanical environment, temperature, mechanical stress and pH.

10.3.2 Antimicrobial QA-based Natural Polymers

PLL is a homopolymer with a primary amine group on each repeating unit, which has been widely used as a coating material for various applications. Under physiological conditions amino groups quaternize resulting in positively charged moieties, thus exhibiting antimicrobial activity against a wide spectrum of yeast, fungi, and both Gram-positive and Gram-negative bacteria species.[72] The correlation between the antibacterial activity of poly(ε-lysine)-based polymers and their molecular weight have been extensively studied and reported. It was found that a chain length of at least nine L-lysine residues is the optimum for inhibition of microbial growth.[72] Another study was reported by Hiraki, who tested the inhibition effect of poly(ε-lysine) on *Bacillus* strains. It was found that inhibitory concentrations for the spore germination of *Bacillus* strains can be varied depending on the strain studied. It was found that the required inhibitory concentrations of poly(ε-lysine) for *Bacillus coagulans*, *Bacillus stearothermophilus* and *Bacillus subtilis* are 12.5 $\mu g\,mL^{-1}$, 2.5 $\mu g\,mL^{-1}$ and 12.5 $\mu g\,mL^{-1}$, respectively.[73]

Chitosan is a linear polysaccharide, biocompatible and biodegradable with tissue-adhesive and antimicrobial properties. These properties mean that chitosan's native structure and modified derivatives as well as blended mixtures with other polymers are very useful materials and an integral part of many compositions for biomedical applications.[88,89] The antimicrobial activities of the chitosan-based polysaccharides have been reported.[90] Chitosan and its derivatives with antimicrobial QA moieties can be synthesized either by direct alkylation reaction with alkyl halide (*e.g.* different alkyl chain lengths) or by reductive amination (Schiff base intermediates) followed by quaternization with alkyl halides.[3,74,90] As a result of quaternization, an increase in water-solubility was reported. This feature plus the resultant cationic polyelectrolyte structure and the varied modification opportunities were all found to increase their potential to be used in biomedical applications and in the field of anti-infection. Over the years, several quaternized structures of chitosan have been reported. A comprehensive review by Honglue *et al.*,[88] details quaternized chitosan antimicrobial activity and their mechanism of action in biomedical applications particularly in orthopaedic implants. In a study reported by Rimondino *et al.*[91] they synthesized a green version of chitosan for biomedical application with improved biocompatibility, biodegradability and functionality. The group prepared amphiphilic and water-soluble versions of chitosan *via* a combination of dendronization modification with a biocompatible dendron followed by a quaternization process with an ammonium salt of glycidyltrimethylammonium chloride (GTMAC), which confers a permanent positive charge to the polymer. However, these materials need to be further evaluated. Guanghua *et al.*[92] reported the preparation of water-superabsorbent gel-based QA chitosan for biomedical application and as a personal hygiene material by modifying chitosan with QA salt at 10–75% (by wt%) and with epichlorohydrin as a crosslinking agent.

Other effective biodegradable chitosan-based QA antimicrobial moieties were reported by Yu *et al.*[93] They prepared them by mixing chitosan QA salt (epoxypropyltrimethylammonium chloride-grafted chitosan and epoxypropyltrimethylammonium chloride-grafted carboxymethyl chitosan) with hydrophilic polymer amphipathic derivatives (1,2-epoxyalkane-grafted CM-cellulose, 1,2-epoxyalkane-grafted alginate salt). KoonGee *et al.*,[94] have reported chitosan and agarose, modifications and functionalization to form stable antimicrobial coatings for specific biomedical applications. Kotelnikova and Panarin[95] have reported the modification of various types of cellulose with biologically active QA compounds, including QA monomers and polymeric versions. QA-based dimethylbenzyl alkyl ammonium chloride, both low and high molecular weight were used to study the adsorption–desorption interaction of cellulose with QA substances. They found that QA substances can be released from the adsorbed complex, pointing to the potential of expanding biomedical application of adsorption complexes of QA derivatives. Also the addition of other natural polymers can potentially expand the utility for different biomedical purposes. Chen *et al.*[96] have reported novel antimicrobial agents by QA-conjugation of poly(*N*,*N*-dimethylaminoethylmethacrylate) with natural rosin as the pendant group (PDMAEMA-g-rosin). The synthesized PDMAEMA-g-rosin copolymers were found to exhibit antimicrobial efficiency against Gram-negative *Escherichia coli* and Gram-positive *Staphylococcus aureus*. The authors suggested that the mechanism of action is a combination of both hydrophobic and ionic interactions between bacterial cells and polymers.

10.3.3 Antimicrobial QA-based Biodegradable Polymers

The use of biodegradable antimicrobial polymers has been considered a promising approach for combating drug-resistant microbes. Garcia-Arguelles *et al.*[97] reported the incorporation of QA salts in the network of biodegradable polyesters *i.e.* poly(octanediol-*co*-citrate) using deep eutectic solvent-assisted synthesis to produce polyesters with antibacterial properties. A wide range of QA compounds were studied (*i.e.* hexadecyltrimethyl ammonium bromide, choline chloride and tetraethyl ammonium bromide) for the inhibition of bacterial growth in the early post-implantation steps. Synthesized materials have been tested on *Escherichia coli* on solid agar plates. The materials exhibited antimicrobial properties with potential application as antimicrobial wound dressings as well as other biomedical applications. Engler *et al.*[98] have studied the effect of charge and hydrophobicity balance of biodegradable antimicrobial polycarbonates on the resultant antimicrobial activity and selectivity toward microbes over mammalian cells. Engler *et al.* reported the manipulation of the hydrophobic/hydrophilic ratio *via* controlling the length of the spacer between the charged QA moiety and the polymer backbone. They tested the synthesized materials against various microbes and they were found to be highly efficient. In addition, they found that hydrophobicity manipulation *via* modifying the length of the hydrophobic side chains, while maintaining the charge on each repeating unit along the polymer structure resulted in higher activity and selectivity.[98] Such polymers were found to act *via* a membrane-disruption mechanism, making them less likely to induce resistance. Adhikari *et al.*[99] reported the preparation of biodegradable eumelanin-inspired aliphatic QA oligomers and polymer conjugates for targeting resistant pathogenic bacteria. Given their biocompatibility, biodegradability and physicochemical properties these polymers were suggested as promising materials for bioelectronics and as coatings for various applications, including for exposed surfaces that are commonly in contact, such as desks, bench tops and doorknobs. Zhengwei and Shaofei[100] reported the synthesis of biodegradable polycation-polyester containing QA under mild conditions by adding and mixing *N,N,N′,N′*-tetramethyl-1,4-butanediamine with 1,2-bis-(chloroacetoxy)ethane at room temperature and for 4–72 h. Chaobo *et al.*[101] reported the preparation of biodegradable blended films of chitosan and *N*-alkylated poly(4-vinylpyridine) with QA salt alkyl modifications (alkyl length = C4–7) or benzyl at a ratio of 60–95 : 5–40 as biodegradable composite materials for preserving food and drug delivery purposes.

10.3.4 Crosslinked Nanoparticles of Antimicrobial QA Polymers

Nanofabrication and nanoformulation of QA polymers is a growing area of research interest aiming to design QA antimicrobial polymers with improved activity *via* the design of nanostructures. By practice, it was found that using tools developed in nanotechnology and nanoscience for the fabrication of QA antimicrobial nanoparticles could expand the utility of already reported QA antimicrobial polymers. In addition to QA polymer performance enhancement, nanofabrication was found to increase their suitability to address several biomedical needs.[3,33–35,74–79] This growing area of research assumes that thinking at the nanoscale would open interesting possibilities for designing efficient antimicrobial surfaces *via* impacting the biology at the interface between bacteria and QA nanoparticles. Accordingly, worldwide, several efforts have been invested in the synthesis of new QA compound structures and nanofabrications for many healthcare areas. One of these areas is dental hygiene, where the preparation of antimicrobial dental restoration composite resins have gained extensive attention in the past few years.[81] In such biomedical applications, insoluble and crosslinked nanoparticles are essential, as the restorative composites are expected to withstand exposure to an aqueous environment and maintain the antimicrobial efficiency.[102]

In the following section, some examples of the synthesis and modification of QA polymeric nanoparticles are presented.

The synthesis of QA-PEI nanoparticles was previously described using three main steps of crosslinking, alkylation and quaternization

(Figure 10.5).[3,23,40–43,74–79] In brief, PEI was reacted with dialkyl halide, *i.e.* diiodopentane at a 1 : 0.04 mol ratio (monomer units of PEI/diiodopentane) under reflux for 24 h to form crosslinked nanoparticles. Then, *N*-alkylation was conducted using alkyl halides with different chain lengths, ranging from butyl to octyl halide, added at a 1 : 0.25 mol ratio (monomer units PEI/alkylhalide) and the reaction carried out under reflux for 24 h followed by neutralization using sodium bicarbonate for an additional 24 h under the same conditions. The final step is *N*-methylation, conducted at 42 °C for 48 h with methyl iodide followed by neutralization with sodium bicarbonate for an additional 24 h. The supernatant obtained was decanted, washed with hexane and DDW and then freeze-dried. The final product was ground to result in fine nanoparticles (80–150 nm); yield ≥80% (mol/mol).[3,23,40–43,74–79] These nanoparticles can be synthesized with high reproducibility and were found to exhibit high chemical and thermal stability.[33,35,78] Crosslinking percentage was found to have a direct effect on the size and thermal properties of the QA-PEI nanoparticles. These nanoparticles were mixed with dental composites at different loading percentages (wt/wt%) ranging from 0.5% to 2% and tested against *Enterococcus faecalis*. It was found that complete inhibition of the bacterial growth can be achieved at 2%, while at lower loading percentages incomplete inhibition was found. Nevertheless, the same octyl-alkylated QA-PEI nanoparticles were found to induce complete bacterial inhibition with reduced loading up to 1% following pre-exposure to H_2O_2.

In another study, C8- and C18-alkylated QA-PEI nanoparticles were studied after anchoring in non-leaching polymeric coatings of polyethylene vinyl acetate (PEVA) and polyethylene methacrylic acid (PEMA) coatings, either by casting or spray coating, as well as tested as aqueous suspension.[33] These coatings were tested against representative bacterial species, *Pseudomonas aeruginosa*, *Staphylococcus aureus*, *Escherichia coli* and *heterotrophic plate count*.

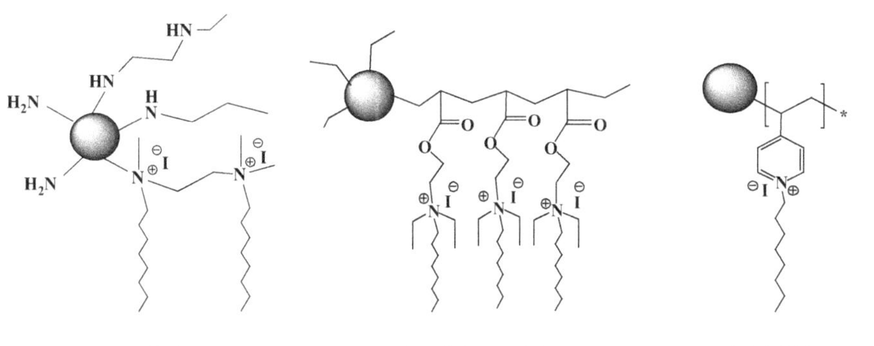

Figure 10.5 Chemical structure of quaternary ammonium (QA)-polyamine nanoparticles: QA-polyethyleneimine (QA-PEI), QA-poly(2-dimethylaminoethyl methacrylate) (QA-PDEAEM) and QA-poly(*N*-alkyl-4-vinylpyridinium) (QA-PVP). For these structures n-octyl iodide was drawn as the alkylating agent and iodide ion as counter ion.

All coatings with the embedded QA-PEI nanoparticles have exhibited strong antimicrobial activity also whenever QA-PEI nanoparticles dispersed in water at concentrations as low as 0.1 ppm.[33]

Crosslinked QA-PEI onto silica particles were synthesized by interpenetrating PEI polymeric networks into silica particles followed by crosslinking, alkylation and quaternization using the same procedure reported for QA-PEI synthesis.[35] By the end of the process, the negatively charged silica particles (−16.7 mV by zeta potential analysis) were shifted to exhibit highly positive charge density of (+50) to (+60) mV following QA-PEI coating resulting in S-QA-PEI particles formation. Antimicrobial activity was tested against representative microbial strains of both Gram-positive and Gram-negative bacteria in plastic coatings and it was found to exhibit a strong antibacterial activity with improved particle distribution in polymeric coatings.[35]

Quaternary ammonium poly(diethylaminoethyl methacrylate) crosslinked nanoparticles (QA-PDEAEM) were prepared by radical polymerization of the quaternized QA-DEAEM monomers with azoisobutyronitrile (AIBN) as free radical initiator under nitrogen plus the addition of 1% either ethylene glycol dimethacrylate (EDGMA) or divinylbenzene (DVB) as crosslinkers (Figure 10.5).[34] 50–70 nm QA-PDEAEM nanoparticles were found to result depending on the crosslinker used. Antimicrobial activity of QA-PDEAEM nanoparticles were tested following embedding in PEVA coating at different loading percentages. These coatings exhibited strong antibacterial activity against four representative Gram-positive and Gram-negative bacteria: *Staphylococcus aureus*, *Escherichia coli*, *heterotrophic plate count* and *Pseudomonas aeruginosa*.

Pyridinium-based particles were prepared using suspension polymerization of 4-vinylpyridine (4-VP) with the crosslinker divinylbenzene (DVB) under reflux condenser with the radical initiator AIBN and polyvinyl alcohol as a dispersing agent (Figure 10.5).[74] The reaction was carried out at 80 °C under nitrogen atmosphere for 17 hours followed by quaternization of the pyridine rings with C6 or C8 under reflux conditions for 48 hours. This resulted in a wide range of particles that varied in size from 400 nm to 1 μm. Although these particles exhibited strong a antibacterial effect against *Streptococcus mutans*, their activity was found to drop following incorporation into restorative composite resins at 1% w/w.[74]

The incorporation of nanoparticles bearing both QA moieties and alkyl modifications within dental composite resins and polymeric surface coatings has been found to be a successful approach for a complete inhibition of bacterial growth. The same factors that affect the antimicrobial properties of linear QA polymers are also valid for crosslinked QA-based nanoparticles and should be taken into account during synthesis and fabrication. On the top of that, other factors related to the nanostructure should be further optimized, including crosslinking percentage, particle chemical stability and aggregation tendency and particle size, shape and distribution in the composite resins or polymeric surface coatings.

10.4 Biomedical Application Summary

Microbial contamination and infections are a continuing source of threat to human life worldwide with each surgical operation, medical implantation and treatment. The antimicrobial QA-based polymers, both linear and crosslinked nanoparticles are considered effective materials that can be used to manage microbial growth and infection development. Given that they have exhibited strong antimicrobial activity and limited microbial resistance, all have paved their way to various biomedical applications.

Several excellent reviews have reported comprehensively the bioactivity of QA polymers and described in detail the potential modifications for addressing various biomedical applications. The main biomedical applications are orthopaedic-related implants,[17,28,88] sutures and wound dressings,[28] dental composites,[40–43,78,79] air filters and water detoxification.[33–35,75] Overall it can be concluded that different QA polymer structures, compositions and moiety modifications are needed to address the various requirements of biomedical applications. Per application, an optimization study is required and should take into account the in-site physiochemical environment of the QA antimicrobial polymers and involved microbes, as well as their growth profile.

The incorporation of the antimicrobial agent as part of a coating or treatment should take into account the mechanism of action of the antimicrobial agent. In some cases, it has been found that some of the active QA compounds lose their bioactivity following their incorporation into surface coatings, *e.g.* C-8 PVP particles in dental composite resins,[74] and accordingly new coating technologies and different structures should be developed.

For challenging biomedical cases such as for medical device-induced infections (*e.g.* catheters) it has been found that monocomponent antibacterial agents are far from meeting requirements. In such cases, dual-action compositions of antibacterial materials or agents were found to be highly effective. One example is cationic polymer poly(4-vinyl-*N*-hexylpyridinium bromide) with embedded silver bromide nanoparticles.[81,103] Another example is *N*-halamine siloxane and QA salt siloxane copolymers for use as biocidal coatings.[81,104] Combination of antimicrobial strategies and bioactive moieties is a growing field of research, giving promising data indicating the potential for reducing microbial resistance in challenging biomedical applications.

10.5 Conclusion and Future Perspectives

Over the past several decades, extensive efforts have been made to control emerging disease infections.[105,106] However, progress has been far from being sufficient and unfortunately lags behind the progress of disease transmission.[3] In this chapter we have reviewed the leading QA-based antimicrobial polymers for biomedical applications, their synthetic methods and modification techniques as well as the recent advances for efficient bioactive QA-based polymeric coatings, including design and manipulation. Emerging progress in other technologies, specifically in nanotechnology, have had significant finger print on designing potent polymeric-based antimicrobial formulations.

For example, the synthesis of antimicrobial polymeric nanoparticles with QA moieties was found to result in antimicrobial agents with improved functionality against well-known challenging microbes, as a result of the high functional QA moieties density per surface area. In addition, such enhancements have been found to lead eventually to better biomedical performance of various biomedical implants including in dental applications as well as various medical catheters, among others.

QA polymers and QA-based biomaterials are effective tools with a proven broad spectrum and long antimicrobial activity; however, there is great need for further investigation of the mechanism of action and developed cytotoxicity. Better understanding of these aspects would definitely assist in designing QA polymers with better performance and eventually to expand their clinical use. In this regard, there is a high demand for the development of advanced QA antimicrobial polymers with the capability of simultaneously targeting multi-species of stubborn pathogenic microbes, *i.e.* challenging resistant bacteria as well as viruses. To do so, further studies focusing on exploring in depth the structure–bioactivity relationship of QA-based biomedical polymers is required. Furthermore, there is a need for further study of QA-based biomedical polymer antimicrobial resistance development and functionality limitation under advanced stages of biofilms.

Lastly, further studies are ongoing, focusing on utilizing the recent progress in polymerization techniques and post-modification strategies for synthesis of advanced QA antimicrobial materials with better control over their structure, moieties composition, stability and functionality. Also, one expected progress is the design of novel QA-polymers having various QA moieties targeting different specific pathogens or QA moieties targeting various stages of microbial infection development. Another expected progress is in developing combined antimicrobial strategies of QA antimicrobial materials. However, further clinical trials are critically needed in order to clinically approve such polymers and eventually commercialize them. Well-designed comparative studies are essential for identifying the most effective antibacterial polymeric QA-based compositions, their limitations and the optimal circumstances for their use.

Acknowledgements

The author would like to acknowledge the Technion for the Technion's presidential grant supporting this work.

References

1. A. E. Madkour, J. M. Dabkowski, K. Nüsslein and G. N. Tew, Fast Disinfecting Antimicrobial Surfaces, *Langmuir*, 2009, **25**, 1060–1067.
2. J. F. Williams and S. D. Worley, Infection-resistant nonleachable materials for urologic devices, *J. Endourol.*, 2000, **14**, 395–400.

3. S. Farah, Bioactive Antimicrobial and Antiproliferative Surfaces, Hebrew University of Jerusalem, 2014.
4. G. J. Verkerke, K. H. Schraffordt, R. P. Veth, J. R. Van Horn, L. Postma and H. J. Grootenboer, First clinical experience with a noninvasively extendable endoprosthesis: a limb-saving procedure in children suffering from a malignant bone tumor, *Artif. Organs*, 1997, **21**, 413–417.
5. K. Page, M. Wilson and I. P. Parkin, Antimicrobial surfaces and their potential in reducing the role of the inanimate environment in the incidence of hospital-acquired infections, *J. Mater. Chem.*, 2009, **19**, 3819–3831.
6. E. R. Kenawy, Biologically Active Polymers IV: Synthesis and Antimicrobial Activity of Polymers Containing 8- HydroxyQuinoline Moiety, *J. Appl. Polym. Sci.*, 2001, **82**, 1364–1374.
7. E. R. Kenawy, S. D. Worley and R. Broughton, The Chemistry and Applications of Antimicrobial Polymers, *Biomacromolecules*, 2007, **8**, 1359–1384.
8. E. R. Kenawy, F. I. Abdel-Hay, A. El-Raheem, R. El-Shanshoury and M. H. El-Newehy, Synthesis and Antibacterial Activity of Modified Glycidyl Methacrylate Polymers Having Quaternary Ammonium and Phosphonium Groups, *J. Controlled Release*, 1998, **50**, 145–152.
9. T. Tashiro, Antibacterial and Bacterium Adsorbing Macromolecules, *Macromol. Mater. Eng.*, 2001, **286**, 63–87.
10. I. L. Shih, M. H. Shen and Y. T. Van, Microbial synthesis of poly (ε-lysine) and its various applications, *Bioresource Technol.*, 2006, **97**, 1148–1159.
11. L. Ferreira and A. Zumbuehl, Non-leaching surfaces capable of killing microorganisms on contact, *J. Mater. Chem.*, 2009, **19**, 7796–7806.
12. N. Beyth, I. Yudovin-Farber, M. Perez-Davidi, A. J. Domb and E. Weiss, Polyethyleneimine nanoparticles incorporated into resin composite cause cell death and trigger biofilm stress in vivo, *PNAS*, 2010, **107**, 22038–22043.
13. L. Timofeeva and N. Kleshcheva, Antimicrobial polymers: mechanism of action, factors of activity, and applications, *Appl. Microbiol. Biotechnol.*, 2011, **89**, 475–492.
14. P. Kurt, L. Wood, D. E. Ohman and K. J. Wynne, Highly Effective Contact Antimicrobial Surfaces via Polymer Surface Modifiers, *Langmuir*, 2007, **23**, 4719–4723.
15. K. Yamamoto, S. Ohashi, M. Aono, T. Kokubo, I. Yamada and J. Yamauchi, Antibacterial activity of silver ions implanted in SiO2 filler on oral streptococci, *Dent. Mater.*, 1996, **12**, 227–229.
16. I. Cakmak, Z. Ulukanli, M. Tuzcu, S. Karabuga and K. Genctav, Synthesis and characterization of novel antimicrobial cationic polyelectrolytes, *Eur. Polym. J.*, 2004, **40**, 2373–2379.
17. V. Sehgal, V. S. Shetty, S. Mogra, G. Bhat, M. Eipe and S. Jacob, Evaluation of antimicrobial and physical properties of orthodontic composite resin modified by addition of antimicrobial agents–an in-vitro study, *Am. J. Orthod. Dentofacial Orthop.*, 2007, **131**, 525–529.

18. S. Imazato, M. Torii, Y. Tsuchitani, J. F. McCabe and R. R. B. Russell, Incorporation of bacterial inhibitor into resin composite, *J. Dent. Res.*, 1994, **73**, 1437–1443.
19. N. Beyth, I. Yudovin-Farber, R. Bahir, A. J. Domb and E. I. Weiss, Antibacterial activity of dental composites containing quaternary ammonium polyethylenimine nanoparticles against Streptococcus mutans, *Biomaterials*, 2006, **27**, 3995–4002.
20. N. Beyth, I. Yudovin-Farber, M. Perez-Davidi, A. J. Domb and E. I. Weiss, Surface antimicrobial activity and biocompatibility of incorporated polyethylenimine nanoparticles, *Biomaterials*, 2008, **29**, 4157–4163.
21. I. Yudovin-Farber, N. Beyth, A. Nyska, E. I. Weiss, J. Golenser and A. J. Domb, Surface Characterization and Biocompatibility of Restorative Resin Containing Nanoparticles, *Biomacromolecules*, 2008, **9**, 3044–3050.
22. I. Yudovin-Farber, N. Beyth, E. I. Weiss and A. J. Domb, Antibacterial effect of composite resins containing quaternary ammonium polyethyleneimine nanoparticles, *J. Nanopart. Res.*, 2009, **12**, 591–603.
23. N. Kawabata and M. Nishiguchi, Antibacterial activity of soluble pyridinium-type polymers, *Appl. Environ. Microbiol.*, 1988, **54**, 2532–2535.
24. K. Talaro and A. Talaro, in *Foundations in Microbiology*, WCB Publishers, Dubuque, IA, 1993, 286.
25. G. J. Li, J. R. Shen and Y. L. Zhu, A study of pyridinium-type functional polymers. III. Preparation and characterization of insoluble pyridinium-type polymers, *J. Appl. Polym. Sci.*, 2000, **78**, 668–675.
26. J. C. Tiller, C. J. Liao, K. Lewis and A. M. Klibanov, Designing surfaces that kill bacteria on contact, *Proc. Natl. Acad. Sci. U. S. A.*, 2001, **98**, 5981–5985.
27. D. G. White, J. Acar, F. Anthony, A. Franklin, R. Gupta, T. Nicholls, Y. Tamura, S. Thompson, E. J. Threlfall, D. Vose, M. V. Vuuren, H. C. Wegener and M. L. Costarrica, Antimicrobial resistance: standardisation and harmonisation of laboratory methodologies for the detection and quantification of antimicrobial resistance, *Rev. Sci. Tech.*, 2001, **20**, 849–858.
28. Y. Jiao, L. Niu, S. Ma, J. Li, F. R. Tay and J.-H. Chen, Quaternary ammonium-based biomedical materials: State-of-the-art, toxicological aspects and antimicrobial resistance, *Prog. Polym. Sci.*, 2017, **71**, 53–90.
29. N. M. Milovic, J. Wang, K. Lewis and A. M. Klibanov, Immobilized N-alkylated polyethylenimine avidly kills bacteria by rupturing cell membranes with no resistance developed, *Biotechnol. Bioeng.*, 2005, **90**, 715–722.
30. M. Ignatova, S. Voccia, B. Gilbert, N. Markova, P. S. Mercuri, M. Galleni, V. Sciannamea, S. Lenoir, D. Cossement, R. Gouttebaron, R. Jerome and C. Jerome, Synthesis of copolymer brushes endowed with adhesion to stainless steel surfaces and antibacterial properties by controlled

nitroxide-mediated radical polymerization, *Langmuir*, 2004, **20**, 10718–10726.

31. J. Huang, H. Murata, R. R. Koepsel, A. J. Russell and K. Matyjaszewski, Antibacterial Polypropylene via Surface-Initiated Atom Transfer Radical Polymerization, *J. Biomacromolecules*, 2007, **8**, 1396–1399.
32. T. Abel, J. I. Cohen, R. Engel, M. Filshtinskaya, A. Melkonian and K. Melkonian, Preparation and investigation of antibacterial carbohydrate-based surfaces, *Carbohydr. Res.*, 2002, **337**, 2495–2499.
33. S. Farah, O. Aviv, N. Laout, S. Ratner, N. Beyth and A. J. Domb, Quaternary ammonium polyethylenimine nanoparticles for treating bacterial contaminated water, *Colloids Surf., B*, 2015, **128**, 614–619.
34. S. Farah, O. Aviv, N. Laout, S. Ratner, N. Beyth and A. J. Domb, Quaternary ammonium poly (diethylaminoethyl methacrylate) possessing antimicrobial activity, *Colloids Surf., B*, 2015, **128**, 608–613.
35. S. Farah, O. Aviv, N. Laout, S. Ratner, N. Beyth and A. J. Domb, Antimicrobial silica particles loaded with quaternary ammonium polyethyleneimine network, *Polym. Adv. Technol.*, 2014, **25**(6), 689–692.
36. K. Yamamoto, S. Ohashi, M. Aono, T. Kokubo, I. Yamada and J. Yamauchi, Antibacterial activity of silver ions implanted in SiO2 filler on oral streptococci, *Dent. Mater.*, 1996, **12**, 227–229.
37. M. J. Wicht, R. Haak, S. Kneist and M. J. Noack, A triclosan-containing compomer reduces Lactobacillus spp. predominant in advanced carious lesions, *Dent. Mater.*, 2005, **21**, 831–836.
38. S. Farah, O. Aviv, M. Daif, K. R. Kunduru, N. Laout, S. Ratner, N. Beyth and A. J. Domb, N-bromo-hydantoin grafted polystyrene beads: Synthesis and nano-micro beads characteristics for achieving controlled release of active oxidative bromine and extended microbial inactivation efficiency, *J. Polym. Sci., Part A: Polym. Chem.*, 2016, **54**, 596–610.
39. S. Farah, O. Aviv, N. Laout, S. Ratner and A. J. Domb, Antimicrobial N-brominated hydantoin and uracil grafted polystyrene beads, *J. Controlled Release*, 2015, **216**, 18–29.
40. O. Aviv, S. Farah, AmirN, N. Laout, S. Ratner and A. J. Domb, *N*-Bromo-dimethylhydantoin Polystyrene Resin for Water Microbial Decontamination, *Biomacromolecules*, 2015, **16**(4), 1442–1447.
41. Y. Sun, T. Y. Chen, S. D. Worley and G. Sun, Novel refreshable N-halamine polymeric biocides containing imidazolidin-4-one derivatives, *J. Polym. Sci., Part A: Polym. Chem.*, 2001, **39**, 3073–3084.
42. J. Liang, Y. Chen, X. Ren, R. Wu, K. Barnes, S. D. Worley, S. D. R. M. Broughton, U. Cho, H. Kocer and T. S. Huang, Fabric treated with antimicrobial N-halamine epoxides, *Ind. Eng. Chem. Res.*, 2007, **46**, 6425–6429.
43. X. Ren, L. Kou, H. B. Kocer, S. D. Worley, R. M. Broughton, Y. M. Tzou and T. S. Huang, Antimicrobial modification of polyester by admicellar polymerization, *J. Biomed. Mater. Res., Part B*, 2009, **89**, 475–480.
44. X. Ren, L. Kou, H. B. Kocer, C. Zhu, S. D. Worley, R. M. Broughton and T. S. Huang, Antimicrobial coating of an N-halamine biocidal

monomer on cotton fibers via admicellar polymerization, *Colloids Surf., A*, 2008, **317**, 711–716.
45. S. J. Grunzinger, P. Kurt, K. M. Brunson, L. Wood, D. E. Ohman and K. J. Wynne, Biocidal activity of hydantoin-containing polyurethane polymeric surface modifiers, *Polymer*, 2007, **48**, 4653–4662.
46. E. R. Kenawy, S. D. Worley and R. Broughton, The chemistry and applications of antimicrobial polymers; A state of the art review, *Biomacromolecules*, 2007, **8**, 1359–1384.
47. G. Sun and S. D. Worley, Chemistry of Durable and Regenerable Biocidal Textiles, *J. Chem. Educ.*, 2005, **82**, 60–64.
48. Y. Chen, S. D. Worley, J. Kim, C. I. Wei, T. Y. Chen, J. Suess, H. Kawai and J. F. Williams, Biocidal polystyrenehydantoin beads. 2. Control of chlorine loading, *Ind. Eng. Chem. Res.*, 2003, **42**, 5715–5720.
49. J. Lin, C. Winkelmann, S. D. Worley, R. M. Broughton and J. F. Williams, Antimicrobial treatment of nylon, *J. Appl. Polym. Sci.*, 2001, **81**, 943–947.
50. D. B. Elrod, J. G. Figlar, S. D. Worley, R. M. Broughton, J. R. Bickert, J. I. Santiago and J. F. Williams, A Novel Biocidal Elastomer, *Rubber Chem. Technol.*, 2001, **74**, 331–337.
51. J. Lin, C. Winkelmann, S. D. Worley, J. Kim, C. I. Wei, U. Cho, R. M. Broughton, J. I. Santiago and J. F. Williams, Biocidal Polyester, *J. Appl. Polym. Sci.*, 2002, **85**, 177–182.
52. X. Ren, L. Kou, J. Liang, S. D. Worley, Y. M. Tzou and T. S. Huang, Antimicrobial efficacy and light stability of N-halamine siloxanes bound to cotton, *Cellulose*, 2008, **15**, 593–598.
53. M. W. Eknoian, S. D. Worley, J. Bickert and J. F. Williams, Novel antimicrobial N-halamine polymer coatings generated by emulsion polymerization, *Polymer*, 1999, **40**, 1367–1371.
54. L. Kou, J. Liang, X. Ren, H. B. Kocer, S. D. Worley, R. M. Broughton and T. S. Huang, Novel N-halaminesilanes, *Colloids Surf., A*, 2009, **345**, 88–94.
55. Y. Chen, S. D. Worley, T. S. Huang, J. Weese, J. Kim, C. I. Wei and J. F. Williams, Biocidal polystyrene beads. III. Comparison of N-halamine and quat functional groups, *J. Appl. Polym. Sci.*, 2004, **92**, 363–367.
56. G. Sun, T. Y. Chen, M. S. Habercom, W. B. Wheatley and S. D. Worley, Performance of A New Polymeric Water Disinfectant, *J. Am. Water. Resour. Assoc.*, 1996, **32**, 793–797.
57. Y. Chen, S. D. Worley, J. Kim, C. I. Wei, T. Y. Chen, J. I. Santiago, J. F. Williams and G. Sun, Biocidal poly(styrenehydantoin) beads for disinfectants, *Ind. Eng. Chem. Res.*, 2003, **42**, 280–284.
58. V. S. Panangala, L. Liu, G. Sun, S. D. Worley and A. Mitra, Inactivation of rotavirus by new polymeric water disinfectants, *J. Virol. Methods*, 1997, **66**, 263–268.
59. J. Liang, R. Wu, J. W. Wang, K. Barnes, S. D. Worley, U. Cho, J. Lee, R. M. Broughton and T. S. Huang, N-halamine biocidal coatings, *J. Ind. Microbiol. Biotechnol.*, 2007, **34**, 157–163.

60. M. W. Eknoian, J. H. Putman and S. D. Worley, Monomeric and Polymeric N-Halamine Disinfectants, *Ind. Eng. Chem. Res.*, 1998, **37**, 2873–2877.
61. W. Yingfeng, Y. Maoli, L. Xinghua, L. Lin, L. Zhiguang, R. Xuehong and S. Yuyu, Tailored synthesis of polymer-brush-grafted mesoporous silicas with N-halamine and quaternary ammoniumgroups for antimicrobial applications, *J. Colloid Interface Sci.*, 2019, **533**, 604–611.
62. J. Zhiqiang, Y. Xiufang, Z. Lianhong, W. Shelby Davis and L. Jie, Eco-friendly synthesis of regenerable antimicrobial polymeric resin with N-halamine and quaternaryammonium salt groups, *RSC Adv.*, 2014, **12**, 6048–6054.
63. R. Xuehong, L. Ying and L. Yin, Halo-amine polymer antimicrobial agent containing quaternary ammonium groups and its preparation method and application, *Faming ZhuanliShenqing*, 2014, CN 103524652 A 20140122.
64. C. Willyard, Drug-resistant bacteria ranked, *Nature*, 2017, **543**, 15.
65. E. Nicolle Lindsay, Catheter-related urinary tract infection, *Drugs Aging*, 2005, **22**(8), 627–639.
66. S. S. Pradeep Kumar, H. V. Easwer and A. Maya Nandkumar, Multiple Drug Resistant Bacterial Biofilms on Implanted Catheters – A Reservoir of Infection, *J. Assoc. Physicians India*, 2013, **61**, 702–707.
67. M. Snoussi, J. P. Talledo, N.-A. Del Rosario, S. Mohammadi, B.-Y. Ha, A. Košmrlj and S. Taheri-Araghi, Heterogeneous absorption of antimicrobial peptide LL37 in Escherichia coli cells enhances population survivability, *eLife*, 2018, **7**, e38174.
68. P. Singha, J. Locklin and H. Handa, A review of the recent advances in antimicrobial coatings for urinary catheters, *ActaBiomaterialia*, 2017, **50**, 20–40.
69. R. Hu, G. Li, Y. Jiang, Y. Zhang, J.-J. Zou, L. Wang and X. Zhang, Silver-zwitterion organic–inorganic nanocomposite with antimicrobial and antiadhesive capabilities, *Langmuir*, 2013, **29**(11), 3773–3779.
70. C. Díaz, A. Miñán, P. Schilardi and M. F. L. de Mele, Synergistic antimicrobial effect against early biofilm formation: micropatterned surface plus antibiotic treatment, *Int. J. Antimicrob. Agents*, 2012, **40**, 221–226.
71. J. Lin, S. Qiu, K. Lewis and A. M. Klibanov, Mechanism of bactericidal and fungicidal activates of textiles covalently modified with alkylated polyethylenimine, *Biotechnol. Bioeng.*, 2003, **83**, 168–172.
72. J. Hiraki, ε-Polylysine, its development and utilisation, *Fine Chem.*, 2000, **29**, 25–28.
73. J. Hiraki, Basic and applied studies on e-polylysine, *J. Antibact. Antifungal Agents*, 1995, **33**, 349–354.
74. I. Yudovin-Farber, Bioactive polyelectrolytes, Hebrew University of Jerusalem, 2008.
75. K. R. Kunduru, M. Nazarkovsky, S. Farah, R. P. Pawar, A. Basu and A. J. Domb, Nanotechnology for water purification: applications of

nanotechnology methods in wastewater treatment, *Water Purification*, 2017, 33–74.
76. A. Ortega, S. Farah, P. Tranque, A. V. Ocaña, S. H. Nam-Cha, N. Beyth, C. Gómez-Roldán, R. Pérez-Tanoira, A. J. Domb, F. C. Pérez-Martínez and J. Pérez-Martínez, Antimicrobial evaluation of quaternary ammonium polyethyleneimine nanoparticles against clinical isolates of pathogenic bacteria, *IET Nanobiotechnol.*, 2015, **9**, 342–348.
77. A. Ortega, F. C. Pérez-Martínez, S. Farah, S. H. Nam-Cha, A. V. Ocaña, N. Beyth, A. J. Domb, G. Martínez, C. Gómez-Roldán and J. Pérez-Martínez, Quaternary ammonium polyethyleneimine nanoparticles: antimicrobial evaluation against bacteria from peritoneal dialysis related peritonitis, *Nephrolo. Dial. Transplanat.*, 2015, 30, Oxford University Press, iii537–iii538.
78. S. Farah, W. Khan, I. Farber, D. Kesler-Shvero, N. Beyth, E. I. Weiss, A. J. Domb and Q. A. Crosslinked, PEI nanoparticles: synthesis reproducibility, chemical modifications, and stability study, *Polym. Adv. Technol.*, 2013, **24**, 446–452.
79. A. J. Domb, N. Beyth and S. Farah, Quaternary Ammonium Antimicrobial Polymers, MRS Online Proceedings Library Archive, Cambridge University Press, 2013, 1569, 97–107.
80. K. De Prijck, N. De Smet, T. Coenye, E. Schacht and H. J. Nelis, *Mycopathologia*, 2010, **170**, 213–221.
81. A. Jain, S. DuvvuriL, S. Farah, N. Beyth, A. J. Domb and W. Khan, Antimicrobial Polymers, *Adv. Healthcare Mater.*, 2014, **3**, 1969–1985.
82. G. Li and J. Shen, *Abstr. Papers Am. Chem. Soc.*, 1998, **216**, U97.
83. E. S. Park, H. S. Kim, M. N. Kim and J. S. Yoon, Antibacterial activities of polystyrene–poly(4-vinyl pyridine) and poly(styrene–4-vinyl pyridine), *Eur. Polym. J.*, 2004, **40**, 2819–2822.
84. S. Imazato, N. Ebi, Y. Takahashi, T. Kaneko, S. Ebisu and R. Russell, Antibacterial activity of bactericide-immobilized filler for resin-based restoratives, *Biomaterials*, 2003, **24**, 3605–3609.
85. S. Imazato, Y. Kinomoto, H. Tarumi, S. Ebisu and F. R. Tay, Antibacterial activity and bonding characteristics of an adhesive resin containing antibacterial monomer MDPB, *Dent. Mater. J.*, 2003, **19**, 313–319.
86. S. Imazato, Bio-active restorative materials with antibacterial effects: new dimension of innovation in restorative dentistry, *Dent. Mater. J.*, 2009, **28**, 11–19.
87. Y. Zhang, J. Jiang and Y. Chen, *Polymer*, 1999, **40**, 6189.
88. T. Honglue, M. Rui, L. Chucheng, L. Ziwei and T. Tingting, Quaternized chitosan as an antimicrobial agent: antimicrobial activity, mechanism of action and biomedical applications in orthopaedics, *Int. J. Mol. Sci.*, 2013, **14**, 1854–1869.
89. S. Farah, J. C. Doloff, D. G. Anderson and R. Langer, Crosslinked chitosan nanoparticles, in *Drug Delivery Techniques, Aldrich Materials Science*, Sigma-Aldrich Co., Milwaukee, WI, 2015, 18–21.

90. H. Sashiwa and S. I. Aiba, Chemically modified chitin and chitosan as biomaterials, *Prog. Polym. Sci.*, 2004, **29**, 887–908.
91. G. N. Rimondino, M. C. Strumia and M. Martinelli, Synthesis and Characterization of Water-Soluble Dendronized Chitosan Using Newkome-Type Dendrons, *ACS Sustainable Chem. Eng.*, 2014, **2**, 2582–2587.
92. H. Guanghua, Q. Xiaoyan, L. Cuifang, Y. Ying, Modified chitosan base water-superabsorbent gel, preparation and application thereof, *Faming ZhuanliShenqing*, 2017, CN 107141519 A 20170908.
93. C. Yu, T. Zongrui and L. Yang, Chitosan-based multifunctional hemostatic microsphere with good biodegradability and hydrophilic and antimicrobial performances, *Faming ZhuanliShenqing*, 2015, CN 105001434 A 20151028.
94. N. KoonGee, L. Min, W. Rong, K. E. Tang and C. Edmund, Antimicrobial coatings for biomedical applications: Challenges and opportunities offered by natural polymers, Abstracts of Papers, 249th ACS National Meeting & Exposition, Denver, CO, United States, 2015, March 22–26, PMSE-204.
95. N. E. Kotelnikova and E. F. Panarin, Cellulose modification by biologically active substances for biomedical applications, *Cell Chem. Technol.*, 2005, **39**, 437–450.
96. Y. Chen, P. A. Wilbon, Chen, P. Yung, J. Zhou, M. Nagarkatti, C. Wang, F. Chu, A. W. Decho and C. Tang, Amphipathic antibacterial agents using cationic methacrylic polymers with natural rosin as pendant group, *RSC Adv.*, 2012, **2**, 10275–10282.
97. S. Garcia-Arguelles, M. C. Serrano, M. C. Gutierrez, M. L. Ferrer, L. Yuste, F. Rojo and F. del Monte, Deep Eutectic Solvent-Assisted Synthesis of Biodegradable Polyesters with Antibacterial Properties, *Langmuir*, 2013, **29**, 9525–9534.
98. A. C. Engler, P. K. Tan Jeremy, Y. Ong Zhan, D. J. Coady, W. L. Ng Victor, Y. Y. Yang and J. L. Hedrick, Antimicrobial polycarbonates: investigating the impact of balancing charge and hydrophobicity using a same-centered polymer approach, *Biomacromolecules*, 2013, **14**, 4331–4339.
99. S. Adhikari, E. Lutter and T. Nelson, Antimicrobial activities of eumelanin-inspired conjugated oligomers and polymers. Abstracts of Papers, 252nd ACS National Meeting & Exposition, Philadelphia, PA, United States, 2016, August 21–25, POLY-333.
100. Y. Zhengwei and W. Shaofei, Manufacture method of biodegradable polycation-polyester containing quaternary ammonium salt at mainchain, *Faming ZhuanliShenqing*, 2016, CN 105482105 A 20160413.
101. X. Chaobo, L. Changhua and H. Aibin, Chitosan/poly(4-vinylpyridinium) quaternary ammonium salt blended film, its preparation and application, *FamingZhuanliShenqing*, 2004, CN 1482164 A 20040317.
102. N. Beyth, S. Farah, A. J. Domb and E. I. Weiss, Antibacterial dental resin composites, *React. Funct. Polym.*, 2014, **75**, 81–88.

103. V. Sambhy, M. M. MacBride, B. R. Peterson and A. Sen, Silver bromide nanoparticle/polymer composites: dual action tunable antimicrobial materials, *J. Am. Chem. Soc.*, 2006, **128**, 9798.
104. J. Liang, Y. Chen, K. Barnes, R. Wu, S. Worley and T.-S. Huang, N-halamine/quat siloxane copolymers for use in biocidal coatings, *Biomaterials*, 2006, **27**, 2495–2501.
105. Z. Chen and Y. Sun, Antimicrobial polymers containing melamine derivatives. II. Biocidal polymers derived from 2-vinyl-4,6-diamino-1,3,5-triazine, *J. Polym. Sci. Part A: Polym. Chem.*, 2005, **43**, 4089–4098.
106. U. Makal, L. Wood, D. E. Ohman and K. J. Wynne, Polyurethane biocidal polymeric surface modifiers, *Biomaterials*, 2006, **27**, 1316–1326.

CHAPTER 11

Polymer–Drug Conjugates for Treating Local and Systemic Fungal Infections

ARUL PRAKASH FRANCIS AND A. JAYAKRISHNAN*

Biomaterials Laboratory, Bhupat and Jyoti Mehta School of Biosciences, Department of Biotechnology, Indian Institute of Technology Madras, Chennai 600036, Tamil Nadu, India
*Email: jayakrishnan1953@gmail.com

11.1 Introduction

Fungal infections commonly found in humans are categorized as superficial, localized, or disseminated. Superficial mycoses confined to the skin and nails are generally not life-threatening and could be treated with topical antifungal agents. However, it is difficult to treat the localized mycoses in gastrointestinal, genitourinary, and respiratory tracts. Disseminated mycoses affecting visceral organs, central nervous system, and generalized fungal septicemia commonly found in immunosuppressed hosts cannot be treated easily.[1] Disseminated fungal infections resulting from the pathogens *Aspergillus*, *Candida*, *Cryptococcus*, and *Pneumocystis* have emerged as a significant problem in immunocompromised patients, such as individuals suffering from HIV/AIDS or diabetes mellitus or following organ transplantations and immunosuppressive chemotherapy during cancer treatment.[2] These pathogens are responsible for the majority of morbidity as well as 90% of lethal fungal-related cases. About 30–50% of patients with invasive

Biomaterials Science Series No. 5
Antimicrobial Materials for Biomedical Applications
Edited by Abraham J. Domb, Konda Reddy Kunduru and Shady Farah

Published by the Royal Society of Chemistry, www.rsc.org

aspergillosis die due to late diagnosis, resistance of existing drugs, and brain dissemination.[3,4]

In immunocompromised hosts, aspergillosis can exist in the respiratory tract in the form of non-invasive aspergilloma or as an invasive disseminated infection, but prognosis in these cases is quite poor.[5] Candidiasis is an opportunistic fungal infection commonly reported in AIDS patients (80%) and is due primarily to *Candida albicans*. Other *Candida* species such as *Candida tropicalis* and *Candida parapsilosis* are also emerging as important opportunistic pathogens, and their importance should not be ignored.[5,6] Localized candidiasis is characterized by mucocutaneous lesions with a whitish, curd-like exudate, substernal burning, and difficulty in swallowing, while disseminated candidiasis results in non-specific symptoms such as fever, fatigue, weight loss, shaking chills, hypotension, tachycardia, and dyspnea. Another common systemic fungal infection in AIDS patients is cryptococcosis, which mainly exists in the form of meningitis due to *Cryptococcus neoformans.* Cryptococcal meningitis progresses rapidly to death and mortality rates remain unacceptably high at ~50% in the treated population.[1]

11.2 Discovery of Antifungal Drugs

The first antifungal agent reported was griseofulvin in 1939, then benzimidazole in 1944, followed by propamidine in 1945.[7–10] The discovery of the first polyene macrolide antifungal, nystatin (NYS), in 1950 by Hazen and Brown, became a lead for the modern era of antifungal therapy.[11,12] In 1951, stilbamidine and its less toxic derivative, 2-hydroxystilbamidine were used in a few human cases of blastomycosis with limited success.[13,14] The discovery of amphotericin B (AmB) in 1955 and its advantages in comparison with existing antifungal drugs in the treatment of blastomycosis led to the development of more antifungal agents such as oral griseofulvin and topical chlormidazole in 1958, intravenous (*iv*) AmB in 1960, 5-fluorocytosine in 1964, and miconazole and clotrimazole in 1969. The imidazole antifungals, with broad-spectrum activity against dermatophytes, *Candida*, and other infections were developed during the 1970s. Econazole for topical use, introduced in 1974, is still used. However, the systemic clotrimazole and miconazole introduced during that decade had limited success.[15–18] Ketoconazole was the only oral agent available for the treatment of systemic fungal infections for a decade from 1981 and became the base in the search for new safer and effective agents. The significance of fluconazole was realized only in 1990, a decade after its discovery.[10] Fluconazole, the first broad-spectrum triazole, overcame the shortcomings of the imidazoles—poor solubility and lack of an *iv* formulation—while itraconazole introduced in 1992 expanded the activity beyond *Candida* spp.[19]

Currently, the three major classes of drugs with different mechanisms of action used as antifungals for systemic infections are (i) polyenes, which exhibit antifungal potential by disrupting the fungal membranes; (ii) azoles, which control fungal growth by inhibiting ergosterol biosynthesis; and (iii) echinocandins, which prevent fungal development by obstructing the synthesis of cell wall β-glucan. However, most of these drugs have serious disadvantages like low specificity (toxicity), narrow spectrum of activity, drug–drug interactions, and high cost. For example, AmB, the gold standard among antifungal drugs, which is highly recommended for systemic infections as well as for the treatment for visceral and mucocutaneous-leishmaniasis, shows limitations in therapy due to its poor water solubility followed by highly toxic side effects, especially high nephrotoxicity. These drawbacks highlight the vital need for the antifungal therapeutics with high potency, selectivity, and improved pharmacokinetics.[3,4]

In order to enhance the potency and selectivity of antifungal agents, researchers attempted to alter the properties of the drugs through physical and chemical modifications. Most of the physical modifications designed for intravenous administration include the micellar suspension of AmB in deoxycholic acid (Fungizone®), non-covalent AmB lipid complexes (ABLC™), liposomal form of AmB (AmBisome®) and AmB colloidal dispersion (Amphocil™). These physically modified forms of AmB possess merits and demerits resulting from their modified pharmacokinetic properties[20] as listed in Table 11.1. All these formulations ensure the smoother release of AmB accompanied by its restricted distribution in the kidney, thereby lowering its nephrotoxicity. However, infusion-related side effects, high cost, instability of liposomes, and other disadvantages still limit their widespread use in clinical practice. New drug delivery systems such as polymeric micelles, nanoparticles, dendrimers, and microspheres have also been used for the formulation of antifungal drugs. Lavasanifar *et al.*[21] encapsulated AmB in poly(ethylene oxide)-*block*-poly(*n*-hexyl stearate-L-aspartamide) micelles and showed that the micelle formation influenced the drug solubility and toxicity. AmB solubility was 250 $\mu g\,mL^{-1}$ in freeze-dried micelles. Moreover, no hemolysis at a drug concentration of 20 $\mu g\,mL^{-1}$ was observed. Vandermeulen *et al.*[22] encapsulated AmB in poly(ethylene glycol)-*block*-poly(caprolactone-*co*-trimethylene carbonate) polymeric micelles. Its solubility increased by two orders of magnitude based on polymer concentration and the aggregation state of the drug was reduced. Chakraborty and Naik[23] complexed AmB with sulfobutyl ether and hydroxypropyl β-cyclodextrin

Table 11.1 Pharmacokinetic properties of AmB preparations.[20]

Parameters[a]	AmBdeoxycholate	Liposomal AmB	AmB lipid complex	AmB colloidal dispersion
C_{max} ($\mu g\,mL^{-1}$)	1.5–2.9	83 ± 35.2	1.7	2.9
AUC ($\mu g\,h\,mL^{-1}$)	17.1–36	555 ± 311	14.0 ± 7.0	36
$t_{1/2}$ (h)	24	8.6 ± 3.1	173.4	28.2
V_d ($L\,kg^{-1}$)	5.0 ± 2.8	0.16	131 ± 57.7	4.1
Cl ($m\,Lh^{-1}\,kg^{-1}$)	38.0 ± 15.0	11.0 ± 6.0	436 ± 188	112

[a]AMB: amphotericin B; AUC: area under the concentration–time curve; Cl: clearance; C_{max}: peak plasma concentration; $t_{1/2}$: half-life; V_d: volume of distribution.

(HPBCD). They showed 15% and 22% lysis of erythrocytes, respectively (at a concentration of 38 $\mu g\,mL^{-1}$), which was milder than the AmBisome®. Lincopan *et al.*[24] evaluated the nephrotoxicity of a novel AmB formulation with dioctadecyl dimethylammonium bromide (DODAB). Nephrotoxicity of the formulation was lower than that of Fungizone®. Dose-dependent cytotoxicity of DODAB/AmB was lower than that exhibited by Fungizone®. DODAB/AmB formulation yielded high efficacy of treatment in a mouse model in which candidiasis had been developed. Jain *et al.*[25] described a targeted delivery of AmB to macrophages. Muramyl dipeptide-conjugated multimeric poly(propyleneimine) dendrimers were loaded with AmB. Cell uptake studies showed appreciable macrophage targeting. The dendrimer formulation showed higher or equivalent antiparasitic activity against parasite-infected macrophage cell lines and *in vivo* infection in balb/c mice. Gupta *et al.*[26] prepared self-assembled ionically cross-linked polymeric nanoparticles loaded with AmB. The *in vivo* experiments showed macrophage uptake and tissue localization towards *Leishmania*-infected organs, *i.e.* spleen and liver, while lower uptake was seen in the kidney. The *in vitro* and *in vivo* results showed better efficacy over AmB alone. Recently, Tutaj *et al.*[27] synthesized silver hybrid nanoparticles using AmB. The effect of AmB–silver nanoparticles on fungal growth was compared with Fungizone®. The inhibitory effects were higher than Fungizone®. Although various methods were used to enhance the efficacy of the antifungal drugs, polymer–drug conjugates of antifungal drugs have received more attention in recent years. This chapter discusses the polymer–drug conjugates of antifungal drugs, their merits, and demerits.

11.3 Polymer–Drug Conjugates

The solubility of small molecules in aqueous media is a major problem in drug development. Recently, it has been estimated that 40–60% of drugs have poor bioavailability because of low aqueous solubility. The solubility of the drugs could be enhanced by conjugating the drug with hydrophilic polymers. A design for polymer conjugation with therapeutic agents was proposed by Ringsdorf.[28] The scheme representing the polymer–drug conjugation consists of three main components, as shown in Figure 11.1: (i) a polymeric carrier, which provides hydrophilicity, ensuring water solubility for the water-insoluble drugs; (ii) therapeutically active agents are then attached to the polymer backbone either through a linker or a functional group present on the therapeutically active agent; (iii) a targeting moiety such as an antibody can be incorporated which would target the drug towards the specific tissue. The polymer–drug conjugates offer several significant advantages over traditional small-molecule therapeutics. A major benefit of the conjugating drug to the polymeric carrier is to modulate the disposition of drug in the body, a process that allows high therapeutic efficacy with low adverse effects. To increase the targeting efficiency of drug through the *iv* route, it is essential to decrease the non-specific removal of

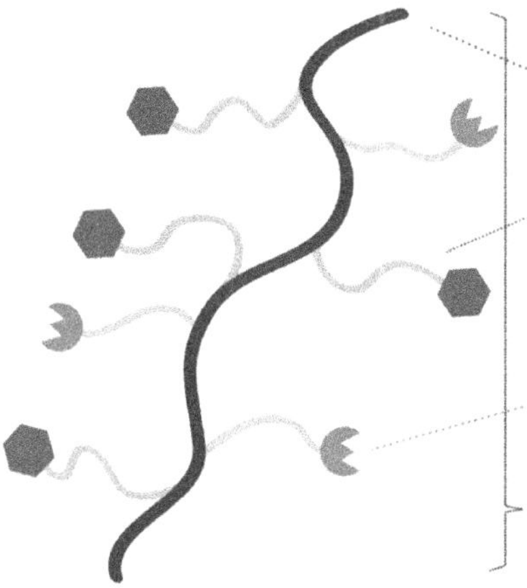

Figure 11.1 The rationale for drug delivery *via* polymer–drug conjugates. Reproduced from www.creative-biostructure.com/polymer-drug-conjugation-service_31.htm with permission from Creative Biostructure.

the drugs by organs such as the kidney and the liver. An obvious benefit expected for drug conjugation to the polymeric carrier is the extended plasma half-life due to decreased kidney excretion. The threshold molar mass for glomerular excretion is approximately 40 000–70 000 g mol^{-1} in the case of natural water-soluble polymers. Conjugation to hydrophilic polymers with neutral or slightly anionic character can attain prolonged circulation. By employing appropriate target moieties, and with the help of spacers, drugs can be targeted to specific sites. The aim of the linker goes further than merely connecting the polymer to the drug; it can also serve to trigger the drug release under certain conditions, such as a change in pH or the presence of enzymes, such as esterases, lipases, or proteases.[29] The stability of the linkage between the drug and polymer determines the rate of release. In addition, a targeting moiety or a solubilizer may also elevate the therapeutic index of the conjugates.[30] Prolonged or controlled drug delivery from a carrier can also help to overcome difficulties with patient compliance in multi-dose therapy. Polymer–drug conjugates retained in the body for a longer period of time can serve as a depot, enabling controlled release of the drug.

11.3.1 Importance of the Polymeric Backbone as Drug Carrier

It is very important to find a suitable polymer carrier for therapeutic application. The choice of carrier for the conjugate has very important roles in altering the pharmacokinetics and pharmacodynamics of the drug.[31] The physical nature of the polymer plays a vital role in the biological environment. The molecular weight, polydispersity, architecture, charge, and hydrophilicity define the drug solubility and loading, its biodistribution, body excretion, and the interaction with the immune system.[32] The polymeric backbone of the conjugate can be biodegradable or non-biodegradable. Various synthetic and natural polymer-based carriers have been developed and utilized to control the release of drug and other active agents for various diseases. Some regularly used polymers in the field of drug delivery are given in Table 11.2. Synthetic polymers such as

Table 11.2 Polymers employed in drug-delivery systems.

Polymer name	Structure of the repeat unit
Poly(ethylene glycol)	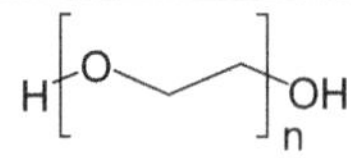
Poly(vinyl alcohol)	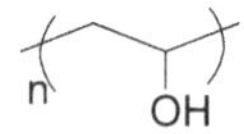
Poly(vinyl pyrrolidone)	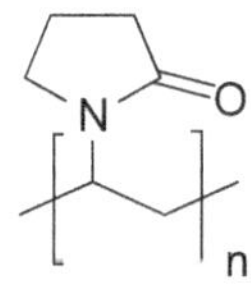
Polylactide	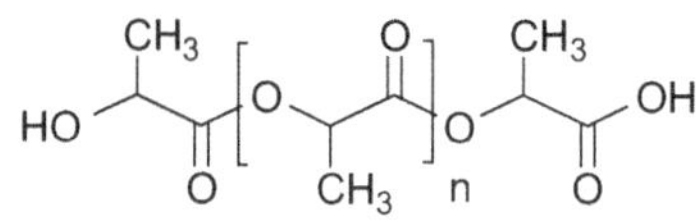
Polyglycolide	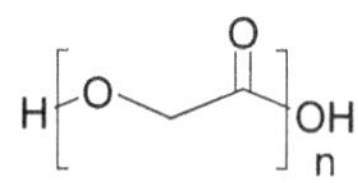
Polyanhydride	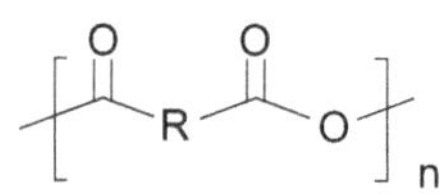
Polyorthoester	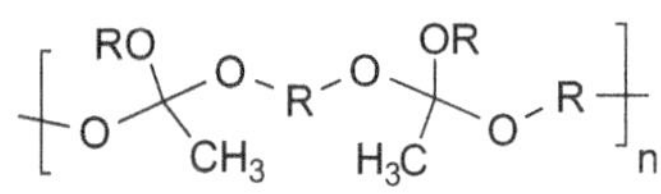
N-(2-hydroxypropyl methacrylamide)	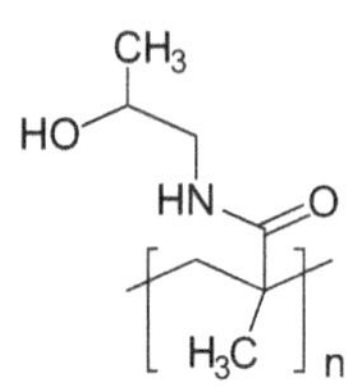
Poly(amino acids)	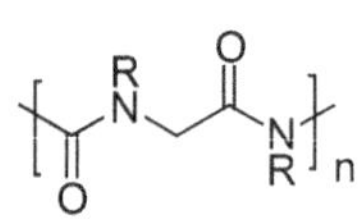

hydroxypropyl methacrylamide (HPMA) and its copolymers, polyethyleneimine (PEI), linear polyamidoamines, poly(vinyl pyrrolidone) (PVP), poly(glycolic acid) (PGA), polyacrylamide (PAM), poly(dimethyl acrylamide) (PDMA), poly(ethylene glycol) (PEG), poly(vinyl alcohol) (PVA),[33] and natural polymers such as chitosan, dextran, hyaluronic acid, pullulan, pectin, and alginic acid[34] have been investigated. The different types of polymer backbones can be divided into two subgroups according to the feature of biodegradability in the main chain, as follows. (i) Synthetic polymers and natural polymers as degradable polymeric biomaterials: the degradation of a polymeric biomaterial after its intended purpose should result in natural by-products, such as carbon dioxide, nitrogen, water, *etc.* The breakdown involves cleavage of hydrolytically or enzymatically susceptible bonds in the polymer, leading to polymer erosion (*e.g.* polysaccharides, polyanhydrides, polyesters, polylactide, and polyglycolide). (ii) Non-degradable polymers that are mostly derived from synthetic polymers. The chance of break-up in these polymeric networks is very rare in any condition (*e.g.* HPMA).

11.3.2 Cell Uptake

Cellular uptake of low molecular weight molecules occurs by the rapid *trans*-membrane passage. The combination of De Duve *et al.*'s realization that the endocytic pathway might be useful for lysosomotropic drug delivery[35] and Ringsdorf's[28] vision of idealized polymer chemistry for the drug conjugation produced the concept of targetable polymer–drug conjugates. With the aid of receptor-mediated endocytosis, a drug can be delivered into a particular cell population for the purpose; a pilot moiety having a specific affinity towards the cellular receptor should be introduced into the conjugates, as polymers without a specific affinity towards plasma membrane are taken up by cells by a process called fluid-phase endocytosis, in which polymers are engulfed into endosomes concomitantly with liquid components. The endocytic vesicles with ingested substances (polymers) are transferred to organelles called endosomes, where they are eventually carried to lysosomes. Lysosomes contain a variety of enzymes capable of degrading most natural macromolecules entering the cell. In the acidic milieu of lysosome (pH 5), substances suffer digestion by lysosomal enzymes (acid hydrolases).[36] Ideally, macromolecules should be stable and pharmacologically inactive during circulation in the bloodstream and active compounds are released after uptake by the cells.

11.3.3 Choice of Linkers

A linker is a bifunctional molecule that connects the drug to polymer. It is selected in such a way that the hydrolysis of the linker–polymer bond is the rate-limiting step in the drug release from the polymer. Esters, carbonates, amides, imines, amines, carbamates, and urethanes are chosen to facilitate aqueous hydrolysis. The rate of the drug release will increase in the order ester > carbonate > urethane > amide > amine and the controlled enzymatic hydrolysis is achieved by using peptidyl linkers.[37] Thus, the choice of a linker between the active agent and the polymer backbone has a significant role in drug delivery. The linker should be stable in the bloodstream and in physiological pH and efficiently cleaved at the target tissue. For some medical applications, the drug is biologically inert when bound to the polymer. Therefore, a drug that would otherwise cause side effects can exist in the bloodstream as a non-active conjugate, thus avoiding any harm to the patient. The linkers in the polymer therapeutics field are either hydrolytically or pH-sensitive or enzymatically cleavable. The release of the free drug can occur extracellularly or intracellularly.[38]

In polymer–drug conjugates, acid-labile spacers are used extensively to cleave the drug from the carrier under low pH conditions. For instance, imines, esters, carbonates, and orthoesters are very labile in nature and cleave the bond between drug and carrier under low pH conditions. This can be explained by pH variation from the physiological pH of 7.2–7.4 in the blood or extracellular space to pH 4.0–6.5 in the various intracellular compartments. In addition, the fact that the extracellular pH surrounding tumor tissue is slightly lower than that of normal tissue supports the rationale for the incorporation of these spacers.[39]

Enzymes play a major role in biological function. In drug delivery, enzymes can facilitate the release of the drug from the carrier. Depending on their functional nature, enzymes are classified into different categories. For instance, nucleases, proteases, phosphatases, and lipases are present in the lysosomes. Because water-soluble polymer–drug conjugates enter the cell *via* endocytosis and then continue to the lysosomes, the presence of an enzymatically cleavable linker between the backbone and the drug enables selective targeting.[40] The peptidyl spacers demonstrate the usefulness of enzymatic cleavage of the drug from carriers only at the tumor site, specifically by cathepsin B or extracellularly by cathepsin K. Cathepsin B-cleavable tetrapeptide, Gly-Phe-Leu-Gly, is used as a pendant group on HPMA copolymer for the delivery of several anticancer drugs, because cathepsin B is overexpressed in many tumors. Cathepsin K-cleavable peptides have been used in polymer–drug conjugates to deliver drugs to bone tissues for the treatment of calcified diseases.[37,41] Proteases, which are overexpressed in tumors and play a critical role in tumor progression, angiogenesis, and metastasis were studied, *e.g.* the lysosomal protease legumain and the matrix metalloproteinases (MMPs) (MMP2 and MMP9), which are active intracellularly and extracellularly, respectively.[40,42]

11.4 Natural Polymers

Polysaccharides are vital biomolecules required for life. Polysaccharides exhibit unique properties such as biocompatibility and biodegradability, which are essential for a material to be used in biomedical applications.[43] An enzymatically biodegradable polysaccharide can be appropriate for administration without the fear of accumulation in the body. Many naturally occurring polysaccharides have been utilized as macromolecular supports in drug delivery, as they have biological similarities and hence may avoid inflammation or immune response. Polysaccharides have numerous advantages such as abundance, ease of isolation, and chemical modification to meet various biomedical applications. Many polysaccharides have been chemically modified to alter physicochemical properties including degradation, mechanical stability, and bioactivity. Polysaccharides can be readily processed into micro and nanoparticles, hydrogels, and porous materials for tissue regeneration applications.[44]

11.4.1 Arabinogalactan Conjugates

Falk *et al.*[45] conjugated AmB to a water-soluble polysaccharide, arabinogalactan (AG). The conjugation was *via* Schiff base formation between the amino group of AmB and carbonyl group of oxidized AG, as shown in Figure 11.2. The conjugate showed similar antifungal activity as AmBisome® and AmB-deoxycholate, but was less toxic. Therapeutic efficacy was studied in mice infected with *Candida albicans*. Total yeast eradication was found in the kidneys of mice treated with the highest dosage of the AG–AmB (8 mg kg^{-1} day^{-1}). Furthermore, these conjugates were found to be effective against leishmanial parasite infections. The AG–AmB conjugate showed altered pharmacokinetic profile and tissue distribution, with persistence of

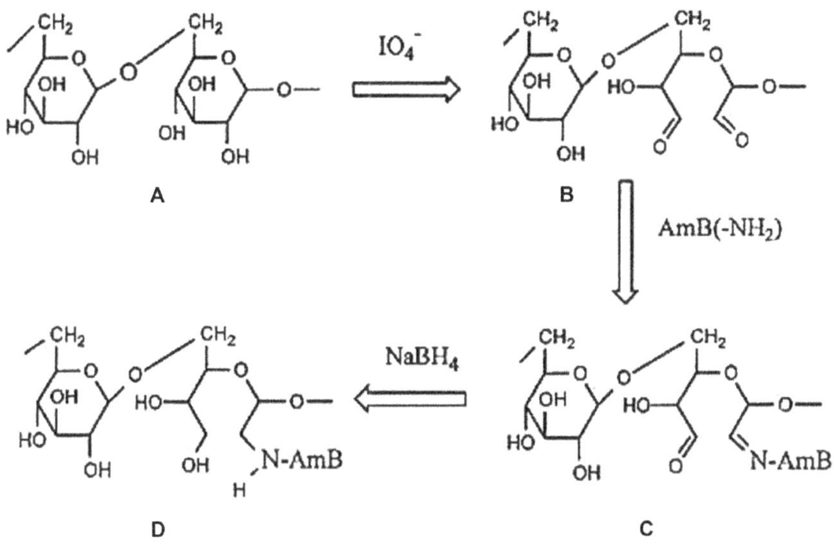

Figure 11.2 Scheme of arabinogalactan (AG)–amphotericin B (AmB) conjugates. (A) AG; (B) oxidized AG; (C) imine conjugates; and (D) amine conjugates. Reproduced from ref. 45 with permission from American Society for Microbiology, Copyright 1999.

drug in the body and its accumulation in lungs, compared to micellar and liposomal AmB formulations. Moreover, AG–AmB conjugate was also found to be effective in treating systemic murine aspergillosis.[46] Oxidation of the polysaccharide chain resulted in the degradation of the polymer and changed its natural structure and biodistribution. The degradation of the polysaccharide can be prevented by conjugating the drug directly to the AG without introducing carbonyl group through prior oxidation. Direct conjugation of drug to the polymer can be achieved through two steps: (i) a leaving functional group introduced into the polymer by reacting with tosyl-chloride or mesyl-chloride; (ii) replacement of the tosylate or mesylate derivatives with the AmB. The AG–AmB conjugates showed similar activity against *Leishmania major* and *Candida albicans* in comparison with free AmB, and were less toxic in erythrocytes and BALB/c mice compared to Fungizone®. However, The AG–AmB conjugates obtained through direct conjugation were expected to be more toxic than the conjugates obtained through reductive amination.[47] The natural and highly water-soluble AG protein (AGP), isolated from groundnut and purified using β-glucosyl Yariv reagent, was used as a carrier for AmB. The conjugation of purified AGP to AmB through Schiff base reaction yielded the water-injectable lesser toxic AGP–AmB conjugate without affecting the antifungal potential of AmB.[48] The testing of AmB and AGP–AmB conjugate against *Candida albicans* clinical isolates showed similar minimum inhibitory concentrations (MICs) and minimum fungicidal concentrations. The AGP–AmB conjugate was less toxic than AmB, and the maximum tolerated dose (MTD) was a high 45 mg kg^{-1} body weight, suggesting AGP as a suitable carrier for AmB formulation. Conjugation of AmB to AG modified its distribution with respect to the molecular weight of the AG when compared to the free AmB. The high molecular weight AG–AmB conjugate was found to be highly soluble in water, with a relatively small volume of distribution and very low clearance, while the free AmB was poorly soluble in water, with the extensive disposition and high clearance rate. The permeability of the conjugates through the blood vessel's wall is limited due to the large molecular dimensions that restrict the distribution within the blood. This is the reason for the higher concentration of conjugated AmB in the blood in comparison with the unmodified AmB. Hence, it was concluded that a novel pharmaceutical entity resulted from the conjugation of AmB with AG with considerably different pharmacokinetic properties compared to the free drug.[49] Scale-up studies on AG–AmB conjugates delivered high conjugation yields and pure product. Similar *in vitro* and *in vivo* antifungal activity was observed in conjugates prepared in the laboratory and pilot scale studies. The conjugates were evaluated for toxicity in animals at high doses for 28 consecutive days and was found to be non-toxic.[50] AmB in the aggregated form is known to develop toxicity by creating pores (polar channels) in cell membranes and disturbing post-endocytic trafficking.

Kagan *et al.*[51] investigated the toxicity mechanism of AmB and reported it to be the same in kidney cell lines as well as in yeast. However, AmB conjugated to AG was non-toxic in mammalian cells and the reduced toxicity of AG–AmB even at high concentrations was attributed to its inability to form pores in the cell membrane. AmB conjugated to oxidized galactomannan showed comparable antifungal activity and low toxicity with respect to AmB.[52] While the hemolytic effect is negligible for the conjugates, even at 2 mg mL^{-1} containing 600 µg of AmB, toxicity studies of galactomannan-AmB conjugates in mice showed a MTD of 200 mg kg^{-1} (60 mg AmB equivalent).

11.4.2 Gum Arabic Conjugates

Gum arabic, a highly branched, water-soluble natural polysaccharide, is obtained from the exudates of an acacia tree. Gum arabic mainly consists of three components: (i) AG (>90%) with a low (0.5%) protein content; (ii) AG (<10%) with a high protein content (10%); and (iii) glycoprotein (<1%) with 50% protein content.[53] The non-toxic, biodegradable, and biocompatible nature of gum arabic enables its use in food, pharma, and in the cosmetic industry. Oxidation of gum arabic results in a molecule with a carbonyl group that facilitates conjugation of AmB through its amine group resulting in the imine conjugate. Nishi *et al.*[54] evaluated gum arabic–AmB conjugates for their toxicity, bioavailability, and antifungal and anti-leishmanial activity. The drug conjugates were stable, non-hemolytic, and non-toxic to the internal organs in a mouse model and showed good antifungal and anti-leishmanial activity *in vitro*. In spite of the large molecular weight of the polysaccharide, AmB from the conjugates showed bioavailability after *iv* injection. The polyene antibiotic NYS conjugated to gum arabic resulted in a Schiff base, *i.e.* imine linkage.[55] The stability of the conjugates was studied in dry form, in solution, and under different pH values, and was found to exhibit higher stability than NYS. The higher aqueous solubility and negligible hemolytic potential of the conjugates suggested that it would be non-toxic at physiological pH and could be used for *iv* administration. The availability of gum arabic at low cost compared to AG could make this a principal carrier for drug conjugation.

11.4.3 Alginate Conjugates

Alginates are natural, anionic linear polysaccharides made up of different proportions of (1-4)-linked β-D-mannuronate (M) and (1-4)-linked α-L-guluronate (G) residues. Alginates are generally isolated from brown seaweeds such as *Laminaria*, *Lessonia*, and *Macrocystis* species. They can also be biosynthesized by the fermentation process. Alginates are used in the preparation of hydrogels, sponges, and fibers for tissue engineering and for drug delivery applications using ionic or covalent cross-linkers. Alginates have attracted much attention as carriers for drug and cell delivery.[56,57] Even though alginates are considered to be biocompatible and non-toxic for medical applications, the higher molecular weight alginates are resistant to biodegradation. Interestingly, oxidized alginates are susceptible to biodegradation and are therefore well suited to applications where biodegradability is a desirable criterion. Thus, oxidized alginates could function as potentially non-toxic and biodegradable cross-linking agents for proteins and polysaccharides in the preparation of hydrogels. The use of oxidized alginate as a potential cross-linking agent has been extensively reported in the preparation of biodegradable hydrogels for wound dressing, tissue engineering and drug delivery applications.[58–62] Sodium alginate (SA)–AmB conjugates with high drug-loading capacity of up to 40 wt% were synthesized and found to be highly soluble in water to an extent of 500 mg mL^{-1}. Such high solubility is expected to provide a potent therapeutic formulation of AmB, suitable for systemic applications. The conjugates were negligibly hemolytic, even with a drug content 100 times that of pure AmB, and less toxic to kidney cells and RAW cells *in vitro*. Moreover, the implantable gel discs fabricated from oxidized alginate were found to release AmB *in vitro*, suggesting that they could be used in treating systemic fungal infections by sustained release of AmB. Overall, the SA–AmB conjugates with their high drug carrying capacity and good aqueous solubility coupled with their non-toxicity and low hemolytic activity could be a cost-effective pro-drug strategy for treating fungal infections.[63]

11.4.4 Dextran Conjugates

Dextran, a non-toxic water-soluble polysaccharide, obtained from *Lactobacillus*, *Leuconostoc*, and *Streptococcus*, consists of α-1,6-linked glucopyranose units. Dextran is commonly used as a plasma expander, in thrombosis prophylaxis as well as in artificial tears. High molecular weight dextran with a longer half-life has been investigated as a drug carrier.[64] Domb *et al.*[65] described a method for producing a stable water-soluble polysaccharide conjugate of NYS with oxidized dextran by a Schiff–base reaction. The conjugates were found to be highly water soluble and suitable for injection. The MIC of the conjugate against *Candida albicans* and *Cryptococcus neoformans* was comparable with the free drug. However, in terms of toxicity in mice, it was about 25 times less toxic than free NYS. Sokolsky-Papkov *et al.*[66] investigated dextran–AmB conjugates for toxicity and anti-leishmanial activity and reported that conjugates wherein the residual aldehyde was end-capped with 2-ethanolamine were at least 15 times less hemolytic than free AmB and more effective in their anti-leishmanial activity. Injectable hydrogels of dextran–AmB conjugate using carboxymethyl cellulose-hydrazide as a cross-linker were developed and investigated for antifungal activity *in vitro* as well as biocompatibility in mice.[67] AmB released from the gel was investigated for *in vitro* antifungal activity and found to be effective for 11 days, while the gel destroyed the *Candida* more effectively for 3 weeks. No apparent tissue toxicity was reported in the mice. It was suggested that the drug-loaded injectable gels could be an alternative for the treatment of local antifungal infections.

11.4.5 Miscellaneous Conjugates

The pectin–AmB conjugates showed reduced toxicity, low hemolytic potential and good antifungal and anti-leishmanial activities.[68] AmB conjugated to bovine serum albumin *via* amide bonds enhanced the water solubility to 150 $mg\,mL^{-1}$; it is negligible with free AmB. The conjugate was non-toxic to HEK293T cells even at a concentration of 500 $\mu g\,mL^{-1}$ (equivalent to ~30 μg AmB), while the IC_{50} of AmB is 0.78 $\mu g\,mL^{-1}$ and showed less than 5% hemolysis to human erythrocytes at a concentration of 200 $\mu g\,mL^{-1}$ (equivalent to ~12 μg AmB). The conjugates showed a very similar release profile in plasma to AmBisome® and enhanced antifungal activity against yeast strains such as *Candida albicans*, *Cryptococcus neoformans* and *Candida parapsilosis* compared to AmBisome®. Human serum albumin could be a potential carrier for AmB, as it was approved by the US Food and Drug Administration (FDA) as a drug carrier for the anticancer drug paclitaxel (Abraxane®).[69] Matsumori *et al.*[70] conjugated AmB to sterols *via* an ethylene carbamate or hexamethylene carbamate linker and examined the membrane permeability. The results suggested that the AmB–ergosterol conjugated with ethylene carbamate linker showed higher channel-open probability than the cholesterol conjugate.

11.5 Synthetic Polymers

Kim *et al.*[71] stated that polymer conjugation creates new chemical entities, which need additional FDA approval although the used drug is already approved. The physical and biological properties of natural polysaccharides depend on their source and method of isolation and purification. For example, the structure and properties, such as the degree of branching, relative amount of particular type of glycosidic links, molecular weight, solubility, optical activity, and physiological action, of dextran derived from different sources are different. Basu *et al.*[31] state, "Making the polysaccharide conjugate system uniform, reproducible, and scalable is challenging. The polysaccharide structure (molecular weight, functional groups, purity) varies from batch to batch and source to source. Therefore, it is difficult to fabricate a reproducible system with appropriate properties". The molecular weight, polydispersity, charge, and structural conformations of the polysaccharide–drug conjugate influence the pharmacokinetics. Therefore, although many naturally occurring polysaccharides have interesting properties for biomedical applications, their clinical use, especially as drug delivery systems, is problematic since regulatory agencies treat the drug–polysaccharide conjugates as new chemical entities and would require consistent physical and chemical properties in the conjugates before regulatory approval.

The aqueous solubility of many therapeutic agents, including AmB, has been increased by conjugating with PEG.[72–75] A number of covalent conjugations have been attempted on AmB to overcome its insolubility and side effects. AmB–methoxy PEG conjugates with antifungal effects and spectrum

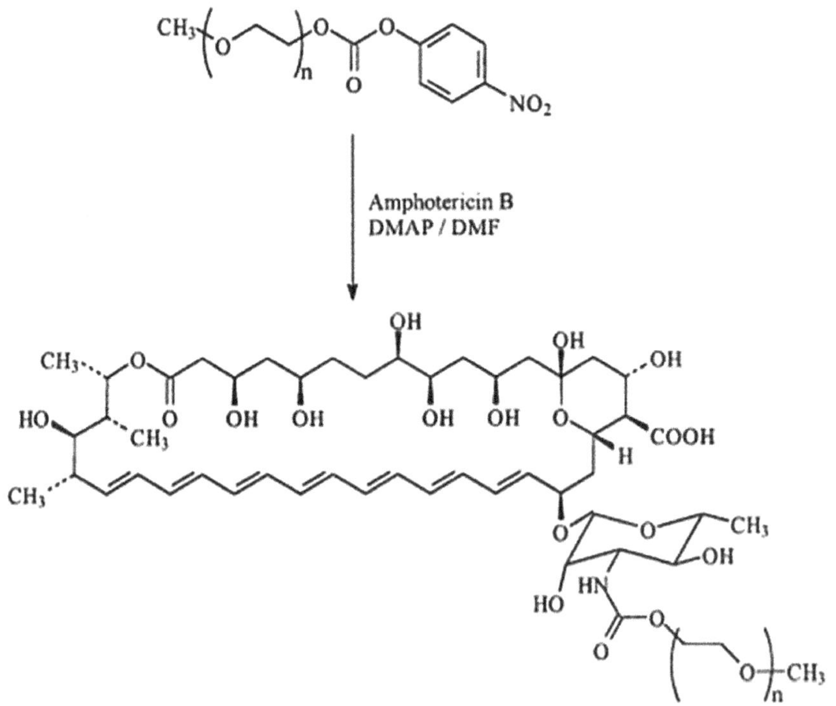

Figure 11.3 Conjugation of amphotericin B to methoxy-poly(ethylene glycol) through carbamate linkage.
Adapted from ref. 72 with permission from Elsevier, Copyright 2001.

of activity comparable to the conventional AmB formulated with sodium deoxycholate was the first one reported by Sedlak *et al.* (Figure 11.3).[72]

The antifungal activity possessed by methoxy-PEG–AmB conjugates is about 10 times higher than that of conventional AmB. Conover *et al.*[73] conjugated PEG to AmB *via* carbamate and carbonate linkages. The PEG–AmB conjugate showed 200 times more solubility than AmB and was six-fold less toxic than AmB–deoxycholate. Sedlák *et al.*[76] synthesized a group of PEG–AmB conjugates with an acid labile linker by keeping the target-specific release of the drug, *i.e.* to the area of the infection as the main objective. It was believed that the acid-labile linker could be cleaved in the acidic environment locally produced by fungal pathogens to release the drug at the site of infection. The drug release kinetics at pH 5.5 revealed the relationship between the rate of drug release and the nature of the linkage. Moreover, the conjugates were reported to be stable at physiological pH and in the presence of blood, plasma and/or serum. The higher lethal dose (LD_{50}) observed in a mice model in comparison with the free drug suggested that the conjugates are a good alternative to free AmB. Parasitic fungal pathogens are known to possess specific hydrolase β-glucosidases in their enzymatic outfit, but not in healthy human tissues. By considering these enzymes,

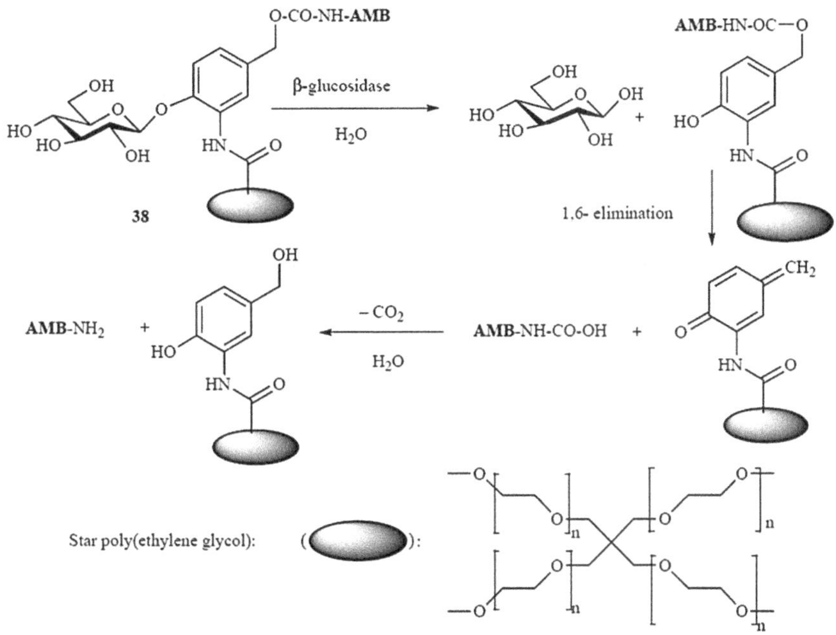

Figure 11.4 Structure and principle of targeted amphotericin B (AmB) release fromβ-glucosidase-sensitive star-poly(ethylene glycol)–AmB. Adapted from ref. 77 with permission from Elsevier, Copyright 2008.

star-PEG–AmB conjugates with β-glucosidase-sensitive β-D-glucopyranoside molecular trigger were synthesized (Figure 11.4). The preliminary evaluation results confirmed the targeted delivery of AmB in the area of the activity of the fungal pathogen. The enzymatic hydrolysis of β-D-glucopyranoside molecular trigger initially provides glucose, and the subsequent 1,6-elimination releases free AmB.[77] Recently, Halperin *et al.*[2] prepared PEG–amide conjugates of AmB that showed antifungal activity in an *in vivo* mouse model of systemic candidiasis. AmB was conjugated to PEG using two different concentrations, one with equal molar ratio and the other with AmB twice the molar ratio with respect to PEG. It was observed that the conjugates with higher molar ratio of AmB possessed more physically bound AmB and had relatively larger particle diameter compared to that formulated using lower concentration of AmB. The solubility of the AmB–PEG conjugates was reported as 8.66 mM and 9.41 mM in PBS–EDTA, whereas the free AmB was practically insoluble. The PEG–AmB conjugates were reported to be less toxic to mammalian cells.[75]

N-(2-Hydroxypropyl)methacrylamide (HPMA)–AmB conjugates were developed by introducing AmB to HPMA with the help of a biodegradable linker Gly-Phe-Leu-Gly. The conjugates were non-toxic for a dose equivalent to 3 mg kg^{-1} AmB, which is higher than the LD_{50} value (2.5 mg kg^{-1}) reported for Fungizone® and showed similar *in vivo* anti-leishmanial activity in comparison with AmBisome®.[78] Nicoletti *et al.*[79] reported polymer–drug combination chemotherapy against visceral leishmaniasis. This combination approach used HPMA as the polymer backbone. AmB and alendronic acid were attached using the amino acid linker. The combinatorial polymer–drug conjugates were effective anti-leishmanial agents *in vitro* and *in vivo.* Charvalos *et al.*[80] synthesized poly(vinyl pyrrolidone)–AmB complexes and showed that they maintained antifungal activity against *Candida* spp. and *Aspergillus* spp. Benincasa *et al.*[81] prepared functionalized carbon nanotube–AmB conjugates. Antifungal activity of these conjugates was tested against clinical fungal strains. Interestingly, the cell viability experiments of conjugates showed no significant toxic effect on Jurkat cells at antifungal concentrations. NYS conjugated to PEG showed an increased solubility which was 10 times higher than NYS with better antifungal activity in comparison with the pure NYS.[82] Dinh *et al.*[83] prepared poly(γ-glutamic acid)–AmB complexes and showed less cytotoxicity compared to Fungizone® and AmBisome® and were more efficacious *in vivo* in experimental murine candidiasis.

Synthetic polysaccharides prepared from their respective monomers would possess reproducible physical and chemical properties especially their molecular weight, branching, glycosidic linkages, and optical properties. Synthetic polysaccharides would be free from extraneous substances such as proteins that are present in many polysaccharides derived from bacterial, animal, and even plant sources, and better suited for many biomedical applications, as they would be expected to have reduced immunogenic character.[32] Glucose, the primary source of energy, was polymerized into polyglucose (PG) by melt condensation reaction using phosphorus acid as the catalyst. AmB was conjugated to oxidized PG through Schiff's base reaction and further reduced to stable amine linkage. The PG–AmB conjugates were highly soluble (four- to five-fold) in water when compared to the AmB conjugates reported with gum arabic or AG. The conjugates showed potent antifungal activity against *Candida albicans* and *Candida parapsilosis* and anti-leishmanial activity against *Leishmania donovani* with negligible hemolytic activity. PG with consistent physical, chemical and biological properties could be an effective drug carrier in terms of enhanced therapeutic efficacy, toxicity, biodistribution, bioavailability, elimination, and targetability.[84] AmB conjugation to polymannose enhanced the aqueous solubility of AmB and thereby increased the antifungal potential against *Candida albicans*, *Candida parapsilosis* and *Cryptococcus neoformans* with reduced toxicity against human erythrocytes and HEK293T cells.[85]

Taming strategy, *i.e.* limiting the entry of toxic drug molecules into the cells and enhancing the retention of drug within the membrane is shown in Figure 11.5. Taming of AmB to the choloyl moiety by covalent attachment reduced its toxicity towards mammalian cells with similar antifungal activity in comparison with free AmB.[86] Janout *et al.*[87] synthesized a tetra-walled and an octa-walled molecular umbrella conjugate of AmB. The conjugates exhibited high water solubility, aggregation, and negligible hemolytic activity. The same research group prepared molecular umbrella from one spermidine

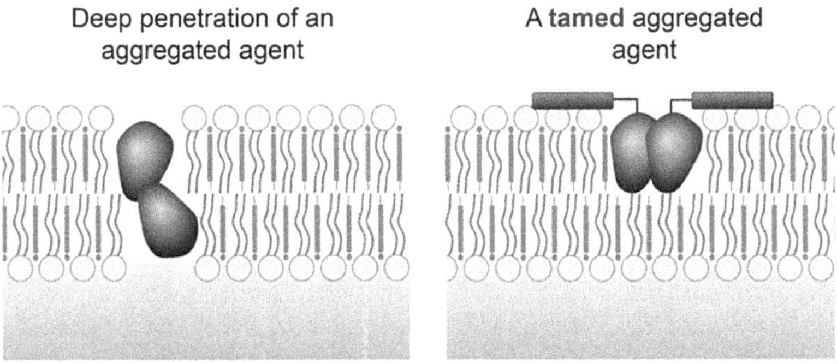

Figure 11.5 Penetration of aggregated amphotericin B (left) and tamed aggregated AmB (right).
Adapted from ref. 86, https://pubs.acs.org/doi/full/10.1021/acs.bioconjchem.6b00629, with permission from American Chemical Society, Copyright 2016.

and two cholic acid molecules and conjugated to AmB. The antifungal activities reported are almost equal to native AmB.[88] A di-walled "molecular umbrella" made up of spermidine-linked deoxycholic or cholic acids was linked to the polyene macrolide antibiotics AmB and NYS through the ω-amino acid linkers. AmB–di-walled molecular umbrella conjugates showed increased cellular selectivity with negligible haemolytic activity and substantial antifungal activity, while the conjugation of NYS (structurally similar to AmB) to a di-walled molecular umbrella showed a decrease in cellular selectivity as well as diminishing antifungal activity in comparison with free NYS due to the decreased affinity of NYS conjugates to ergosterol.[89]

Tween 20, an additive in food and pharmaceutical preparations, has been approved by the US FDA. The non-ionic, non-toxic surfactant can be conjugated to the drugs through carbamate linkages that are susceptible to breakdown by hydrolysis in aqueous media releasing the drug molecule. The resulting conjugates improved the aqueous solubility and reduced the toxicity of many drugs. AmB was conjugated to Tween 20 *via* carbamate linkage by the activation of hydroxyl groups of Tween 20 using *p*-nitrophenylchloroformate. AmB covalently linked to Tween 20 was less toxic to kidney cells with improved aqueous solubility and negligible hemolytic potential when compared to AmB. Furthermore, AmB bound to Tween 20 showed good antifungal and anti-leishmanial activity *in vitro*.[90] As polysorbate-coated nanocarriers were reported to cross the blood–brain barrier,[91] the conjugates with potent antifungal activity could be investigated for their ability to treat fungal infections in the central nervous system.

11.6 Conclusions

Increased fungal infections in immunocompromised patients and the toxicity induced by conventional antifungal agents have led to the modification of existing drug molecules to enhance their potency and reduce their toxicity. Physical modifications of AmB reduce the nephrotoxicity by limiting its distribution in the kidney and enhancing the duration of action. However, infusion-related side effects, high cost, instability of liposomes, and other disadvantages in physically modified formulations limit their widespread use in clinical practice. Polymer–drug conjugates of antifungal drugs have emerged as potential drug delivery systems for antifungal therapy in recent years due to their unique properties such as improved aqueous solubility, stability, non-toxicity, and the possibility to release the drug at the site of action. A number of biocompatible natural and synthetic polymers have been investigated to prepare the polymer–drug conjugates of antifungal drugs. Although many naturally occurring polymers are biocompatible and biodegradable, their use as drug delivery systems appear to be problematic due to the difficulty involved in ensuring consistent physical and structural properties. Synthesis of polysaccharides from their respective monomers (*e.g.* PG) appears to be an interesting option to ensure consistent and reproducible properties in the polymeric carrier.

References

1. A. M. Clark, The need for new antifungal drugs, in *New Approaches for Antifungal Drugs*, ed. P. B. Fernandes, Birkhäuser Boston, Boston, MA, 1992, p. 1–19.
2. A. Halperin, Y. Shadkchan, E. Pisarevsky, A. M. Szpilman, H. Sandovsky, N. Osherov and I. Benhar, *J. Med. Chem.*, 2016, **59**, 1197.
3. M. V. Castelli, E. Butassi, M. C. Monteiro, L. A. Svetaz, F. Vicente and S. A. Zacchino, *Expert Opin. Ther. Pat.*, 2014, **24**, 323.
4. D. W. Denning and M. J. Bromley, *Science*, 2015, **347**, 1414.
5. J. W. Rippon, Therapy in the Immunosuppressed Host, in *Antifungal Drug Therapy*, ed. P. H. Jacobs and L. Nall, Marcel Dekker, Inc, New York, 1990, p. 247–256.
6. M. R. Cairns, *J. Electron Microsc. Technol.*, 1988, **8**, 115.
7. A. E. Oxford, H. Raistrick and P. Simonart, *Biochem. J.*, 1939, **33**, 240.
8. D. W. Woolley, *J. Biol. Chem.*, 1944, **152**, 225.
9. W. O. Elson, *J. Infect. Dis.*, 1945, **76**, 193.
10. D. J. Sheehan, C. A. Hitchcock and C. M. Sibley, *Clin. Microbiol. Rev.*, 1999, **12**, 40.
11. E. L. Hazen and R. Brown, *Science*, 1950, **112**, 423.
12. E. L. Hazen and R. Brown, *Exp. Biol. Med.*, 1951, **76**, 93.
13. E. B. Schoenbach, J. M. Miller, M. Ginsberg and P. H. Long, *J. Am. Med. Assoc.*, 1951, **146**, 1317.
14. I. Snapper and L. V. McVay, *Am. J. Med.*, 1953, **15**, 603.
15. M. A. Burgess and G. P. Bodey, *Antimicrob. Agents Chemother.*, 1972, **2**, 423–426.
16. P. R. Sawyer, R. N. Brogden, R. M. Pinder, T. M. Speight and G. S. Avery, *Drugs*, 1975, **9**, 424.

17. R. C. Heel, R. N. Brogden, G. E. Pakes, T. M. Speight and G. S. Avery, *Drugs*, 1980, **19**, 7.
18. V. Fainstein and G. P. Bodey, *Ann. Intern. Med.*, 1980, **93**, 432.
19. A. Espinel-Ingroff, S. Shadomy and R. J. Gebhart, *Antimicrob. Agents Chemother.*, 1984, **26**, 5.
20. R. J. Hamill, *Drugs*, 2013, **73**, 919.
21. A. Lavasanifar, J. Samuel and G. S. Kwon, *Colloids Surf., B.*, 2001, **22**, 115.
22. G. Vandermeulen, L. Rouxhet, A. Arien, M. E. Brewster and V. Préat, *Int. J. Pharm.*, 2006, **309**, 234.
23. K. K. Chakraborty and S. R. Naik, *J. Pharm. Pharm. Sci.*, 2003, **6**, 231.
24. N. Lincopan, E. M. Mamizuka and A. M. Carmona-Ribeiro, *J. Antimicrob. Chemother.*, 2005, **55**, 727.
25. K. Jain, A. K. Verma, P. R. Mishra and N. K. Jain, *Nanomedicne*, 2015, **11**, 705.
26. P. K. Gupta, A. K. Jaiswal, S. Asthana, A. Verma, V. Kumar, P. Shukla, P. Dwivedi, A. Dube and P. R. Mishra, *Pharm. Res.*, 2015, **32**, 1727.
27. K. Tutaj, R. Szlazak, K. Szalapata, J. Starzyk, R. Luchowski, W. Grudzinski, M. Osinska-Jaroszuk, A. Jarosz-Wilkolazka, A. Szuster-Ciesielska and W. I. Gruszecki, *Nanomedicine*, 2016, **12**, 1095.
28. H. Ringsdorf, *J. Polym. Sci., Polym. Symp.*, 1975, **51**, 135.
29. R. Duncan and M. J. Vicent, *Adv. Drug Delivery Rev.*, 2013, **65**, 60.
30. N. Larson and H. Ghandehari, Polymeric conjugates for drug delivery, *Chem. Mater.*, 2012, **24**, 840.
31. A. Basu, K. R. Kunduru, E. Abtew and A. J. Domb, *Bioconjugate Chem.*, 2015, **26**, 1396.
32. X. Liu, S. Siegrist, M. Amacker, R. Zurbriggen, G. Pluschke and P. H. Seeberger, *ACS Chem. Biol.*, 2006, **1**, 161.
33. R. Duncan, S. Gac-Breton, R. Keane, R. Musila, Y. N. Sat, R. Satchi and F. Searle, *J. Controlled Release*, 2001, **74**, 135.
34. R. Pawar, W. Jadhav, S. Bhusare, R. Borade, S. Farber, D. Itzkowitz and A. Domb, Polysaccharides as carriers of bioactive agents for medical applications, in *Natural-based Polymers for Biomedical Applications*, ed. R. R. L. Reis, N. M. Neves, J. F. Mano, M. E. Gomes, A. P. Marques and H. S. Azevedo, CRC Press, 2008, p. 3–54.
35. C. De Duve, T. De Barsy, B. Poole, A. Trouet, P. Tulkens and F. Van Hoof, *Biochem. Pharmacol.*, 1974, **23**, 2495.
36. M. C. Garnett, *Adv. Drug Delivery Rev.*, 2001, **53**, 171.
37. R. Duncan, I. C. Hume, P. Kopečková, K. Ulbrich, J. Strohalm and J. Kopeček, *J. Controlled Release*, 1989, **10**, 51.
38. G. Leriche, L. Chisholm and A. Wagner, *Bioorg. Med. Chem.*, 2012, **20**, 571.
39. F. Kratz, U. Beyer and M. T. Schütte, *Crit. Rev. Ther. Drug Carrier Syst.*, 1999, **16**, 245.
40. R. Reents, D. A. Jeyaraj and H. Waldmann, *Drug Discovery Today*, 2002, **7**, 71.

41. A. J. M. D'Souza, R. L. Schowen and E. M. Topp, *J. Controlled Release*, 2004, **94**, 91.
42. F. Kratz, I. A. Müller, C. Ryppa and A. Warnecke, *Chem. Med. Chem.*, 2008, **3**, 20.
43. K. J. Yarema, *Handbook of Carbohydrate Engineering*, Taylor & Francis, 2005.
44. T. Heinze, T. Liebert, B. Heublein and S. Hornig, Functional polymers based on dextran, in *Advanced Polymer Science 205 (Polysaccharides II)*, ed. D. Klemm, Springer, Berlin Heidelberg, 2006, p. 199–291.
45. R. Falk, A. J. Domb and I. Polacheck, *Antimicrob. Agents Chemother.*, 1999, **43**, 1975.
46. R. Falk, J. Grunwald, A. Hoffman, A. J. Domb and I. Polacheck, *Antimicrob. Agents Chemother.*, 2004, **48**, 3606.
47. T. Ehrenfreund-Kleinman, J. Golenser and A. J. Domb, *Biomaterials*, 2004, **25**, 3049.
48. S. Parveen, A. D. Gupta and R. Prasad, *Int. J. Pharm.*, 2007, **333**, 79.
49. A. Elgart, S. Farber, A. J. Domb, I. Polacheck and A. Hoffman, *Biomacromolecules*, 2010, **11**, 1972.
50. D. E. Ickowicz, S. Farber, E. Sionov, S. Kagan, A. Hoffman, I. Polacheck and A. J. Domb, *Biomacromolecules*, 2014, **15**, 2079.
51. S. Kagan, D. Ickowicz, M. Shmuel, Y. Altschuler, E. Sionov, M. Pitusi, A. Weiss, S. Farber, A. J. Domb and I. Polacheck, *Antimicrob. Agents Chemother.*, 2012, **56**, 5603.
52. S. Farber, D. Ickowicz, E. Sionov, S. Kagan, I. Polacheck and A. J. Domb, *Polym. Adv. Technol.*, 2011, **22**, 119.
53. D. Verbeken, S. Dierckx and K. Dewettinck, *Appl. Microbiol. Biotechnol.*, 2003, **63**, 10.
54. K. K. Nishi, M. Antony, P. V. Mohanan, T. V. Anilkumar, P. M. Loiseau and A. Jayakrishnan, *Pharm. Res.*, 2007, **24**, 971.
55. J. Stefanovic, D. Jakovljevic, G. Gojgic-Cvijovic, M. Lazic and M. Vrvic, *J. Appl. Polym. Sci.*, 2013, **127**, 4736.
56. R. Pandey and G. K. Khuller, Alginate as a Drug Delivery Carrier, in *Handbook of Carbohydrate Engineering*, ed. K. J. Yarema, CRCPress, Boca Raton, 2005, p. 799–816.
57. T. K. Giri, D. Thakur, A. Alexander, Ajazuddin, H. Badwaik and D. K. Tripathi, *Curr. Drug Delivery*, 2012, **9**, 539.
58. B. Balakrishnan and A. Jayakrishnan, *Biomaterials*, 2005, **26**, 3941.
59. B. Balakrishnan, M. Mohanty, P. R. Umashankar and A. Jayakrishnan, *Biomaterials*, 2005, **26**, 6335.
60. S. Manju, C. V. Muraleedharan, A. Rajeev, A. Jayakrishnan and R. Joseph, *J. Biomed. Mater. Res., Part B*, 2011, **98B**, 139.
61. A. Jejurikar, X. T. Seow, G. Lawrie, D. Martin, A. Jayakrishnan and L. Grøndahl, *J. Mater. Chem.*, 2012, **22**, 9751.
62. B. Balakrishnan, N. Joshi, A. Jayakrishnan and R. Banerjee, *Acta Biomater.*, 2014, **10**, 3650.

63. V. Ravichandran and A. Jayakrishnan, *Int. J. Biol. Macromol.*, 2018, **108**, 1101.
64. S. R. Van Tomme and W. E. Hennink, *Expert Rev. Med. Devices*, 2007, **4**, 147.
65. A. J. Domb, G. Linden, I. Polacheck and S. Benita, *J. Polym. Sci., Part A: Polym. Chem.*, 1996, **34**, 1229.
66. M. Sokolsky-Papkov, A. J. Domb and J. Golenser, *Biomacromolecules*, 2006, **7**, 1529.
67. S. P. Hudson, R. Langer, G. R. Fink and D. S. Kohane, *Biomaterials*, 2010, **31**, 1444.
68. G. P. Kothandaraman, V. Ravichandran, C. Bories, P. M. Loiseau and A. Jayakrishnan, *J. Drug Delivery Sci. Technol.*, 2017, **39**, 1.
69. S. Gurudevan, A. P. Francis and A. Jayakrishnan, *Eur. J. Pharm. Sci.*, 2018, **115**, 167.
70. N. Matsumori, N. Eiraku, S. Matsuoka, T. Oishi, M. Murata, T. Aoki and T. Ide, *Chem. Biol.*, 2004, **11**, 673.
71. S. Kim, J.-H. Kim, O. Jeon, I. C. Kwon and K. Park, *Eur. J. Pharm. Biopharm.*, 2009, **71**, 420.
72. M. Sedlák, V. Buchta, L. Kubicová, P. Šimůnek, M. Holčapek and P. Kašparová, Synthesis and characterisation of a new amphotericin b–methoxy poly(ethylene glycol) conjugate, *Bioorg. Med. Chem. Lett.*, 2001, **11**, 2833.
73. C. D. Conover, H. Zhao, C. B. Longley, K. L. Shum and R. B. Greenwald, *Bioconjugate Chem.*, 2003, **14**, 661.
74. F. M. Veronese and G. Pasut, *Drug Discovery Today*, 2005, **10**, 1451.
75. T. R. M. Tan, K. M. Hoi, P. Zhang and S. K. Ng, *PLoS One*, 2016, **11**, 1.
76. M. Sedlák, M. Pravda, F. Staud, L. Kubicová, K. Týčová and K. Ventura, *Bioorg. Med. Chem.*, 2007, **15**, 4069.
77. M. Sedlák, P. Drabina, E. Bílková, P. Šimůnek and V. Buchta, *Bioorg. Med. Chem. Lett.*, 2008, **18**, 2952.
78. S. Nicoletti, K. Seifert and I. H. Gilbert, *Int. J. Antimicrob. Agents*, 2009, **33**, 441.
79. S. Nicoletti, K. Seifert and I. H. Gilbert, *Bioorg. Med. Chem.*, 2010, **18**, 2559.
80. E. Charvalos, M. N. Tzatzarakis, F. Van Bambeke, P. M. Tulkens, A. M. Tsatsakis, G. N. Tzanakakis and M.-P. Mingeot-Leclercq, *J. Antimicrob. Chemother.*, 2006, **57**, 236.
81. M. Benincasa, S. Pacor, W. Wu, M. Prato, A. Bianco and R. Gennaro, *ACS Nano*, 2011, **5**, 199.
82. M. Spulber, A. Fifere, D. A. Anamaria and N. Fifere, *J. Incl. Phenom. Macrocycl. Chem.*, 2011, **71**, 87.
83. T. Dinh, Q. Zia, S. Zubair, P. Stapleton, R. Singh, M. Owais and S. Somavarapu, *Nanomedicine*, 2017, **13**, 1773.
84. V. Ravichandran, G. P. Kothandaraman, C. Bories, P. M. Loiseau and A. Jayakrishnan, *J. Nanosci. Nanotechnol.*, 2018, **18**, 2405.

85. A. P. Francis, S. Gurudevan and A. Jayakrishnan, *J. Biomater. Sci., Polym. Ed.*, 2018, **29**, 1529.
86. Y. Yu, M. J. Sabulski, W. A. Schell, M. M. Pires, J. R. Perfect and S. L. Regen, *Bioconjugate Chem.*, 2016, **27**, 2850.
87. V. Janout, C. Bienvenu, W. Schell, J. R. Perfect and S. L. Regen, *Bioconjugate Chem.*, 2014, **25**, 1408.
88. V. Janout, W. A. Schell, D. Thévenin, Y. Yu, J. R. Perfect and S. L. Regen, *Bioconjugate Chem.*, 2015, **26**, 2021.
89. A. S. Skwarecki, K. Skarbek, D. Martynow, M. Serocki, I. Bylińska, M. J. Milewska and S. Milewski, *Bioconjugate Chem.*, 2018, **29**, 1454.
90. V. Ravichandran, V. Kesavan, S. Cojean, P. M. Loiseau and A. Jayakrishnan, *Curr. Drug Delivery*, 2018, **15**, 1028.
91. X.-H. Tian, X.-N. Lin, F. Wei, W. Feng, Z.-C. Huang, P. Wang, L. Ren and Y. Diao, *Int. J. Nanomed.*, 2011, **6**, 445.

CHAPTER 12

Methods for Sterilization of Biopolymers for Biomedical Applications

CHAU CHUN BEH,*[a,b] SHADY FARAH,[b,c,d,e] ROBERT LANGER[b,c,d,f,g] AND ANA JAKLENEC*[b]

[a] Western Australia School of Mines: Minerals, Energy and Chemical Engineering, Curtin University, Bentley 6102, WA, Australia; [b] David H Koch Institute for Integrative Cancer Research, Massachusetts Institute of Technology, 500 Main Street, Cambridge, MA 02139, USA; [c] Department of Chemical Engineering, Massachusetts Institute of Technology, 77 Massachusetts Avenue, Cambridge, MA 02139, USA; [d] Department of Anesthesiology, Boston Children's Hospital, Harvard Medical School, 300 Longwood Ave, Boston, MA 02115, USA; [e] The Wolfson Faculty of Chemical Engineering and The Russell Berrie Nanotechnology Institute, Technion-Israel Institute of Technology, Haifa 3200003, Israel; [f] Institute for Medical Engineering and Science, Massachusetts Institute of Technology, 77 Massachusetts Avenue, Cambridge, MA 02139, USA; [g] Harvard-MIT Division of Health Science and Technology, Massachusetts Institute of Technology, 77 Massachusetts Avenue, Cambridge, MA 02139, USA
*Email: jane.beh@curtin.edu.au; jaklenec@mit.edu

12.1 Introduction

Biopolymers are formed naturally in living organisms or synthesized by using natural biological materials and are commonly used for many

Biomaterials Science Series No. 5
Antimicrobial Materials for Biomedical Applications
Edited by Abraham J. Domb, Konda Reddy Kunduru and Shady Farah

Published by the Royal Society of Chemistry, www.rsc.org

biomedical applications such as tissue engineering, vascular implants, wound care products, surgical sutures, fixation rods, screws and clips, and controlled drug delivery systems. Biopolymers are suitable for biomedical applications because of their low toxicity, biocompatibility, controlled degradability, and reasonable thermal processability. There are two main categories of biopolymers: (i) bioresorbable biopolymers and (ii) biodegradable biopolymers.[1,2] Bioresorbable biopolymers are defined as polymers that reabsorb into the body or blood plasma over a predetermined duration while fulfilling a specific purpose; biodegradable biopolymers are polymers that degrade into low-molecular-weight compounds in the body, which subsequently are metabolized.[1] Several common examples of biopolymers include hyaluronic acid, polypeptides, polysaccharides, chitosan, chitin, cellulose, polylactic acid (PLA), poly-L-lactic acid (PLLA), polyglycolic acid, poly(lactide-*co*-glycolide) (PLGA), polycaprolactone (PCL), and polyhydroxybutyrate.[1]

The characteristics of biopolymers such as their biocompatibility, precisely engineered rate of degradability, thermal processability, relatively high strength, low toxicity, controlled crystallinity, and hydrophilicity have made biopolymers exceptionally useful in biomedical applications such as medical devices, implantation components, tissue engineering, and drug delivery systems. As biopolymers are formed naturally in living organisms, or are synthesized using natural biological materials, they are generally biocompatible with low cytotoxicity, immune toxicity, and sensitive and allergic reactions.[3–5] The rate of biodegradability of the biopolymers depends on the water permeability, water absorption, crystallinity, physical size and shape, molecular weight, and chemical environment such as acidic, basic and enzymatic.[1] The main factors that affect the rate of degradability of biopolymers are summarized in Table 12.1. The rates of degradation of several examples of biopolymers have been tabulated in Table 12.2.[1,6,7]

Biopolymers have to be appropriately sterilized before being utilized as medical devices, due to infection risks associated with unsterilized biomaterials being used in the human body. Sterilization is a process of elimination and inactivation of all microorganisms such as bacteria, spores, fungus, and viruses. A sterility assurance level of 10^{-6} is regarded as acceptable. The sterilization efficiency (SE) of a sterilization process is a

Table 12.1 The main factors that determine the rate of degradation of biopolymers.

Factors affecting degradation	To achieve fast degradation
Water absorption	High
Water vapor permeability	High
Chemical environment (acidic, basic)	Depends on chemical and physical form of the biopolymers
Crystallinity	Amorphous materials
Molecular weight	Low

Table 12.2 Several examples of biopolymers and their rates of degradation.[1,6,7]

Biopolymers	Rate of degradation (months)
Polylactic acid (PLA)	3–24
Poly-L-lactic acid (PLLA)	18–60
Polyglycolic acid (PGA)	0.5–2
Poly(lactide-*co*-glycolide) (PLGA)	1–6
Polycaprolactone (PCL)	24
Poly(ethylene glycol) (PEG)	Varied
Polyhydroxybutyrate (PHB)	2–18

measurement of the reduction factor or degree of inactivation of microorganisms:

$$SE = -\log 10 \frac{\text{number of microorganisms after sterilization}}{\text{number of microorganisms before sterilization}}$$

12.2 Sterilization Methods

Sterilization involves removing or inactivating all living microorganisms such as bacteria, fungi, and viruses. Various sterilization methods have been adapted to sterilize medical devices for over 100 years.[8] The most commonly used sterilization techniques are heat and steam-autoclaving sterilization. At present, different sterilization methods are used to sterilize biopolymers such as ethylene oxide treatment, irradiation (γ-rays, X-rays, electron beam, and ultraviolet), gas plasma, and supercritical fluid technology. A comparison of the pros and cons of various sterilization techniques is tabulated in Table 12.3.[8–10]

12.2.1 Steam-autoclaving

Steam-autoclaving, also known as moist heat sterilization, is a simple, easy to apply, safe, non-toxic, and inexpensive method that is widely in use in hospitals and healthcare facilities. In this approach objects/devices/biomedical polymers are placed in an autoclave and exposed to saturated steam under high temperatures (121–134 °C) and usually for short period of a few minutes up to 15 min, depending on heating profile and applied pressure.[11] These conditions were found to cause the destruction of microorganisms *via* the irreversible denaturation of both enzyme and structural protein components.[12] Given method's simplicity and numerous advantages, steam-autoclaving found its way to biomedical polymers for various applications, including hematic circuit (*i.e.* polyvinyl chloride), catheters (*i.e.* polyurethanes) and vascular prosthesis (*i.e.* polytetrafluoroethylene).[11] In a recent study, Venault and coworkers studied the effect of steam sterilization

Table 12.3 A comparison of the most commonly used sterilization techniques.[8–10]

Sterilization method	Pro(s)	Con(s)
Steam-autoclaving	✓ Economical ✓ Short processing time ✓ Non-toxic ✓ Safe for the environment	✗ Not recommended for materials that are sensitive to high heat and moisture ✗ Not suitable for oil-based materials and electric devices
Dry heat	✓ Fast ✓ Simple ✓ No toxic residues ✓ High penetration ✓ Effective ✓ Can be used for oil-based materials (*e.g.* glycerine, soft paraffin)	✗ Not ideal for thermal-labile materials ✗ High temperature ✗ Long duration ✗ Change in physicochemical properties of biopolymers
Ethylene oxide	✓ Low temperature ✓ Effective ✓ No limit for lumen	✗ Toxic residues left in products ✗ Carcinogenic ✗ Flammable ✗ Explosive ✗ Change in structural properties of biopolymers
Gamma irradiation	✓ High penetration ✓ Low temperature ✓ No residue ✓ Easy control ✓ Effective	✗ Change in structural properties of biopolymers ✗ Long duration
X-rays	✓ High penetration ✓ Suitable for in-line sterilization (low-energy X-ray): decreased lead time, increased dosage control	✗ Extra shielding required ✗ Not suitable for continuous bulk sterilization
Electron beam	✓ Low temperature ✓ No residue ✓ Fast ✓ Easy control	✗ Low penetration ✗ Change in structural properties of biopolymers ✗ Limited to low density and small products
Ultraviolet irradiation	✓ Fast ✓ Low temperature ✓ No toxic residues ✓ Low cost	✗ Not effective ✗ Prolonged exposure causes a change in structural and biochemical properties of biopolymers
Plasma	✓ Low temperature ✓ Fast ✓ Increase wettability on the surface of biopolymers	✗ Reactive species left in products ✗ Change in chemical and mechanical properties of biopolymers
Supercritical fluid (CO_2)	✓ No toxic residue ✓ No biochemical property change ✓ No chemical reagents involved	✗ The possibility of change in morphology and porosity of biopolymers

on the stability of antifouling zwitterionic biopolymeric gels of poly(4-vinylpyridine propylsulfobetaine) (4VPPS) and explored their biofouling capability following 1 h steam sterilization at 121 °C in comparison to poly(sulfobetaine methacrylate) (SBMA).[13] Differently to SBMA, unstable gels, the 4VPPS gels maintained their structure, interactions, and antifouling properties, even following multiple sterilization cycles.[13]

However, this method should be avoided whenever the object is made of either degradable biomedical polymers or oil-based materials, as well whenever electric device components are involved. Several studies have indicated that elevated moisture and pressure can induce a dramatic change in a biopolymer's mechanical properties, due to hydrolysis and degradation.[11] Accordingly, the effect of the sterilization profile on a biopolymer's mechanical properties should be deeply studied. One example has been reported by Zanelli *et al.*, were they studied the effect of steam-autoclaving sterilization process practised in healthcare facilities on the mechanical properties of two block copolymers in use for manufacturing biomedical devices: thermoplastic polyurethane and a poly(ether-*block*-amide).[14] In this study, the authors determined the functional dependence between material constitutive parameters, in order to obtain an optimal constitutive model using the uniaxial tensile tests, before and after the sterilization process. The authors recommended the appliance of such an approach to different biomedical biopolymers under various tensile tests, as well for other sterilization processes that may induce a change in biopolymer's mechanical properties.[14]

12.2.2 Dry-heat Sterilization

The dry-heat sterilization process is one of the earliest practised sterilization methods, having simplicity and feasibility, using an oven with elevated temperature to kill microorganisms and bacterial spores. Different from the steam-autoclaving sterilization method, higher temperatures and longer exposure times are needed for decontaminating an object during a dry-heat process, as illustrated in Figure 12.1.[11]

In the process, first, the heat is absorbed by the object's exterior surface and then passed inward to the next layer until the object reaches the desired sterilization temperature. This approach is highly advantageous and desirable for the effective sterilization of complex designs/biomedical devices with closed cavities.[11] These conditions have been found to kill microorganisms by deep dehydration and protein denaturation.[11] Due to lack of moisture (or minimal presence) this method can be used to sterilize oil-based materials. Overall, the need for high temperature and for prolonged time limit the use of this method to thermo-resistant materials such as glass and steel and less for biopolymers.[11] In general, dry-heat sterilization is operated within the temperature range of 150 to 170 °C, and a duration of 60 to 150 min.[11]

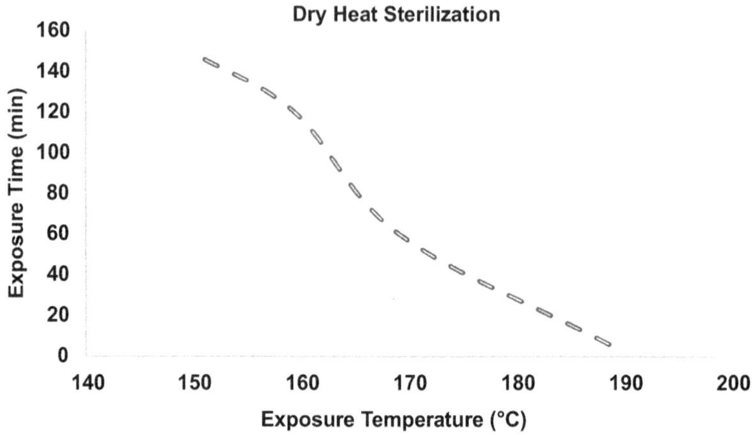

Figure 12.1 The dry heat sterilization method, exposure time *vs.* exposure temperature (graph plotted based on the data from Tessarolo and Nollo).[11]

12.2.3 Chemical Treatment—Ethylene Oxide

Ethylene oxide (EtO) is a commonly used gaseous agent for the chemical sterilization process. Ethylene oxide is known to be an effective, flexible, and robust sterilization tool at low operating temperatures. Ethylene oxide offers a better sterilization option to thermally labile, radiation- and moisture- sensitive biopolymers, as well in the medical device industry. The decontamination mechanism of action of EtO against a wide range of microorganisms is attributed to chemical alkylation following interaction with the EtO gas. Alkylation of carboxylic, hydroxylic, and sulphydrilic groups in nucleic acids causes proteins to change their shape, eventually leading to cell death.[11,15] The EtO sterilization process is composed of two steps: first, a vacuum is applied to the autoclave/chamber and then EtO gas is flushed in at concentrations of 600–1200 $mg\,L^{-1}$. Depending on the sterilized biopolymer and sterilizer, load the following parameters are usually followed: humidity of 40–50%, temperature 30–50 °C, and sterilization cycle duration 2–8 h.[11] Zhu and co-workers have studied key factors impacting the sterilization efficiency, *i.e.* temperature, humidity, concentration, reaction time, and loading method. They found that the mechanisms of each factor was very variable, and all should be considered and optimized for achieving effective sterilization.[16]

The compatibility of EtO with a wide range of biopolymers/materials as well as the fact that the low temperature at which the process is handled offers a huge advantage with extra applicability.[11] However, chemical sterilization can leave toxic residues potentially either on the surface and/or within the biopolymer/medical implant.[17] Giving the fact that EtO is a colorless and flammable gas causing extreme skin irritation, extra safety steps are required to ensure that all gas is removed from the biopolymer/device before final packaging. Typically, an extremely long vacuum aeration time is applied to fully remove EtO gas toxic residues.[17]

Exposure of biopolymers to the EtO sterilization method may affect their molecular weight, leading eventually to changes in their mechanical properties and their degradation rate, as well as changing their surface chemistry.[11] Although detrimental effects of EtO sterilization on the mechanical properties of PLA-based biopolymers have rarely been reported in the literature, other studies have reported EtO acting as a plasticizer, leading to changes in the biopolymer structure.[17] Letzner *et al.* have reported modified biopolymers of poly-3-hydroxybutyric acid and poly-L-lactide with the incorporation of metronidazole for coating of dental implants. The sterilization with EtO showed no damage to the biopolymers without any influence of coating thickness.[18]

Chemical composition has an important role when selecting a sterilization method. Kohn's group have reported a comparison study of sterilization methods on PEG-based bioresorbable biopolymers with different terminal chemical groups. They have found that EtO esterifies the carboxylic acid groups in desaminotyrosol-tyrosine and causes significant degradation.[19] Alternatively, multiblock polyesters have been found less sensitive to sterilization, as reported by El Fray *et al.*[20] Two sterilization methods of EtO and electron-beam radiation were applied and were not found to affect the structure and properties of this biopolymer.[20]

12.2.4 Gas Plasma—Hydrogen Peroxide

Sterilization by gas plasma was first introduced as an alternative for EtO treatment.[8] The procedure involves a combination of hydrogen peroxide (H_2O_2) vapor and low-temperature gas plasma to perform sterilization without toxic chemical residues such as EtO.[21] The gas-plasma sterilization is operated at a relatively low range of temperatures (25–70 °C).[9] Hence, sterilization by gas plasma is beneficial for thermally labile materials. The plasma technique is efficient in achieving a sterility assurance level of 10^{-6}. In general, reactive gas mixtures with high oxygen contents are used for gas-plasma sterilization to increase the efficiency of sterilization.[22] It has been reported that reactive gas plasma is effective to inactivate highly resistant microorganisms, in particular spores.[23]

Firstly, a biopolymer is placed within a chamber. After the chamber undergoes a process of removal of moisture content under vacuum, the chamber is sealed with a set pressure and 60% H_2O_2 vaporized into the chamber. As the vapor diffuses throughout the chamber, inactivation of microorganisms takes place.[21] Plasma is generated as H_2O_2 vapor is converted into reactive and biocidal free radicals by an electric field or a magnetic field.[11] Microorganisms such as vegetative bacteria, mycobacteria, spores, bacterial endospores, fungi, yeasts, and viruses can be inactivated by gas plasma H_2O_2.[21,22] However, plasma is not recommended for products that have long lumens, as gas penetration can be difficult, as well as porous materials, due to the fact that materials may absorb H_2O_2 before the gas becomes an active plasma form.[9]

12.2.5 Radiation Process

Sterilization can be achieved by ionizing radiation such as gamma (γ), X-ray and electron beam (E-beam), and non-ionizing radiation like ultraviolet (UV) radiation. Ionizing radiation involves the use of high intensity and short wavelength radiation to achieve sterilization. Ionizing radiation has a profoundly harmful effect due to the inhibition of cell division of microorganisms and genetic damage. Therefore, ionizing radiation is useful in inactivating and eliminating bacteria, fungi, and viruses.[24] In contrast, non-ionizing radiation of long wavelength and low energy is more useful to sterilize surfaces of materials, because it is not capable of penetrating substances. Inactivation of microorganisms by radiation process can be achieved *via* direct ionization or indirect reactions of the free radicals generated during the procedure.

The radiation process exhibits advantages, particularly during the sterilization of multi-component biomaterials with a porous structure, and thermally labile materials. The radiation process has control over the penetration and intensity of the high-energy radiation, with precise and uniform dosage distribution. For instance, the radiation process can be useful in the synthesis and modification of polymers for biomedical applications without involving toxic additives. At times, free radicals generated from the radiation can be chemically active and used during the process of modification, such as crosslinking of polymers and the polymerization process.[25] The radiation process involves specialized equipment, and the process can be more expensive than the other conventional sterilization methods such as dry heat and steam-autoclaving treatment.

Gamma irradiation uses radioisotope cobalt 60 (Co-60) or caesium 137 (Cs-137) as an energy source.[24] Cobalt 60 is the most commonly used energy source as it is a non-flammable, non-soluble, non-dispersible, and non-fissionable metal.[26] Cobalt 60 has a half-life of 5.3 years with a decay process of emitting electrons and γ-rays to convert into a non-radioactive nickel-60 (Ni-60). The energy of γ-rays is relatively low and does not result in radioactive effects on the sterilized substances. Gamma irradiation is a useful tool to sterilize moderately dense or sealed materials. The dose of radiation required is dependent on the density and size of the substances, temperature, and their water content. A single standard dose of γ-irradiation used to sterilize a substance is 25 000 Gray (25 kGy).[24] High doses of γ-rays are required for substances that may include stubborn microorganisms such as viruses, parasites, and helminths.

X-ray radiation is a form of electromagnetic energy that can be generated by various sources such as an X-ray tube, fast-protons, and electron beam accelerators.[27] X-rays are formed when the electrons are accelerated by a high energy source and interact with high atomic number nuclei of metal atoms such as tungsten, molybdenum, copper, gallium, indium, and silver. The accelerated electrons slow down as they interact with the metal atoms, which results in a release of energy in the form of X-rays. Metals with high atomic numbers produce high conversion efficiency of X-rays, while elements with low atomic numbers generate low X-rays conversion. The energy of X-rays is determined by the energy of electrons. The wavelengths of X-rays are shorter than those of UV rays, but longer than γ-rays. X-rays have been demonstrated to possess higher penetrating ability than γ-rays and electron beams.[26] X-rays have advantages such as clean process without toxic residues, and they do not cause thermal damage to biopolymers as the procedures are operated at a relatively low temperature.

Electron beam radiation is also used in sterilization with tunability of the penetration depth to decrease the risk of damaging biological components. E-beam is generated by electron beam accelerator. As the penetration of E-beam is lower than γ-rays,[24] E-beam is limited to sterilization of low density and small products. In addition, the dose distribution of E-beam is less uniform than γ-irradiation, due to the penetration depth.[24] Nevertheless, E-beam has advantages such as shorter treatment time with higher dosages, higher throughput with lesser damaging impacts on the sterilized products, and lower cost than γ-irradiation.[9] As electrons are negatively charged, the scan pattern and direction of an E-beam can be controlled by strong magnets.

Ultraviolet irradiation is generally used to sterilize surfaces of materials and biopolymer implants. Ultraviolet radiation is electromagnetic with wavelengths shorter than visible light.[28] The electromagnetic spectrum comprises energies with both electrical and magnetic properties that can be categorized based on wavelength and photonic interaction with matter.[29] The UV wavelength ranges between the high-energy X-rays (<100 nm) and the lower-energy visual spectrum (>400 nm). Ionization with a change in the atomic charge of matter results from the interactions between the energy and the matter under a wavelength of <100 nm. Increasing the wavelengths could increase electron excitation and reduce ionization from the interaction between the energy and the matter.[29] Ultraviolet wavelengths can be classified in different ranges:

1. "Vacuum UV", the most energetic wavelengths (<200 nm) that interact with oxygen atoms and organic molecules at low doses
2. "UV C", the "germicidal" spectrum (200–280 nm), which has biocidal effects on bacteria
3. "UV B", the synthesis of Vitamin D and the "sun burning" effect (280–315 nm);
4. "UV A", the light generated by black light fixtures (315–400 nm)

Ultraviolet irradiation results in the excitation of electrons and the accumulation of photoproducts.[9] Various microorganisms react differently to the sterilization process of UV irradiation. For instance, vegetative bacteria can be inactivated easily by UV irradiation.[30] Comparatively, bacterial spores are more resistant to UV sterilization.[30] Besides, prions and numerous viruses are difficult to inactivate by UV irradiation.[9,31,32] The most-used UV wavelength for sterilization is within "UV C" range, which is from 200 nm to 280 nm; 260 nm is reported to be the most effective.[9,29,33,34] Sterilization by UV can be optimized based on the wavelength and the duration of exposure.[35,36] However, it has been reported that PLGA showed a significant decrease in molecular weight and tensile strength after 1 hour of UV exposure.[35] Hence, the operating conditions of UV sterilization have to be further optimized and investigated before being applied to biopolymers.

12.2.6 Supercritical Fluid

Supercritical fluid (SCF) is a pure substance at a pressure and a temperature that is beyond its critical point on a three-phase diagram. Supercritical fluid generally appears as a single-phase while possessing intermediate properties of a liquid and a gas. Specifically, SCF dissolves solutes like a liquid and SCF is also able to diffuse through solids like a gas. For instance, SCF is commonly used as a solvent to dissolve polymers and active pharmaceutical ingredients (APIs);[37–40] SCF can be used as an anti-solvent to saturate a solution to precipitate the solutes; and SCF can be used as an extraction tool to extract organic compounds from oil extracts or to eliminate undesired residual organic solvents from precipitated polymers and APIs. Among all the SCF types, carbon dioxide (CO_2) is the most commonly used SCF as it is widely available, non-toxic, non-flammable, inexpensive, and environmentally friendly. In addition, supercritical CO_2 has been demonstrated to achieve a sterility effect.[41–49] Supercritical CO_2 ($scCO_2$) has been used as an alternative to conventional sterilization processes such as dry heat treatment, steam-autoclaving, and irradiation. The benefits of using $scCO_2$ are that CO_2 does not contaminate substances with toxic chemical residues like EtO treatment does, CO_2 is unreactive to polymers and APIs, and CO_2 offers milder sterilization conditions for thermally labile materials than heat treatment or steam-autoclaving.

Supercritical fluid sterilization has been demonstrated to be an effective method of inactivation of several types of microorganisms such as bacteria, fungi, and yeast. Supercritical CO_2 possesses the property of antimicrobial activity, which has been recognized and applied for biomedical applications. Sterilization by SCF takes place when the cytoplasmic pH of the microorganism cell is decreased by forming carbonic acid.[50,51] Hence, the cell membrane is damaged, and the main enzymes are inactivated. Furthermore, the diffusivity of CO_2 and the permeability of cells can be increased by elevating the operating temperature, which leads to cell death.[49] A high diffusivity of SCF decreases the duration of the sterilization process. Dillow *et al.* have demonstrated inactivation of several microorganisms by $scCO_2$ at 205 bar pressure and a range of operating temperatures, which are listed in Table 12.4.[43] The microorganisms that were investigated included both Gram-positive (*Staphylococcus aureus*, *Bacillus cereus*, and *Listeria innocua*) and Gram-negative (*Salmonella salford*, *Psoriasis vulgaris*, *Legionella dunnifii*, *Pseudomonas aeruginosa*, and *Escherichia coli*). A continuous cycle of

Table 12.4 The sterilization of several microorganisms by supercritical carbon dioxide ($scCO_2$) at 205 bar.[43]

Microorganisms		Temperature, °C	Duration, min	Number of cycles	The degree of inactivation, log
Gram-positive	*Staphylococcus aureus*	34	36	3	3
		34	36	6	7
		40	120	6	6
		40	240	6	9
	Bacillus cereus	34	36	3	2
		34	120	6	1
		60	120	6	5
		60	240	6	8
	Listeria innocua	34	36	3	3
		34	36	6	9
Gram-negative	*Salmonella salford*	34	36	3	3
		34	36	6	3
		40	120	6	6
		40	240	6	9
	Proteus vulgaris	34	36	3	8
	Legionella dunnifii	40	90	6	4
	Pseudomonas aeruginosa	34	36	3	6
		40	90	6	6
		40	240	6	8
	Escherichia coli	34	30	3	8

depressurization and repressurization with a differential pressure of <100 bar (approximately five cycles per hour) was applied during the sterilization process by Dillow *et al.*[43] to provide a driving force for mass transport. The degree of inactivation of microorganisms is determined by the log of the ratio of the number of active microorganisms after the sterilization process to the initial total number of microorganisms before the sterilization process.

More defects were found in Gram-negative than in the Gram-positive microorganisms, which could be because the cell wall of Gram-negative microorganisms is thinner than those of Gram-positive species.[52] However, the literature suggests that the mechanism of $scCO_2$ sterilization is neither by rupture of cells caused by an increased internal pressure of the cell nor an extraction of cell wall lipids.[43] On the contrary, the mechanism of $scCO_2$ sterilization is initiated by the diffusion of CO_2 into the cell and the change of pH within the cell.[43] Generally, a sterility assurance level of 6 logs or 10^{-6} can be achieved by $scCO_2$ at lower temperatures and process duration, as compared with the conventional heat treatment and steam-autoclaving.[43]

In some cases, the addition of low molecular volatile additives can help with the sterilization process when $scCO_2$ alone is not enough to inactivate bacterial endospores at ambient temperature.[49] Studies involving water as an additive in $scCO_2$ sterilization have been demonstrated to increase the

Table 12.5 Some successful examples of a combination of additives with supercritical carbon dioxide ($scCO_2$) sterilization to achieve a sterility assurance level of 6 logs.[49]

Microorganisms	Additives	Operating conditions	Ref.
Bacillus subtilis	Peracetic acid (0.002%), water (0.15%), trifluoracetic acid	96.5 bar, 35 °C, 60 min	55
Bacillus cereus	Ethanol (2%)	100 bar, 60 °C, 90 min	101
Bacillus pumilus	Hydrogen peroxide (0.0002%)	275 bar, 60 °C, 240 min	102
	Water (3.3%), methanol (0.33%), formic acid (0.033%)	1101.3 bar, 50 °C, 45 min	103
	Water (3.3%), hydrogen peroxide (0.1%)	81 bar, 50 °C, 30 min	104
Bacillus atrophaeus	Hydrogen peroxide (0.6%)	304 bar, 40 °C, 240 min	105
Geobacillus stearothermophilus	Peracetic acid (0.002%), water (0.15%), trifluoracetic acid	96.5 bar, 35 °C, 60 min	55
	Hydrogen peroxide (0.6%)	304 bar, 40 °C, 240 min	105
	Hydrogen peroxide (0.0002–0.0006%)	270 bar, 40 °C, 30–90 min	106

inactivation of microorganisms.[43,53,54] Water has been added to increase the permeability of the cell wall, which subsequently facilitates the diffusion of $scCO_2$.[43] In the presence of water, the sterilizing effect of $scCO_2$ can be significantly enhanced because CO_2 reacts and forms carbonic acid that can further decrease the pH of the cell.[43] Besides water, peracetic acid has been demonstrated to inactivate bacteria endospores effectively, followed by hydrogen peroxide.[55] Other effective combinations of additives with $scCO_2$ sterilization are tabulated in Table 12.5.[49]

12.3 Sterilization of Biopolymers

In general, steam-autoclaving sterilization can be applied when it does not cause thermal resistance and hydrolytic instability of the biopolymers. Ethylene oxide treatment is used when it does not result in significant physical property changes of biopolymers. Irradiation has the potential to cause significant degradation of polymers at doses of radiation ≥25 kGy. A careful selection of types of polymer subjected to irradiation is essential. A general selection guide to sterilization methods for the most commonly used biopolymers is simplified in Table 12.6.

Poly-L-lactic acid (PLLA) exhibits useful applications in biomedical areas with its combination of biocompatibility and mechanical and hydrolytic degradation properties. PLLA is commonly applied in orthopaedics for implants. Several sterilization techniques such as EtO, γ-irradiation, steam-autoclaving, and gas plasma treatment have been investigated to inactivate microorganisms from PLLA. Studies showed that PLLA experienced an extensive reduction in molecular weight and in thermal and mechanical properties after γ-irradiation treatment.[56–58] The high operating temperature

Table 12.6 A general selection guide to sterilization techniques for some commonly used biopolymers.[1,43,100]

Sterilization method	Biopolymers[a] PLA	PLLA	PLGA	PGA	PHB	PCL
Steam-autoclaving	✗	○	✗	✓	✗	○
Dry heat	○	✓	✗	✓	✗	✓
Ethylene oxide	✓	✓	✓	✓	✓	✓
Gas plasma: hydrogen peroxide	○	○	✗	○	✗	✗
Gamma radiation	✓	✓	○	○	○	✓
Electron beam	✓	✓	○	✓	○	✓
Supercritical carbon dioxide	✓	✓	✓	N/A	N/A	N/A

[a]✓: good; ○: fair; ✗: poor; N/A: data not available; PCL: polycaprolactone; PGA: polyglycolic acid; PHB: polyhydroxybutyrate; PLA: polylactic acid; PLGA: poly(lactide-*co*-glycolide; PLLA: poly-L-lactic acid.

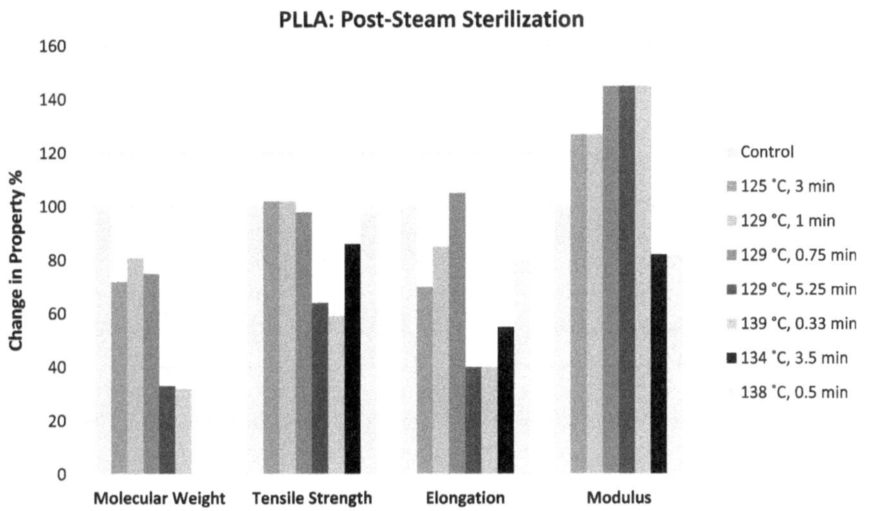

Figure 12.2 Change in properties of poly-L-lactic acid (PLLA) after steam-autoclaving sterilization (graph plotted based on the data from Rozema *et al.*).[59]

and moist environment of the steam-autoclaving sterilization method can result in significant hydrolytic degradation of PLLA. However, steam-autoclaving sterilization can be operated at a relatively low temperature within a short duration of exposure without impairing the properties of polymers.[59,60] A range of operating temperatures and duration of exposure for steam-autoclaving sterilization of PLLA was investigated by Rozema *et al.*, as illustrated in Figure 12.2.[59] Ethylene oxide treatment and H_2O_2 gas plasma sterilization for PLLA were investigated by Peniston and Choi.[58] Both EtO treatment and gas plasma by H_2O_2 have been demonstrated to have insignificant changes on the molecular weight of PLLA and throughout the 12 week *in vitro* degradation profile, as illustrated in Figure 12.3.

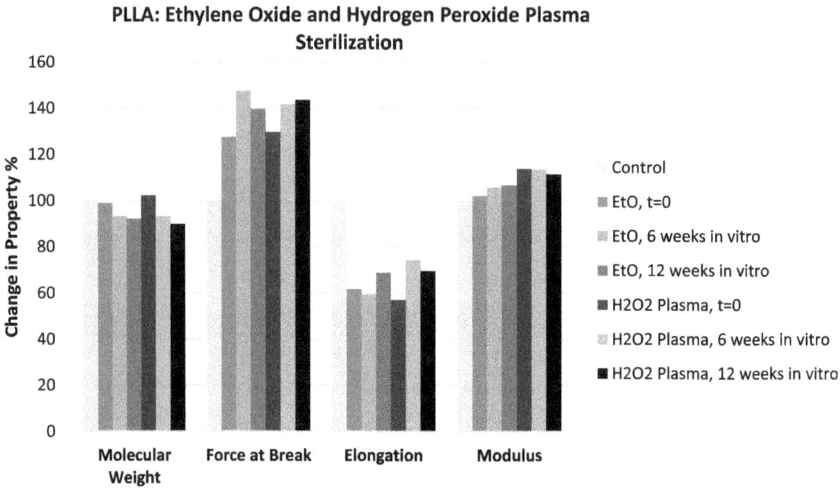

Figure 12.3 Change in properties of poly-L-lactic acid (PLLA) after an ethylene oxide (EtO) treatment and a hydrogen peroxide (H_2O_2) gas plasma sterilization (graph plotted based on the data from Peniston and Choi).[58]

A crosslinking process of the polymer can cause a slight increase of molecular weight in PLLA after H_2O_2 plasma.[58] The changes in properties of PLLA such as molecular weight, force at break, elongation, and modulus after EtO treatment and H_2O_2 plasma sterilization are illustrated in Figure 12.3, which is plotted based on the data from Peniston and Choi.[58] Supercritical CO_2 was used to sterilize PLLA scaffolds by Lanzalaco *et al.*[61] *E. coli* bacteria were inactivated under 100 bar of CO_2 pressure and 40 °C for 5 min.[61] *Streptomuyces coelicolor* spores were inactivated under 300 bar of CO_2 pressure with 200 ppm of H_2O_2 and 30 °C for 6 hours or 300 bar of CO_2 pressure without the presence of H_2O_2 and 40 °C for 6 hours.[61] Lanzalaco *et al.* have verified no changes in biocompatibility, morphology, and crystallinity of PLLA after $scCO_2$ sterilization.[61] The range of operating conditions of $scCO_2$ with the respective changes to PLLA is shown in Figure 12.4 based on the study by Lanzalaco *et al.*[61]

Polylactic acid (PLA) is a leading biopolymer in biomedical and non-medical applications. However, most sterilization techniques are not suitable for PLA. The high operating temperature of steam-autoclaving technique has been demonstrated to affect PLA by softening, melting, deforming, and causing hydrolysis of the polymer.[62,63] Ultraviolet irradiation was investigated to sterilize PLA by Janorkar *et al.*[64] The molecular weight of PLA was reduced significantly after UV-irradiation under atmospheric conditions. However, the degradation of PLA during the UV-irradiation can be minimized by irradiating through a Pyrex container.[64] Supercritical CO_2 was used to sterilize PLA by Dillow *et al.*[43] Supercritical CO_2 has been demonstrated to not have significant changes in morphology and rate of degradation of PLA.[43]

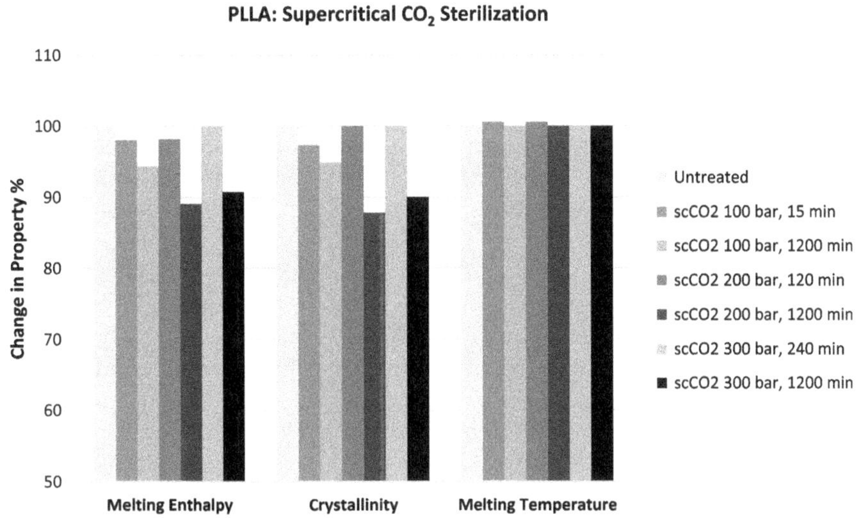

Figure 12.4 Effect on poly-L-lactic acid (PLLA) after supercritical carbon dioxide ($scCO_2$) sterilization (graph plotted based on the data from Lanzalaco *et al.*).[61]

A commonly used biodegradable polymer for biomedical applications, namely poly(lactide-*co*-glycolide) (PLGA) is popular due to its biocompatibility and controlled rate of degradation.[65,66] Applications include tissue engineering, particularly for bone, skin, nerve, and vascular regeneration.[67–70] Steam-autoclaving sterilization is not suitable for PLGA due to the high operating temperatures, and pressure conditions could deform and cause hydrolysis of the polymer.[71] γ-irradiation was applied to PLGA for sterilization.[72,73] The γ-irradiated PLGA fiber diameter is reduced by approximately 6% from unsterilized PLGA fibers. Selim *et al.* compared the PLGA fiber diameter with those following other sterilization methods such as chemical treatments woth ethanol and peracetic acid. The resulting PLGA fiber diameters from ethanol and peracetic acid treatment were reduced by more than 50%,[72] which was due to the hydrophobicity nature of PLGA and the aqueous conditions necessary for ethanol and peracetic acid sterilization procedures. The aqueous environment introduces breakage of polymer chains, which subsequently results in structure rearrangement of polymer that causes reduction of fiber diameter.[74] In contrast, γ-irradiation was conducted in dry conditions that did not cause changes in the diameter of PLGA fibers. The γ-irradiated PLGA fibers have also shown a reduction in tensile strength and molecular weight of the PLGA scaffolds.[72,73] Nevertheless, the γ-irradiated PLGA scaffolds have been demonstrated to retain sterility for more than 3 months when incubated in antibiotic-free culture medium.[72]

Sterilization of PLGA by UV radiation has been investigated.[35,75] Literature reported that UV radiation causes a significant change of morphology and reduction of molecular weight of PLGA at 254 nm within 30–480 min.[35,75] The degradation induced by UV radiation could influence PLGA applications in tissue engineering. Hence, optimization is essential before conducting UV radiation. Ethylene oxide has been used to sterilize PLGA foam scaffolds with a resulting change of dimensions and volume. Holy *et al.* demonstrated that EtO-treated PLGA has a reduction of 50% in volume and a reduction of 12% in molecular weight, even though a sterilization efficiency of 100% was achieved.[73] Gas plasma sterilization by using an inert gas (argon) was investigated by Holy *et al.*, with a 36% increase in molecular weight of the PLGA scaffolds.[73]

Polycaprolactone (PCL) is popularly known for its applications such as surgical implants and tissue engineering scaffolds to replace human bone tissue.[76,77] The degradation rate of PCL is more than a year, which could be a major advantage for bone tissue engineering. Long degradation time of scaffolds can allow sufficient time for human bone cells to grow and replace the scaffold before the scaffold fully degrades. PCL was sterilized under γ-irradiation of 30.8 kGy by Cottam *et al.*[76] γ-irradiation has resulted in crosslinking within PCL and increases in tensile strength and yield stress[†].[76] The findings were consistent with the other reports on sterilization of PCL with γ-irradiation.[78,79] Besides γ-irradiation, by Oláh *et al.* investigated E-beam sterilization of PCL scaffolds with varied porosity, ranging from 70% to 90%.[77] There is a slight increment of the modulus of elasticity (MPa) of PCL with respect to the increment of the E-beam radiation dose. However, the changes are insignificant at high porosity (85–90%) of PCL. In general, the modulus of elasticity is a linear function of the dose that is between 25 kGy and 150 kGy at constant porosity.[77] The yield strength (MPa) is found to be dependent on the porosity, but not on the radiation dose.[77]

Collagen is a major part of the extracellular matrix (ECM) of tissues. The ECM plays an important role in the regulation of essential cellular processes such as wound healing, vessel formation, and maintaining differentiated cell phenotypes.[80,81] Hence, it is a popular natural scaffold material for tissue engineering applications. Sterilization of collagen-based scaffolds was investigated *via* several sterilization methods such as scCO_2, γ-irradiation, and EtO treatment by Bernhardt *et al.*[49] Steam-autoclaving is not a suitable sterilization method for collagen-based scaffolds because the elevated temperatures can denaturize the scaffolds. It was found that γ-irradiation did not compromise the compressive modulus of the collagen scaffolds. On the other hand, the EtO-treated collagen scaffolds experienced a decrease in the compressive modulus while scCO_2-treated collagen scaffolds had an increase in the compressive modulus. Bernhardt *et al.* have demonstrated scCO_2 sterilization with a combination of additives such as water, H_2O_2, and acetic anhydride has effectively inactivated vegetative bacteria, fungi, bacteriophages, and bacteria spores.[49]

The operating conditions of several sterilization techniques for commonly used biopolymers are summarized in Table 12.7.

[†]Yield stress is the stress level where the material ceases to respond elastically and becomes plastic.[38]

Table 12.7 The operating conditions of sterilization techniques for commonly used biopolymers for tissue engineering.[a]

Biopolymer	Sterilization method	Operating conditions	Outcomes	Ref.
PLLA	γ-irradiation	39 kGy	Decreased MW and mechanical strength, increased degradation rate	107
	Steam	129 °C, 1 min	Effective sterilization for contaminations $<10^4$ microbiological death kinetics, increased mechanical strength, decreased MW, promising for thermal- and moisture-labile polymers	59
	Ethylene oxide	< 50 °C	Insignificant changes in MW	58
	Plasma—H_2O_2	43 °C, 99 min	Increased crystallinity, brittleness, melting temperature, and glass transition temperature	
	scCO_2	100–400 bar, 40 °C, 15–1200 min	No change in biocompatibility, morphology, and crystallinity of PLLA	61
PLA	UV irradiation	245–365 nm, 720 min	Inactivated microorganisms, decreased MW, increased degradation rate	64
	scCO_2	205 bar, 25–40 °C, 36–240 min	Complete inactivation of microorganisms, no morphology changes	43
PLGA	γ-irradiation	3 kGy (Caesium)	Sterility remained more than three months, decreased tensile strength	72
	γ-irradiation	25 kGy	Decreased MW, accelerated degradation rate	73
	UV irradiation	254 nm, 30–120 min	Decreased MW	75
PLGA	UV irradiation	254 nm, 30–480 min	Decreased MW and tensile strength, increased degradation rate, change in morphology	35
	Plasma—argon	33 W, 2–10 min; 100 W, 4 min	Increased MW, change in chemical structure and degradation rate	73
	Ethylene oxide	57 °C, 120 min	Decreased MW, shrinkage, change in brittleness and stiffness	73
	scCO_2	205 bar, 25–40 °C, 36–240 min	Complete inactivation of microorganisms, no morphology changes	43
PCL	γ-irradiation	30.8 kGy	Increased yield point and maximum stress, decreased MW, change in mechanical structure	76
	Electron beam	25–150 kGy, room temperature (ambient)	Increased modulus of elasticity, crosslinked, chain scission	77
Collagen-based scaffolds	scCO_2	85 bar, 38 C, 0.25% water/0.15% H_2O_2/0.5% acetic anhydride	Inactivation of vegetative bacteria, fungi, bacteriophages, bacteria spores, increased compressive modulus, no cytotoxic effects of the additives detected *in vitro*	49
	Steam	121 °C, 20 min	Denaturized by elevated temperatures	
	γ-irradiation	25–30 kGy	No change in compressive modulus of collagen	
	Ethylene oxide	Room temperature	Decreased compressive modulus of collagen	

[aa]H_2O_2: hydrogen peroxide; MW: molecular weight; PCL: polycaprolactone; PLA: polylactic acid; PLGA: poly(lactide-*co*-glycolide; PLLA: poly-L-lactic acid; sCO_2: supercritical CO_2; UV: ultraviolet.

12.3.1 Other Natural Biopolymers

Hyaluronic acid or hyaluronan is a biopolymer from the disaccharides family. Hyaluronic acid can be synthesized from integral membrane proteins, which are membrane proteins attached to biological membranes. Hyaluronic acid is essential in tissue repair of skin, as the lubricant in muscular connective tissues, articular cartilage, and extracellular matrices. Hyaluronic acid has a wide range of uses for biomedical application,[82–87] for instance, in artificial tears for dry eyes,[82] lotion for dry skin,[83] osteoarthritis knee treatment that has been approved by the US Food and Drug Administration,[84] and as a dermal filler for cosmetic surgery.[85] Hyaluronic acid is sensitive to thermal sterilization procedures. The heating process can cause rapid oxidation or hydrolysis of hyaluronic acid, resulting in a significant decrease of molecular weight of the biopolymer. Chemical sterilization could cause contamination with residues remaining in hyaluronic acid with potential risks if the hyaluronic acid is used in a medical device. Steam-autoclaving was applied to sterilize solid hyaluronic acid under the exposure of saturated or superheated steam at about 121–126 °C and a positive gauge pressure of at least 1 bar for 15–30 min was demonstrated with positive outcomes.[88,89] A pre-filled sodium hyaluronate solution was sterilized using steam-autoclaving at 115 °C, 121 °C, and 130 °C for 15 min with the conclusion that degradation can be minimized, given that the exposure time was low regardless of the application of a high temperature.[90] A long exposure time at lower temperature seemed to cause more damage than a high temperature with a short exposure time.[90]

Cellulose and chitin are the essential biomass resources in the family of polysaccharides. Chitin can be obtained from animals while cellulose is produced from plants. Chitosan is the most important derivative of chitin that is produced from deacetylation of chitin under alkaline conditions. During the deacetylation, chitin becomes soluble in aqueous acidic media.[91] The solubilization takes place due to a protonation process of the NH_2 functional group, where the polysaccharides are transformed to polyelectrolyte in an acidic media. The NH_2 functional groups in chitosan have made chitosan preferable to chitin. Chitosan exhibits advantages such as antimicrobial effects, biocompatibility, biodegradability, and low toxicity, and is inexpensive. There is a wide range of biomedical applications that chitosan is used for, which include biological adhesives for soft tissues, wound healing, tissue engineering, antioxidants, biofilms, biosensors, drug delivery systems, hydrogels, and biomedicine.[91–93] Several sterilization methods have been investigated to sterilize chitosan including the steam-autoclaving, dry heat, γ-irradiation, and EtO treatment.[94] Studies indicated that saturated steam could cause chain scission of the chitosan polymer gels that subsequently induced a 20–50% reduction of viscosity and a 30% decrease in molecular weight.[95] In contrast, the molecular weight of chitosan was unchanged after steam-autoclaving when the chitosan powder was dispersed in water before the steam sterilization.[92] γ-irradiation has been used to sterilize chitosan with outcomes such as main chain scissions,[96,97] a significant decrease in molecular weight, lower water sorption capacity,[98] and an increase in tensile strength from polymer chain rearrangements.[99] Ethylene oxide treatment has been demonstrated to show minor changes to chitosan while achieving sterilization.[96,99]

12.4 Conclusion

Biopolymers are particularly useful in biomedical applications such as medical devices, tissue engineering, implants, and drug delivery systems. It is essential to ensure the sterility of the biopolymers before they are applied *in vivo*. Various sterilization techniques such as steam-autoclaving, dry heat, radiation processes, EtO treatment, gas plasma, and supercritical fluid technology have been demonstrated to inactivate microorganisms from biopolymers. The pros and cons of several sterilization methods have been discussed and are summarized in Table 12.3. As biodegradability, functionality, and the physicochemical properties of biopolymers affect their applications, sterilization methods should be carefully selected and optimized to achieve sterility while minimizing their side effects influencing biodegradability, functionality, and physicochemical properties of the end products. Among the sterilization methods, $scCO_2$ is non-toxic and is able to achieve inactivation of microorganisms at a low range of operating temperatures of $scCO_2$, which is less damaging to the thermally labile biopolymers. In addition, processing by $scCO_2$ does not leave any toxic chemical residues in the products. Hence, sterilization by $scCO_2$ seems to be the most promising of the sterilization techniques discussed.

References

1. V. R. Sastri, in *Plastics in Medical Devices*, ed. V. R. Sastri, William Andrew Publishing, Oxford, 2nd edn, 2014, pp. 215–261.
2. U. Congress, *Office of Technological Assessement*, 1993, 51–62.
3. G.-Q. Chen and Q. Wu, *Biomaterials*, 2005, **26**, 6565–6578.
4. X. Yang, K. Zhao and G.-Q. Chen, *Biomaterials*, 2002, **23**, 1391–1397.
5. Y.-W. Wang, Q. Wu and G.-Q. Chen, *Biomaterials*, 2003, **24**, 4621–4629.
6. H. Nayar, J. P. Rubin and K. G. Marra, in *Stem Cell Biology and Tissue Engineering in Dental Sciences*, ed. A. Vishwakarma, P. Sharpe, S. Shi and M. Ramalingam, Academic Press, Boston, 2015, pp. 603–609.
7. M. Lim and N. Sultana, *Malaysian J. Anal. Sci.*, 2017, **21**, 627–632.
8. Medical Device Developments, http://www.medicaldevice-developments.com/features/featureleave-nothing-to-chance-sterilising-medical-devices-different-techniques-pros-cons/, 2013.
9. Z. Dai, J. Ronholm, Y. Tian, B. Sethi and X. Cao, *J. Tissue Eng.*, 2016, 7, 2041731416648810.
10. M. Silindir and A. Y. Özer, *FABAD J. Pharm. Sci.*, 2009, **34**, 43.

11. F. Tessarolo and G. Nollo, *Encycl. Biomater. Biomed. Eng.*, 2008, 2501–2510.
12. R. Galante, T. J. Pinto, R. Colaço and A. P. Serro, *J. Biomed. Mater. Res., B*, 2018, **106**, 2472–2492.
13. A. Venault, M.-W. Lai, J.-F. Jhong, C.-C. Yeh, L.-C. Yeh and Y. Chang, *ACS Appl. Mater. Interfaces*, 2018, **10**, 17771–17783.
14. L. Zanelli, S. Todros, E. L. Carniel, P. G. Pavan and A. N. Natali, *Acta Bioeng. Biomech.*, 2017, 19.
15. D. J. Dempsey and R. R. Thirucote, *J. Biomater. Appl.*, 1988, **3**, 454–523.
16. J. Zhu, D. Zhou and X. Zhang, *Zhongguo Yaoye*, 2016, **25**, 6–8.
17. S. Farah, D. G. Anderson and R. Langer, *Adv. Drug Delivery Rev.*, 2016, **107**, 367–392.
18. G. Letzner, H. V. Schwanewede, K. Sternberg and D. Behrend, *Deutsche Zahnartliche Zeitschrift*, 2005, **60**, 83.
19. D. Bhatnagar, K. Dube, V. B. Damodaran, G. Subramanian, K. Aston, F. Halperin, M. Mao, K. Pricer, N. S. Murthy and J. Kohn, *Macromol. Mater. Eng.*, 2016, **301**, 1211–1224.
20. M. El Fray, A. Bartkowiak, P. Prowans and J. Slonecki, *J. Mater. Sci.: Mater. Med.*, 2000, **11**, 757–762.
21. K. Sekhar, *Indian J. Anaesth.*, 2005, **49**, 389.
22. S. Lerouge, M. Wertheimer and L.'H Yahia, *Plasmas Polym.*, 2001, **6**, 175–188.
23. D. Vicoveanu, S. Popescu, Y. Ohtsu and H. Fujita, *Plasma Process. Polym.*, 2008, **5**, 350–358.
24. R. Singh, D. Singh and A. Singh, *World J. Radiol.*, 2016, **8**, 355.
25. A. N. Chakoli, J. He, M. A. Chayjan, Y. Huang and B. Zhang, *RSC Adv.*, 2015, **5**, 55544–55549.
26. Gamma Industry Processing Alliance (GIPA) and International Irradiation Association (iia), 31 August 2017 edn, 2017, pp. 1–49.
27. B. P. Fairand, *Radiation Sterilization for Health Care Products: X-ray, Gamma, and Electron Beam*, CRC Press, 2001.
28. W. Kowalkski, in *Ultraviolet Germicidal Irradiation Handbook*, Springer, Heidelberg, 2009, pp. 17–50.
29. T. D. Cutler and J. J. Zimmerman, *Anim. Health Res. Rev.*, 2011, **12**, 15–23.
30. P. Setlow, *J. Appl. Microbiol.*, 2006, **101**, 514–525.
31. C. Dean, P. Feldschreiber and J. Lett, *Nature*, 1966, **209**, 49.
32. Y. Watanabe, H. Miyata and H. Sato, *Exp. Anim.*, 1989, **38**, 305–311.
33. M. Prescott, P. Harley and D. Klein, New York, USA, 2005, 910.
34. C. Meulemans, *Ozone: Sci. Eng.*, 1987, **9**, 299–313.
35. D. Yixiang, T. Yong, S. Liao, C. K. Chan and S. Ramakrishna, *Tissue Eng. Part A*, 2008, **14**, 1321–1329.
36. C. Fischbach, J. Tessmar, A. Lucke, E. Schnell, G. Schmeer, T. Blunk and A. Göpferich, *Surf. Sci.*, 2001, **491**, 333–345.
37. N. Foster, R. Mammucari, F. Dehghani, A. Barrett, K. Bezanehtak, E. Coen, G. Combes, L. Meure, A. Ng, H. L. Regtop and A. Tandya, *Ind. Eng. Chem. Res.*, 2003, **42**, 6476–6493.

38. A. Tabernero, E. M. Martín del Valle and M. A. Galán, *Chem. Eng. Process.*, 2012, **60**, 9–25.
39. N. R. Foster, R. Mammucari, L. T. Danh and W. H. Teoh, *Dense Phase Carbon Dioxide: Food and Pharmaceutical Applications*, 2012, 199–226.
40. A. Tabernero, A. González-Garcinuño, M. A. Galán and E. M. M. Del Valle, *Rev. Chem. Eng.*, 2016, **32**, 507–532.
41. M. Perrut, *J. Supercrit. Fluids*, 2012, **66**, 359–371.
42. C. C. Beh, R. Mammucari and N. R. Foster, *Chem. Eng. J.*, 2012, **188**, 1–14.
43. A. K. Dillow, F. Dehghani, J. S. Hrkach, N. R. Foster and R. Langer, *Proc. Natl. Acad. Sci. U. S. A.*, 1999, **96**, 10344–10348.
44. C. C. Beh, R. Mammucari and N. R. Foster, IOP Conference Series: Materials Science and Engineering, 2017.
45. S. Spilimbergo, F. Dehghani, A. Bertucco and N. R. Foster, *Biotechnol. Bioeng.*, 2003, **82**, 118–125.
46. R. B. Yoganathan, R. Mammucari and N. R. Foster, *Polym. Rev.*, 2010, **50**, 144–177.
47. C. C. Beh, R. Mammucari and N. R. Foster, *Langmuir*, 2014, **30**, 11046–11054.
48. C. C. Beh, R. Mammucari and N. R. Foster, *Chem. Eng. J.*, 2015, **266**, 320–328.
49. A. Bernhardt, M. Wehrl, B. Paul, T. Hochmuth, M. Schumacher, K. Schütz and M. Gelinsky, *PLoS One*, 2015, **10**, e0129205.
50. J. Zhang, T. A. Davis, M. A. Matthews, M. J. Drews, M. LaBerge and Y. H. An, *J. Supercrit. Fluids*, 2006, **38**, 354–372.
51. S. Spilimbergo and A. Bertucco, *Biotechnol. Bioeng.*, 2003, **84**, 627–638.
52. L. McKane and J. Kandel, McGraw-Hill, New York, 1996.
53. A. N. Sabirzyanov, A. P. Il'in, A. R. Akhunov and F. M. Gumerov, *High Temp.*, 2002, **40**, 203–206.
54. A. W. Francis, *J. Phys. Chem.*, 1954, **58**, 1099–1114.
55. A. White, D. Burns and T. W. Christensen, *J. Biotechnol.*, 2006, **123**, 504–515.
56. P. Nugroho, H. Mitomo, F. Yoshii and T. Kume, *Polym. Degrad. Stab.*, 2001, **72**, 337–343.
57. Ö. Kantoğlu and O. Güven, Nuclear Instruments and Methods in Physics Research Section B: Beam Interactions with Materials and Atoms, 2002, 197, 259–264.
58. S. J. Peniston and S. J. Choi, *J. Biomed. Mater. Res., B*, 2007, **80**, 67–77.
59. F. Rozema, R. Bos, G. Boering, J. Van Asten, A. Nijenhuis and A. Pennings, *J. Appl. Biomater.*, 1991, **2**, 23–28.
60. N. Weir, F. Buchanan, J. Orr, D. Farrar and A. Boyd, *Biomaterials*, 2004, **25**, 3939–3949.
61. S. Lanzalaco, S. Campora, V. Brucato, F. Carfì Pavia, E. R. Di Leonardo, G. Ghersi, O. Scialdone and A. Galia, *J. Supercrit. Fluids*, 2016, **111**, 83–90.
62. B. K. Lee, Y. Yun and K. Park, *Adv. Drug Delivery Rev.*, 2016, **107**, 176–191.

63. R. Jain, N. H. Shah, A. W. Malick and C. T. Rhodes, *Drug Dev. Ind. Pharm.*, 1998, **24**, 703–727.
64. A. V. Janorkar, A. T. Metters and D. E. Hirt, *J. Appl. Polym. Sci.*, 2007, **106**, 1042–1047.
65. L. Lu and A. G. Mikos, *MRS Bull.*, 1996, **21**, 28–32.
66. S. Li, *J. Biomed. Mater. Res.*, 1999, **48**, 342–353.
67. J. Stitzel, J. Liu, S. J. Lee, M. Komura, J. Berry, S. Soker, G. Lim, M. Van Dyke, R. Czerw and J. J. Yoo, *Biomaterials*, 2006, **27**, 1088–1094.
68. B.-M. Min, Y. You, J.-M. Kim, S. J. Lee and W. H. Park, *Carbohydr. Polym.*, 2004, **57**, 285–292.
69. T. Bini, S. Gao, X. Xu, S. Wang, S. Ramakrishna and K. W. Leong, *J. Biomed. Mater. Res., A*, 2004, **68**, 286–295.
70. K. Kim, M. Yu, X. Zong, J. Chiu, D. Fang, Y.-S. Seo, B. S. Hsiao, B. Chu and M. Hadjiargyrou, *Biomaterials*, 2003, **24**, 4977–4985.
71. G. Kumar, N. Shafiq and S. Malhotra, *Critical Reviews™ in Therapeutic Drug Carrier Systems*, 2012, 29.
72. M. Selim, A. J. Bullock, K. A. Blackwood, C. R. Chapple and S. MacNeil, *BJU Int.*, 2011, **107**, 296–302.
73. C. E. Holy, C. Cheng, J. E. Davies and M. S. Shoichet, *Biomaterials*, 2000, **22**, 25–31.
74. J. K. Perron, H. E. Naguib, J. Daka, A. Chawla and R. Wilkins, *J. Biomed. Mater. Res., B*, 2009, **91**, 876–886.
75. D. I. Braghirolli, D. Steffens, K. Quintiliano, G. A. X. Acasigua, D. Gamba, R. A. Fleck, C. L. Petzhold and P. Pranke, *J. Biomed. Mater. Res., B*, 2014, **102**, 700–708.
76. E. Cottam, D. W. Hukins, K. Lee, C. Hewitt and M. J. Jenkins, *Med. Eng. Phys.*, 2009, **31**, 221–226.
77. L. Oláh, K. Filipczak, T. Czvikovszky, T. Czigány and L. Borbás, *Radiat. Phys. Chem.*, 2007, **76**, 1430–1434.
78. M. Narkis, S. Sibony-Chaouat, A. Siegmann, S. Shkolnik and J. P. Bell, *Polymer*, 1985, **26**, 50–54.
79. V. Masson, F. Maurin, H. Fessi and J. P. Devissaguet, *Biomaterials*, 1997, **18**, 327–335.
80. M. J. Bissell, H. G. Hall and G. Parry, *J. Theor. Biol.*, 1982, **99**, 31–68.
81. S. W. Tsai, R. L. Liu, F. Y. Hsu and C. C. Chen, *Biopolymers*, 2006, **83**, 381–388.
82. A. D. Pucker, S. M. Ng and J. J. Nichols, *The Cochrane Library*, 2016.
83. P. E. Bermann, *Nurs. Clin. North Am.*, 2007, **42**, 485–500.
84. T. Gower, Arthritis Foundation, https://www.arthritis.org/living-with-arthritis/treatments/medication/drug-types/other/hyaluronic-acid-injections.php, 2018.
85. S. N. A. Bukhari, N. L. Roswandi, M. Waqas, H. Habib, F. Hussain, S. Khan, M. Sohail, N. A. Ramli, H. E. Thu and Z. Hussain, *Int. J. Biol. Macromol.*, 2018, **120**, 1682–1695.
86. S. Farah, *ACS Appl. Mater. Interfaces*, 2018, **10**, 9010–9022.
87. S. Farah and A. J. Domb, *J. Controlled Release*, 2018, **271**, 107–117.

88. D. A. Swann, J.-w. Kuo and V. Pinsky, *US Pat.*, US5621093A, 1997.
89. A. A. M. Shimojo, I. C. de Souza Brissac, L. M. Pina, C. S. Lambert and M. H. A. Santana, *Bio-Med. Mater. Eng.*, 2015, **26**, 183–191.
90. M. L. Bernuzzi and A. Giori, 2016.
91. H. Honarkar and M. Barikani, *Monatshefte für Chem. – Chem. Monthly*, 2009, **140**, 1403.
92. A. San Juan, A. Montembault, D. Gillet, J. Say, S. Rouif, T. Bouet, I. Royaud and L. David, IOP Conference Series: Materials Science and Engineering, 2012.
93. S. Farah, J. C. Dolof, D. G. Anderson and R. Langer, Crosslinked Chitosan Nanoparticles and Chemical Modifications for Drug Delivery Applications, 2015.
94. S. B. Rao and C. P. Sharma, *J. Biomater. Appl.*, 1995, **10**, 136–143.
95. C. Jarry, C. Chaput, A. Chenite, M. A. Renaud, M. Buschmann and J. C. Leroux, *J. Biomed. Mater. Res.*, 2001, **58**, 127–135.
96. Y. M. Yang, Y. H. Zhao, X. H. Liu, F. Ding and X. S. Gu, *J. Appl. Polym. Sci.*, 2007, **104**, 1968–1972.
97. C. Jarry, J.-C. Leroux, J. Haeck and C. Chaput, *Chem. Pharm. Bull.*, 2002, **50**, 1335–1340.
98. E. Szymańska and K. Winnicka, *Mar. Drugs*, 2015, **13**, 1819–1846.
99. P. R. Marreco, P. D. L. Moreira, S. C. Genari and Â. M. Moraes, *J. Biomed. Mater. Res., B*, 2004, **71**, 268–277.
100. A. K. Dillow, R. S. Langer, N. Foster and J. S. Hrkach, *US Pat.*, US6149864A, 2000.
101. H. S. Park, H. J. Choi, M. D. Kim and K. H. Kim, *Int. J. Food Microbiol.*, 2013, **166**, 207–212.
102. J. Zhang, S. Burrows, C. Gleason, M. A. Matthews, M. J. Drews, M. LaBerge and Y. H. An, *J. Microbiol. Methods*, 2006, **66**, 479–485.
103. E. Shieh, A. Paszczynski, C. M. Wai, Q. Lang and R. L. Crawford, *J. Microbiol. Methods*, 2009, **76**, 247–252.
104. A. Checinska, I. A. Fruth, T. L. Green, R. L. Crawford and A. J. Paszczynski, *J. Microbiol. Methods*, 2011, **87**, 70–75.
105. J. D. Hemmer, M. J. Drews, M. LaBerge and M. A. Matthews, *J. Biomed. Mater. Res., B*, 2007, **80**, 511–518.
106. I. Donati, M. Benincasa, M. P. Foulc, G. Turco, M. Toppazzini, D. Solinas, S. Spilimbergo, I. Kikic and S. Paoletti, *Biomacromolecules*, 2012, **13**, 1152–1160.
107. K. A. Hooper, J. D. Cox and J. Kohn, *J. Appl. Polym. Sci.*, 1997, **63**, 1499–1510.

CHAPTER 13

Recent Advances in Antimicrobial Hydrogels

KONDA REDDY KUNDURU[a] AND ABRAHAM J. DOMB*[b]

[a] School of Chemistry, University of Hyderabad, Hyderabad, India;
[b] School of Pharmacy-Faculty of Medicine, The Hebrew University of Jerusalem, Jerusalem, Israel
*Email: avid@ekmd.huji.ac.il

13.1 Introduction

Despite major advances in healthcare, infectious diseases caused by various pathogens such as bacteria, viruses, parasites, or fungi remain major challenges. These pathogens may indirectly cause socio-economic problems. Microbial infection is a global challenge. According to the World Health Organization 2008 report, microbial infection is the second largest cause of death.[1] Antibiotics play a vital role in the treatment of various pathogens. However, the excessive use of conventional antibiotics has led to resistance to these pathogens. β-Lactamase is the enzyme released by bacteria, which can cleave the β-lactam ring of the antibiotics and cause resistance to their antimicrobial action.[2]

Furthermore, the residues of these conventional antibiotics in the environment are a serious health hazard. To overcome the problems associated with the use of conventional antibiotics and drug-resistant microbes, there is a need to develop new antimicrobial materials. The relevant formulation of these new antimicrobial materials is also necessary. Various classes of materials have been reported to possess antimicrobial activity. Among these materials, antimicrobial peptides (AMPs),[3] cationic polymers,[4] and antimicrobial nanoparticles[5] have been reported as potential alternatives to conventional antibiotics.

Biomaterials Science Series No. 5
Antimicrobial Materials for Biomedical Applications
Edited by Abraham J. Domb, Konda Reddy Kunduru and Shady Farah

Published by the Royal Society of Chemistry, www.rsc.org

A hydrogel is a three-dimensional (3-D) matrix that holds water in large quantities within the molecular network *via* hydrophilic interactions. It is formed from natural and synthetic polymers.[6] Hydrogels consist of polymer chains that are formed either by physical or chemical crosslinking. The term 'hydrogel' was used for the first time in 1894, but its first biological application was reported in 1960 by Wichterle and Lim.[7] Hydrogels have been synthesized from natural and synthetic polymers. Varying classes of natural polymers, such as polysaccharides or proteins and synthetic polymers, such as poly(vinyl alcohol), polyethylene oxide, and poly(acrylic acid) were used to prepare hydrogels. Synthesis and modification of various types of hydrogel have been reported elsewhere.[8,9] Hydrogels have numerous applications in biomedical research, such as medical devices, drug delivery, protein delivery, wound dressings, tissue engineering, immunoisolation, and as coatings.[10–23] The large amount of water in hydrogels causes a soft consistency, which is similar to natural tissue; no other class of biomaterial has this quality.[24,25] There has been an ever-growing demand for the development of hydrogel materials in recent years, especially for physical or chemical crosslinking processes.[26]

Hydrogels have been studied as an alternative for antimicrobial activities instead of conventional antibiotics. Careful selection of monomers and crosslinkers can provide hydrogels with desirable properties that are certainly useful for antimicrobial applications. Some of the materials having inherent antimicrobial action can be used in the synthesis of hydrogels.[27] Hydrogels can show responsiveness towards various stimuli, such as temperature, pH, light, enzymes, *etc.* They can be used as a coating on urinary and central venous catheters. They can also be used as joint and dental implants and in contact lenses. They may be used as a local injection for drug release and in wound healing. In addition, certain hydrogels possess inherent antimicrobial action. Nanoantimicrobials can be combined with hydrogels and used in doses lower than systemic administration. By doing so, resistance to pathogens and side effects associated with the drugs can be avoided.[28] These remarkable features make hydrogels versatile materials for applications in the pharmaceutical and medical fields, particularly for antimicrobial applications. In this chapter, we discuss recent advances in the different classes of antimicrobial hydrogels. Antimicrobial hydrogels are classified into two main categories: (1) hydrogels with inherent antimicrobial activity; and (2) hydrogels loaded with biocides. Antimicrobial hydrogels inherently contain active antimicrobial species; in the second category, antimicrobial active molecules are inserted into hydrogel matrices.

13.2 Classification of Hydrogels Based on their Fabrication Strategies

Hydrogels are categorized based on their fabrication into three types: (1) physically crosslinking hydrogels or non-covalent hydrogels; (2) chemically crosslinking hydrogels; and (3) dual network hydrogels. Physical hydrogels are generated *via* non-covalent crosslinks and are reversible. The non-covalent crosslinks are due mainly to hydrophobic interactions, hydrogen bonds, electrostatic interactions, or even crystallized segments. Chemically crosslinking hydrogels are generated *via* the reaction of functionalized monomers with a crosslinking agent. However, after this reaction, a purification step is required to remove unreacted monomers. There is an alternative route to preparing chemically crosslinking hydrogels. Instead of using monomers, treatment of polymers with γ-irradiation, X-rays, UV light, or heat generates crosslinking hydrogels. These irradiated processes for synthesis of chemically crosslinking hydrogels are inexpensive, safe, and do not require purification. The third option is dual-network hydrogels that are formed by the combination of both physically and chemically crosslinking networks.[29–31]

13.3 Hydrogels with Inherent Antimicrobial Activity

Inherently antimicrobial hydrogels contain active antimicrobial species. Synthesis of these hydrogel networks involves either chemical or physical crosslinking, depending on the polymer type. These hydrogels kill microbes upon contact due to the active moiety of the polymer. In most cases, the active moiety of the polymer will be from the quaternary ammonium group.

13.3.1 Natural Polymeric Hydrogels

Living organisms usually produce natural biopolymers, such as polypeptides, polynucleotides, and polysaccharides. Among these natural polymers, polysaccharides play an important role in the development of polymeric hydrogels possessing antimicrobial activity. Among polysaccharides, chitosan is an abundantly available linear polysaccharide that possesses biodegradation, low toxicity, and a biocompatible nature. It is also less expensive. The excellent features of this natural polymer and its derivatives have generated various biomedical applications. The amino groups of this polymer are protonated under acidic conditions to enhance water solubility, thus creating the antimicrobial property of chitosan. The biomaterials prepared from chitosan and its derivatives have been applied to wound dressing materials.[32] Synthesis of chitosan hydrogels can be possible by either physical or chemical crosslinking methods *via* blending or with a suitable crosslinker.

Mixing of chitosan and γ-poly(glutamic acid) has been gelated by polyelectrolyte complexation of cationic chitosan and anionic γ-poly(glutamic acid). This hydrogel works against *Staphylococcus aureus* and *Escherichia coli*. This hydrogel is a promising wound-healing material.[33,34] Aniline-grafted quaternized chitosan hydrogels were prepared by using the polydextran aldehyde crosslinker. This hydrogel shows antimicrobial activity against *S. aureus* and *E. coli*. The hydrogel has been electroactive with proliferated myoblasts. Thus, the hydrogel matrix can be used in tissue engineering.[35]

Instead of polydextran aldehyde, a benzaldehyde group functionalized crosslinker has been used to prepare aniline-grafted quaternized chitosan hydrogels (Figure 13.1). The hydrogel shows a self-healing, radical scavenging antioxidant property along with the antimicrobial activity. This hydrogel can be used for wound-healing applications.[36]

An antibacterial hydrogel based on chitosan and polydextran aldehyde has been synthesized. The hydrogel matrix displays efficacy against *S. aureus*,

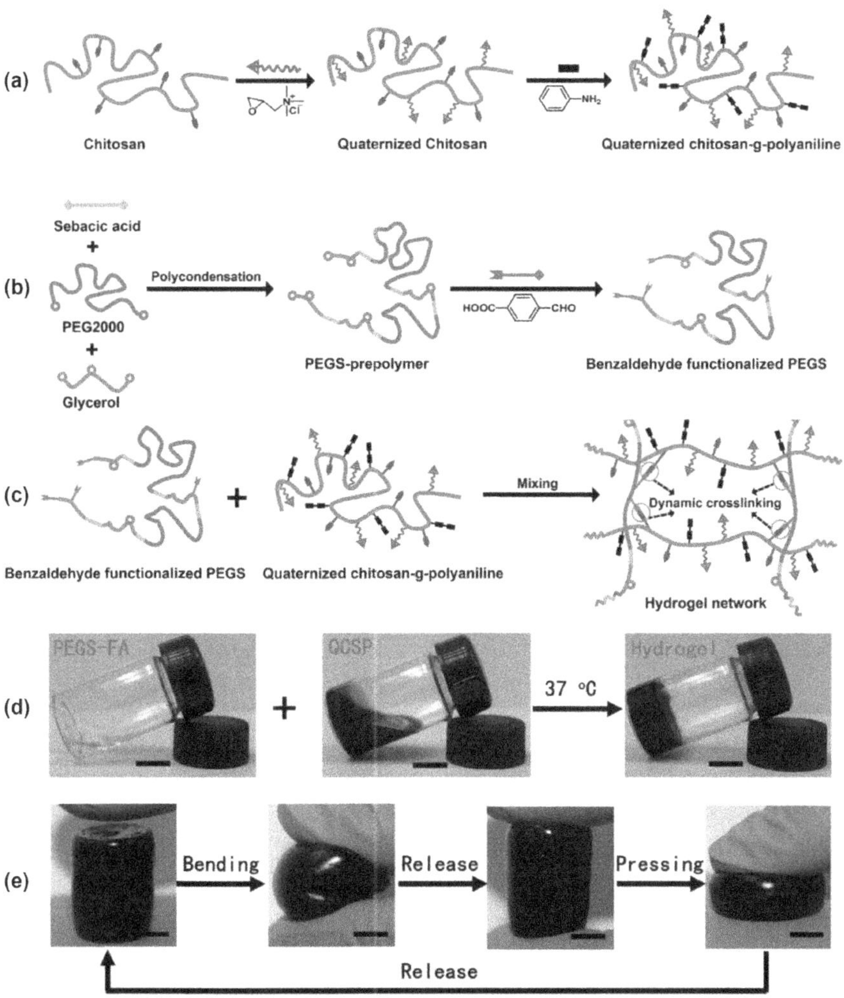

Figure 13.1 Synthesis of hydrogels based on aniline-grafted chitosan and benzaldehyde-functionalized polymer (a, b, and c). Quaternized chitosan-g-polyaniline (QCSP) and benzaldehyde functionalized poly(ethylene glycol)-*co*-poly(glycerol sebacate) grafted 4-formylbenzoic acid (PEGS-FA) are mixed to form a hydrogel (d). The bending and pressing shapes of the hydrogel (e).[36]
Reproduced from ref. 36 with permission from Elsevier, Copyright 2017.

E. coli, and *Pseudomonas aeruginosa*, along with drug-resistant bacteria methicillin-resistant *S. aureus* (MRSA), vancomycin-resistance *Enterococcus* (VRE), and β-lactam-resistant *Klebsiella pneumoniae*. This system has been applied successfully in wound healing and is a biocompatible hydrogel.[37] Natural polymeric hydrogels have tremendous growth in the past and now are common antimicrobial materials used to kill microbes. However, the variation in the molecular weights of natural polymers may affect their physical properties, thereby possibly affecting the efficacy of the hydrogel materials as antimicrobials. Impurities in the natural polymers may cause immunogenicity, a considerable constraint when designing natural antimicrobial hydrogels.[26]

13.3.2 Synthetic Polymer-based Hydrogels

Synthetic cationic polymers are used as an alternative to antibiotics in the treatment of microbial infections. These cationic group containing polymers are prepared using a variety of synthetic methods. However, an adaptation of living radical polymerization techniques, such as atom transfer radical polymerization (ATRP), reversible addition–fragmentation chain transfer (RAFT), *etc.*, in the synthesis of cationic group containing synthetic polymers has emerged in the development of antimicrobial materials. Various cationic synthetic polymers, include poly(acrylate), poly(norbornene), poly(ethyleneimine)s, poly(arylamide)s, poly-β-lactams, poly-α-amino acids, and polycarbonates.[38–45]

Polycarbonate-based antimicrobial hydrogel was prepared by non-covalent interaction and shows antimicrobial activity against VRE, *Acinetobacter baumannii*, and *K. pneumoniae*.[46] In another study, polycarbonate-based biodegradable hydrogels containing vitamin E were prepared. This hydrogel system shows activity against various microbes, such as *S. aureus*, *E. coli*, and *Candida albicans*.[47] In addition, antimicrobial hydrogels were prepared by thiol–ene reaction. Condensation of fumaryl chloride, dodecylbis(2-hydroxyethyl) methylammonium chloride, and oligo(ethylene glycol) generated multi 'ene' functionality. Polymers with multi 'thiol' groups were generated by a condensation reaction between mercaptosuccinic acid and oligo(ethylene glycol). Mixing 'ene' and 'thiol' functionalities forms a hydrogel matrix by covalent crosslinking. The hydrogel system shows activity against Gram-negative and Gram-positive bacteria.[48]

An injectable hydrogel was prepared by reacting polydextran aldehyde with branched polyethyleneimine covalently. The hydrogel shows activity against Gram-negative and Gram-positive bacteria. This hydrogel possessing bioadhesive properties can be used for wound-healing applications.[49] An injectable ABA-type triblock copolymer hydrogel containing catechol and PEG has been prepared (Figure 13.2). The hydrogel shows activity against *E. coli*. It has thermosensitive properties and a self-healing nature, certainly useful in bioengineering applications.[50]

A double-network hydrogel possessing antimicrobial activity was prepared by using 3,4en-ionene. A hydrogel network was generated by

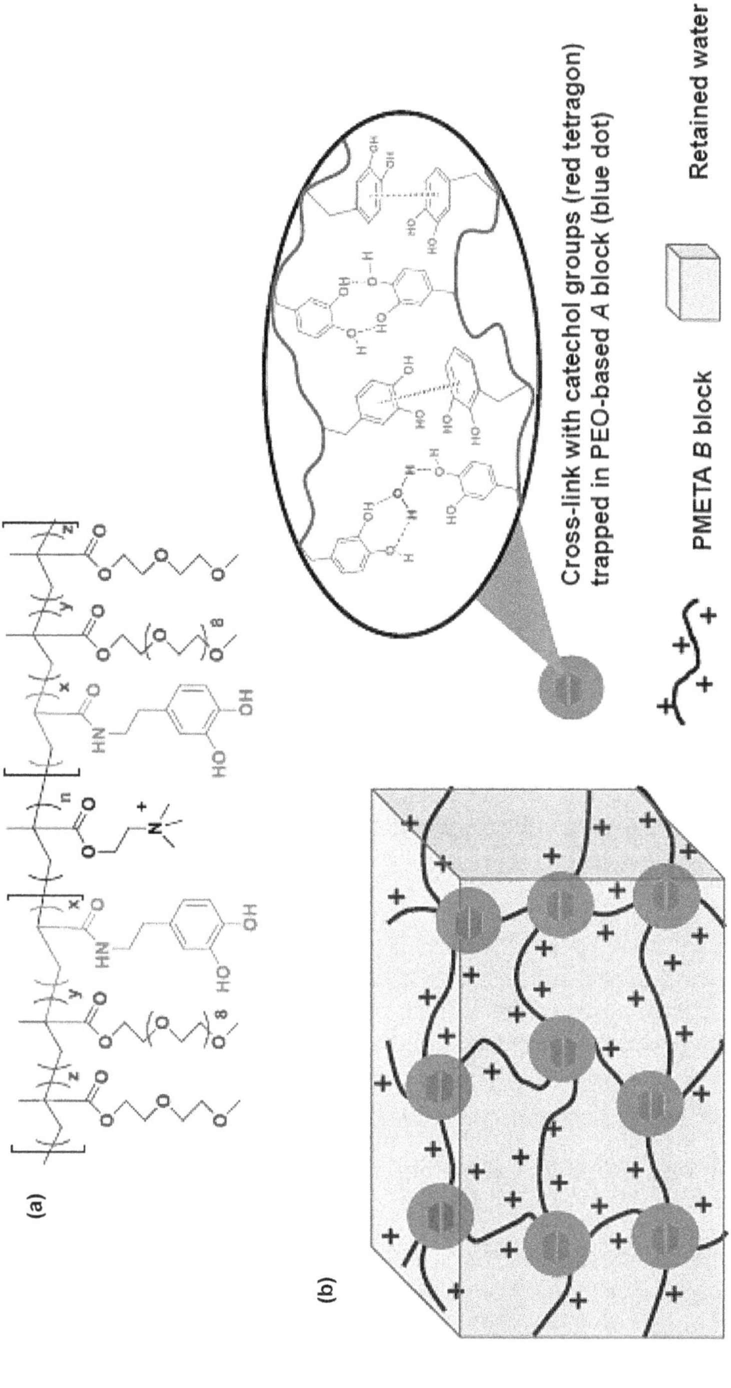

Figure 13.2 (a) Structure of the ABA tri-block copolymer. (b) Proposed structure of the hydrogel and the mussel-inspired self-healing mechanism.[50]
Reproduced from ref. 50 with permission from the American Chemical Society, Copyright 2017.

photopolymerization of 3,4en-ionene with 2-hydroxyethyl acrylate and glycerol dimethacrylate. This double-network hydrogel shows activity against *S. aureus*, *E. coli*, and *P. aeruginosa*. Even after washing the double network hydrogels with water, antimicrobial activity is retained for 4 weeks.[51]

13.3.3 Polypeptide-based Hydrogels

Recently, peptide-based hydrogel systems have gained importance for antimicrobial activity research. A self-assembling peptide-based hydrogel system was prepared. The surface of the hydrogel shows a broad spectrum of antimicrobial activity against both Gram-positive and Gram-negative bacteria. This hydrogel system can be used for tissue regeneration, even in non-sterile environments.[52] In another study, a peptide-based hydrogel was prepared by a self-assembling reaction to external stimuli. This hydrogel shows activity against *E. coli*. This hydrogel system can be used in drug delivery, tissue engineering, and regenerative medicine.[53] The dopamine-modified ε-poly-L-lysine-PEG-based hydrogel was prepared. Dopamine moieties were involved in crosslinking to generate a hydrogel matrix. This system shows activity against *S. aureus* and *E. coli*. As the dopamine substitution in the polylysine backbone increases, fewer free amino groups are available, and reduced antimicrobial activity is observed. This hydrogel can be useful for the development of wound dressings.[54] In another study ε-poly-L-lysine-based self-healing hydrogels were prepared (Figure 13.3). This hydrogel system exhibits antibacterial activity against *S. aureus* and *E. coli*. This hydrogel is also effective as a wound-closure gel. The system may be useful in tissue engineering applications.[55]

13.3.4 Mechanism of Action of Hydrogels Possessing Antimicrobial Activity

Unlike other cationic polymers, the mechanism of the antimicrobial activity of hydrogels has yet to be fully understood. However, from the literature, it would appear that lysis of the bacterial cell wall is most likely key.[26,56,57] The anions on the cell walls of the bacteria are attracted to the cationic moieties in the hydrogel matrix. This leads to disruption of the cell wall followed by leakage of cellular material.[26] The antimicrobial activity of the hydrogels depends mainly on the amphiphilicity and cationic charge on the polymers. The activity also depends on the porosity of the hydrogel matrix and power penetrating into the microbial membrane.[26]

13.4 Hydrogels Loaded with Biocides

Various biocide molecules have been inserted into the hydrogel matrix for the development of sustainable antimicrobial materials. Among biocides metal nanoparticles, conventional antibiotics, metal ions, small molecules, and antimicrobial peptides have been inserted into the hydrogel matrix.

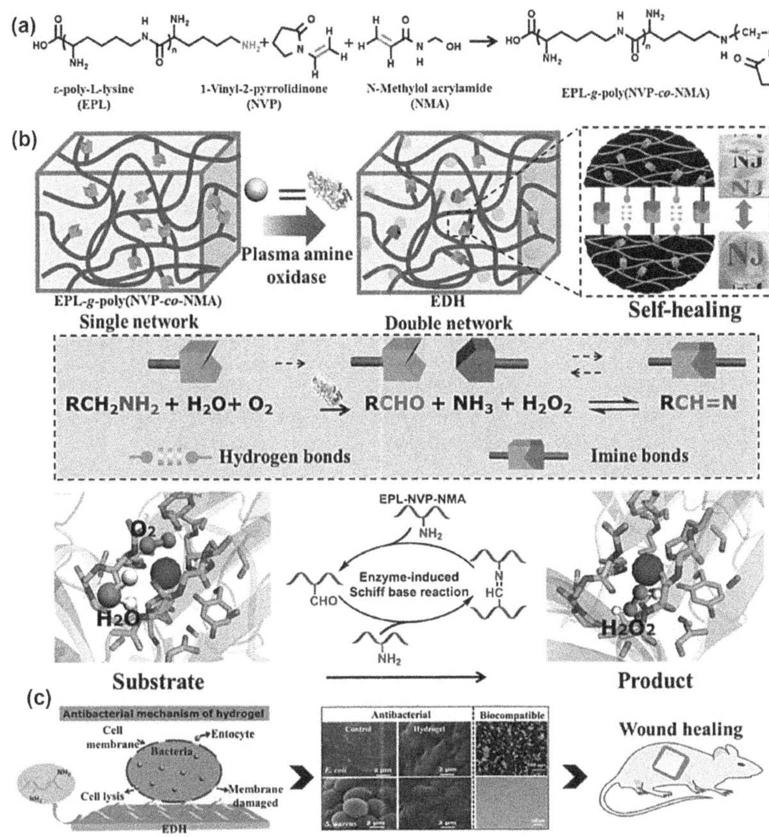

Figure 13.3 Enzyme-assisted dual-network self-healing hydrogel synthesis (a, b) and *in vitro* and *in vivo* applications (c).[55]
Reproduced from ref. 55 with permission from the Royal Society of Chemistry.

Biocide molecules released from the hydrogel matrix not only kill the microbe, but they also cure the infected region, since most of the hydrogels are biocompatible. In the following sections, various recently developed biocide-loaded hydrogel matrices are discussed.

13.4.1 Metal Ions and Nanoparticle-loaded Hydrogels

Metal ions have been used in the treatment of microbial infections since ancient times. Among various others, silver, copper, zinc, and gold are the prominent metal ions reported in the literature. Silver has shown tremendous antimicrobial activity against various pathogens with less toxicity than other metal ions. Other metal ions have also shown antimicrobial activities with varying mechanisms, but they possess advantages over other pathogens (Table 13.1).

Table 13.1 Antimicrobial mechanisms of various metal nanoparticles (Reproduced from ref. 28 with the permission from Dove Press).[a]

Metal nanoparticle	Mechanism of action	Ref.
AgNPs	1. Silver ions (Ag^+) dissolved from AgNPs interact with cell wall and plasma membrane of bacteria. Binding to negatively charged parts of the membrane creates holes in the membrane, allowing plasma contents (including K^+) to leak out of the cell, thus dissipating the H^+ gradient across the membrane. 2. Inside the microbial cell, AgNPs exert several antimicrobial effects: (a) inhibiting cytochromes of the electron transport chain of microbes; (b) causing damage to DNA and RNA of microbes; (c) inducing the formation of ROS, which is also toxic to host cells; (d) inhibiting cell wall synthesis in Gram-positive bacteria. 3. After the AgNPs are leaked from dead microbes, AgNPs can kill other microbial cells.	58–63
AuNPs	AuNPs can attach to the bacterial membrane, which leads to leakage of bacterial contents or penetration of the outer membrane and peptidoglycan layer, and results in bacterial death.	64
AuNP-Amp	The presence of multiple Amp molecules on the surface of AuNP allows the AuNP-Amp to overwhelm high concentrations of β-lactamase expressed by these bacteria. Then, AuNP-Amp inhibits the transmembrane pump that catalyzes drug efflux from the bacterial cell.	65
ZnONPs	1. ZnONPs bind to bacterial cell membranes and destroy their lipids and proteins. 2. ZnONPs can cause formation of Zn^{2+} ions and ROS, which damage the bacterial cell. 3. When coated with PVA, ZnONPs increase membrane permeability and enter the cytoplasm of the bacterial cell.	66–70
CuO/CuNPs	Copper interacts with amine and carboxyl groups on the surfaces of microbial cells. Therefore, microbes with a higher density of the two functional groups have a higher sensitivity to CuO/CuNPs.	71–73

[a]AgNPs: silver nanoparticles; Amp: ampicillin; AuNPs: gold nanoparticles; CuO/CuNPs: copper-containing NPs; PVA: polyvinyl alcohol; ROS: reactive oxygen species; ZnONPs, zinc oxide nanoparticles.

Free metal ions have been used as antimicrobial materials. But, the toxicity towards mammalian cells made their application limited. This toxicity was addressed by developing antimicrobial metal ion-loaded hydrogel systems.[74–76] These hydrogel matrices can be used as wound care materials with less toxicity. Various metal ion nanoparticles were developed for use as antimicrobial materials; these antimicrobial nanoparticles generate reactive oxygen species. The metal nanoparticles may also have toxic effects towards mammalian cells, which can be minimized with the loading of these metal nanoparticles into hydrogel systems.[77–80]

13.4.1.1 Silver Nanoparticle-loaded Hydrogels

Silver has been used as an antimicrobial agent for thousands of years, even before the existence of microorganisms was known. Silver vessels have been used for food and water storage.[81,82] Silver has been employed in various biomedical applications, such as wound dressings, textiles, and bone implants. Silver nanoparticles show activity against various pathogens due to their multiple mechanisms against microbes.[5,58,67]

Natural polymer-based hydrogel systems, such as chitosan hydrogels loaded with nanosilver have been prepared for wound dressings. As the concentration of nanosilver in the hydrogel system for low molecular weight chitosan increases, activity against *P. aeruginosa* and *S. aureus* increases.[83] Covalently crosslinked alginate-based hydrogels have been prepared for loading silver nanoparticles. Silver nanoparticle-loaded alginate hydrogels can be used in wound-healing materials.[84] Carboxymethyl cellulose-based hydrogels loaded with silver nanoparticles have been prepared and studied for their antimicrobial activity. The hydrogel shows activity against both Gram-positive and Gram-negative bacteria. It can be used as a potential antimicrobial gel to treat pathogens in the medical field.[85] In another study, synthesis of hydrogel based on chitosan/polyvinyl alcohol/graphene was carried out with silver ions embedded into the hydrogel by electrochemical reduction. This hydrogel shows activity against *S. aureus* and *E. coli.* The hydrogel does not show toxicity towards human and mouse fibroblast cells.[86] Along with these examples, several other natural polymer-based hydrogel systems loaded with silver nanoparticles have been reported.[28]

Synthetic polymers have also been used to fabricate hydrogel systems to load silver nanoparticles. The most commonly used polymers are poly(acrylamide), poly(acrylic acid), and poly(vinyl alcohol). A hydrogel nanocomposite based on polyacrylamide/starch was prepared with silver nanoparticles. The hydrogel system showed activity against both bacteria and fungi.[87] Other synthetic polymer-based hydrogel matrices possessing silver nanoparticles have been reported.[28]

13.4.1.2 Gold Nanoparticle-loaded Hydrogels

Gold is a bioinert metal, but in nanoparticles it performs various biological functions.[88] Gold nanoparticles can be functionalized with various polymers depending on the application. The mechanism of action of gold nanoparticles is presented in Table 13.1. There are fewer studies of the antimicrobial action of gold nanoparticles than there are for silver nanoparticles.[28] Progress regarding gold nanoparticles as antimicrobials has been reviewed.[89] Layered poly(methacrylic acid) hydrogel capsules have been prepared and can generate gold nanoparticles *in situ* in the walls of the hydrogel matrix.[90] This layered hydrogel possessing gold nanoparticles can be used for antimicrobial applications. In another report, hydrogel was used to formulate pH-responsive gold nanoparticle-stabilized liposomes for topical antimicrobial delivery. This formulation showed no toxicity in a 7 day study on mouse skin. Therefore, this can be used as a potential formulation for the treatment of skin diseases.[91] Silk fibroin/nanohydroxyapatite-based hydrogels have been synthesized and loaded with silver and gold nanoparticles. These hydrogels show activity against both Gram-positive and Gram-negative bacteria. They can potentially be used for bone tissue engineering.[92] The antimicrobial action of gold nanoparticles is weaker than silver nanoparticles. The gold nanoparticles, however, possess their own advantages, for example, in clinical orthopaedic surgery.[28] Along with these nanoparticulate hydrogel systems, other metal nanoparticle-based hydrogel systems possessing zinc, copper, and others have been reported.[93–99]

Hydrogels possessing metal nanoparticles can counter antibiotic resistance, so they can be a good replacement for antibiotics. Microbial resistance is the modern cause of mortality with microbes that are resistant to existing antibiotic molecules. Microbial resistance to metal nanoparticles is rare, due to their multiple mechanisms of action towards pathogens. Another advantage is that the small size of these nanoparticulate systems enables them to penetrate easily through the cell walls of the microbes, to disrupt them as needed. They are also sufficiently stable to pass on to other microbial cells after being released from dead cells. All these benefits indicate that hydrogel systems possessing metal nanoparticles may solve the present day problem of antimicrobial resistance.[28]

13.4.2 Antibiotic-loaded Hydrogel Systems

The most commonly used and effective antimicrobial treatment is antibiotics. However, the biggest obstacle is the resistance of pathogens to antibiotics.[100] Almost all the antibiotics available on the market have shown resistance towards pathogens except teixobactin, which does not show any resistance towards bacteria.[101] This drug is more effective towards Gram-positive than Gram-negative bacteria.[102] To overcome bacterial resistance, new antibiotics must be manufactured, or else usage must be minimized. The most effective way to reduce the usage of antibiotics is to load the antibiotic molecule into a system that can directly deliver molecules to the infected site. The hydrogel system perfectly delivers these antibiotic molecules to the infected site in controlled and prolonged release of the antibiotics to reduce antibiotic toxicity. This system possesses a hydrophilic nature with crosslinked voids, wherein small molecule antibiotics can be loaded. Important antibiotics that have been used to load into the hydrogel matrix include gentamicin, ciprofloxacin, amoxicillin.

Synthesis of hydrogel matrices loaded with ciprofloxacin has been achieved using electrosynthesis on a titanium surface to release the antibiotic.

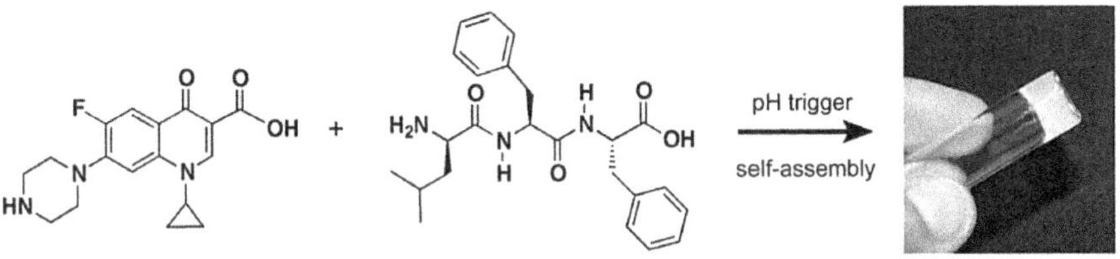

Figure 13.4 Structures of ciprofloxacin and peptide ^{D}Leu-Phe-Phe, which self-assemble into a hydrogel following a pH trigger.
Reproduced from ref. 104 with permission from Elsevier, Copyright 2013.

This system effectively inhibits MRSA. This hydrogel system possessing ciprofloxacin can be used to treat orthopaedic infections.[103] A pH-responsive tripeptidic hydrogel matrix loaded with ciprofloxacin has been prepared. The antibiotic molecule is bound with the hydrogel matrix by non-covalent interactions (Figure 13.4). This hydrogel matrix shows activity against *S. aureus*, *E. coli*, and a clinical strain of *K. pneumoniae.*[104]

pH-sensitive hydrogels containing a chitosan/poly-γ-glutamic acid nanoparticle system have been prepared to deliver the antibiotic amoxicillin. Since the hydrogel is pH-responsive, the nanoparticles are protected from the gastric juices of the intestine. This hydrogel system shows activity against *Helicobacter pylori.*[105]

13.4.3 Antimicrobial-agent Loaded Hydrogels

Various other strategies have been proposed to incorporate antimicrobial agents into the hydrogel matrix for delivery. AMPs are promising antimicrobial agents loaded into hydrogel matrices. AMPs show activity against a wide spectrum of targets, such as viruses, bacteria, fungi, and other parasites.[106] Poly(2-hydroxyethyl methacrylate)-based hydrogel systems loaded with AMPs were studied for their release and antimicrobial actions. This hydrogel system shows activity against *Staphylococcus epidermidis*. The release of AMPs from the hydrogel matrix works against biomaterial associated infections.[107]

Hydrogels loaded with antifungal agents have also been reported. Amphotericin-B, an antifungal agent, generally used to treat medical device derived infections, has been incorporated into hydrogel matrices. Amphogel is a hydrogel matrix prepared from dextran possessing amphotericin-B. This hydrogel system works well against *C. albicans* in less than 2 hours. It can be reused for at least 53 days without losing effectiveness. This antifungal hydrogel matrix can be coated onto medical devices.[108] In another study, carboxymethylcellulose containing hydrazide functionality reacts with dextran aldehyde, forming a hydrogel loaded with amphotericin-B (Figure 13.5).

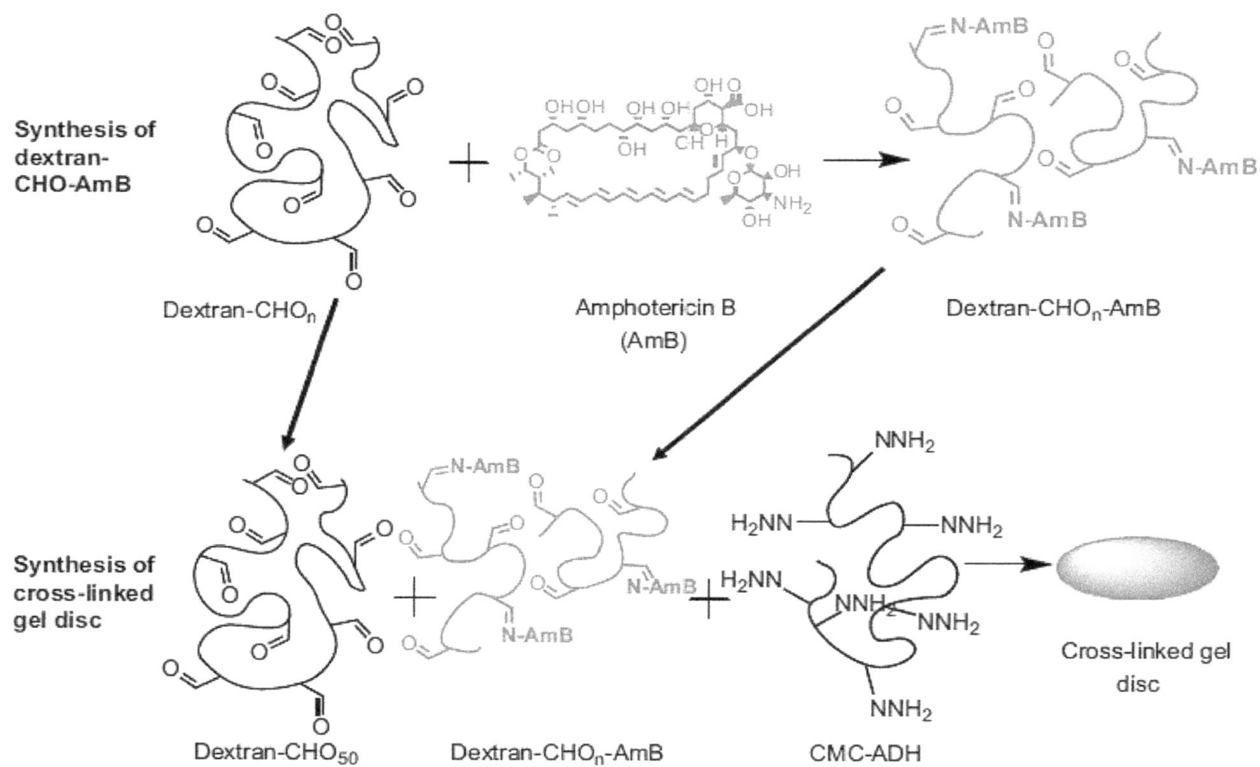

Figure 13.5 Conjugation of amphotericin B to oxidized dextran and the incorporation of dextran-CHO–AmB into a CMC–dextran gel. Reproduced from ref. 109 with permission from Elsevier, Copyright 2010.

This hydrogel system releases the antifungal agent for 11 days to treat *Candia* fungal strains. It can be used as a potential injectable gel for sustained local antifungal therapy.[109]

13.5 Conclusions

This chapter provides an overview of antimicrobial hydrogel systems. Antimicrobial hydrogels are suitable for various biomedical applications, such as wound care materials, catheter infections, *etc.* Hydrogels are loaded for the delivery of antimicrobial materials, such as small molecules, antibiotics, antimicrobial peptides, metal nanoparticles, antimicrobial polymers, *etc.* These molecules are released from the hydrogels in a controlled manner in the treatment of infectious diseases. Biodegradable polymers or peptide-loaded hydrogels are promising antimicrobial candidates compared to hydrogels loaded with antibiotics due to antibiotic drug resistance towards pathogens. Other antimicrobial hydrogels possess intrinsic antimicrobial moieties. This type of hydrogel contains mostly quaternary ammonium moieties that can react with the negatively charged cell surfaces of the pathogens and kill them effectively.

Although antimicrobial hydrogels have made enormous progress to reduce resistance associated with antibiotics, the delivery of loaded biocides may not be accurate. Hydrogels sometimes may degrade too quickly to release the drug in a sustained manner. A few hydrogels may react with loaded biocides, limiting their application. These problems can be solved with collaboration in research: innovative strategies in combating microbial resistance with antimicrobial hydrogels can be discovered by the successful collaboration of polymer chemists, microbiologists, pharmacists, and toxicologists.

References

1. C. Mathers, D. M. Fat and J. Boerma, *The Global Burden of Disease: 2004 Update*, World Health Organization, 2008.
2. W. Li, K. Dong, J. Ren and X. Qu, A β-Lactamase-Imprinted Responsive Hydrogel for the Treatment of Antibiotic-Resistant Bacteria, *Angew. Chem., Int. Ed.*, 2016, **55**, 8049–8053.
3. J. L. Fox, Antimicrobial peptides stage a comeback, *Nat. Biotechnol.*, 2013, **31**, 379.
4. A. Muñoz-Bonilla and M. Fernández-García, Polymeric materials with antimicrobial activity, *Prog. Polym. Sci.*, 2012, **37**, 281–339.
5. A. J. Huh and Y. J. Kwon, "Nanoantibiotics": a new paradigm for treating infectious diseases using nanomaterials in the antibiotics resistant era, *J. Controlled Release*, 2011, **156**(2), 128–145.
6. J. Rodríguez-Hernández, Antimicrobial Hydrogels, in *Polymers against Microorganisms*, Springer, Cham, 2017, pp. 179–204.

7. O. Wichterle and D. Lim, Hydrophilic gels for biological use, *Nature*, 1960, **185**, 117.
8. N. A. Peppas, J. Z. Hilt, A. Khademhosseini and R. Langer, Hydrogels in Biology and Medicine: From Molecular Principles to Bionanotechnology, *Adv. Mater.*, 2006, **18**, 1345–1360.
9. F. Ullah, M. B. H. Othman, F. Javed, Z. Ahmad and H. M. Akil, Classification, processing and application of hydrogels: A review, *Mater. Sci. Eng., C*, 2015, **57**, 414–433.
10. N. Beyth, I. Yudovin-Farber, R. Bahir, A. J. Domb and E. I. Weiss, Antibacterial activity of dental composites containing quaternary ammonium polyethylenimine nanoparticles against Streptococcus mutans, *Biomaterials*, 2006, **27**, 3995–4002.
11. K. Y. Lee and D. J. Mooney, Hydrogels for Tissue Engineering, *Chem. Rev.*, 2001, **101**, 1869–1880.
12. T.-Y. Cheng, M.-H. Chen, W.-H. Chang, M.-Y. Huang and T.-W. Wang, Neural stem cells encapsulated in a functionalized self-assembling peptide hydrogel for brain tissue engineering, *Biomaterials*, 2013, **34**, 2005–2016.
13. X. Shi, J. Zhou, Y. Zhao, L. Li and H. Wu, Gradient-Regulated hydrogel for interface tissue engineering: steering simultaneous Osteo/Chondrogenesis of stem cells on a chip, *Adv. Healthcare Mater.*, 2013, **2**, 846–853.
14. N. Q. Tran, Y. K. Joung, E. Lih and K. D. Park, In Situ Forming and Rutin-Releasing Chitosan Hydrogels As Injectable Dressings for Dermal Wound Healing, *Biomacromolecules*, 2011, **12**, 2872–2880.
15. C. Gong, Q. Wu, Y. Wang, D. Zhang, F. Luo, X. Zhao, Y. Wei and Z. Qian, A biodegradable hydrogel system containing curcumin encapsulated in micelles for cutaneous wound healing, *Biomaterials*, 2013, **34**, 6377–6387.
16. K.-C. Yang, C.-C. Wu, F.-H. Lin, Z. Qi, T.-F. Kuo, Y.-H. Cheng, M.-P. Chen and S. Sumi, Chitosan/gelatin hydrogel as immunoisolative matrix for injectable bioartificial pancreas, *Xenotransplantation*, 2008, **15**, 407–416.
17. W. Zhu, L. Gao, Q. Luo, C. Gao, G. Zha, Z. Shen and X. Li, Metal and light free "click" hydrogels for prevention of post-operative peritoneal adhesions, *Polym. Chem.*, 2014, **5**, 2018–2026.
18. H. Zhang, A. Patel, A. K. Gaharwar, S. M. Mihaila, G. Iviglia, S. Mukundan, H. Bae, H. Yang and A. Khademhosseini, Hyperbranched Polyester Hydrogels with Controlled Drug Release and Cell Adhesion Properties, *Biomacromolecules*, 2013, **14**, 1299–1310.
19. Y. Qiu and K. Park, Environment-sensitive hydrogels for drug delivery, *Adv. Drug Delivery Rev.*, 2012, **64**, 49–60.
20. K. Chaturvedi, K. Ganguly, M. N. Nadagouda and T. M. Aminabhavi, Polymeric hydrogels for oral insulin delivery, *J. Controlled Release*, 2013, **165**, 129–138.

21. L. Gao, G. Zha, Y. Wang, Q. Luo, W. Zhu, Z. Shen and X. Li, An injectable drug-loaded hydrogel using a "clickable" amphiphilic triblock copolymer as a precursor, *Polym. Chem.*, 2015, **6**, 8240–8243.
22. L. Xiong, Q. Luo, Y. Wang, X. Li, Z. Shen and W. Zhu, An injectable drug-loaded hydrogel based on a supramolecular polymeric prodrug, *Chem. Commun.*, 2015, **51**, 14644–14647.
23. W. Zhu, L. Xiong, H. Wang, G. Zha, H. Du, X. Li and Z. Shen, Sustained drug release from an ultrathin hydrogel film, *Polym. Chem.*, 2015, **6**, 7097–7099.
24. T. Vermonden, R. Censi and W. E. Hennink, Hydrogels for Protein Delivery, *Chem. Rev.*, 2012, **112**, 2853–2888.
25. D. Seliktar, Designing Cell-Compatible Hydrogels for Biomedical Applications, *Science*, 2012, **336**, 1124–1128.
26. V. W. Ng, J. M. Chan, H. Sardon, R. J. Ono, J. M. García, Y. Y. Yang and J. L. Hedrick, Antimicrobial hydrogels: A new weapon in the arsenal against multidrug-resistant infections, *Adv. Drug Delivery Rev.*, 2014, **78**, 46–62.
27. S. Li, S. Dong, W. Xu, S. Tu, L. Yan, C. Zhao, J. Ding and X. Chen, Antibacterial Hydrogels, *Adv. Sci.*, 2018, **5**, 1700527.
28. K. Yang, Q. Han, B. Chen, Y. Zheng, K. Zhang, Q. Li and J. Wang, Antimicrobial hydrogels: promising materials for medical application, *Int. J. Nanomed.*, 2018, **13**, 2217–2263.
29. C. M. González-Henríquez, M. A. Sarabia-Vallejos and J. Rodriguez-Hernandez, Advances in the Fabrication of Antimicrobial Hydrogels for Biomedical Applications, *Materials*, 2017, **10**, 232.
30. E. S. Place, J. H. George, C. K. Williams and M. M. Stevens, Synthetic polymer scaffolds for tissue engineering, *Chem. Soc. Rev.*, 2009, **38**, 1139–1151.
31. F. Ullah, M. B. H. Othman, F. Javed, Z. Ahmad and H. M. Akil, Classification, processing and application of hydrogels: A review, *Mater. Sci. Eng., C*, 2015, **57**, 414–433.
32. R. Jayakumar, M. Prabaharan, P. T. Sudheesh Kumar, S. V. Nair and H. Tamura, Biomaterials based on chitin and chitosan in wound dressing applications, *Biotechnol. Adv.*, 2011, **29**, 322–337.
33. C. T. Tsao, C. H. Chang, Y. Y. Lin, M. F. Wu, J.-L. Wang, J. L. Han and K. H. Hsieh, Antibacterial activity and biocompatibility of a chitosan–γ-poly(glutamic acid) polyelectrolyte complex hydrogel, *Carbohydr. Res.*, 2010, **345**, 1774–1780.
34. C. T. Tsao, C. H. Chang, Y. Y. Lin, M. F. Wu, J. L. Wang, T. H. Young, J. L. Han and K. H. Hsieh, Evaluation of chitosan/γ-poly(glutamic acid) polyelectrolyte complex for wound dressing materials, *Carbohydr. Polym.*, 2011, **84**, 812–819.
35. X. Zhao, P. Li, B. Guo and P. X. Ma, Antibacterial and conductive injectable hydrogels based on quaternized chitosan-graft-polyaniline/oxidized dextran for tissue engineering, *Acta Biomater.*, 2015, **26**, 236–248.

36. X. Zhao, H. Wu, B. Guo, R. Dong, Y. Qiu and P. X. Ma, Antibacterial anti-oxidant electroactive injectable hydrogel as self-healing wound dressing with hemostasis and adhesiveness for cutaneous wound healing, *Biomaterials*, 2017, **122**, 34–47.
37. J. Hoque, R. G. Prakash, K. Paramanandham, B. R. Shome and J. Haldar, Biocompatible Injectable Hydrogel with Potent Wound Healing and Antibacterial Properties, *Mol. Pharmaceutics*, 2017, **14**, 1218–1230.
38. M. Mizutani, E. F. Palermo, L. M. Thoma, K. Satoh, M. Kamigaito and K. Kuroda, Design and Synthesis of Self-Degradable Antibacterial Polymers by Simultaneous Chain- and Step-Growth Radical Copolymerization, *Biomacromolecules*, 2012, **13**, 1554–1563.
39. E. F. Palermo, S. Vemparala and K. Kuroda, Cationic Spacer Arm Design Strategy for Control of Antimicrobial Activity and Conformation of Amphiphilic Methacrylate Random Copolymers, *Biomacromolecules*, 2012, **13**, 1632–1641.
40. S. Colak, C. F. Nelson, K. Nüsslein and G. N. Tew, Hydrophilic Modifications of an Amphiphilic Polynorbornene and the Effects on its Hemolytic and Antibacterial Activity, *Biomacromolecules*, 2009, **10**, 353–359.
41. N. Pasquier, H. Keul, E. Heine and M. Moeller, From Multi-functionalized Poly(ethylene imine)s toward Antimicrobial Coatings, *Biomacromolecules*, 2007, **8**, 2874–2882.
42. H. D. Thaker, A. Som, F. Ayaz, D. Lui, W. Pan, R. W. Scott, J. Anguita and G. N. Tew, Synthetic Mimics of Antimicrobial Peptides with Immunomodulatory Responses, *J. Am. Chem. Soc.*, 2012, **134**, 11088–11091.
43. R. Liu, X. Chen, Z. Hayouka, S. Chakraborty, S. P. Falk, B. Weisblum, K. S. Masters and S. H. Gellman, Nylon-3 Polymers with Selective Antifungal Activity, *J. Am. Chem. Soc.*, 2013, **135**, 5270–5273.
44. A. C. Engler, A. Shukla, S. Puranam, H. G. Buss, N. Jreige and P. T. Hammond, Effects of Side Group Functionality and Molecular Weight on the Activity of Synthetic Antimicrobial Polypeptides, *Biomacromolecules*, 2011, **12**, 1666–1674.
45. A. C. Engler, J. P. K. Tan, Z. Y. Ong, D. J. Coady, V. W. L. Ng, Y. Y. Yang and J. L. Hedrick, Antimicrobial Polycarbonates: Investigating the Impact of Balancing Charge and Hydrophobicity Using a Same-Centered Polymer Approach, *Biomacromolecules*, 2013, **14**, 4331–4339.
46. Y. Li, K. Fukushima, D. J. Coady, A. C. Engler, S. Liu, Y. Huang, J. S. Cho, Y. Guo, L. S. Miller and J. P. Tan, Broad-Spectrum Antimicrobial and Biofilm-Disrupting Hydrogels: Stereocomplex-Driven Supramolecular Assemblies, *Angew. Chem., Int. Ed.*, 2013, **52**, 674–678.
47. A. L. Z. Lee, V. W. L. Ng, W. Wang, J. L. Hedrick and Y. Y. Yang, Block copolymer mixtures as antimicrobial hydrogels for biofilm eradication, *Biomaterials*, 2013, **34**, 10278–10286.
48. H. Du, G. Zha, L. Gao, H. Wang, X. Li, Z. Shen and W. Zhu, Fully biodegradable antibacterial hydrogels via thiol–ene "click" chemistry, Polymer, *Chemistry*, 2014, **5**, 4002–4008.

49. M. C. Giano, Z. Ibrahim, S. H. Medina, K. A. Sarhane, J. M. Christensen, Y. Yamada, G. Brandacher and J. P. Schneider, Injectable bioadhesive hydrogels with innate antibacterial properties, *Nat. Commun.*, 2014, **5**, 4095.
50. L. Li, B. Yan, J. Yang, W. Huang, L. Chen and H. Zeng, Injectable Self-Healing Hydrogel with Antimicrobial and Antifouling Properties, *ACS Appl. Mater. Interfaces*, 2017, **9**, 9221–9225.
51. A. Strassburg, J. Petranowitsch, F. Paetzold, C. Krumm, E. Peter, M. Meuris, M. Köller and J. C. Tiller, Cross-Linking of a Hydrophilic, Antimicrobial Polycation toward a Fast-Swelling, Antimicrobial Superabsorber and Interpenetrating Hydrogel Networks with Long Lasting Antimicrobial Properties, *ACS Appl. Mater. Interfaces*, 2017, **9**, 36573–36582.
52. D. A. Salick, J. K. Kretsinger, D. J. Pochan and J. P. Schneider, Inherent Antibacterial Activity of a Peptide-Based β-Hairpin Hydrogel, *J. Am. Chem. Soc.*, 2007, **129**, 14793–14799.
53. Y. Liu, Y. Yang, C. Wang and X. Zhao, Stimuli-responsive self-assembling peptides made from antibacterial peptides, *Nanoscale*, 2013, **5**, 6413–6421.
54. R. Wang, J. Li, W. Chen, T. Xu, S. Yun, Z. Xu, Z. Xu, T. Sato, B. Chi and H. Xu, A Biomimetic Mussel-Inspired ε-Poly-L-lysine Hydrogel with Robust Tissue-Anchor and Anti-Infection Capacity, *Adv. Funct. Mater.*, 2017, **27**, 1604894.
55. R. Wang, Q. Li, B. Chi, X. Wang, Z. Xu, Z. Xu, S. Chen and H. Xu, Enzyme-induced dual-network ε-poly-l-lysine-based hydrogels with robust self-healing and antibacterial performance, *Chem. Commun.*, 2017, **53**, 4803–4806.
56. C. Zhou, P. Li, X. Qi, A. R. M. Sharif, Y. F. Poon, Y. Cao, M. W. Chang, S. S. J. Leong and M. B. Chan-Park, A photopolymerized antimicrobial hydrogel coating derived from epsilon-poly-l-lysine, *Biomaterials*, 2011, **32**, 2704–2712.
57. Y. Li, K. Fukushima, D. J. Coady, A. C. Engler, S. Liu, Y. Huang, J. S. Cho, Y. Guo, L. S. Miller, J. P. K. Tan, P. L. R. Ee, W. Fan, Y. Y. Yang and J. L. Hedrick, Broad-Spectrum Antimicrobial and Biofilm-Disrupting Hydrogels: Stereocomplex-Driven Supramolecular Assemblies, *Angew. Chem., Int. Ed.*, 2013, **52**, 674–678.
58. H. H. Lara, N. V. Ayala-Núñez, L. D. C. I. Turrent and C. R. Padilla, Bactericidal effect of silver nanoparticles against multidrug-resistant bacteria, *World J. Microbiol. Biotechnol.*, 2010, **26**, 615–621.
59. K. M. Hindi, A. J. Ditto, M. J. Panzner, D. A. Medvetz, D. S. Han, C. E. Hovis, J. K. Hilliard, J. B. Taylor, Y. H. Yun, C. L. Cannon and W. J. Youngs, The antimicrobial efficacy of sustained release silver–carbene complex-loaded l-tyrosine polyphosphate nanoparticles: Characterization, in vitro and in vivo studies, *Biomaterials*, 2009, **30**, 3771–3779.
60. A. Ivask, A. ElBadawy, C. Kaweeteerawat, D. Boren, H. Fischer, Z. Ji, C. H. Chang, R. Liu, T. Tolaymat, D. Telesca, J. I. Zink, Y. Cohen,

P. A. Holden and H. A. Godwin, Toxicity Mechanisms in Escherichia coli Vary for Silver Nanoparticles and Differ from Ionic Silver, *ACS Nano*, 2014, **8**, 374–386.
61. A. Taglietti, Y. A. Diaz Fernandez, E. Amato, L. Cucca, G. Dacarro, P. Grisoli, V. Necchi, P. Pallavicini, L. Pasotti and M. Patrini, Antibacterial Activity of Glutathione-Coated Silver Nanoparticles against Gram Positive and Gram Negative Bacteria, *Langmuir*, 2012, **28**, 8140–8148.
62. L. Guo, W. Yuan, Z. Lu and C. M. Li, Polymer/nanosilver composite coatings for antibacterial applications, *Colloids Surf. A Physicochemi. Eng. Asp.*, 2013, **439**, 69–83.
63. M. L. W. Knetsch and L. H. Koole, New Strategies in the Development of Antimicrobial Coatings: The Example of Increasing Usage of Silver and Silver Nanoparticles, *Polymers*, 2011, **3**, 340.
64. Y. Zhao and X. Jiang, Multiple strategies to activate gold nanoparticles as antibiotics, *Nanoscale*, 2013, **5**, 8340–8350.
65. A. N. Brown, K. Smith, T. A. Samuels, J. Lu, S. O. Obare and M. E. Scott, Nanoparticles functionalized with ampicillin destroy multiple-antibiotic-resistant isolates of Pseudomonas aeruginosa and Enterobacter aerogenes and methicillin-resistant Staphylococcus aureus, *Appl. Environ. Microbiol.*, 2012, **78**, 2768–2774.
66. R. Y. Pelgrift and A. J. Friedman, Nanotechnology as a therapeutic tool to combat microbial resistance, *Adv. Drug Delivery Rev.*, 2013, **65**, 1803–1815.
67. A. Nasir and A. Friedman, The growing role of nanotechnology in combating infectious disease AU – Blecher, Karin, *Virulence*, 2011, **2**, 395–401.
68. A. J. Huh and Y. J. Kwon, "Nanoantibiotics": A new paradigm for treating infectious diseases using nanomaterials in the antibiotics resistant era, *J. Controlled Release*, 2011, **156**, 128–145.
69. M. J. Hajipour, K. M. Fromm, A. Akbar Ashkarran, D. Jimenez de Aberasturi, I. R. D. Larramendi, T. Rojo, V. Serpooshan, W. J. Parak and M. Mahmoudi, Antibacterial properties of nanoparticles, *Trends Biotechnol.*, 2012, **30**, 499–511.
70. P. T. Sudheesh Kumar, V.-K. Lakshmanan, T. V. Anilkumar, C. Ramya, P. Reshmi, A. G. Unnikrishnan, S. V. Nair and R. Jayakumar, Flexible and Microporous Chitosan Hydrogel/Nano ZnO Composite Bandages for Wound Dressing: In Vitro and In Vivo Evaluation, *ACS Appl. Mater. Interfaces*, 2012, **4**, 2618–2629.
71. T. Kruk, K. Szczepanowicz, J. Stefańska, R. P. Socha and P. Warszyński, Synthesis and antimicrobial activity of monodisperse copper nanoparticles, *Colloids Surf., B*, 2015, **128**, 17–22.
72. A. P. Ingle, N. Duran and M. Rai, Bioactivity, mechanism of action, and cytotoxicity of copper-based nanoparticles: a review, *Appl. Microbiol. Biotechnol.*, 2014, **98**(3), 1001–1009.
73. T. Zhong, G. S. Oporto, J. Jaczynski and C. Jiang, Nanofibrillated Cellulose and Copper Nanoparticles Embedded in Polyvinyl Alcohol Films for Antimicrobial Applications, *BioMed Res. Int.*, 2015, **2015**, 8.

74. Y. Li, Y. Han, X. Wang, J. Peng, Y. Xu and J. Chang, Multifunctional Hydrogels Prepared by Dual Ion Cross-Linking for Chronic Wound Healing, *ACS Appl. Mater. Interfaces*, 2017, **9**, 16054–16062.
75. C. Hu, M. X. Wang, L. Sun, J. H. Yang, M. Zrínyi and Y. M. Chen, Dual-Physical Cross-Linked Tough and Photoluminescent Hydrogels with Good Biocompatibility and Antibacterial Activity, *Macromol. Rapid Commun.*, 2017, **38**, 1600788.
76. S. McMahon, R. Kennedy, P. Duffy, J. M. Vasquez, J. G. Wall, H. Tai and W. Wang, Poly(ethylene glycol)-Based Hyperbranched Polymer from RAFT and Its Application as a Silver-Sulfadiazine-Loaded Antibacterial Hydrogel in Wound Care, *ACS Appl. Mater. Interfaces*, 2016, **8**, 26648–26656.
77. W. Gao, D. Vecchio, J. Li, J. Zhu, Q. Zhang, V. Fu, J. Li, S. Thamphiwatana, D. Lu and L. Zhang, Hydrogel Containing Nanoparticle-Stabilized Liposomes for Topical Antimicrobial Delivery, *ACS Nano*, 2014, **8**, 2900–2907.
78. A. GhavamiNejad, C. H. Park and C. S. Kim, In Situ Synthesis of Antimicrobial Silver Nanoparticles within Antifouling Zwitterionic Hydrogels by Catecholic Redox Chemistry for Wound Healing Application, *Biomacromolecules*, 2016, **17**, 1213–1223.
79. K. Baek, J. Liang, W. T. Lim, H. Zhao, D. H. Kim and H. Kong, In Situ Assembly of Antifouling/Bacterial Silver Nanoparticle-Hydrogel Composites with Controlled Particle Release and Matrix Softening, *ACS Appl. Mater. Interfaces*, 2015, **7**, 15359–15367.
80. M. I. González-Sánchez, S. Perni, G. Tommasi, N. G. Morris, K. Hawkins, E. López-Cabarcos and P. Prokopovich, Silver nanoparticle based antibacterial methacrylate hydrogels potential for bone graft applications, *Mater. Sci. Eng., C*, 2015, **50**, 332–340.
81. J. García-Barrasa, J. M. López-de-Luzuriaga and M. Monge, Silver nanoparticles: synthesis through chemical methods in solution and biomedical applications, *Cent. Eur. J. Chem.*, 2011, **9**, 7–19.
82. N. Sahiner, Soft and flexible hydrogel templates of different sizes and various functionalities for metal nanoparticle preparation and their use in catalysis, *Prog. Polym. Sci.*, 2013, **38**, 1329–1356.
83. H.-W. Chang, Y.-S. Lin, Y.-D. Tsai and M.-L. Tsai, Effects of chitosan characteristics on the physicochemical properties, antibacterial activity, and cytotoxicity of chitosan/2-glycerophosphate/nanosilver hydrogels, *J. Appl. Polym. Sci.*, 2013, **127**, 169–176.
84. K. Neibert, V. Gopishetty, A. Grigoryev, I. Tokarev, N. Al-Hajaj, J. Vorstenbosch, A. Philip, S. Minko and D. Maysinger, Wound-Healing with Mechanically Robust and Biodegradable Hydrogel Fibers Loaded with Silver Nanoparticles, *Adv. Healthcare Mater.*, 2012, **1**, 621–630.
85. A. Hebeish, M. Hashem, M. M. A. El-Hady and S. Sharaf, Development of CMC hydrogels loaded with silver nano-particles for medical applications, *Carbohydr. Polym.*, 2013, **92**, 407–413.

86. K. Nešović, A. Janković, V. Kojić, M. Vukašinović-Sekulić, A. Perić-Grujić, K. Y. Rhee and V. Mišković-Stanković, Silver/poly(vinyl alcohol)/chitosan/graphene hydrogels – Synthesis, biological and physicochemical properties and silver release kinetics, *Composites, Part B*, 2018, **154**, 175–185.
87. E. S. Abdel-Halim and S. S. Al-Deyab, Antimicrobial activity of silver/starch/polyacrylamide nanocomposite, *Int. J. Biol. Macromol.*, 2014, **68**, 33–38.
88. E. Faoucher, P. Nativo, K. Black, J. B. Claridge, M. Gass, S. Romani, A. L. Bleloch and M. Brust, In situ preparation of network forming gold nanoparticles in agarose hydrogels, *Chem. Commun.*, 2009, 6661–6663.
89. Y. Zhao and X. Jiang, Multiple strategies to activate gold nanoparticles as antibiotics, *Nanoscale*, 2013, **5**, 8340–8350.
90. V. Kozlovskaya, E. Kharlampieva, S. Chang, R. Muhlbauer and V. V. Tsukruk, pH-Responsive Layered Hydrogel Microcapsules as Gold Nanoreactors, *Chem. Mater.*, 2009, **21**, 2158–2167.
91. W. Gao, D. Vecchio, J. Li, J. Zhu, Q. Zhang, V. Fu, J. Li, S. Thamphiwatana, D. Lu and L. Zhang, Hydrogel Containing Nanoparticle-Stabilized Liposomes for Topical Antimicrobial Delivery, *ACS Nano*, 2014, **8**, 2900–2907.
92. M. Ribeiro, M. P. Ferraz, F. J. Monteiro, M. H. Fernandes, M. M. Beppu, D. Mantione and H. Sardon, Antibacterial silk fibroin/nanohydroxyapatite hydrogels with silver and gold nanoparticles for bone regeneration, *Nanomedicine*, 2017, **13**, 231–239.
93. A. Mohandas, P. T. S. Kumar, B. Raja, V.-K. Lakshmanan and R. Jayakumar, Exploration of alginate hydrogel/nano zinc oxide composite bandages for infected wounds, *Int. J. Nanomed.*, 2015, **10**(Suppl 1), 53–66.
94. J. Wang, H. Hu, Z. Yang, J. Wei and J. Li, IPN hydrogel nanocomposites based on agarose and ZnO with antifouling and bactericidal properties, *Mater. Sci. Eng., C*, 2016, **61**, 376–386.
95. Z. Zare-Akbari, H. Farhadnejad, B. Furughi-Nia, S. Abedin, M. Yadollahi and M. Khorsand-Ghayeni, PH-sensitive bionanocomposite hydrogel beads based on carboxymethyl cellulose/ZnO nanoparticle as drug carrier, *Int. J. Biol. Macromol.*, 2016, **93**, 1317–1327.
96. T. Kruk, K. Szczepanowicz, J. Stefańska, R. P. Socha and P. Warszyński, Synthesis and antimicrobial activity of monodisperse copper nanoparticles, *Colloids Surf., B*, 2015, **128**, 17–22.
97. A. P. Ingle, N. Duran and M. Rai, Bioactivity, mechanism of action, and cytotoxicity of copper-based nanoparticles: a review, *Appl. Microbiol. Biotechnol.*, 2014, **98**, 1001–1009.
98. T. Zhong, G. S. Oporto, J. Jaczynski and C. Jiang, Nanofibrillated Cellulose and Copper Nanoparticles Embedded in Polyvinyl Alcohol Films for Antimicrobial Applications, *BioMed Res. Int.*, 2015, **2015**, 8.
99. M. E. Villanueva, A. M. D. R. Diez, J. A. González, C. J. Pérez, M. Orrego, L. Piehl, S. Teves and G. J. Copello, Antimicrobial Activity of Starch

Hydrogel Incorporated with Copper Nanoparticles, *ACS Appl. Mater. Interfaces*, 2016, **8**, 16280–16288.
100. P. A. Norowski Jr. and J. D. Bumgardner, Biomaterial and antibiotic strategies for peri-implantitis: A review, *J. Biomed. Mater. Res. B*, 2009, **88B**, 530–543.
101. C. A. Arias and B. E. Murray, A new antibiotic and the evolution of resistance, *N. Engl. J. Med.*, 2015, **372**, 1168–1170.
102. H. Marshall, New antibiotic discovered for Gram-positive bacteria, *Lancet Respir. Med.*, 2015, **3**, 104.
103. E. De Giglio, S. Cometa, M. A. Ricci, D. Cafagna, A. M. Savino, L. Sabbatini, M. Orciani, E. Ceci, L. Novello, G. M. Tantillo and M. Mattioli-Belmonte, Ciprofloxacin-modified electrosynthesized hydrogel coatings to prevent titanium-implant-associated infections, *Acta Biomater.*, 2011, 7, 882–891.
104. S. Marchesan, Y. Qu, L. J. Waddington, C. D. Easton, V. Glattauer, T. J. Lithgow, K. M. McLean, J. S. Forsythe and P. G. Hartley, Self-assembly of ciprofloxacin and a tripeptide into an antimicrobial nanostructured hydrogel, *Biomaterials*, 2013, **34**, 3678–3687.
105. C.-H. Chang, Y.-H. Lin, C.-L. Yeh, Y.-C. Chen, S.-F. Chiou, Y.-M. Hsu, Y.-S. Chen and C.-C. Wang, Nanoparticles Incorporated in pH-Sensitive Hydrogels as Amoxicillin Delivery for Eradication of Helicobacter pylori, *Biomacromolecules*, 2010, **11**, 133–142.
106. A. A. Bahar and D. Ren, Antimicrobial Peptides, *Pharmaceuticals*, 2013, **6**, 1543.
107. G. Laverty, S. P. Gorman and B. F. Gilmore, Antimicrobial peptide incorporated poly(2-hydroxyethyl methacrylate) hydrogels for the prevention of Staphylococcus epidermidis-associated biomaterial infections, *J. Biomed. Mater. Res., Part A*, 2012, **100A**, 1803–1814.
108. A. Zumbuehl, L. Ferreira, D. Kuhn, A. Astashkina, L. Long, Y. Yeo, T. Iaconis, M. Ghannoum, G. R. Fink, R. Langer and D. S. Kohane, Antifungal hydrogels, *Proc. Natl. Acad. Sci. U. S. A.*, 2007, **104**, 12994–12998.
109. S. P. Hudson, R. Langer, G. R. Fink and D. S. Kohane, Injectable in situ cross-linking hydrogels for local antifungal therapy, *Biomaterials*, 2010, **31**, 1444–1452.

CHAPTER 14

Catheters with Antimicrobial Surfaces

SHADY FARAH,*[a] MALIA MCAVOY[b,c] AND AMANI JAHJAA[d,e]

[a] The Wolfson Faculty of Chemical Engineering and The Russell Berrie Nanotechnology Institute, Technion-Israel Institute of Technology, Haifa, 3200003, Israel; [b] David H Koch Institute for Integrative Cancer Research, Massachusetts Institute of Technology, 500 Main Street, Cambridge, Massachusetts 02139, USA; [c] Harvard-MIT Division of Health Science Technology, Massachusetts Institute of Technology, 77 Massachusetts Avenue, Cambridge, MA 02139, USA; [d] Clinical Trials Designer, Boston, Massachusetts 02128, USA; [e] The Ruth and Bruce Rappaport Faculty of Medicine-Technion, Haifa, 31096, Israel
*Email: sfarah@technion.ac.il

14.1 Introduction

14.1.1 Catheters and Catheterization

Catheters are tubular medical devices that vary in their type, length, size, shape, design, purpose and function. Usually they are made of polymeric materials, mainly silicones, polyurethanes, polytetrafluoroethylene (PTFE), plastic (polyvinyl chloride) and rubber, among others, with different characteristics (*e.g.* porosity) and flexibility profiles. Catheters are useful and convenient tools for addressing varied biomedical needs, such as to keep a passage open, drainage of excess fluids, withdrawal and injection of fluids, insertion of medical devices, as diagnostic tools and for hemodynamic

Biomaterials Science Series No. 5
Antimicrobial Materials for Biomedical Applications
Edited by Abraham J. Domb, Konda Reddy Kunduru and Shady Farah

Published by the Royal Society of Chemistry, www.rsc.org

monitoring, as well as for delivery of therapeutic formulations into body cavities, vessels and canals.

The low cost and wide applicability of catheters have paved the way for hundreds of different catheter designs and uses in the medical field. Applications include vascular/intravascular catheter, central venous catheter (CVC), peripheral venous/intravenous catheter, percutaneous nephrostomy catheter (PCN), urinary catheter, urethral catheter, bladder catheter, indwelling catheter, hemodialysis catheter, hydrocephalus shunt catheter, external ventricular drainage catheter, peritoneal catheter, intrasynovial/intra-articular catheter and pulmonary artery catheters. A comprehensive list of various catheters with detailed descriptions of their use, insertion protocol and characteristics have been reported.[1]

Catheterization is considered an elegant process, minimally invasive and an integral part of modern medicine. Catheter insertion should be undertaken under strict sterilization techniques, although different types of catheters may have varied protocols of insertion depending on catheter's dimensions, with differences in insertion and final destination sites. However, despite the biomedical utility of catheters, their performance is limited due to microbial infection and implant rejection. There is a direct correlation between catheter-associated infections and prolonged hospitalization as well as increased morbidity and mortality.

14.1.2 Infection Problem

Bacterial growth has been reported as a major concern limiting wider usage of catheters, dental composites and implanted devices.[2] As foreign bodies, biomedical catheters are associated with a substantial risk of microbial infections. Catheter infection is a major contributor to the increasing problem of worldwide nosocomial infections.[3] Microbial accumulation and contamination on the catheter's surface have led to loss of function while simultaneously endangering the patient's life. In addition, it was found that infection starts on the catheter's surface with the adherence of a few microorganisms, followed by rapid development of therapeutic-resistant microbial biofilm and spread of infection.[4–6]

Although the incidence of catheter-associated infections is increasing worldwide,[7] different bacterial populations were found to be involved, depending on catheter type, duration of the insertion, patient health background, microenvironment of the insertion and destination sites. Two major types of catheter have been found contribute to the overall catheter-related/-associated infections, morbidity and mortality; thus, in this chapter we give them special attention: catheter-related blood infections (CRBIs) and catheter-associated urinary tract infections (CAUTIs).

CRBIs are defined as the presence of bacteremia originating from an intravenous catheter,[8,9] with an occurrence rate of 5 per 1000 catheter days in the intensive care unit.[10] CRBIs were found to be a major cause of morbidity and mortality, with 250 000 cases being reported annually in the

United States alone, and an overall \$2.3 billion estimated healthcare expenditure and attributable mortality of up to 35%.[11–16] Following intravascular catheterization, CRBI should be suspected if the patient develops clinical or laboratory criteria of systemic inflammatory response syndrome (SIRS) which include heart rate >90 beats per minute, temperature <36 °C or >38 °C, respiratory rate >20 breaths per minute or peripheral white blood cell count <4000 cells per μL or >12 000 cells per μL.[17] In CRBIs, the interaction of blood proteins, *i.e.* plasma proteins, and their binding to the catheter's surface was found to promote bacterial adhesion, increasing the risk of infection complications both locally and systemically.[11] Also, the majority of CRBI cases have been reported to have SIRS.[16] Most CRBIs are attributed to three types of organisms: *Candida* species, *Staphylococcus aureus* and coagulase-negative *Staphylococci* (usually *Staphylococcus epidermidis*).[18] CVCs account for an estimated 90% of all CRBIs.[18] Moreover, CVCs were found to pose the greatest risk as well as considered to be the main source of bacteremia and septicemia in hospitalized patients, and eventually the main cause of morbidity and mortality.[8,16] Prospective studies indicated that the relative risk for CRBI cases with CVCs is up to 64 times greater than with peripheral venous catheters.[8] In a retrospective study of 321 patients undergoing hemodialysis by Chun-Xiao *et al.*, Gram-positive bacteria were among the highest percentage among pathogens with 50.76% for CVC-induced infection, compared to 31.06% for Gram-negative bacteria and 18.18% for fungi, while 13 patients have been reported to have mixed infections.[19]

CAUTIs have been defined by the Centers for Disease Control and Prevention as the existence of either funguria or bacteriuria induced by the invasion and colonization of pathogens through the route of the urinary catheter at elevated concentrations ($>10^5$ CFU mL^{-1}).[20] While urinary tract infections (UTIs) are defined as invasions of phathogenic fungi or bacteria specimens to any part of the urinary system,[20–22] a strong link exists between urinary catheters and subsequent UTIs, the most common hospital-acquired infection in the United States.[23] Urinary catheters are widely used in hospitals, with an estimated 15–25% of hospitalized patients having them placed at some point during the course of the hospital stay and they are associated with 75% of hospital-acquired or nosocomial UTIs.[20] In the United States alone, up to 2009, 450 000 CAUTI cases were reported annually with treatment costs exceeding \$350 million.[24] Today, the estimation in the United States is well over a milion CAUTI cases and much greater annual treatment costs.[20] Worldwide, the incidence of CAUTIs is increasing, leading to greater hospital costs,[25] and they account for more than 40% of all healthcare-associated infections.[26] Several types of bacteria and fungi were found to be involved in CAUTIs, forming single- or multi-species biofilms. The common organisms that cause CAUTIs are diverse and include Gram-negative bacteria: *Escherichia coli*, *Proteus mirabilis*, *Klebsiella pneumoniae* and *Pseudomonas aeruginosa*; Gram-positive bacteria: *Staphylococcus epidermidis*, *Staphylococcus aureus* and *Enterococcus faecalis*; as well as fungi: *Candida albicans*.[20,27–29] For indwelling urinary catheters, several microbial organisms were found to be involved in the infection, including yeast species; however, *Escherichia coli* was found to be the most common infecting pathogen, with a total infection rate of 5% per day.[30] The indwelling urinary catheters are widely used among older populations.

In hemodialysis, it was reported that the rate of infection, morbidity and mortality is much higher when patients dialyzed using catheters compared to fistulas or grafts.[31] Catheters were found to induce infection complications, leading to higher a percentage of morbidity and mortality in this patient population, and thus they are more favored for short-term use.[32] Despite these complications, worldwide use of tunneled hemodialysis catheters *via* vascular access is becoming more widespread, encompassing more than 25% of all dialysis patients.[31] Also, hemodialysis catheters through the neck, chest and groin offer advantages such as capability to function under emergency conditions or potential usage until other methods of access mature.

For shunt catheterization, the Gram-positive *Propionibacterium acnes* which is an abundant bacterium on human skin was recognized as a shunt pathogen, causing up to 14% of related infections. The infections related to shunt catheters are considered a serious issue and were found to affect ~10% ofprocedures.[33]

For the majority of catheter-induced infections, microorganisms at an advanced stage of the infection were found to form biofilm/biofilms, in which the pathogenic contiminations are more challenging to treat. At this stage, the microbes have been reported to be more protected, less receptive to treatment and even exhibit resistance to treatments. Resistance was found both to antibiotic treatment and the host's own immune system. Biofilm formation was found to induce the spread of dangerous infections, leading to several health issues, including death.

The fact that microbial contamination can spread rapidly and the capability at later stages to become more resistant to treatments *via* challenging biofilm formation have increased public awareness of pathogenic risk and increased the need for the development of protective/therapeutic technologies from which patients can gain the medical benefit from the medical devices, *i.e.* catheters, with minimal risk to life.

14.1.3 The Need for Antimicrobial Catheters

The catheter serves as a nidus for microbes during its insertion or implantation or through repeated exposure to the environment, such as during blood draws from a CVC.[34,35] In one animal study, contamination of the skin entry site during insertion resulted in catheter tip colonization within 1 hour of insertion.[36] The microbes survive by attaching to solid substrates in structured communities, *i.e.* biofilms, where they may persist for extended periods of time, during which their pathways of transmission multiply.[37] Microbes grown in biofilms are more resistant to systemic antimicrobial treatments and can withstand host immune responses.[38] Additionally, the majority of nosocomial infections are caused by Gram-negative bacteria, among which there is a growing resistance to antibiotics.[39] Biofilms form readily on catheters typically composed of materials such as polyurethane and silicone and the current clinical management of a catheter-related infection is to remove the device completely.[40]

Accordingly, there is a great need to develop novel and efficient antimicrobial materials as well as coating technologies for achieving antimicrobial coatings to which microbes are less resistant throughout the entire duration of catheter implantation (Figure 14.1). In order to achieve this, materials with antimicrobial properties must be prepared with the capability to prevent complex epidemiological situations as well as infection risks. More importantly, these materials and coatings should be effectively functional in the healthcare environment and physiological microenvironment of the catheters, taking into account the bacteria species at both the insertion and final destination sites.[4,41] These requirements are valid for a wide range of biomedical implants for various applications, including orthopaedics, surgical armaments and dental composites, among many others.[4,41]

Antimicrobial coatings and materials are rapidly emerging to mitigate catheter-associated infections. Recent advances in materials science and developments in biotechnology methodologies have led to an extensive variety of surfaces/coatings with antimicrobial properties. Worldwide, initial efforts have been invested to develop various coating strategies where bioactive materials can be applied as a coating or part of the coating of medical implants, *e.g.* catheters. Initially, a simple "preventative strategy" was implemented, in which coatings capable of preventing microbial

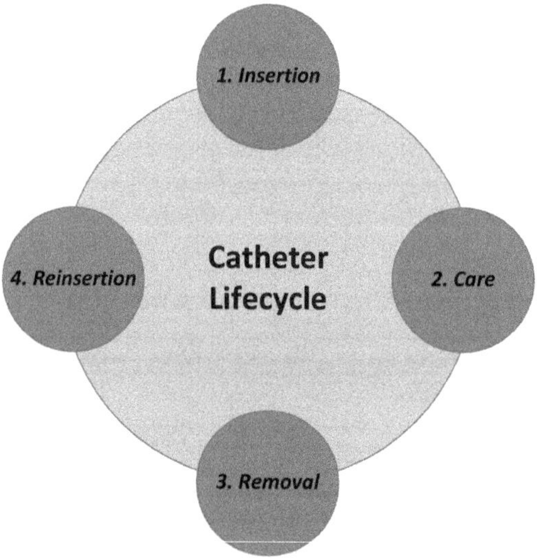

Figure 14.1 Lifecycle of a catheter.

adhesion to the surface were developed. Few examples have been reported using poly(ethylene glycol) coatings and diamond-like carbon films.[41] However, bioactivity results have indicated that such a strategy is limited and underline further the need for the development of both potent antimicrobial materials and antimicrobial coatings.

Intensive efforts worldwide have been carried out to overcome the microbial challenge, aiming to develop antimicrobial catheters with functional antimicrobial coatings. Current strategies to prevent microbial infection include the following surface modification approaches: antibiotic agent release, contact killing and repelling or anti-adhesive formulations.[42] Here we discuss the recent advances in antimicrobial materials and development of coatings for catheters that have been clinically tested, as well as catheter agents/materials currently being researched. We also highlight the challenges and opportunities in altering the surface properties of catheters to limit microbial adhesion, growth and biofilm formation.

14.2 Antimicrobial Materials

The fast rate of microbial spread, resistance to treatment and growing public health awareness of catheter-associated infections have all contributed to the increased need to develop novel efficient antimicrobial materials, including antibiotics with minimal resistance applied to surfaces coating medical devices. Accordingly, these efforts have yielded various families of antimicrobial materials. The antimicrobial materials range from small molecules up to high molecular weight polymer-based antimicrobial agents and particles, and include organic and inorganic compounds (Figure 14.2). The following antimicrobial agents have been widely reported and associated with antimicrobial catheter design.

14.2.1 Chlorhexidine

Chlorhexidine (*N*,*N*‴-1,6-hexanediylbis[*N*′-(4-chlorophenyl)(imidodicarbonimidic diamide)]) is a cationic bisbiguanide which acts as an antiseptic agent by binding to negatively charged bacterial walls, causing membrane disruption. Figure 14.2 shows the chemical structure of chlorhexidine. This agent is widely used in dental applications such as gelatin for the treatment of periodontal infection[43] and in mouthwash due to its broad spectrum of antimicrobial action and lower risk of drug resistance compared with conventional antibiotics, because of its non-specific mechanism of action.[44] Chlorhexidine has been used for a long time in drainage bags, as well as forming part of the coating of urinary catheters.[20,45,46]

14.2.2 Silver

Silver (Ag) has been approved by the US Food and Drug Administration and is known to kill microbes non-specifically. Even low concentrations of silver

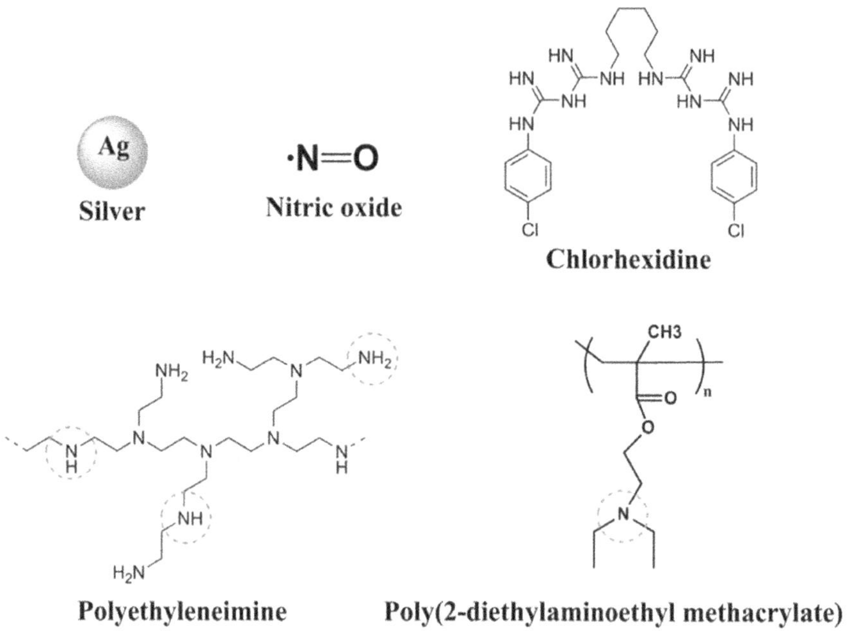

Figure 14.2 Representative examples of antimicrobial agents in use for catheters. Dashed circles highlight potential sites for alkylation and/or quaternization.

ions are sufficient to kill microbes through the following mechanisms: (1) loss of membrane potential; (2) destruction of Fe–S clusters in proteins; and (3) oxidative stress.[20,47–51]

The combination of these three mechanisms for antimicrobial activity makes silver a potent biocide and, consequently, one of the most popular antimicrobial agents used in medical device coating formulations. Pure silver coatings alone deteriorate quickly and do not demonstrate enough antimicrobial efficacy, so silver ion-releasing coatings of catheters in the clinic have primarily been silver alloys with gold or palladium.[52–54] A few silver-coated urinary catheters have been developed and tested clinically to resist CAUTIs. Multiple studies have supported the efficacy, safety and cost-effectiveness of the use of a CAUTI-reducing coating consisting of a silver alloy and hydrogel layer.[52,53,55] This coating has been integrated into clinical practice in the United States. However, there is considerable variation of the CAUTI-reducing effect observed in these studies, and there is high dependence on patient group characteristics, geographic region, catheterization time period and varying definitions of CAUTI. One multicenter randomized controlled superiority trial showed almost no difference when comparing silver alloy urinary catheters with standard polytetrafluoroethylene catheters.[56] Furthermore, silver ions are poorly soluble, which diminishes their antimicrobial efficacy. One potential solution to this problem is silver nanoparticles (AgNPs), which may allow for sustained release of silver ions.

14.2.3 Nitric Oxide

Nitric oxide is a colorless gas, with highly reactive small molecular structure (formula •NO). Nitric oxide was found to exhibit antimicrobial activity by inducing detrimental damage to microbial cellular targets *via* its free-radical form.[57,58] It was found to metabolize rapidly, thus promoting an innate immune response, which is also considered an effective tool to target pathogens associated with medical implants.[59] In order to maintain its antimicrobial activity, a continuous elution of nitric oxide is required from the implanted catheter's surface/coating; this mode of action has driven research into the design of nitric oxide eluting catheters. To achieve this purpose, polymeric catheters and their surface coatings are loaded with nitric oxide, either by embedding in nitric oxide forming molecules/complexes and/or loading with nitric oxide gas by simply exposing the catheter for extended time under high pressure. The incorporation of nitric oxide forming molecules/complexes (Figure 14.3) to polymeric resins of the catheter prior to an extrusion step was found to significantly enhance nitric oxide loading and eventually antimicrobial activity.[60]

14.2.4 Antibiotics

Impregnation of catheters with antibiotics is considered an efficient approach for prevention of catheter-associated infections.[61] Most antibiotics are small molecules with a relatively low molecular weight (Figure 14.4), varying in their antifungal, antibacterial and antiparasitic activity *via* different mechanisms of action, ranging from slowing microbial growth to significant microbial contamination killing.[20] Examples of efficient antibiotics have been reported in the literature and extensively studied, such as sparfloxacin, nitrofural, minocycline and rifampin.[20,62–75] Although antibiotics have proven to be more efficient than many other antimicrobial agent based coatings such as silver, serious concerns over resistance have

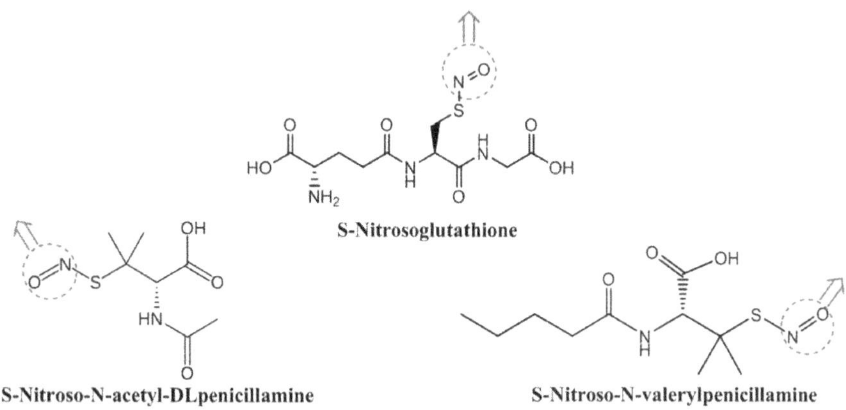

Figure 14.3 Chemical structures of commonly studied nitric oxide (NO) donors. Dashed circles highlight functional groups releasing NO.

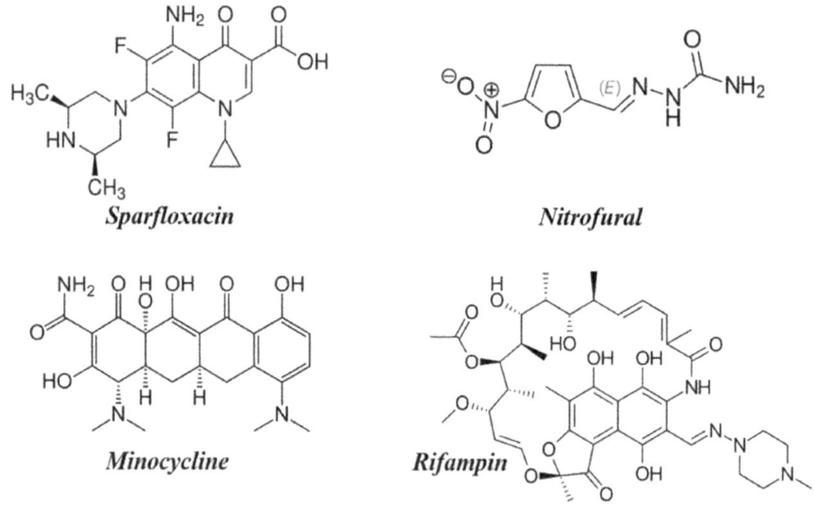

Figure 14.4 Chemical structures of commonly studied antibiotic compounds for catheters.

been reported. Increasing numbers of studies clearly indicate that the frequent and long-term use of antibiotics can lead to the formation of antibiotic-resistant bacteria species, thus emphasizing the need to examine other antibiotics and therapeutic combinations.

14.2.4.1 Nitrofural

Nitrofural (5-nitro-2-furaldehyde semicarbazone) is a low molecular weight antibiotic and considered an effective antibacterial agent against Gram-positive and Gram-negative bacteria species. Nitrofural exhibited strong antimicrobial activity which paved the way for its use against CAUTI microbes. Nitrofural was found to exhibit excellent properties, particularly against bacterial adherence and biofilm formation. The mechanism of action of nitrofural was reported to be *via* inhibition of DNA replication in minutes. Nitrofural-coated catheters were found to be more effective than silver-coated/-impeded catheters. Initially, nitrofural-impregnated catheters were commercially available and in wide use; however, nitrofural-coated catheters have been discontinued in the US, due to serious side effects, as well as reported discomfort shortly following catheterization, thus need for the development of safer antibiotics.[20]

14.2.4.2 Minocycline–Rifampin

*The use of minocycline–rifampin in dual-*antibiotic impregnated CVCs was found to cover a broad spectrum of activity against many Gram-positive and

Gram-negative bacteria including *Candida albicans*, *Pseudomonas aeruginosa*, *Enterobacter* species and *Escherichia coli*.[76] According to a recent systemic review and network meta-analysis of randomized clinical trials, the minocycline–rifampin impregnated CVCs appeared to be the most effective in preventing catheter colonization; however, colonization reduction by minocycline–rifampin impregnated CVCs is not accompanied by a significant reduction in catheter-related bloodstream infections (CRBSIs).[77]

The embedding of low molecular weight antimicrobial agents in catheter coatings was considered for many years to be an efficient tool for minimizing catheter-associated infections. However, later studies indicated residual toxicity and several side effects, thus limiting their widespread usage as well as increasing the need for development of other materials and coating techniques.[78] Macromolecular antimicrobial agents, specifically polymers known as antimicrobial polymers, affording chemically stable non-volatile structures have exhibited reduced toxicity. The reduced toxicity was attributed to their minimal capability to permeate through biological membranes due to their macromolecular structure and high molecular weight.[79]

In addition, antimicrobial polymeric-based agents exhibited further promise due to the ease of applying them as a coating *via* relatively simple techniques and their wide applicability to various catheter surfaces. On top of that, some coatings were found to maintain the catheter's flexibility, thus increasing the coating's design and manipulation opportunities. Lastly, antimicrobial polymers can offer a high surface density of active antimicrobial functional groups that can be controlled by their structure and design, thus potentially resulting in less microbial resistance as well as eventually an increased overall treatment efficiency.[80]

A potent class of high molecular weight antimicrobial agents are polycationic polymers with a leading subgroup, namely quaternary ammonium (QA)-based polymers. Their incorporation in coatings was found to provide long-term antimicrobial and anti-adhesion properties, thus improving the efficiency of antimicrobial coatings. In biomedical applications including catheters, several polycations have been synthesized and tested. Bioactive polymers including homo-, co- and random polymers with various architecture structures—linear, branched and dendritic—were synthesized. Poly(2-diethylaminoethyl methacrylate) (PDEAEM) and its acrylate derivatives, poly(*N*-alkyl-4-vinylpyridinium) (PVP) salts, polyethyleneimine (PEI), polyamidoamine (PAMAM) dendrimers, polyguanidines and poly(L-lysine) (PLL) are the most commonly used and studied QA-based polyamines for biomedical applications. Multiple fundamental studies that discuss the antimicrobial functionality, therapeutic effect and physiochemical properties of these polymers have been extensively reported and summarized in several reviews and books.[81] Next, we cover the strategies for the fabrication of catheters with long-term antimicrobial activity.

14.3 Strategies for the Development of Antimicrobial Catheters

Different strategies have been developed to inhibit initial bacterial colonization at the catheter surface. The systematic application of antibiotics has been suboptimal for catheter-associated infections, because of the difficulty of penetrating biofilms, as well as side effects associated with systemic delivery.[82] Local delivery of antibiotics from the catheter's surface has been developed to kill the bacteria directly before biofilm formation. One disadvantage of this approach is that the release of antimicrobial agents is finite, limiting their efficacy in cases where more long-term implanted catheters are required. Eventually, the concentration of the release of antimicrobial agent will drop below the minimal inhibitory concentration. Various methods have been developed to control and prolong the release kinetics, including the development and incorporation of nanoparticles (organic and inorganic), multilayer releasing systems (*e.g.* polyelectrolyte) and environment-dependent release (stimulus-responsive). Another approach to address this issue of finite antimicrobial loading is the development of contact-killing coatings. By this strategy, antimicrobial agents are permanently anchored to the catheter surface by utilizing various hydrophobic and flexible polymeric chains.[42] Cationic compounds are often attached to the polymeric chains, killing the bacteria due to disruption of their cellular membrane. One additional approach involves prevention of bacterial adhesion to the surface of the catheter, preventing the earliest step of biofilm production using non-cytotoxic molecules. This requires modification of the catheter surface, primarily with hydrophilic polymeric surface coatings, which minimize protein adsorption and subsequent microbial adhesion.[83,84] In this section, we overview the recent advances and emerging strategies to prevent and combat the proliferation of bacteria on the catheter surface (Figure 14.5). Strategies are divided into three categories: (1) antimicrobial-releasing catheters; (2) catheters with contact killing surface modifications/coatings; and (3) bacterial-repelling and anti-adhesive catheter surfaces.

14.3.1 Release-based Antimicrobial Catheters

Engineering antimicrobial surfaces and coatings capable of releasing agents formed an effective approach for various medical devices and implant applications, including several types of catheters. Antimicrobial-releasing coatings are preferable in cases where is limited direct contact between the surface and the microbe and/or when a short time is available for such interactions.[85–88] This mode of action was also found to be potent whenever the antimicrobial agent required diffuse coverage within the microenvironment of the catheter at the final destination site. Accordingly, small therapeutic molecules and nanoparticles (both non-polymeric and

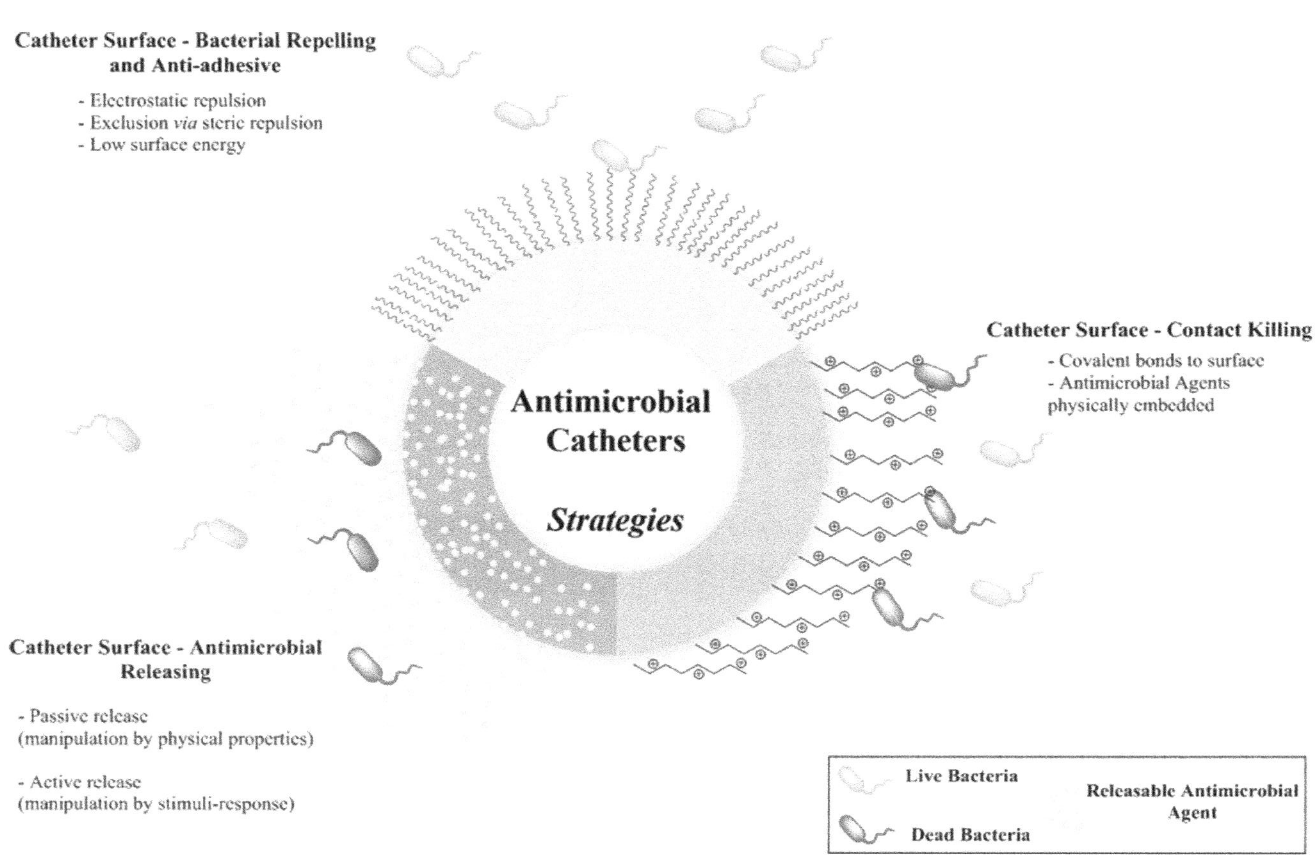

Figure 14.5 Strategies for the prevention of colonization of bacteria on catheter surfaces: antimicrobial-releasing catheters; catheters with contact-killing surface modifications/coatings; and bacteria-repelling and anti-adhesive catheter surfaces.

polymer-based) became the preferred and dominant agents. Given that therapeutic coatings must answer specific medical needs, several parameters should be taken into account in designing the catheter's surface coating in order to control both the kinetics profile and the overall duration of antimicrobial agent release. However, it is always desired that the levels of released antimicrobial agents stay in the therapeutic window; simultaneously sufficient to kill the microbes and low enough to avoid/with minimal cytotoxicity to patient cells and surrounding tissue.

A wide range of antimicrobial compounds have been developed over the past several decades using release-based kinetics (Table 14.1).[42]

Antibiotic compounds such as minocycline–rifampin, as well as silver- and chlorhexidine-coated catheters have displayed significant activity against a range of microorganisms such as *Candida albicans* and *Escherichia coli*.[105] The use of antimicrobial peptides (AMPs) for biomaterial applications has become popular due to several advantages over release of antibiotics, elements and organic cationic compounds, including increased biocompatibility, minimal immunogenicity, availability of moieties for additional functionalization, chemical versatility and biodegradability.[106] AMPs are a component of the innate immune response and can induce antimicrobial effects by disrupting the membranes or by passing through the membranes and targeting intracellular processes.[107] Certain peptides released from the surface of catheters have been developed to prevent *Candida albicans* colonization.[103]

The oldest and most common method to coat catheters under this strategy consists of simple impregnation by soaking a porous material in effective agent solution or by direct coating with the antimicrobial compound using fast evaporation techniques. These methods create an amorphous coating with fast release kinetics. In order to achieve an elongated effect, various delivery systems have been developed to control the release of antimicrobial agents from coatings. The most frequently employed carriers range from synthetic polymers such as poly(lactic-*co*-glycolic acid) (PLGA), polyurethane (PU), polyacrylic acid (PAA), poly(methacrylic acid) (PMMA) hydroxyapatite and natural polymers such as chitosan and hyaluronic acid as well as their blends and mixtures. Agent releasing strategies have been progressed and further developed to have both passive (without active triggers) and active (with environmental triggers) release kinetics.

14.3.1.1 Controlled Release Kinetics: Passive Approaches

Bioactive antimicrobial agents may be passively released from the catheter/catheter's coating. The release profile can be tuned mainly by manipulating the physical properties of (1) antimicrobial agent properties (*e.g.* size, distribution, polarity/charge, loading); (2) antimicrobial agent carrier (*e.g.* chemical moieties, hydrophilicity/hydrophobicity, crosslinking/porosity); and (3) therapeutic coating characteristics (*e.g.* thickness).[42,108–110] In addition, the appliance of a thin polymeric top layer, mostly using either plasma

Table 14.1 Antimicrobial compounds in release-based catheter coatings. Adapted from *Trends in Biotechnology*, **33**, M. Cloutier, D. Mantovani, F. Rosei, Antibacterial Coatings: Challenges, Perspectives, and Opportunities, 637–652, Copyright 2015, with permission from Elsevier.[42]

Antimicrobial type	Released compounds	Mechanism of action	Catheter type	Comments	References
Antibiotics	Aminoglycosides (tobramycin, gentamycin)	Inhibit protein synthesis by binding to the bacterial 30S ribosomal subunit	Urinary catheter		89
	Quinolones (norfloxacin, ciprofloxacin)	Inhibit DNA replication and transcription by targeting DNA topisomerases II and IV	Urinary catheter		90, 91
	Glycoproteins (vancomycin)	Disrupt cell wall peptidoglycan synthesis by binding to amino acids	Ventriculostomy catheters	Effective against mycobacteria and Gram-positive bacteria Not approved for coatings, since this antibiotic is the drug of choice for treatment of catheter-related bloodstream infections	92
	Tetracyclines (minocycline)	Inhibit protein synthesis	Central venous catheters, percutaneous nephrostomy		93–96
	Rifamycins (rifampin)	Inhibit transcription by binding to RNA polymerase	Central venous catheters, percutaneous nephrostomy	Effective against Gram-positive bacteria and mycobacteria	93, 94, 96
Elements (metals and non-metals)	Silver	Impaired membrane function, protein dysfunction and oxidative stress	Urinary catheters	Most used antibacterial metal/nanometal Reported to have toxicity in humans at high doses	20, 97, 98

Table 14.1 (*Continued*)

Antimicrobial type	Released compounds	Mechanism of action	Catheter type	Comments	References
	Copper	Induce lipid peroxidation in bacterial membrane by generating reactive oxygen species and depletion of antioxidants	Intravenous catheters	Similarly to other heavy metals (*e.g.* zinc), can induce several metal-catalyzed oxidation reactions that damage proteins, membranes or DNA	99
	Iodine	Penetrate the cell wall and disrupt protein and nucleic acids structure and synthesis	Central venous catheter		100
Organic cationic compounds	Chlorhexidine	Causing bacterial membrane disruption, by binding to bacterial walls			43, 45, 46
	Quaternary ammonium compounds	Vital ion leakage induces bacterial death by destabilizing the negatively charged bacterial cell wall			101
Antimicrobial peptides	Charged peptides	Depends on the specific tested antimicrobial peptide	Urinary catheters, central venous catheters		102–104
	Common examples includenisin, magainin	Usually by metabolic inhibition and transmembrane pore formation			

polymerization or spray coating, can be used as an effective tool to either delay, slow or sustain the release of the antimicrobial agent by acting as physical barrier. Top-coating porosity/crosslinking, thickness and hydrophobicity were shown to be the dominant parameters affecting the release profile.

14.3.1.2 Controlled Release Kinetics: Active Approaches

In contrast to passive release, active approaches can be triggered by external stimuli in response to changes at the microenvironment of the implant, *i.e.* catheter, either during the insertion process or at both short- and long-term stays at the desired destination. Stimuli-responsive materials have been studied for several decades for biomedical applications as part of drug delivery systems and self-healing coatings, and found to have great potential for use as bioactive coatings.[42,111,112] Today, they are classified as smart materials which can be utilized for immediate response to changes in the medical situation, such as bacterial infection development or progress. As such, stimuli-responsive coatings were found to be efficient tools for increasing the ability to extend the useful lifetime of implants and therapeutic coatings as well as to produce ‘on demand’ antimicrobial effects (Figure 14.6). Antimicrobial release systems in response to magnetic, electrical, thermal, mechanical and ultrasonic triggers have been reported.[42,113–118]

Smart coating-based nanoparticles (NPs) are an emerging field which hold the potential for precise and timed active release of antibacterial agents,[42,119,120] *e.g.* silver NPs embedded in polyelectrolyte multilayers (PEMs), where the active release can be triggered following exposure to near-infrared light by inducing polymer degradation.[121] More importantly, bacteria-triggered coatings are an exciting approach and represent the ultimate solution for catheter infections. Examples of pH-responsive coatings were reported in the literature, in which the release of the antimicrobial agent is a function of bacterial metabolism and the involvement of acidic substances (*e.g.* acetic and lactic acid).[122,123] Zhuk *et al.* have reported the combination of positively charged antibiotics, polymyxin B, tobramycin and gentamicin, with a counterpart of polyanionic tannic acid, forming pH-responsive antimicrobial PEM systems that burst-release antibiotics as a function of lowering pH.[123] Another example is gentamicin sulfate bonded to NPs through pH-sensitive imine bonds; the NPs are attached with stable amide bonds to titanium substrates.[124] The uniqueness of such systems is that with the absence of bacterial stimuli and/or pH lowering the antimicrobial agents can remain within the coatings for several months. Accordingly, such coatings are more successful than passive drug-eluting coatings in fighting delayed catheter-induced infections.

In contrast to the pH-trigger approach, using bacteria-generated enzymes to degrade or cleave bound antimicrobial agents from a specific coating offers a huge advantage. Such enzymatic pathways offer the possibility of controlled release triggered by specific microbial species. Several examples

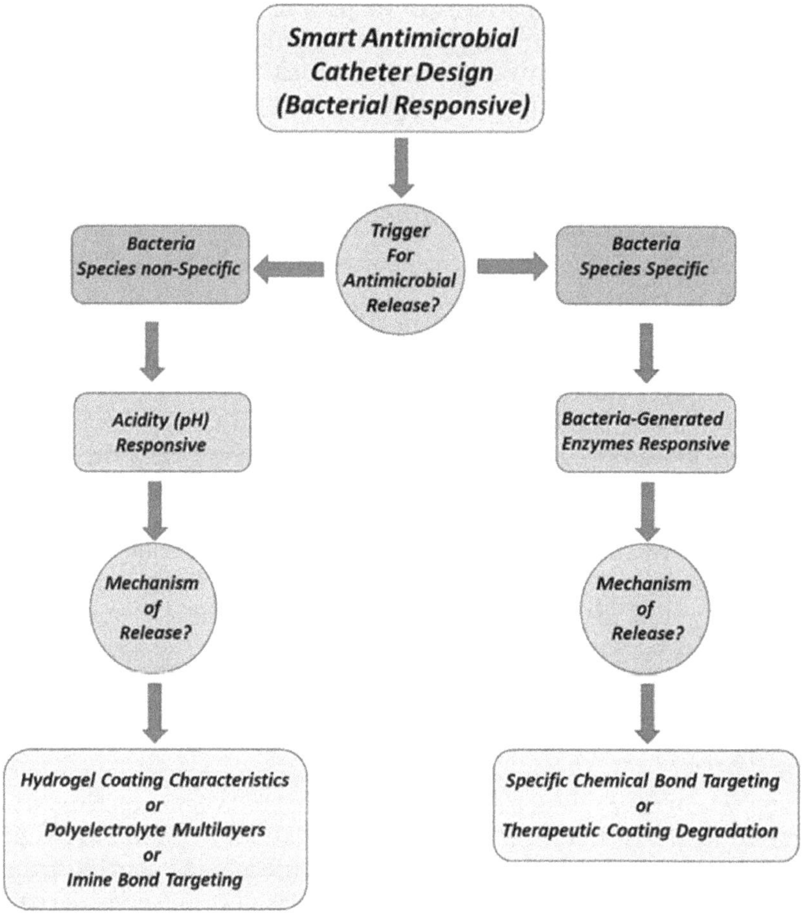

Figure 14.6 Smart antimicrobial catheter design with a bacterial-responsive system to the bacterial triggers: secreted enzymes or bacterial metabolism.

have been reported, such as hydrolyzable anhydride bonds by lipase[125] or thrombin-sensitive peptide linkers.[126] Cado *et al.* have reported the release of an antimicrobial peptide (*i.e.* cateslytin) from a chitosan/hyaluronic acid multilayer system *via* the cleavage/enzyme pathway by hyaluronidase, an enzyme secreted by pathogens.[127] The antimicrobial coatings were found to be effective against *Candida albicans* for three cycles, but failed after the first cycle to fully inhibit *Staphylococcus aureus*.[127]

14.3.1.3 Nanoparticles

Antimicrobial-loaded NPs are an emerging method allowing for sustained antimicrobial release. Nanospheres have been used as drug delivery

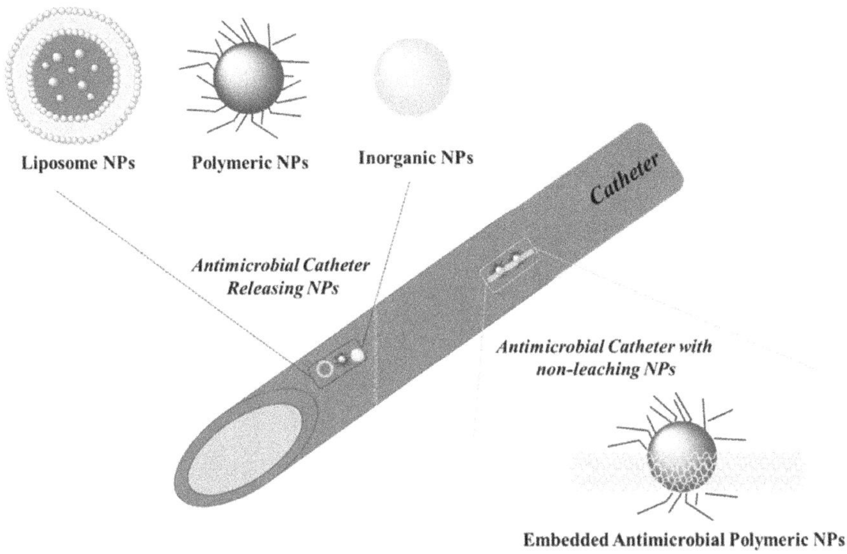

Figure 14.7 Nanoparticle (NP)-based catheters: antimicrobial-releasing NP agents (liposomal (encapsulated antimicrobial agents), polymeric or inorganic) or non-releasing, with polymeric NPs embedded on the catheter's surface.

systems to prolong the release of hydrophobic agents.[128–130] Here, we present briefly three antimicrobial NP-based releasing systems from catheters (Figure 14.7).

14.3.1.3.1 Liposome Nanospheres. Liposomes are spherical lipid vesicles consisting of one or more phospholipid bilayers. These liposomes are prepared mainly *via* two steps: (1) lipid materials dissolved in an organic solvent which is removed, forming a clear lipid film; and (2) lipid film dispersed in an aqueous medium. Upon agitation in an aqueous medium, the hydrophobic tails turn inward and the hydrophilic heads face the aqueous medium, forming concentric layers of phospholipids.[131] Various sizes of these vesicles may be prepared through different mechanical dispersion techniques. Liposomes may carry hydrophilic, hydrophobic and amphipathic drugs that can be protected from external disturbances for sustained release. Therefore, catheters coated with liposomes have the advantages of being highly customizable while allowing protection from degradation or non-specific reactions increasing the rate of drug delivery. In a study by the Khoury group, liposomes containing ciprofloxacin were prepared and embedded within a hydrogel coating on a urinary catheter. Using a rabbit model, the developed coating was evaluated for efficiency against *Escherichia coli*, a common pathogen involved in CAUTIs. When compared with hydrogel-coated silicone catheters, the time to positive urine cultures for control catheters (blank) was prolonged from 3.5 to

5.3 days ($P = 0.04$).[20,132] For optimization purposes, liposome surfaces can be modified to be anionic or cationic.[133] Furthermore, liposome-based catheter coatings are highly biocompatible compared to other coatings.[134]

14.3.1.3.2 Polymeric Nanospheres. Polymeric nanospheres have demonstrated more stable storage, higher drug loading and better controlled release than other colloidal formulations.[135] For example, biodegradable polycaprolactone (PCL) was used to prepare chlorhexidine-loaded polymeric NPs and applied by spray-coating onto silicone catheter surfaces.[20,136] This coating based on PCL NPs showed antibacterial activity against *Staphylococcus aureus*, *Streptococcus epidermidis* and *Escherichia coli* for up to 15 days, a duration three times longer than that of chlorhexidine polymer blend coatings used in the clinic.[136]

14.3.1.3.3 Inorganic Nanoparticles. The evolution of bacterial strains and the increasing resistance to multiple antibiotics has prompted researchers to consider AgNPs due to the broad-spectrum antimicrobial coverage of silver.[137] Furno *et al.* were the first to demonstrate the potential of AgNPs as antimicrobial agents in medical devices *via* the impregnation of silicone with AgNPs.[138] Although NPs showed promise as antimicrobial agents, surface fixation is required to prolong the efficacy. Another issue of AgNPs is their safety. Several reports have demonstrated that AgNPs initiate non-specific killing of cells, risking the patient's own cells when there are high concentrations of AgNPs.[139]

14.3.2 Contact Killing

Sub-lethal doses of antimicrobial compounds, as occur in release-based coatings, have been shown to accelerate the development of resistance in bacteria species, making biofilms even more difficult to eradicate.[140] In contrast to antimicrobial release methods, contact-killing surfaces retain their activity for extended durations.[141]

The majority of non-leaching/non-releasing long-lasting antimicrobial surfaces are prepared *via* surface modification of medical implants with hydrophobic polycationic polymers and mainly based upon QA compounds (QACs). QACs have been found to induce microbial death, where the positively charged QA moieties attract the partially negatively charged bacterial cell membranes and induce microbial cell wall destabilization and, later, leakage of vital ions.[142] QACs have been found to exhibit strong antimicrobial activity against a broad spectrum of Gram-negative and Gram-positive bacteria, fungi and viruses.[81] These modifications result in increased surface density of the antimicrobial functional moieties (*i.e.* QA) per coating area, thus less microbial resistance and increased antimicrobial activity. These long-lasting antimicrobial surfaces were prepared by either chemical or physical incorporation of the active agents which caus damage to bacteria upon contact. In comparison to small QA molecules, a high density of QA moieties following QA-polymer surface incorporation was found to urge the active hydrophobic chains to be erected from the surface, thus increasing the antimicrobial functionality.[143,144]

The bioactivity of these QA polymers were found to be related to QA moieties (density and charge) and the four attached alkyl chains. Their hydrophilicity/hydrophobicity balance, length and QA counter-ion have a significant effect on the observed antimicrobial activity of QA polymers.[80] One alkyl chain should be long enough, at least C4–C8, to disturb the microbe cell wall, while shorter or longer alkyl chain-substituted QA polymers are less effective.[6,81,143,145] Other parameters that affect activity are chemical structure linearity and branching, degree of quaternization and alkylation, and QA polymer molecular weight. In order to maximize their antimicrobial efficiency and suitability for multiple biomedical applications, including the various types of catheters, each of the aforementioned parameters should be independently studied and optimized while taking into consideration the microbial species involved in each case.

Several examples have reported the incorporation of high molecular weight QACs or their crosslinked NPs structures into top coatings for different antimicrobial applications.[6,145–152] Given the importance of QACs and the fact that the majority of contact-killing materials employ QACs, we provide an overview of synthetic and natural-based antimicrobial QA polymers for surface coatings.

14.3.2.1 Polymeric Coating Modifications—Synthetic Cationic Polymers

The fact that the nature of the organic substituents attached to the functional groups (*i.e.* QA moieties) and the polymer structure have a direct impact on their antimicrobial activity[153] has provoked research worldwide continuing effort to synthesize active QA polymers with various QA substituents, varying in their chemical composition (R1, R2, R3, R4 = H-, alkyl-, aryl- and heterocyclic-; Figure 14.8), length and hydrophobicity/hydrophilicity ratio.[142,153–157] Intensive efforts over the past four decades have yielded a wide spectrum of QA polymers with active and stable QA moieties.[79,80,157–161] Several QA polymers were identified with prolonged shelf-lives, high efficiency and selectivity, as well as reduced toxicity, thus increasing their potential application in medicine.[80,161,162] Ferreira and Zumbuehlhave comprehensively discussed QA antimicrobial surface modification methods as well as immobilizing them on different surfaces for various applications.[143]

For medical applications, QA-PVP, quaternary ammonium diethylaminoethyl methacrylate (QA-DEAEM) and its copolymerized acrylate-based derivatives, and QA-PEI and its modified versions are among the most studied QA polymers for contact-killing antimicrobial surface coatings. Due to their low cytotoxicity and ease of application to medical devices, these polymers have gained increasing interest for orthopaedic implants,[157]

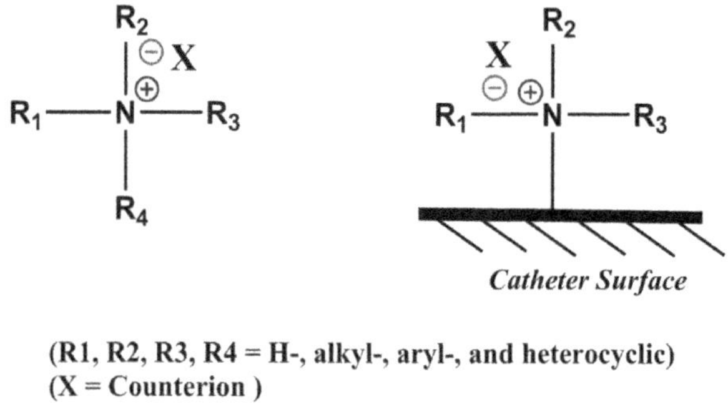

Figure 14.8 Chemical structure of antimicrobial agents/compounds based on quaternary ammonium (QA): left: free structure; right: connected to the catheter surface. The R groups may be the same or different.

catheter coatings[163,164] and dental composites,[146–149,165–167] as well as water disinfection.[150–152,168,169] Here we highlight the synthesis and antimicrobial activity of leading polyamine-polycationic polymers.

Pyridinium-based polymers were found to exhibit strong antimicrobial activity after heterocyclic ring-*N*-alkylation.[170] Also, the polymeric backbone was found to have a direct effect on the observed antimicrobial activity. QA-polystyrene-*block*-PVP in comparison to QA-polystyrene-*random*-PVP was found to exhibit higher potency against both *Staphylococcus aureus* and *Pseudomonas aeruginosa*. Following its addition to bonding resin it was confirmed that these particles can prevent bacterial growth on the surface without affecting resin bonding characteristics.[171,172]

PEI is a branched polyamine well known for its abundance of amino groups (primary, secondary and tertiary amine groups) along the structure, thus making it easily accessible for chemical modifications such as alkylation followed by quaternarization, among others, resulting in an active antimicrobial polymer with a high density of QA moieties. Furthermore, its antimicrobial activity was found to be affected by PEI's molecular weight, percentage of *N*-alkylation and alkyl chain length.[159] QA-PEI, both as polymers and 3D crosslinked NPs have been reported to have superior properties in the inhibition of the growth of a broad spectrum of bacterial species following embedding in different surfaces and for various medical needs.[6,145–150,152,159,162–166,169,173] In a recent study reported by Ortega, Farah and co-workers, they examined the antimicrobial activity of QA-PEI NPs against clinical isolates of the pathogenic bacteria involved in catheter-induced recurrent bacterial peritonitis, a global public health problem.[163,164] In this disease, patients are often infected because of their weakened immune system and as a result of procedures such as the introduction of catheters into the peritoneum. Moreover, it is reported that the bacterial strains involved have become resistant to many antibiotics. In this study, they evaluated the activity of QA-PEI NPs against *Stenotrophomonas maltophilia*, *Escherichia coli* and *Streptococcus viridians* and have found that the minimum inhibitory concentrations are 12.5, 25 and 100 μg ml^{-1}, respectively. Moreover, at a concentration of 12.5 μg ml^{-1}, QA-PEI NPs were found to inhibit significantly the bacterial growth for the studied bacterial isolates. Furthermore, QA-PEI NPs at concentrations up to 50 μg ml^{-1} and exposure for 48 h had no significant toxicity on human embryonic kidney 293 (HEK293 T) cells. This study highlights the antimicrobial efficiency of QA-PEI against catheter-related antibiotic-resistant Gram-negative bacteria.[163,164] These QA-PEI NPs were embedded and tested in non-leaching dental composites and polyethylenemethacrylic acid and polyethylenevinyl acetate coatings at different loading percentages (wt/wt %) against *Enterococcus faecalis* (streptomycin-resistant), *Staphylococcus aureus*, *Pseudomonas aeruginosa*, *Escherichia coli* and *Staphylococcus epidermidis* were found to induce complete inhibition of the bacterial growth at low loading.[150]

Cationic compound-based QA moieties attached to catheter surfaces have been reported by applying on QA polymers. On the surface, the QA functional groups were found to interact with the negatively charged cell wall of bacteria and disrupt the normal functions of the membrane while remaining non-cytotoxic to human cells, *i.e.* tissue adjacentto the catheter.[79,174] Prior literature has demonstrated important antimicrobial activity of QA salts on biofilm formation.[174–178] Several studies have described various processes of catheter surface modification with QA. One novel method utilizes a vapor phase plasma-induced graft polymerization of acrylic acid onto polyurethane catheters to which create a negative charge at neutral pH, readily adsorbing positively charged 3-(trimethyoxysilyl)-propyldimethyloctadecylammonium chloride (QAS).[179] Other processes allow for more rapid surface modification with QA, such as using thiol–ene "click" chemistry to attach a series of QA thiol compounds possessing hydrocarbon tails (8–14 carbons) to an allyl ether side chain.[180] QA-based coatings were found to be promising for medical devices such as catheters and prostheses.[173,181] Coatings of dimethylaminoethyl methacrylate and PEI covalently bound to polymethyl methacrylate or polydimethylsiloxane followed by quaternization were reported by De Prijck *et al.*[181] These coatings were examined for prevention of *Candida albicans* biofilm formation—which is a major challenge for the safety of prosthesis deterioration in laryngectomized patients—and were found to exhibit up to 92% reduction in bacterial growth.

14.3.2.2 Polymeric Coating Modifications—Natural Cationic Polymers

Several natural polymers were found to exhibit antimicrobial properties following chemical quaternization or *via* QAC modifications or simply due to the existence/formation of QA moieties in the physiological environment. For instance, chitosan is a biocompatible polysaccharide with natural antimicrobial properties and tissue-adhesive capabilities, all of which have contributed to their wide usage in biomedical applications.[182–184] Antimicrobial chitosan and modified derivatives with QA moieties can be prepared either by reductive amination followed by alkyl halide quaternization[6,162,184] or simply by direct alkylation reaction *via* alkyl halide. Over the past few decades, various QA-chitosan preparations have been reported and tested for antimicrobial properties. Tan *et al.*[182] have reviewed the activity and mechanism of action of recently synthesized antimicrobial QA-chitosans for biomedical applications, particularly in orthopaedic implants. Reaction of chitosan with QACs was also found to be an effective approach for conferring a permanent QA moiety. One example was reported by Rimondino *et al.*, who synthesized chitosan with improved functionality and biocompatibility for biomedical applications *via* dendronization modification followed by quaternization with glycidyltrimethylammonium chloride.[185]

PLL is a linear polypeptide in which each repeating unit has a free primary amine side arm. Primary amines have been widely utilized for coating materials for various applications, where in the physiological environment these group quaternize, forming QA-PLL. QA-PLL was found to exhibit strong antimicrobial activity against both Gram-negative and -positive bacteria species, yeast and fungi.[160] Furthermore, these activities were found to correlate with molecular weight of PLL: at least nine L-lysine units are required for various microbial growth inhibition.[160] Moreover, QA-PLL activity was found to vary against different strains, as reported by Hiraki *et al.*, who studied QA-PLL required inhibitory concentrations for *Bacillus* strains.[161] It was found that for *Bacillus stearothermophilus* the least required concentration was 2.5 $\mu g\ ml^{-1}$, while five times more concentrated levels were required for *Bacillus subtilis* and *Bacillus coagulans.*[161]

Modification of natural polymers and materials with QACs was found to expand their utility and to produce natural-based antimicrobial polymers for biomedical applications. One example reported by Neoh *et al.*[186] is the functionalization of chitosan and agarose used for fabrication of stable antimicrobial coatings to address biomedical applications. In another study, the modification of cellulose with various types of biologically active QA compounds of different molecular weight, including QA monomers and polymer-based QA-alkyldimethylbenzylammonium chloride were reported by Kotelnikova *et al.*,[187] thus expanding the biomedical application of cellulose. Chen *et al.*,[188] reported the synthesis of novel active antimicrobial materials *via* conjugation of natural rosin with poly(*N*,*N*-dimethylaminoethyl methacrylate) forming PDMAEMA-g-rosin copolymers. These copolymers were found to exhibit strong antimicrobial activity against *Staphylococcus aureus* (Gram-positive) and *Escherichia coli* (Gram-negative) through the combination of ionic and hydrophobic interactions between the copolymers and bacteria cells.

Overall, quaternization modifications of both synthetic and natural polymers resulting in antimicrobial QA-polymers (cationic) was found to impact their water solubility. Enhanced and increasing solubility properties under some circumstances can increase their potential usage in the antimicrobial field as well as other biomedical applications.

14.3.3 Bacteria-repelling and Anti-adhesive Surfaces

While both strategies of antimicrobial-releasing and contact-killing coatings were found to be effective tools for producing catheters with active antimicrobial surfaces, several issues surrounding the emergence of resistant microorganisms and toxicity to host cells still restrict their widespread use in the clinic. These facts and the recent progress in surface modification technologies urges the further expansion of bacterial repelling and anti-adhesive surfaces strategies by developing new materials and coating techniques. As a consequence, anti-fouling catheters have been developed, which do not kill microbes directly, but instead prevent bacterial attachment to catheter surfaces through mechanisms such as steric and electrostatic repulsion. Here we describe the leading materials under this category.

14.3.3.1 Hydrogel-coated Catheters

Hydrogels consist of crosslinked, insoluble, hydrophilic polymers that trap water. Hydrogel-coated catheters allow for improved patient comfort, prevention of microbial adherence and reduced encrustation.[189] However, there is some evidence to suggest that hydrogel catheters may obstruct more rapidly than uncoated silicone catheters. An *in vitro* study of the migratory capacity of *Proteus mirabilis* showed that catheters coated with hydrogels such as polyhexamethylenebiguanide and polyvinylpyrrolidine-iodine complex became occluded after 2 days, limiting bacteriostatic activity early after catheterization. However, with the addition of triclosan, hydrogel-coated catheters resisted encrustation for more than 7 days. The conclusion was that hydrogel coatings should be applied in conjunction with antimicrobial agents to counter the negative effects of hydrogel.[190] There are currently insufficient data to suggest the routine use of hydrogel-coated catheters.

14.3.3.2 Polyethylene glycol-based coatings

Surface modification with hydrophilic polymers, including polyethylene glycol (PEG), have demonstrated promising *in vitro* anti-fouling ability.[191] The tightly water-bound hydrophilic layer (*i.e.* hydration shell) features water molecules in a directional configuration, as directed by PEG hydrogen bonds on the catheter surface. This hydration shell, along with the dynamic movement of the polymer chains projecting from the surface, hinder attachment of bacterial cells acting as both physical and steric barriers.[192] For urinary catheters, surface coatings *via* grafting of PEG-based polymers were found to be an efficient tool for producing catheter coatings with good anti-fouling activity.[20,193]

14.3.3.3 Polyzwitterion-coated Catheters

Polymers with zwitterionic head groups can also be applied as surface coatings to inhibit fouling of the catheter surface.[97] With a net neutral charge, zwitterions carrying equal numbers (one or more) of both positive and negative moieties were found to exhibit antifouling properties *via* both steric repulsion/steric hindrance and electrostatic interactions.[20] Zwitterionic polymer coatings were found to be superior to PEG-based coatings given that electrostatic interactions were found to form a tight/high-density hydration shell around the polymer, thus preventing proteins, as well as microbes, from accumulating and adhering to the surface of medical device (*i.e.* catheters).[20,194,195] Zwitterion-coated surfaces are both biocompatible and non-thrombogenic, since the polymer head group mimics that found in the lipid bilayers of biological membranes.[20] Usually, zwitterionic polymers are represented as zwitterionic moieties conjugated to polymethacrylate or polyacrylamide backbones; however, the fact that zwitterionic coatings are easily functionalizable have expanded their potential applications to include a variety of medical device coatings (Figure 14.9).[196,197] The leading zwitterionic coatings studied thus far are poly(sulfobetaine methacrylate) (PSBMA) and Poly(methacryloyloxylethylphosphorylcholine) (PMPC).

PSBMA is one of the earliest reported zwitterionic anti-fouling polymers examined as a coating for various applications and implantable medical devices. In one study it was found that silicone-based catheters, following

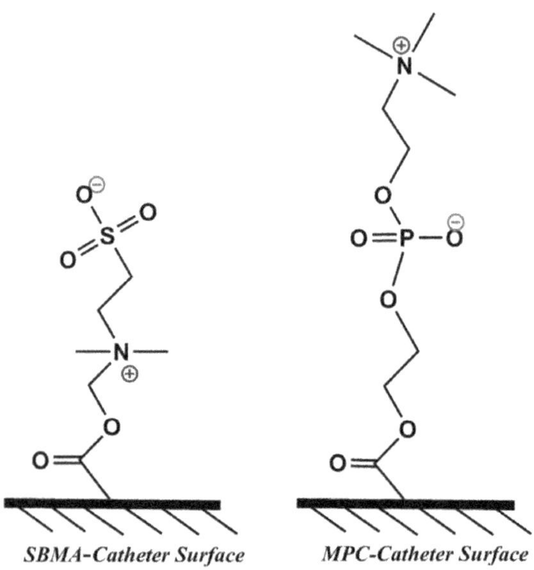

Figure 14.9 Chemical structures of the commonly studied zwitterionic compounds for catheters, either *via* conjugation to the surface or as part of a polymeric coating. SBMA: sulfobetaine methacrylate; MPC: methacryloyloxylethylphosphorylcholine.

functionalization of polydimethylsiloxane coatings with PSBMA, exhibited strong anti-biofilm properties against *Staphylococcus aureus* and *Pseudomonas aeruginosa*.[198] The functionalization step was through plasma activation followed by radical polymerization of zwitterionic SBMA monomers on the surface. As result, biofilm formation decreased by 80% for coated catheters compared to uncoated catheters.

PMPC has been tested by different groups as an anti-fouling coating. In one study PMPC coatings were tested for urinary catheters. PMPC-coated catheters in comparison to uncoated catheters exhibited anti-adhesion properties for both *Escherichia coli* and fibrinogen.[199] In addition, in this study various catheters made of polyurethanes, latex, polyvinyl chloride and silicone were PMPC coated and tested in presence an artificial medium of urine with *Proteus mirabilis* and found to decrease encrustation. PMPC coatings were clinically tested as coatings for ureteral stents; following the PMPC coatings the biofilm coverage decreased from 70% to 30% in 3 months.[200] These reports clearly indicate the potency of PMPC coatings with the potential for further expanding their utility evaluation to other medical devices in the field of urology, among others.

14.4 Clinically Tested Antimicrobial Catheters

A wide range of catheters have been reported worldwide for various applications, as have their clinical trials, testing various coating strategies and mechanisms of action, surface functionalization and antimicrobial polymer grafting. Here, we report clinical trials of antimicrobial catheters for the two leading categories of catheter-induced infections: CRBIs and CAUTIs. Selected commercial vascular, urinary and ventricular catheters with antimicrobial claims are listed in Table 14.2.[9]

Overall, we found that the antimicrobial activity of most of these catheters is based on either releasing antibiotic/anticancer agents or impregnation with antiseptic compounds such as silver. While most of these catheters have demonstrated clinical antimicrobial outcomes including reduced CRBIs and catheter colonization, some were not effective across all blood infections.[201]

14.5 Challenges and Future Approaches

14.5.1 Antimicrobial Resistance

Worldwide, ∼700 000 people die per year due to antibiotic resistance, while experts are warning that these numbers are rapidly increasing, thus estimating an increase in death cases to 10 million by 2050, given the lack of cutting-edge antimicrobial technologies and new antibiotics development.[218] Despite the urgent need for new antibiotics, only 40 new antibiotics were in clinical development in US by 2016.[218] Accordingly, microbial resistance to antimicrobial treatment is a source of concern for therapeutic

Table 14.2 Randomized clinical trials of antimicrobial catheters.[a] Adapted by permission from Springer Nature: Springer eBook, *Antimicrobial Modifications on Critical Care Implants* by Z. Zhang, V. E. Wagner, J. C. Victor, Copyright 2017.[9]

Antimicrobial modifications and agents	Catheter type(s), modification description	Reference
Modification with antibiotics		
Minocycline–rifampin (Spectrum® Cook)	Applied to: PICCs** (Silicone, Cook Spectrum Glide®) CVCs* (Silicones, Polyurethanes) Modification was applied to enhance the insertion step, MR was impregnated with hydrophilic coating consisting of polyvinulpyrrolidone and polyacrylamide (EZ-Pass® Cook)	93, 94, 202–205
Miconazole–rifampin (Vygon, Multistar+)	Applied to: Acute CVCs (Vygon, Multistar+)	206
Cefazolin	Applied to: CVCs and arterial catheters Modification was applied to bond cefazolin to the surface in the following order: (1) Cationic tridodecylmethylammonium chloride (2) Anionic antibiotic cefazolin	207
Modification with anticancer drugs		
5-Fluorouracil (Angiotech)	Applied to: 5-FU CVCs (Angiotech) Modification was applied by coating CVC surface with 5-FU	208
Modification with antiseptics		
Chlorhexidine–silver sulfadiazine (ARROW +gard Blue®, Teleflex)	Applied to: Acute hemodialysis catheters Pressure injectable CVCs Modification was applied to swelling surfaces	209, 210
Chlorhexidine–silver sulfadiazine (ARROW +gard Blue PLUS®, Teleflex)	Applied to: Pressure injectable CVCs Modifications were applied inside and outside the catheters, as well as the extension lines and hubs	211–213
Silver sulfadiazine coating (BioBloc®, CR Bard)	Applied to: Tunneled cuffed catheters Tunneled long-term hemodialysis catheters (HemoGlide®, HemoSplit and HemoStar)	214

Table 14.2 (*Continued*)

Antimicrobial modifications and agents	Catheter type(s), modification description	Reference
Silver ion sleeve with/or without heparin coating (Covidien-Medtronic)	Applied to: Chronic hemodialysis catheters (Palindrome™ HIS, Palindrome™ SI)	214
	Modification was applied to the catheter with a silver-impregnated sleeve covering the device's outer surface cuff-to-hub	
Inorganic silver powder and inert ceramic zeolite (AgION®, Sciessent)	Applied to: Temporary dialysis catheters (Vygon) CVCs	215
Silver/slatinum/carbon black (Oligon, Edwards Lifesciences)	Applied to: Silver CVCs (OligonVantex®)	216
	Modification was applied to catheter made of carbon black plus metal alloys (platinum and silver) with a polyurethane and heparin coating	
Benzalkonium chloride	Applied to: BZK-impregnated CVCs (Multi-Med®, Baxter-Edwards)	217
	Modification was applied to by impregnation of BZK with hydrophilic coating or without	
Silver-alloy and hydrogel-coated	Applied to: Urinary catheter	52

[a]BZK: benzalkonium chloride; CVCs: central venous catheters; 5-FU: 5-fluorouracil; PICCs: peripherally inserted central catheters; MR: minocycline–rifampin.

approaches including catheter usage and is considered a challenging issue into which more efforts should be invested. Following catheterization, catheters with antibiotic release as mechanism of action are among the most reported for antimicrobial resistance. Given the importance of this issue, and to increase the attention given to it, recently and for the first time, the World Health Organization (WHO) released a list ranking the 12 drug-resistant bacteria/bacterial species found to pose the greatest threat to human health, thus highlighting that new antibiotics and antimicrobial treatments are desperately needed.[218] The list ranks the bacteria starting with the greatest threat to human health in the following order: *Acinetobacter baumannii*, *Pseudomonas aeruginosa*, and Enterobacteriaceae—carbapenem-resistant, *Enterococcus faecium*—vancomycin-resistant, *Staphylococcus aureus*—resistant to both methicillin and vancomycin, *Helicobacter pylori*—clarithromycin-resistant, *Campylobacter* and *Salmonella* species—fluoroquinolone-resistant, *Neisseria gonorrhoeae*—resistant to both fluoroquinolone and cephalosporin,

Streptococcus pneumoniae—penicillin-non-susceptible, *Haemophilus influenzae*—ampicillin-resistant and *Shigella* species—fluoroquinolone-resistant.

Use of implantable medical devices such as catheters is associated with foreign body related infections starting with bacterial adhesion on their surfaces and later biofilm formation.[219] Bacterial biofilms are the major cause of antibiotic resistance in which bacteria are more protected, compared to planktonic forms, and thus are less receptive to the antimicrobial treatments and host immune response, which could affect the clinical use of catheters. Interestingly, a recent study led by a group from Princeton University has revealed that under some circumstances inhibited bacteria can still assist the survival and growth of nearby bacteria by absorption of large amounts of antibiotic.[220] Also, with repeated antimicrobial courses (*e.g.* antibiotic-containing catheters), it was found that bacteria are more likely to show increased resistance.[30] In such situations, the catheter is usually recommended to be removed in order to resolve the infection.[219] However, in some cases such as catheter-related *Staphylococcus aureus* bacteremia, it was found that the bacteremia can continue, even after catheter removal plus initiation of suitable antimicrobial therapy.[221]

Several studies have examined catheter-related antibiotic resistance by examining the prevalence of bacterial biofilms as well as antibiotic-resistant strains on catheters retrieved from patients. In a comprehensive study, analysis was performed on 86 and 141 retrieved Foley and vascular catheters, respectively.[219] It was found that different microbial species at varied presence percentages were involved in each case of biofilm formation and were tested for antibiotic resistance. For Foley catheters, the isolates were found to exhibit multiple drug resistance (*e.g.* quinolones, gentamycin, penicillin); *Enterococcus faecalis* was found to be the major isolate, followed by *Escherichia coli*, and other various species of *Enterococcus*, *Staphylococcus*, *Enterobacter*, *Klebsiella*, *Citrobacter*, *Pseudomonas aeruginosa*, and fungus—yeast. In comparison, many of the vascular catheter-related isolates were found to exhibit methicillin resistance, while a small portion exhibited vancomycin resistance.The major isolates were *Staphylococcus* species followed by *Escherichia coli* and species of *Enterococcus*, *Streptococcus*, *Klebsiella*, *Micrococcus*, *Pseudomonas* and fungus—yeast. Accordingly, the authors have crystallized the recommendation to remove catheters as early as possible to prevent the development of bacterial biofilm on them.[219] In another study focused on CAUTIs, overall 166 isolates of pathogenic bacteria causing infections were identified.[222] Out of these, *Escherichia coli*, *Pseudomonas aeruginosa* and *Staphylococcus aureus* were found to share the major portion of pathogens involved, with contribution of 39.8%, 13.9% and 13.3%, respectively. Also, the vast majority of the identified pathogenic bacteria were found to develop drug resistance to commonly used antibiotics in response, and at varied levels.[222] Another group, focused on urethral CAUTIs, have identified 159 strains of pathogenic bacteria, with similar percentages of the aforementioned leading pathogens, as well as antibiotic resistance.[223] Both studies recommended that in order to raise the clinical recovery rate as well as enhance the control on the increasing tendency of bacterial resistance, clinicians should use antibiotics wisely, taking into account drug susceptibility testing results.[222,223]

Pathogen species involved, as well as antimicrobial resistance were subjects for a study for other types of catheters, such as closed suction catheters in the case of ventilator-associated pneumonia (VAP).[29] In a study involving 30 patients, it was found that Gram-negative bacteria were the dominant portion, with 78.2% of VAP pathogenic bacteria led by *Acinetobacter baumanii* and *Pseudomonas aeruginosa* with 28.2% and 19.6%, respectively, while *Escherichia coli* and *Klebsiella pneumoniae* each formed 10.9% of pathogenic bacteria. Higher drug resistance was found for most chain cocci against the tested antibiotics (aminoglycosides, cephalosporins and penicillins).[29]

Given the above facts about the challenging bacterial resistance and its complexity, it is currently well understood that for effective treatment against catheter-related infections, a better design of effective antimicrobial coatings is needed. One way is by designing multimechanism-targeting antimicrobial coatings and thus reducing pathogen's ability to resist the antimicrobial agent. Another effective strategy is by applying synergistic approaches into one antimicrobial coating, for example, urinary catheters with anti-fouling surface coating plus one or two releasable antimicrobial agents.[20,224,225] While the anti-fouling surface coating would aim to prevent bacteria attachment and accumulation on the catheter surface, the other released antimicrobial agents would target and kill pathogens in the nearby microenvironment (see Section 14.5.2).

14.5.2 Multi-approach Antimicrobial Catheters

Monocomponent antibacterial agents for catheter-induced infections as well as infections induced by other medical devices are still far from meeting all the requirements, and yet implant-induced infections remain very challenging and represent health threats. As such, coupling multiple antimicrobial approaches to target pathogens represents a promising area of research, thus avoiding the disadvantages of each individual approach. Li *et al.*[226] were among the first groups of researchers to design an antibacterial coating with both contact-killing and release-killing capabilities. This coating consisted of two distinct layered functional regions. The bottom layer (polyelectrolyte multilayer—PEM reservoir) consisted of a reservoir region for the loading and release of two bactericidal chemicals—poly(allylamine hydrochloride) (PAH) and poly(acrylic acid) (PAA)—and AgNPs. The top layer (cap region) was made of bilayers of PAH and SiO_2NPs, and finally, the SiO_2NP cap was modified with a QA silane, [3-(trimethoxysilyl)propyl]octadecyldimethylammonium chloride (OQAS) (Figure 14.10).[226]

The design of this coating has since been modified in various ways.[227,228] One multi-approach antimicrobial strategy specifically for catheter surfaces was proposed by Dayyoub *et al.*[90] for urinary catheters. This design

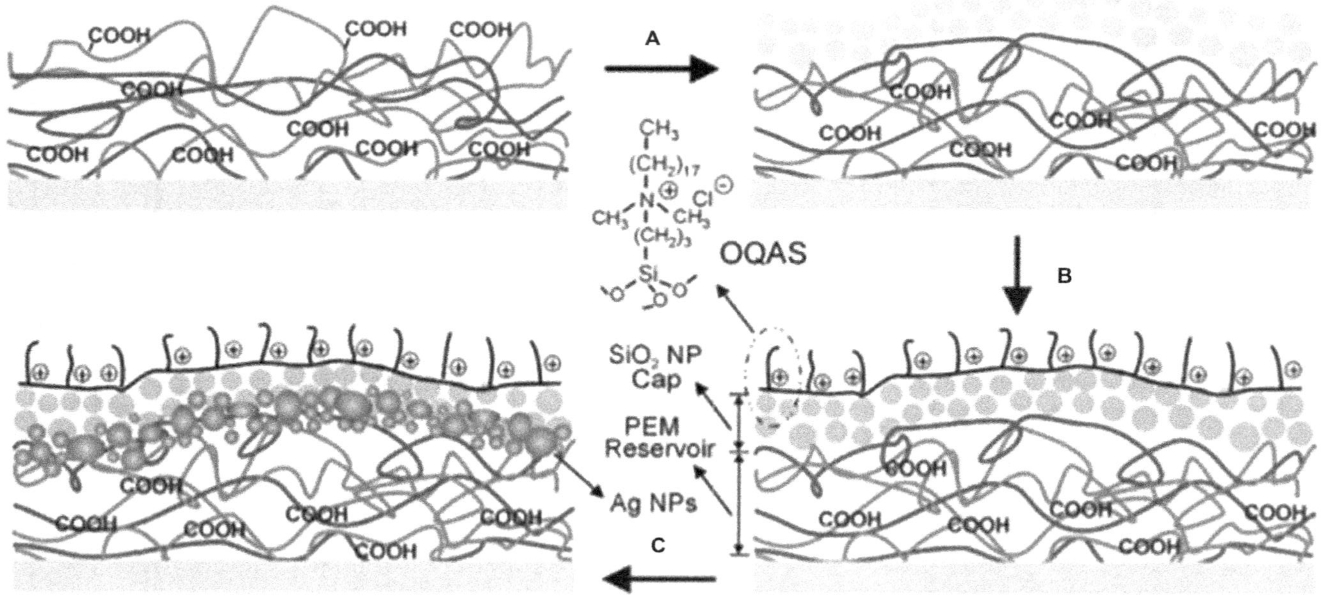

Figure 14.10 Scheme showing the preparation of two-level dual-functional antibacterial coating based on silver nanoparticles (AgNPs), quaternary ammonium compounds (QACs) and layer-by-layer (LbL) deposition of poly(allylamine hydrochloride) (PAH) and poly(acrylic acid) (PAA). (A) A cap region made of bilayers of PAH and SiO_2NPs is added to the top. (B) The SiO_2NP cap is modified with a QAsilane, OQAS. (C) Ag^+ loaded inside the coating using the available unreacted carboxylic acid groups in the LbL multilayers and AgNPs are created *in situ* using the nanoreactor.
Reprinted with permission from Zhi Li, Daeyeon Lee, Xiaoxia Sheng, Robert E. Cohen, and Michael F. Rubner, Two-Level Antibacterial Coating with Both Release-Killing and Contact-Killing Capabilities, *Langmuir*, 2006, **22**, 9820–9823, Copyright 2006 American Chemical Society.[226]

combined the antibacterial effects of norfloxacin and lipid-coated AgNPs with the anti-adhesive property of PLGA. Other multi-approaches have combined release of both AgNPs and antibiotics. One study showed significantly decreased microbial attachment from the surface of urinary catheters using combined AgNP and antibiotic release compared to catheters with antibiotic impregnation only.[229] However, the use of AgNPs and chlorhexidine on central venous catheters was not associated with significant antimicrobial activity, due to the loss of AgNPs within 48 hours of implantation.[230]

14.6 Summary, Concluding Remarks and Future Perspectives

In this chapter we have discussed the different approaches and strategies for addressing catheter-induced microbial infections. By providing an overview of the recent literature on antimicrobial materials and coating development for various catheter applications, a clear conclusion can be made that significant and promising progress has been achieved. However, given the complexity of microbial contaminations, infectious complications remain a significant cause of morbidity and mortality in the patient population.

Taking into account the pros and cons of the developed strategies, it might appear impossible to create catheter surfaces or coatings with no bacterial colonization or host immune foreign body response (FBR), as *in vivo* catheters are rapidly covered by connective tissue proteins and plasma. Alternatively, we see both stimuli-responsive and multi-approach antimicrobial catheter as game-changing strategies in fighting catheter-induced microbial infection, giving their "smartness" producing "on-demand" active agents as well as their multitargeting approach.

Furthermore, in turn, investment of more attention and effort into addressing the following significant challenges, as well as expanding prior studies, should contribute to better design and eventually enhance the performance and efficiency of antimicrobial coatings for catheters:

- Coating stability during catheter manufacture and the catheterization process.
- Live clinical feedback *vs* kinetics and timeframe of antimicrobial agent release, thus encouraging application–patient dependency.
- Maintaining the release levels of the antibacterial agents within the therapeutic window and with limited cytotoxicity toward eukaryotic host cells.
- Stimuli-triggered releasing coatings: minimize non-triggered background leaching or undesired release, and achievement of meaningful release doses over multiple cycles.[42]
- Destabilization and loss of efficiency of zwitterionic anti-fouling coating hydration layers over time. We suggest focusing on designing superhydrophilic zwitterionic coatings.

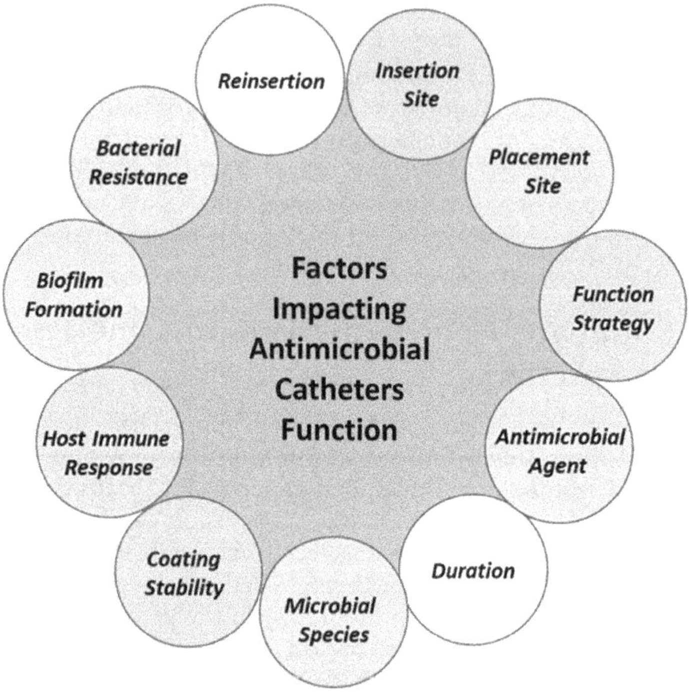

Figure 14.11 Catheter- and non-catheter-related factors that impact antimicrobial catheter function and efficiency. Deep understanding of the role of these factors will have a direct impact on designing efficient antimicrobial catheters and clinical performance.

- Design of bacteria- and bacteria species-specific antimicrobial coatings.
- Simultaneous multi-approach microbial targeting *vs* developed microbial resistance.

Lastly, we realize that there are also non-catheter related factors that could significantly contribute to both rates of infection and the development of microbial resistance, such as (among others): site of insertion, treatment duration and infection control standards (Figure 14.11). Better understanding their role with the help of clinicians worldwide sharing their clinical observations regarding these aspects could be a useful approach in fighting both microbial resistance and catheter-related infections.

Review Criteria

The PubMed database was searched using the terms "urinary catheter", "vascular catheter", "shunt catheter", "biomaterial", "nanoparticles", "antimicrobial" and "infection" to identify papers published at least in the past two or three decades. In addition, the major practice guidelines were identified and specific seminal reference articles published earlier than the review period were also included in the manuscript. We reviewed clinical trials, editorials, letters, meta-analyses, practice guidelines, randomized controlled trials and review and original articles in the English language published in the past 20–30 years and selected the relevant articles.

Acknowledgements

Shady Farah would like to acknowledge the Technion for the Technion's presidential grant supporting this work.

References

1. Catheter (n.d.), *Miller-Keane Encyclopedia and Dictionary of Medicine, Nursing, and Allied Health*, 7th edn., 2003, Retrieved February 25 2019 from https://medical-dictionary.thefreedictionary.com/Catheter.
2. G. J. Verkerke, K. H. Schraffordt, R. P. Veth, J. R. Van Horn, L. Postma and H. J. Grootenboer, First clinical experience with a noninvasively extendable endoprosthesis: a limb-saving procedure in children suffering from a malignant bone tumor, *Artif. Organs*, 1997, **21**, 413–417.
3. C. Von Eiff, W. Kohnen, K. Becker and B. Jansen, Modern strategies in the prevention of implant-associated infections, *Int. J. Artif. Organs*, 2005, **28**(11), 1146–1156.
4. A. E. Madkour, J. M. Dabkowski, K. Nüsslein and G. N. Tew, Fast Disinfecting Antimicrobial Surfaces, *Langmuir*, 2009, **25**, 1060–1067.
5. J. F. Williams and S. D. Worley, Infection-resistant nonleachable materials for urologic devices, *J. Endourol.*, 2000, **14**, 395–400.
6. S. Farah, *Bioactive Antimicrobial and Antiproliferative Surfaces*, Hebrew University of Jerusalem, 2014.
7. H. Saini, S. Chhibber and K. Harjai, Antimicrobial and antifouling efficacy of urinary catheters impregnated with a combination of macrolide and fluoroquinolone antibiotics against *Pseudomonas aeruginosa*, *Biofouling*, 2016, **32**(5), 511–522.
8. R. Gahlot, *et al.*, Catheter-related bloodstream infections, *Int. J. Crit. Illn. Inj. Sci.*, 2014, **4**(2), 162–167.
9. Z. Zhang, *Antimicrobial Coatings and Modifications on Medical Devices*, Springer Science + Business Media, New York, NY, 2017, pages cm p.
10. P. Ramritu, K. Halton, D. Cook, M. Whitby and N. Graves, Catheter-related bloodstream infections in intensive care units: a systematic review with meta-analysis, *J. Adv. Nurs.*, 2008, **62**(1), 3–21.
11. J. Pant, M. J. Goudie, S. M. Chaji, B. W. Johnson and H. Handa, Nitric oxide releasing vascular catheters for eradicating bacterial infection, *J. Biomed. Mater. Res., Part B: Appl. Biomater.*, 2018, **106**(8), 2849–2857.
12. L. A. Mermel, B. M. Farr, R. J. Sherertz, I. I. Raad, N. O'Grady, J. S. Harris and D. E. Craven, Guidelines for the management of

intravascular catheter-related infections 2018 U.S. National Library of Medicine, *J. Intravenous Nurs.*, 2001, **24**(3), 180–205.
13. P. Ramritu, K. Halton, P. Collignon, D. Cook, D. Fraenkel, D. Battistutta, M. Whitby and N. Graves, A systematic review comparing the relative effectiveness of antimicrobial-coated catheters in intensive care units, *Am. J. Infect. Control*, 2008, **36**(2), 104–117.
14. D. Marciante Kristin, L. Veenstra David, A. Lipsky Benjamin and S. Sanjay, Which antimicrobial impregnated central venous catheter should we use? Modeling the costs and outcomes of antimicrobial catheter use, *Am. J. Infect. Control*, 2003, **31**(1), 1–8.
15. A. Novikov, M. Y. Lam, L. A. Mermel, A. L. Casey and T. S. Elliott, Nightingale Peter Impact of catheter antimicrobial coating on species-specific risk of catheter colonization: a meta-analysis, *Antimicrob. Resist. Infect. Control*, 2012, **1**(1), 40.
16. G. Donelli, Vascular catheter-related infection and sepsis, *Surg. Infect.*, 2006, 7(Suppl 2S), 25–27.
17. H. Shah, W. Bosch, K. M. Thompson and W. C. Hellinger, Intravascular catheter-related bloodstream infection, *Neurohospitalist.*, 2013, **3**(3), 144–151.
18. A. Marvaso, Central venous catheter – related infections Le infezioni in medicina: rivistaperiodica di eziologia, epidemiologia, diagnostica, *clinica e terapia delle patologie infettive*, 2000, **8**(4), 202–210.
19. C.-X. Lu, T. Cui, S.-J. Yu and C.-P. Chai, Analysis of pathogen test results for infections of central venous catheter for blood purification, *Zhonghua Yiyuanganranxue Zazhi*, 2015, **25**(7), 1587–1589.
20. P. Singha, J. Locklin and H. Handa, A review of the recent advances in antimicrobial coatings for urinary catheters, *Acta Biomater.*, 2017, **50**, 20–40.
21. A. Bowen, *Urinary Tract Infections: A Primer for Clinicians.* <http://www.medscape.org/viewarticle/556040>, 2007.
22. F. M. E. Wagenlehner and K. G. Naber, Treatment of bacterial urinary tract infections: presence and future, *Eur. Urol.*, 2006, **49**(2), 235–244.
23. S. Saint, C. P. Kowalski, S. R. Kaufman, T. P. Hofer, C. A. Kauffman, R. N. Olmsted, J. Forman, J. Banaszak-Holl, L. Damschroder and S. L. Krein, Preventing hospital-acquired urinary tract infection in the United States: a national study, *Clin. Infect. Dis.*, 2008, **46**(2), 243–250.
24. J. Lo, D. Lange and B. Chew, Ureteral stents and Foley catheters-associated urinary tract infections: the role of coatings and materials in infection prevention, *Antibiotics*, 2014, **3**(1), 87–97.
25. H. Saini, C. Sanjay and H. Kusum, Antimicrobial and antifouling efficacy of urinary catheters impregnated with a combination of macrolide and fluoroquinolone antibiotics against *Pseudomonas aeruginosa*, *Biofouling*, 2016, **32**(5), 511–522.
26. C. E. Chenoweth and S. Sanjay, Urinary tract infections, *Infect. Dis. Clin. North Am.*, 2011, **25**(1), 103–115.

27. P. J. Nowatzki, R. R. Koepsel, P. Stoodley, K. Min, A. Harper, H. Murata, J. Donfack, E. R. Hortelano, G. D. Ehrlich and A. J. Russell, Salicylic acid-releasing polyurethane acrylate polymers as antibiofilm urological catheter coatings, *Acta Biomater.*, 2012, **8**(5), 1869–1880.
28. A. J. Bleyer and M. Murea, Antimicrobial catheter locks: searching for the ideal solution, *J. Am. Soc. Nephrol.*, 2011, **22**(10), 1781–1782.
29. X. Ma, G. Hou and Y. Zhang, Clinical observation of using closed suction catheter for pathogen in ventilator-associated pneumonia and antimicrobial resistance, *Zhonghua Yiyuanganranxue Zazhi*, 2011, **21**(7), 1447–1448.
30. E. Nicolle Lindsay, Catheter-related urinary tract infection, *Drugs & Aging*, 2005, **22**(8), 627–639.
31. S. Jorge, T. e Costa, B. Alexandre, R. Aura and P. Pedro, Catheter-related bacteremia in hemodialysis: which preventive measures to take?, *Nephron. Clini. pract.*, 2008, **110**(4), c251–c257.
32. C. Manierski and A. Besarab, Antimicrobial locks: putting the lock on catheter infections, *Adv. Chronic Kidney Dis.*, 2006, **13**(3), 245–258.
33. A. D. Szvalb, H. El Haddad, K. V. Rolston, Y. Jiang, I. I. Raad, G. M. Viola and S. H. Sabir, Risk factors for recurrent percutaneous nephrostomy catheter-related infections, *Infection*, 2019, **47**(2), 239–245.
34. J. D. Bryers, Medical biofilms, *Biotechnol. Bioeng.*, 2008, **100**(1), 1–18.
35. R. M. Donlan, Biofilms: microbial life on surfaces, *Emerg. Infect. Dis.*, 2002, **8**(9), 881–890.
36. G. L. Cooper, A. L. Schiller and C. C. Hopkins, Possible role of capillary action in pathogenesis of experimental catheter-associated dermal tunnel infections, *J. Clin. Microbiol.*, 1988, **26**(1), 8–12.
37. J. W. Costerton, P. S. Stewart and E. P. Greenberg, Bacterial biofilms: a common cause of persistent infections, *Science*, 1999, **284**(5418), 1318–1322.
38. K. Glinel, P. Thebault, V. Humblot, C. M. Pradier and T. Jouenne, Antibacterial surfaces developed from bio-inspired approaches, *Acta Biomater.*, 2012, **8**(5), 1670–1684.
39. J. L. Fox, Antimicrobial peptides stage a comeback, *Nat. Biotechnol.*, 2013, **31**(5), 379–382.
40. R. M. Donlan and J. W. Costerton, Biofilms: survival mechanisms of clinically relevant microorganisms, *Clin. Microbiol. Rev.*, 2002, **15**(2), 167–193.
41. K. Page, M. Wilson and I. P. Parkin, Antimicrobial surfaces and their potential in reducing the role of the inanimate environment in the incidence of hospital-acquired infections, *J. Mater. Chem.*, 2009, **19**, 3819–3831.
42. M. Cloutier, D. Mantovani and F. Rosei, Antibacterial Coatings: Challenges, Perspectives, and Opportunities, *Trends Biotechnol.*, 2015, **33**(11), 637–652.
43. F. R. Dametto, C. C. Ferraz, B. P. Gomes, A. A. Zaia, F. B. Teixeira and F. J. de Souza-Filho, *In vitro* assessment of the immediate and

prolonged antimicrobial action of chlorhexidine gel as an endodontic irrigant against *Enterococcus faecalis*, *Oral. Surg., Oral. Med., Oral. Pathol., Oral. Radiol., Endod.*, 2005, **99**(6), 768–772.
44. L. Zhao, P. K. Chu, Y. Zhang and Z. Wu, Antibacterial coatings on titanium implants, *J. Biomed. Mater. Res. B: Appl. Biomater.*, 2009, **91**(1), 470–480.
45. H. Phuengkham and N. Nasongkla, Development of antibacterial coating on silicone surface via chlorhexidine-loaded nanospheres, *J. Mater. Sci.-Mater. Med.*, 2015, **26**(2), 78.
46. N. J. Wood, S. E. Maddocks, H. J. Grady, A. M. Collins and M. E. Barbour, Functionalization of ethylene vinyl acetate with antimicrobial chlorhexidine hexametaphosphate nanoparticles, *Int. J. Nanomed.*, 2014, **9**, 4145–4152.
47. R. Kumar and H. Munstedt, Silver ion release from antimicrobial polyamide/silver composites, *Biomaterials*, 2005, **26**(14), 2081–2088.
48. J. S. Kim, E. Kuk, K. N. Yu, J. H. Kim, S. J. Park, H. J. Lee, S. H. Kim, Y. K. Park, Y. H. Park, C. Y. Hwang, Y. K. Kim, Y. S. Lee, D. H. Jeong and M. H. Cho, Antimicrobial effects of silver nanoparticles, *Nanomedicine*, 2007, **3**(1), 95–101.
49. A. B. Lansdown, Silver. I: Its antibacterial properties and mechanism of action, *J. Wound Care*, 2002, **11**(4), 125–130.
50. Q. L. Feng, J. Wu, G. Q. Chen, F. Z. Cui, T. N. Kim and J. O. Kim, A mechanistic study of the antibacterial effect of silver ions on *Escherichia coli* and *Staphylococcus aureus*, *J. Biomed. Mater. Res.*, 2000, **52**(4), 662–668.
51. J. A. Lemire, J. J. Harrison and R. J. Turner, Antimicrobial activity of metals: mechanisms, molecular targets and applications, *Nat. Rev. Microbiol.*, 2013, **11**(6), 371–384.
52. J. W. Lederer, W. R. Jarvis, L. Thomas and J. Ritter, Multicenter cohort study to assess the impact of a silver-alloy and hydrogel-coated urinary catheter on symptomatic catheter-associated urinary tract infections, *J. Wound Ostomy Continence Nurs.*, 2014, **41**(5), 473–480.
53. M. E. Rupp, T. Fitzgerald, N. Marion, V. Helget, S. Puumala, J. R. Anderson and P. D. Fey, Effect of silver-coated urinary catheters: efficacy, cost-effectiveness, and antimicrobial resistance, *Am. J. Infect. Control*, 2004, **32**(8), 445–450.
54. A. M. Leuck, J. R. Johnson, M. A. Hunt, K. Dhody, K. Kazempour, P. Ferrieri and S. Kline, Safety and efficacy of a novel silver-impregnated urinary catheter system for preventing catheter-associated bacteriuria: a pilot randomized clinical trial, *Am. J. Infect. Control*, 2015, **43**(3), 260–265.
55. K. Stenzelius, S. Persson, U. B. Olsson and M. Stjarneblad, Noble metal alloy-coated latex versus silicone Foley catheter in short-term catheterization: a randomized controlled study, *Scand. J. Urol. Nephrol.*, 2011, **45**(4), 258–264.
56. R. Pickard, T. Lam, G. MacLennan, K. Starr, M. Kilonzo, G. McPherson, K. Gillies, A. McDonald, K. Walton, B. Buckley, C. Glazener, C. Boachie,

J. Burr, J. Norrie, L. Vale, A. Grant and J. N'Dow, Antimicrobial catheters for reduction of symptomatic urinary tract infection in adults requiring short-term catheterisation in hospital: a multicentre randomised controlled trial, *Lancet.*, 2012, **380**(9857), 1927–1935.

57. G. M. Viola, J. Rosenblatt and I. I. Raad, Drug eluting antimicrobial vascular catheters: Progress and promise, *Adv. Drug Delivery Rev.*, 2017, **112**, 35–47.
58. F. C. Fang, Perspectives series: host/pathogen interactions. Mechanisms of nitric oxide-related antimicrobial activity, *J. Clin. Invest.*, 1997, **99**, 2818–2825.
59. D. S. Bredt and S. H. Snyder, Nitric oxide: a physiologic messenger molecule, *Annu. Rev. Biochem.*, 1994, **63**, 175–195.
60. K. A. Amoako, C. Archangeli, H. Handa, T. Major, M. E. Meyerhoff, G. M. Annich and R. H. Bartlett, Thromboresistance characterization of extruded nitric oxide-releasing silicone catheters, *ASAIO J.*, 2012, **58**, 238–246.
61. C. von Eiff, W. Kohnen, K. Becker and B. Jansen, Modern strategies in the prevention of implant-associated infections, *Int. J. Artif. Organs*, 2005, **28**(11), 1146–1156.
62. J. Lo, D. Lange and B. Chew, Ureteral stents and Foley catheters-associated urinary tract infections: the role of coatings and materials in infection prevention, *Antibiotics*, 2014, **3**(1), 87–97.
63. D. Campoccia, L. Montanaro and C. R. Arciola, A review of the biomaterials technologies for infection-resistant surfaces, *Biomaterials*, 2013, **34**(34), 8533–8554.
64. J. R. Johnson, B. Johnston and M. A. Kuskowski, *In vitro* comparison of nitrofurazone- and silver alloy-coated Foley catheters for contact-dependent and diffusible inhibition of urinary tract infection-associated microorganisms, *Antimicrob. Agents Chemother.*, 2012, **56**(9), 4969–4972.
65. D. Kowalczuk, G. Ginalska, T. Piersiak and M. Miazga-Karska, Prevention of biofilm formation on urinary catheters: comparison of the sparfloxacin- treated long-term antimicrobial catheters with silver-coated ones, *J. Biomed. Mater. Res. B Appl. Biomater.*, 2012, **100**(7), 1874–1882.
66. R. Pickard, T. Lam, G. Maclennan, K. Starr, M. Kilonzo, G. McPherson, K. Gillies, A. McDonald, K. Walton, B. Buckley, C. Glazener, C. Boachie, J. Burr, J. Norrie, L. Vale, A. Grant and J. N'Dow, Types of urethral catheter for reducing symptomatic urinary tract infections in hospitalised adults requiring short-term catheterisation: multicenter randomised controlled trial and economic evaluation of antimicrobial- and antiseptic-impregnated urethral catheters (the CATHETER trial), *Health Technol. Assess.*, 2012, **16**(47), 1–197.
67. R. Pickard, T. Lam, G. MacLennan, K. Starr, M. Kilonzo, G. McPherson, K. Gillies, A. McDonald, K. Walton, B. Buckley, C. Glazener, C. Boachie, J. Burr, J. Norrie, L. Vale, A. Grant and J. N'Dow, Antimicrobial catheters

for reduction of symptomatic urinary tract infection in adults requiring short-term catheterisation in hospital: a multicenter randomised controlled trial, *Lancet*, 2012, **380**(9857), 1927–1935.
68. O. Akre, H. Thulin and M. Bottai, Assessing catheter-associated urinary tract infection, *Lancet*, 2013, **381**(9877), 1535.
69. G. Regev-Shoshani, M. Ko, A. Crowe and Y. Av-Gay, Comparative efficacy of commercially available and emerging antimicrobial urinary catheters against bacteriuria caused by E. coli in vitro, *Urology*, 2011, **78**(2), 334–339.
70. M. Leone, Prevention of CAUTI: simple is beautiful, *Lancet*, 2012, **380**(9857), 1891–1892.
71. C. Y. Loo, P. M. Young, W. H. Lee, R. Cavaliere, C. B. Whitchurch and R. Rohanizadeh, Superhydrophobic, nanotextured polyvinyl chloride films for delaying *Pseudomonas aeruginosa* attachment to intubation tubes and medical plastics, *ActaBiomater.*, 2012, **8**(5), 1881–1890.
72. M. Kilonzo, L. Vale, R. Pickard, T. Lam, J. N'Dow and Catheter Trial Group, Cost effectiveness of antimicrobial catheters for adults requiring short-term catheterisation in hospital, *Eur. Urol.*, 2014, **66**(4), 615–618.
73. L. E. Fisher, A. L. Hook, W. Ashraf, A. Yousef, D. A. Barrett, D. J. Scurr, X. Chen, E. F. Smith, M. Fay, C. D. Parmenter, R. Parkinson and R. Bayston, Biomaterial modification of urinary catheters with antimicrobials to give long-term broad spectrum antibiofilm activity, *J. Controlled Release*, 2015, **202**, 57–64.
74. A. Salvarci, M. Koroglu and T. Gurpinar, Evaluation of antimicrobial activities of minocycline and rifampin-impregnated silicone surfaces in an *in vitro* urinary system model,*J. Pak. Med. Assoc.*, 2015, **65**(2), 115–119.
75. L. E. Nicolle, Catheter-related urinary tract infection: practical management in the elderly, *Drugs Aging*, 2014, **31**(1), 1–10.
76. J. M. Schierholz, C. Fleck, J. Beuth and G. Pulverer, The antimicrobial efficacy of a new central venous catheter with long-term broad-spectrum activity,*J. Antimicrob. Chemother*, 2000, **46**(1), 45–50.
77. B. J. Rijnders, E. Van Wijngaerden and W. E. Peetermans, Catheter-tip colonization as a surrogate end point in clinical studies on catheter-related bloodstream infection: how strong is the evidence?, *Clin. Infect. Dis*, 2002, **35**(9), 1053–1058.
78. E. R. Kenawy, Biologically Active Polymers IV: Synthesis and Antimicrobial Activity of Polymers Containing 8- HydroxyQuinoline Moiety, *J. Appl. Polym. Sci.*, 2001, **82**, 1364–1374.
79. E. R. Kenawy, S. D. Worley and R. Broughton, The Chemistry and Applications of Antimicrobial Polymers, *Biomacromolecules*, 2007, **8**, 1359–1384.
80. E. R. Kenawy, F. I. Abdel-Hay, A. El-Raheem, R. El-Shanshoury and M. H. El-Newehy, Synthesis and Antibacterial Activity of Modified Glycidyl Methacrylate Polymers Having Quaternary Ammonium and Phosphonium Groups,*J. Controlled Release*, 1998, **50**, 145–152.

81. I. L. Shih, M. H. Shen and Y. T. Van, Microbial synthesis of poly (ε-lysine) and its various applications, *Bioresource Technol.*, 2006, **97**, 1148–1159.
82. N. Hoiby, T. Bjarnsholt, M. Givskov, S. Molin and O. Ciofu, Antibiotic resistance of bacterial biofilms, *Int. J. Antimicrob. Agents*, 2010, **35**(4), 322–332.
83. D. Pavithra and M. Doble, Biofilm formation, bacterial adhesion and host response on polymeric implants–issues and prevention, *Biomed. Mater.*, 2008, **3**(3), 034003.
84. C. Sousa, P. Teixeira and R. Oliveira, Influence of Surface Properties on the Adhesion of *Staphylococcus epidermidis* to Acrylic and Silicone, *Int. J. Biomater.*, 2009, **2009**, 718017.
85. S. Farah, O. Aviv, M. Daif, K. R. Kunduru, N. Laout, S. Ratner, N. Beyth and A. J. Domb, N-bromo-hydantoin grafted polystyrene beads: Synthesis and nano-micro beads characteristics for achieving controlled release of active oxidative bromine and extended microbial inactivation efficiency, *Polym. Sci., Part A: Polym. Chem*, 2016, **54**, 596–610.
86. S. Farah, O. Aviv, N. Laout, S. Ratner and A. J. Domb, Antimicrobial N-brominated hydantoin and uracil grafted polystyrene beads, *J. Controlled Release*, 2015, **216**, 18–29.
87. S. Farah, A. J. Domb, Controlled release N-Halamine polymers for water decontamination and detoxification. U.S. Application No. 15/767,108.
88. O. Aviv, S. Farah, N. Amir, N. Laout, S. Ratner and A. J. Domb, N-bromo-dimethylhydantoin polystyrene resin for water microbial decontamination, *Biomacromolecules*, 2015, **16**, 1442–1447.
89. L. E. Fisher, A. L. Hook, W. Ashraf, A. Yousef, D. A. Barrett, D. J. Scurr, X. Chen, E. F. Smith, M. Fay, C. D. Parmenter, R. Parkinson and R. Bayston, Biomaterial modification of urinary catheters with antimicrobials to give long-term broadspectrum antibiofilm activity, *J. Controlled Release*, 2015, **202**, 57–64.
90. E. Dayyoub, M. Frant, S. R. Pinnapireddy, K. Liefeith and U. Bakowsky, Antibacterial and anti-encrustation biodegradable polymer coating for urinary catheter, *Int. J. Pharm*, 2017, **531**(1), 205–214.
91. J. H. Park, Y. W. Cho, Y. H. Cho, J. M. Choi, H. J. Shin, Y. H. Bae, H. Chung, S. Y. Jeong and I. C. Kwon, Norfloxacin-releasing urethral catheter for long-term catheterization, *J. Biomater. Sci., Polym. Ed.*, 2003, **14**(9), 951–962.
92. A. J. Hamilton, J. Orozco, P. Narotam and T. Bowersock, Efficacy of vancomycin/tri-iododecyclemethyl ammonium chloride-coated ventriculostomy catheters in reducing infection, *Neurosurgery*, 1997, **40**(5), 1043–10439.
93. R. O. Darouiche, J. A. Smith, H. Hanna, C. B. Dhabuwala, M. S. Steiner, R. J. Babaian, T. B. Boone, P. T. Scardino, J. I. Thornby and I. I. Raad, Efficacy of antimicrobial-impregnated bladder catheters in reducing catheter-associated bacteriuria: a prospective, randomized, multicenter clinical trial, *Urology*, 1999, **54**(6), 976–981.

94. R. E. Gilbert, Q. Mok, K. Dwan, K. Harron, T. Moitt, M. Millar, P. Ramnarayan, S. M. Tibby, D. Hughes and C. Gamble, Impregnated central venous catheters for prevention of bloodstream infection in children (the CATCH trial): a randomised controlled trial, *Lancet*, 2016, **387**(10029), 1732–1742.
95. I. Raad, A. Buzaid, J. Rhyne, R. Hachem, R. Darouiche, H. Safar, M. Albitar and R. J. Sherertz, Minocycline and ethylenediaminetetraacetate for the prevention of recurrent vascular catheter infections, *Clin. Infect. Dis.*, 1997, **25**(1), 149–151.
96. N. Vargas-Cruz, R. A. Reitzel, J. Rosenblatt, M. Jamal, A. D. Szvalb, A. M. Chaftari, R. Hachem and I. Raad, In Vitro Study of Antimicrobial Percutaneous Nephrostomy Catheters for Prevention of Renal Infections, *Antimicrob. Agents Chemother*, 2017, **61**(6).
97. K. Page, M. Wilson and I. P. Parkin, Antimicrobial surfaces and their potential in reducing the role of the inanimate environment in the incidence of hospital-acquired infections, *J. Mater. Chem.*, 2009, **19**, 3819–3831.
98. K. Chaloupka, Y. Malam and A. M. Seifalian, Nanosilver as a new generation of nanoproduct in biomedical applications, *Trends Biotechnol.*, 2010, **28**(11), 580–588.
99. M. K. Ballo, S. Rtimi, C. Pulgarin, N. Hopf, A. Berthet, J. Kiwi, P. Moreillon, J. M. Entenza and A. Bizzini, In Vitro and In Vivo Effectiveness of an Innovative Silver-Copper Nanoparticle Coating of Catheters To Prevent Methicillin-Resistant *Staphylococcus aureus* Infection, *Antimicrob. Agents Chemother.*, 2016, **60**(9), 5349–53456.
100. B. Jansen, K. G. Kristinsson, S. Jansen, G. Peters and G. Pulverer, In-vitro efficacy of a central venous catheter complexed with iodine to prevent bacterial colonization, *J. Antimicrob. Chemother.*, 1992, **30**(2), 135–139.
101. A. M. Carmona-Ribeiro and L. D. de Melo Carrasco, Cat-ionic antimicrobial polymers and their assemblies, *Int. J. Mol. Sci.*, 2013, **14**, 9906–9946.
102. K. Lim, R. Saravanan, K. K. L. Chong, S. H. M. Goh, R. R. Y. Chua, P. A. Tambyah, M. W. Chang, K. A. Kline and S. S. J. Leong, Anhydrous polymer-based coating with sustainable controlled release functionality for facile, efficacious impregnation, and delivery of antimicrobial peptides, *Biotechnol Bioeng.*, 2018, **115**(8), 2000–2012.
103. N. Raman, K. Marchillo, M. R. Lee, A. L. Rodriguez Lopez, D. R. Andes, S. P. Palecek and D. M. Lynn, Intraluminal Release of an Antifungal beta-Peptide Enhances the Antifungal and Anti-Biofilm Activities of Multilayer-Coated Catheters in a Rat Model of Venous Catheter Infection, *ACS Biomater. Sci. Eng.*, 2016, **2**(1), 112–121.
104. R. N. Dave, H. M. Joshi and V. P. Venugopalan, Novel biocatalytic polymer-based antimicrobial coatings as potential ureteral biomaterial: preparation and *in vitro* performance evaluation, *Antimicrob. Agents Chemother*, 2011, **55**(2), 845--8453.

105. T. A. Gaonkar and S. M. Modak, Comparison of microbial adherence to antiseptic and antibiotic central venous catheters using a novel agar subcutaneous infection model, *J. Antimicrob. Chemother.*, 2003, **52**(3), 389–396.
106. A. M. Dagsgupta, J. H. Mondal and D. Das, Peptide hydrogels, *RSC Adv.*, 2013, **3**, 9117–9149.
107. R. M. Epand and H. J. Vogel, Diversity of antimicrobial peptides and their mechanisms of action, *Biochim. Biophys. Acta*, 1999, **1462**(1–2), 11–28.
108. D. Campoccia, *et al.*, A review of the biomaterials technologies for infection-resistant surfaces, *Biomaterials*, 2013, **34**, 8533–8554.
109. E. S. Zarie, *et al.*, Solvent free fabrication of micro and nanostructured drug coatings by thermal evaporation for controlled release and increased effects, *PLoS One*, 2012, 7, e40746.
110. H. Cheng, *et al.*, Long-lasting *in vivo* and *in vitro* antibacterial ability of nanostructured titania coating incorporated with silver nanoparticles, *J. Biomed. Mater. Res. A*, 2014, **102**, 3488–3499.
111. C. D. L. H. Alarcon, *et al.*, Stimuli responsive polymers for biomedical applications, *Chem. Soc. Rev.*, 2005, **34**, 276–285.
112. M. A. C. Stuart, *et al.*, Emerging applications of stimuli-responsive polymer materials, *Nat. Mater.*, 2010, **9**, 101–113.
113. D. Esrafilzadeh, *et al.*, Multifunctional conducting fibres with electrically controlled release of ciprofloxacin, *J. Controlled Release*, 2013, **169**, 313–320.
114. B. Hu, *et al.*, Core-shell-shell nanorods for controlled release of silver that can serve as a nanoheater for photothermal treatment on bacteria, *Acta Biomater.*, 2015, **11**, 511–519.
115. P. Norris, *et al.*, Ultrasonically controlled release of ciprofloxacin from self-assembled coatings on poly(2-hydroxyethyl methacrylate) hydrogels for *Pseudomonas aeruginosa* biofilm prevention, *Antimicrob. Agents Chemother.*, 2005, **49**, 4272–4279.
116. M. L. Noble, *et al.*, Digital drug delivery: on-off ultrasound controlled antibiotic release from coated matrices with negligible background leaching, *Biomater. Sci.*, 2014, **2**, 893–902.
117. H. Wang, *et al.*, Magnetic/NIR-thermally responsive hybrid nanogels for optical temperature sensing, tumor cell imaging and triggered drug release, *Nanoscale*, 2014, **6**, 13001–13011.
118. D. G. Shchukin and H. Möhwald, Self-repairing coatings containing active nanoreservoirs, *Small*, 2007, **3**, 926–943.
119. J. Gallo, *et al.*, Antibacterial surface treatment for orthopaedic implants, *Int. J. Mol. Sci.*, 2014, **15**, 13849–13880.
120. D. Shchukin and H. Möhwald, A coat of many functions, *Science*, 2013, **341**, 1458–1459.
121. J. Borges, *et al.*, Layer-by-layer assembly of lightresponsive polymeric multilayer systems, *Adv. Funct. Mater.*, 2014, **24**, 5624–5648.
122. C. Traba and J. F. Liang, Bacteria responsive antibacterial surfaces for indwelling device infections, *J. Controlled Release*, 2015, **198**, 18–25.

123. I. Zhuk, *et al.*, Self-defensive layer-by-layer films with bacteria-triggered antibiotic release, *ACS Nano*, 2014, **8**, 7733–7745.
124. L. Pichavant, *et al.*, pH-controlled delivery of gentamicin sulfate from orthopedic devices preventing nosocomial infections, *J. Controlled Release*, 2012, **162**, 373–381.
125. V. V. Komnatnyy, *et al.*, Bacteria-triggered release of anti-microbial agents, *Angew. Chem.*, 2014, **126**, 449–451.
126. M. Tanihara, *et al.*, A novel microbial infection-responsive drug release system, *J. Pharm. Sci.*, 1999, **88**, 510–514.
127. G. Cado, *et al.*, Self-defensive biomaterial coating against bacteria and yeasts: polysaccharide multilayer film with embed-ded antimicrobial peptide, *Adv. Funct. Mater.*, 2013, **23**, 4801–4809.
128. H. Phuengkham, V. Teeranachaideekul, M. Chulasiri and N. Nasongkla, Preparation and optimization of chlorophene-loaded nanospheres as controlled release antimicrobial delivery systems, *Pharm. Dev. Technol.*, 2016, **21**(1), 8–13.
129. G. Eke, A. M. Kuzmina, A. V. Goreva, E. I. Shishatskaya, N. Hasirci and V. Hasirci, *In vitro* and transdermal penetration of PHBV micro/nanoparticles, *J. Mater. Sci.: Mater. Med.*, 2014, **25**(6), 1471–1481.
130. N. Duran, A. F. De Oliveira and M. M. De Azevedo, In vitro studies on the release of isoniazid incorporated in poly(epsilon-caprolactone), *J. Chemother.*, 2006, **18**(5), 473–479.
131. A. Akbarzadeh, R. Rezaei-Sadabady, S. Davaran, S. W. Joo, N. Zarghami, Y. Hanifehpour, M. Samiei, M. Kouhi and K. Nejati-Koshki, Liposome: classification, preparation, and applications, *Nanoscale Res. Lett.*, 2013, **8**(1), 102.
132. J. L. Pugach, V. DiTizio, M. W. Mittelman, A. W. Bruce, F. DiCosmo and A. E. Khoury, Antibiotic hydrogel coated Foley catheters for prevention of urinary tract infection in a rabbit model, *J. Urol.*, 1999, **162**(3 Pt 1), 883–887.
133. S. Tamilvanan, N. Venkateshan and A. Ludwig, The potential of lipid- and polymer-based drug delivery carriers for eradicating biofilm consortia on device-related nosocomial infections, *J. Controlled Release*, 2008, **128**(1), 2–22.
134. D. M. Siddiq and R. O. Darouiche, New strategies to prevent catheter-associated urinary tract infections, *Nat. Rev. Urol.*, 2012, **9**(6), 305–314.
135. K. S. Soppimath, T. M. Aminabhavi, A. R. Kulkarni and W. E. Rudzinski, Biodegradable polymeric nanoparticles as drug delivery devices, *J. Controlled Release*, 2001, **70**(1–2), 1–20.
136. H. Phuengkham and N. Nasongkla, Development of antibacterial coating on silicone surface *via* chlorhexidine-loaded nanospheres, *J. Mater. Sci.: Mater. Med.*, 2015, **26**(2), 78.
137. M. L. W. Knetsch and L. H. Koole, New Strategies in the Development of Antimicrobial Coatings: The Example of Increasing Usage of Silver and Silver Nanoparticles, *Polymers*, 2011, **3**(4), 340–366.

138. F. Furno, K. S. Morley, B. Wong, B. L. Sharp, P. L. Arnold, S. M. Howdle, R. Bayston, P. D. Brown, P. D. Winship and H. J. Reid, Silver nanoparticles and polymeric medical devices: a new approach to prevention of infection?, *J. Antimicrob. Chemother.*, 2004, **54**(6), 1019–1024.
139. H. J. Johnston, G. Hutchison, F. M. Christensen, S. Peters, S. Hankin and V. Stone, A review of the in vivo and in vitro toxicity of silver and gold particulates: particle attributes and biological mechanisms responsible for the observed toxicity, *Crit. Rev. Toxicol.*, 2010, **40**(4), 328–346.
140. N. Bagge, M. Schuster, M. Hentzer, O. Ciofu, M. Givskov, E. P. Greenberg and N. Hoiby, *Pseudomonas aeruginosa* biofilms exposed to imipenem exhibit changes in global gene expression and beta-lactamase and alginate production, *Antimicrob. Agents Chemother.*, 2004, **48**(4), 1175–1187.
141. M. A. Kohanski, D. J. Dwyer and J. J. Collins, How antibiotics kill bacteria: from targets to networks, *Nat. Rev. Microbiol.*, 2010, **8**(6), 423–435.
142. N. Kawabata and M. Nishiguchi, Antibacterial activity of soluble pyridinium-type polymers, *Appl. Environ. Microbiol.*, 1988, **54**, 2532–2535.
143. L. Ferreira and A. Zumbuehl, Non-leaching surfaces capable of killing microorganisms on contact, *J. Mater. Chem.*, 2009, **19**, 7796–7806.
144. L. Timofeeva and N. Kleshcheva, Antimicrobial polymers: mechanism of action, factors of activity, and applications, *Appl. Microbiol. Biotechnol.*, 2011, **89**, 475–492.
145. N. Beyth, I. Yudovin-Farber, M. Perez-Davidi, A. J. Domb and E. Weiss, Polyethyleneimine nanoparticles incorporated into resin composite cause cell death and trigger biofilm stress in vivo, *PNAS*, 2010, **107**, 22038–22043.
146. N. Beyth, I. Yudovin-Farber, M. Perez-Davidi, A. J. Domb and E. I. Weiss, Surface antimicrobial activity and biocompatibility of incorporated polyethylenimine nanoparticles, *Biomaterials*, 2008, **29**, 4157–4163.
147. I. Yudovin-Farber, N. Beyth, A. Nyska, E. I. Weiss, J. Golenser and A. J. Domb, Surface Characterization and Biocompatibility of Restorative Resin Containing Nanoparticles, *Biomacromolecules*, 2008, **9**, 3044–3050.
148. I. Yudovin-Farber, N. Beyth, E. I. Weiss and A. J. Domb, Antibacterial effect of composite resins containing quaternary ammonium polyethyleneimine nanoparticles, *J. Nanopart. Res.*, 2009, **12**, 591–603.
149. S. Farah, W. Khan, I. Farber, D. Kesler-Shvero, N. Beyth, E. I. Weiss and A. J. Domb, Crosslinked QA-PEI nanoparticles: synthesis reproducibility, chemical modifications, and stability study, *Polym. Adv. Technol.*, 2013, **24**(5), 446–452.
150. S. Farah, O. Aviv, N. Laout, S. Ratner, N. Beyth and A. J. Domb, Quaternary ammonium polyethylenimine nanoparticles for treating bacterial contaminated water, *Colloids Surf., B*, 2015, **128**, 614–619.

151. S. Farah, O. Aviv, N. Laout, S. Ratner, N. Beyth and A. J. Domb, Quaternary ammonium poly (diethylaminoethyl methacrylate) possessing antimicrobial activity, *Colloids Surf., B*, 2015, **128**, 608–613.
152. S. Farah, O. Aviv, N. Laout, S. Ratner, N. Beyth and A. J. Domb, Antimicrobial silica particles loaded with quaternary ammonium polyethyleneimine network, *Polym. Adv. Technol.*, 2014, **25**(6), 689–692.
153. K. Talaro, A. Talaro, in *Foundations in Microbiology*, WCB Publishers, Dubuque, IA, 1993, 286.
154. G. J. Li, J. R. Shen and Y. L. Zhu, A study of pyridinium-type functional polymers. III. Preparation and characterization of insoluble pyridinium-type polymers, *J. Appl. Polym. Sci.*, 2000, **78**, 668–675.
155. J. C. Tiller, C. J. Liao, K. Lewis and A. M. Klibanov, Designing surfaces that kill bacteria on contact, *Proc. Natl. Acad. Sci. U. S. A.*, 2001, **98**, 5981–5985.
156. D. G. White, J. Acar, F. Anthony, A. Franklin, R. Gupta, T. Nicholls, Y. Tamura, S. Thompson, E. J. Threlfall, D. Vose, M. V. Vuuren, H. C. Wegener and M. L. Costarrica, Antimicrobial resistance: standardisation and harmonisation of laboratory methodologies for the detection and quantification of antimicrobial resistance, *Rev. Sci. Tech. OIE.*, 2001, **20**, 849–858.
157. Y. Jiao, L.-n. Niu, S. Ma, J. Li, F. R. Tay and J.-h. Chen, Quaternary ammonium-based biomedical materials: State-of-the-art, toxicological aspects and antimicrobial resistance, *Prog. Polym. Sci.*, 2017, **71**, 53–90.
158. T. Tashiro, Antibacterial and Bacterium Adsorbing Macromolecules, *Macromol. Mater. Eng.*, 2001, **286**, 63–87.
159. J. Lin, S. Qiu, K. Lewis and A. M. Klibanov, Mechanism of bactericidal and fungicidal activates of textiles covalently modified with alkylated polyethylenimine, *Biotechnol. Bioeng.*, 2003, **83**, 168–172.
160. J. Hiraki, ε-Polylysine, its development and utilisation, *Fine Chem.*, 2000, **29**, 25–28.
161. J. Hiraki, Basic and applied studies on e-polylysine, *J. Antibact Antifungal Agents*, 1995, **33**, 349–354.
162. I. Yudovin-Farber, Bioactive polyelectrolytes, Hebrew University of Jerusalem, 2008.
163. A. Ortega, S. Farah, P. Tranque, A. V. Ocaña, S. H. Nam-Cha, N. Beyth, C. Gómez-Roldán, R. Pérez-Tanoira, A. J. Domb, F. C. Pérez-Martínez and J. Pérez-Martínez, Antimicrobial evaluation of quaternary ammonium polyethyleneimine nanoparticles against clinical isolates of pathogenic bacteria, *IET Nanobiotechnol.*, 2015, **9**(6), 342–348.
164. A. Ortega, F. C. Pérez-Martínez, S. Farah, S. H. Nam-Cha, A. V. Ocaña, N. Beyth, A. J. Domb, G. Martínez, C. Gómez-Roldán and J. Pérez-Martínez, Quaternary ammonium polyethyleneimine nanoparticles: antimicrobial evaluation against bacteria from peritoneal dialysis related peritonitis, *Nephrol. Dial. Transplant.*, 2015, **30**, iii537–iii538. Oxford University Press.

165. N. Beyth, I. Yudovin-Farber, R. Bahir, A. J. Domb and E. I. Weiss, Antibacterial activity of dental composites containing quaternary ammonium polyethylenimine nanoparticles against *Streptococcus mutans*, *Biomaterials*, 2006, **27**, 3995–4002.
166. A. J. Domb, N. Beyth, S. Farah, *Quaternary Ammonium Antimicrobial Polymers*, MRS Online Proceedings Library Archive, Cambridge University Press, 2013, 1569, 97–107.
167. N. Beyth, S. Farah, A. J. Domb and E. I. Weiss, Antibacterial dental resin composites, *React. Funct. Polym.*, 2014, **75**, 81–88.
168. T. Abel, J. I. Cohen, R. Engel, M. Filshtinskaya, A. Melkonian and K. Melkonian, Preparation and investigation of antibacterial carbohydrate-based surfaces, *Carbohydr. Res.*, 2002, **337**, 2495–2499.
169. K. R. Kunduru, M. Nazarkovsky, S. Farah, R. P. Pawar, A. Basu and A. J. Domb, Nanotechnology for water purification: applications of nanotechnology methods in wastewater treatment, *Water Purif.*, 2017, 33–74.
170. G. Li and J. Shen, *Abstr. Papers Am. Chem. Soc.*, 1998, **216**, U97.
171. S. Imazato, Y. Kinomoto, H. Tarumi, S. Ebisu and F. R. Tay, Antibacterial activity and bonding characteristics of an adhesive resin containing antibacterial monomer MDPB, *Dent. Mater. J.*, 2003, **19**, 313–319.
172. S. Imazato, Bio-active restorative materials with antibacterial effects: new dimension of innovation in restorative dentistry, *Dent. Mater. J.*, 2009, **28**, 11–19.
173. A. Jain, L. S. Duvvuri, S. Farah, N. Beyth, A. J. Domb and W. Khan, Antimicrobial Polymers, *Adv. Healthcare Mater.*, 2014, **3**, 1969–1985.
174. M. Lindstedt, S. Allenmark, R. A. Thompson and L. Edebo, Antimicrobial activity of betaine esters, quaternary ammonium amphiphiles which spontaneously hydrolyze into nontoxic components, *Antimicrob. Agents Chemother.*, 1990, **34**(10), 1949–1954.
175. L. Furi, M. L. Ciusa, D. Knight, V. Di Lorenzo, N. Tocci, D. Cirasola, L. Aragones, J. R. Coelho, A. T. Freitas, E. Marchi, L. Moce, P. Visa, J. B. Northwood, C. Viti, E. Borghi, G. Orefici, B. Consortium, I. Morrissey and M. R. Oggioni, Evaluation of reduced susceptibility to quaternary ammonium compounds and bisbiguanides in clinical isolates and laboratory-generated mutants of *Staphylococcus aureus*, *Antimicrob. Agents Chemother.*, 2013, **57**(8), 3488–3497.
176. C. J. Ioannou, G. W. Hanlon and S. P. Denyer, Action of disinfectant quaternary ammonium compounds against *Staphylococcus aureus*, *Antimicrob. Agents Chemother.*, 2007, **51**(1), 296–306.
177. M. Crismaru, L. A. Asri, T. J. Loontjens, B. P. Krom, J. de Vries, H. C. van der Mei and H. J. Busscher, Survival of adhering staphylococci during exposure to a quaternary ammonium compound evaluated by using atomic force microscopy imaging, *Antimicrob. Agents Chemother.*, 2011, **55**(11), 5010–5017.
178. J. Stefanska, A. Pietruczuk-Padzik, M. Struga, M. Borkowski and S. Tyski, Biological evaluation of quaternary bis ammonium salt and

cetylpyridinum bromide against *S. epidermidis* biofilm, *Pol. J. Microbiol.*, 2013, **62**(4), 359–364.
179. S. Zanini, A. Polissi, E. A. Maccagni, E. C. Dell'Orto, C. Liberatore and C. Riccardi, Development of antibacterial quaternary ammonium silane coatings on polyurethane catheters, *J. Colloid Interface Sci.*, 2015, **451**, 78–84.
180. Z. K. Zander, P. Chen, Y. H. Hsu, N. Z. Dreger, L. Savariau, W. C. McRoy, A. E. Cerchiari, S. D. Chambers, H. A. Barton and M. L. Becker, Post-fabrication QAC-functionalized thermoplastic polyurethane for contact-killing catheter applications, *Biomaterials*, 2018, **178**, 339–350.
181. K. De Prijck, N. De Smet, T. Coenye, E. Schacht and H. J. Nelis, *Mycopathologia*, 2010, **170**, 213.
182. H. Tan, R. Ma, C. Lin, Z. Liu and T. Tang, Quaternized chitosan as an antimicrobial agent: antimicrobial activity, mechanism of action and biomedical applications in orthopaedics, *Int. J. Mol. Sci.*, 2013, **14**, 1854–1869.
183. S. Farah, J. C. Doloff, G. Anderson, R. Langer, Crosslinked chitosan nanoparticles, in *Drug Delivery Techniques; Aldrich Materials Science*, Sigma-Aldrich Co., Milwaukee, WI, 2015, pp. 18–21.
184. H. Sashiwa and S. I. Aiba, Chemically modified chitin and chitosan as biomaterials, *Prog. Polym. Sci.*, 2004, **29**, 887–908.
185. G. N. Rimondino, M. C. Strumia and M. Martinelli, Synthesis and Characterization of Water-Soluble Dendronized Chitosan Using Newkome-Type Dendrons, *ACS Sustainable Chem. Eng.*, 2014, **2**, 2582–2587.
186. K. G. Neoh; M. Li; R. Wang; E. T. Kang; E. Chiong, Antimicrobial coatings for biomedical applications: Challenges and opportunities offered by natural polymers, Abstracts of Papers, 249th ACS National Meeting & Exposition, Denver, CO, United States, March 22–26, 2015, PMSE-204.
187. N. E. Kotelnikova and E. F. Panarin, Cellulose modification by biologically active substances for biomedical applications, *Cellul. Chem. Technol.*, 2005, **39**(5–6), 437–450.
188. Y. Chen, P. A. Wilbon, Y. P. Chen, J. Zhou, M. Nagarkatti, C. Wang, F. Chu, A. W. Decho and C. Tang, Amphipathic antibacterial agents using cationic methacrylic polymers with natural rosin as pendant group, *RSC Adv.*, 2012, **2**(27), 10275–10282.
189. D. T. Beiko, B. E. Knudsen, J. D. Watterson, P. A. Cadieux, G. Reid and J. D. Denstedt, Urinary tract biomaterials, *J. Urol.*, 2004, **171**(6 Pt 1), 2438–2444.
190. K. A. Kazmierska, R. Thompson, N. Morris, A. Long and T. Ciach, In vitro multicompartmental bladder model for assessing blockage of urinary catheters: effect of hydrogel coating on dynamics of *Proteus mirabilis* growth, *Urology.*, 2010, **76**(2), 515 e15–515 e20.
191. Q. Wei, T. Becherer, P. L. Noeske, I. Grunwald and R. Haag, A universal approach to crosslinked hierarchical polymer multilayers as stable and

highly effective antifouling coatings, *Adv. Mater.*, 2014, **26**(17), 2688–2693.

192. I. Francolini and G. Donelli, Prevention and control of biofilm-based medical-device-related infections, *FEMS Immunol. Med. Microbiol.*, 2010, **59**(3), 227–238.
193. L. M. Hamming and P. B. Messersmith, Fouling resistant biomimetic poly(ethylene glycol)based grafted polymer coatings, *Mater. Matters*, 2008, **3.3**, 52.
194. S. Chen, J. Zheng, L. Li and S. Jiang, Strong resistance of phosphorylcholine self assembled monolayers to protein adsorption: insights into nonfouling properties of zwitterionic materials, *J. Am. Chem. Soc.*, 2005, **127**(41), 14473–14478.
195. R. Bernstein, S. Belfer and V. Freger, Bacterial attachment to RO membranes surface-modified by concentration-polarization-enhanced graft polymerization, *Environ. Sci. Technol.*, 2011, **45**(14), 5973–5980.
196. J. L. Harding and M. M. Reynolds, Combating medical device fouling, *Trends Biotechnol.*, 2014, **32**(3), 140–146.
197. L. Mi and S. Jiang, Integrated antimicrobial and nonfouling zwitterionic polymers, *Angew. Chem., Int. Ed.*, 2014, **53**(7), 1746–1754.
198. C. Diaz Blanco, A. Ortner, R. Dimitrov, A. Navarro, E. Mendoza and T. Tzanov, Building an antifouling zwitterionic coating on urinary catheters using an enzymatically triggered bottom-up approach, *ACS Appl. Mater. Interfaces*, 2014, **6**(14), 11385–11393.
199. J. C. Russell, Bacteria, biofilms, and devices: the possible protective role of phosphorylcholine materials, *J. Endourol.*, 2000, **14**(1), 39–42.
200. D. J. Stickler, A. Evans, N. Morris and G. Hughes, , Strategies for the control of catheter encrustation, *Int. J. Antimicrob. Agents*, 2002, **19**(6), 499–506.
201. N. M. Lai, N. Chaiyakunapruk, N. A. Lai, E. O'Riordan, W. S. Pau and S. Saint, Catheter impregnation, coating or bonding for reducing central venous catheter-related infections in adults, *Cochrane Database Syst. Rev*, 2016, 3. CD007878.
202. I. Raad, R. Darouiche, J. Dupuis, D. Abi-Said, A. Gabrielli, R. Hachem, M. Wall, R. Harris, J. Jones, A. Buzaid, C. Robertson, S. Shenaq, P. Curling, T. Burke and C. Ericsson, Central venous catheters coated with minocycline and rifampin for the prevention of catheter-related colonization and bloodstream infections. A randomized, double-blind trial. The Texas Medical Center Catheter Study Group, *Ann. Intern. Med.*, 1997, **127**(4), 267–274.
203. H. Hanna, R. Benjamin, I. Chatzinikolaou, B. Alakech, D. Richardson, P. Mansfield, T. Dvorak, M. F. Munsell, R. Darouiche, H. Kantarjian and I. Raad, Long-term silicone central venous catheters impregnated with minocycline and rifampin decrease rates of catheter-related bloodstream infection in cancer patients: a prospective randomized clinical trial, *J. Clin. Oncol.*, 2004, **22**(15), 3163–3171.

204. M. E. Falagas, K. Fragoulis, I. A. Bliziotis and I. Chatzinikolaou, Rifampicin-impregnated central venous catheters: a meta-analysis of randomized controlled trials, *J. Antimicrob. Chemother*, 2007, **59**(3), 359–369.
205. R. W. Allan, H. Alnuaimat, W. D. Edwards and H. D. Tazelaar, Embolization of hydrophilic catheter coating to the lungs: report of a case mimicking granulomatous vasculitis, *Am. J. Clin. Pathol.*, 2009, **132**(5), 794–797.
206. A. Yousif, A. M. Chaftari, M. Michael, M. Jordan, Z. Al Hamal, A. Hussain, N. Elizabeth, Y. Jiang, R. Hachem and I. Raad, The influence of using antibiotic-coated peripherally inserted central catheters on decreasing the risk of central line-associated bloodstream infections, *Am. J. Infect. Control.*, 2016, **44**(9), 1037–1040.
207. G. D. Kamal, M. A. Pfaller, L. E. Rempe and P. J. Jebson, Reduced intravascular catheter infection by antibiotic bonding. A prospective, randomized, controlled trial, *JAMA*, 1991, **265**(18), 2364–2368.
208. J. M. Walz, R. L. Avelar, K. J. Longtine, K. L. Carter, L. A. Mermel, S. O. Heard and Group FUCS, Anti-infective external coating of central venous catheters: a randomized, noninferiority trial comparing 5-fluorouracil with chlorhexidine/silver sulfadiazine in preventing catheter colonization, *Crit. Care Med.*, 2010, **38**(11), 2095–2102.
209. D. L. Veenstra, S. Saint, S. Saha, T. Lumley and S. D. Sullivan, Efficacy of antiseptic-impregnated central venous catheters in preventing catheter-related bloodstream infection: a meta-analysis, *JAMA*, 1999, **281**(3), 261–267.
210. M. N. Carrasco, A. Bueno, C. de las Cuevas, S. Jimenez, I. Salinas, A. Sartorius, T. Recio, M. Generelo and F. Ruiz-Ocana, Evaluation of a triple-lumen central venous heparin-coated catheter versus a catheter coated with chlorhexidine and silver sulfadiazine in critically ill patients, *Intensive Care Med.*, 2004, **30**(4), 633–638.
211. M. E. Rupp, S. J. Lisco, P. A. Lipsett, T. M. Perl, K. Keating, J. M. Civetta, L. A. Mermel, D. Lee, E. P. Dellinger, M. Donahoe, D. Giles, M. A. Pfaller, D. G. Maki and R. Sherertz, Effect of a second-generation venous catheter impregnated with chlorhexidine and silver sulfadiazine on central catheter-related infections: a randomized, controlled trial, *Ann. Intern. Med.*, 2005, **143**(8), 570–580.
212. C. Brun-Buisson, F. Doyon, J. P. Sollet, J. F. Cochard, Y. Cohen and G. Nitenberg, Prevention of intravascular catheter-related infection with newer chlorhexidine-silver sulfadiazine-coated catheters: a randomized controlled trial, *Intensive Care Med.*, 2004, **30**(5), 837–843.
213. K. Yorganci, C. Krepel, J. A. Weigelt and C. E. Edmiston, Activity of antibacterial impregnated central venous catheters against *Klebsiella pneumoniae*, *Intensive Care Med.*, 2002, **28**(4), 438–442.
214. S. K. Kakkos, G. K. Haddad, R. K. Haddad and M. M. Scully, Effectiveness of a new tunneled catheter in preventing catheter malfunction: a comparative study, *J. Vasc. Interv. Radiol.*, 2008, **19**(7), 1018–1026.

215. P. Kalfon, C. de Vaumas, D. Samba, E. Boulet, J. Y. Lefrant, D. Eyraud, T. Lherm, F. Santoli, W. Naija and B. Riou, Comparison of silver-impregnated with standard multi-lumen central venous catheters in critically ill patients, *Crit. Care Med.*, 2007, **35**(4), 1032–1039.
216. L. Corral, M. Nolla-Salas, J. Ibanez-Nolla, M. A. Leon, R. M. Diaz, M. Cruz Martin, R. Iglesia and R. Catalan, A prospective, randomized study in critically ill patients using the Oligon Vantex catheter, *J. Hosp. Infect*, 2003, **55**(3), 212–219.
217. K. Jaeger, A. Osthaus, J. Heine, H. Ruschulte, C. Kuhlmann, H. Weissbrodt, A. Ganser and M. Karthaus, Efficacy of a benzalkonium chloride-impregnated central venous catheter to prevent catheter-associated infection in cancer patients, *Chemotherapy*, 2001, **47**(1), 50–505.
218. C. Willyard, Drug-resistant bacteria ranked, *Nature*, 2017, **543**, 15.
219. S. S. Pradeep Kumar, H. V. Easwer and A. Maya Nandkumar, Multiple Drug Resistant Bacterial Biofilms on Implanted Catheters – A Reservoir of Infection, *J. Assoc. Physicians India*, 2013 Oct, **61**(10), 702–707.
220. M. Snoussi, J. P. Talledo, N.-A. Del Rosario, S. Mohammadi, B.-Y. Ha, A. Košmrlj and S. Taheri-Araghi, Heterogeneous absorption of antimicrobial peptide LL37 in *Escherichia coli* cells enhances population survivability, *eLife*, 2018, 7, e38174.
221. K.-H. Park, Y.-M. Lee, H.-L. Hong, T. Kim, H. J. Park, S.-Y. Park, S. M. Moon and Y. P. Chong, Persistent catheter-related *Staphylococcus aureus* bacteremia after catheter removal and initiation of antimicrobial therapy, *PLoS One*, 2012, 7(10), e46389.
222. L. Liu, W. Wu, D. Chen and Y. Lu, Distribution and drug resistance of pathogens causing catheter-related urinary tract infections, *Zhonghua Yiyuanganranxue Zazhi*, 2012, **22**(7), 1347–1348.
223. G. Cheng, L. Jiang and X. Yao, Distribution and antimicrobial resistance of pathogenic bacteria causing urethral catheter-associated urinary, *Zhonghua Yiyuanganranxue Zazhi*, 2012, **22**(4), 854–855.
224. R. Hu, G. Li, Y. Jiang, Y. Zhang, J.-J. Zou, L. Wang and X. Zhang, Silver-zwitterion organic–inorganic nanocomposite with antimicrobial and antiadhesive capabilities, *Langmuir*, 2013, **29**(11), 3773–3779.
225. C. Díaz, A. Miñán, P. Schilardi and M. F. L. de Mele, Synergistic antimicrobial effect against early biofilm formation: micropatterned surface plus antibiotic treatment, *Int. J. Antimicrob. Agents*, 2012, **40**(3), 221–226.
226. Z. Li, D. Lee, X. Sheng, R. E. Cohen and M. F. Rubner, Two-Level Antibacterial Coating with Both Release-Killing and Contact-Killing Capabilities, *Langmuir*, 2006, **22**, 9820–9823.
227. J. A. Lichter, K. J. Van Vliet and M. F. Rubner, Design of antibacterial surfaces and interfaces: polyelectrolyte multilayers as a multifunctional platform, *Macromolecules*, 2009, **42**(22), 8573–8586.
228. B. V. Worley, D. L. Slomberg and M. H. Schoenfisch, Nitric oxide-releasing quaternary ammonium-modified poly(amidoamine)

dendrimers as dual action antibacterial agents, *Biconjugate Chem.*, 2014, **25**(5), 918–927.

229. R. Mala, A. Annie Aglin, A. S. Ruby Celsia, S. Geerthika, N. Kiruthika, C. VazagaPriya and K. Srinivasa Kumar, Foley catheters functionalised with a synergistic combination of antibiotics and silver nanoparticles resist biofilm formation, *IET Nanobiotechnol.*, 2017, **11**(5), 612–620.
230. Y. J. Choi, J. K. Lim, J. J. Park, H. Huh, D. J. Kim, C. H. Gong and S. Z. Yoon, Chlorhexidine and silver sulfadiazine coating on central venous catheters is not sufficient for protection against catheter-related infection: Simulation-based laboratory research with clinical validation, *J. Int. Med. Res.*, 2017, **45**(3), 1042–1053.

CHAPTER 15

Dendrimers and Hyperbranched Polymers as Antimicrobial Agents

CHANDRAKALA UMMADISETTI,[a,†]
KONDA REDDY KUNDURU*[b,†] AND ABRAHAM J. DOMB*[c]

[a] Centre for Lipid Research, CSIR-Indian Institute of Chemical Technology, Hyderabad, India; [b] School of Chemistry, University of Hyderabad, Hyderabad, India; [c] School of Pharmacy-Faculty of Medicine, The Hebrew University of Jerusalem, Jerusalem, Israel
*Email: kkondareddy@gmail.com; avid@ekmd.huji.ac.il

15.1 Introduction

Since the 20th century polymer science and technology has focused mainly on the development of linear polymer chains. Various techniques have been developed in which individual repeating units are joined by polymerization with their covalent bonds. This opened up the concept of macromolecules. Since then various types of macromolecules with different architectures have been effectively synthesized. Among them, dendrimers and hyperbranched polymers merit special attention from both academic and industrial points of view, due to their multifaceted properties and applications as catalysts and as materials for various applications in biology.[1–4]

[†]Equal contribution to this chapter.

Biomaterials Science Series No. 5
Antimicrobial Materials for Biomedical Applications
Edited by Abraham J. Domb, Konda Reddy Kunduru and Shady Farah

Published by the Royal Society of Chemistry, www.rsc.org

Dendrimers and hyperbranched polymers are a group of macromolecules that have a novel, globular, and highly branched architecture, unlike linear and cross-linked polymers (Table 15.1).[5] Among these, dendrimers are characterized by their perfectly branched structure and absolutely symmetrical globular shape and mono-dispersity, having all the end-groups at the surface. Whereas hyperbranched polymers resemble dendrimers in structure and shape with multiple branching and end-groups, they are different with respect to their less perfect globular shape and dispersity (*Đ*). Hyperbranched polymers have many structural similarities to dendrimers and have attracted increasingly more attention than dendrimers. This is because of their unique properties, easy production on a large scale, and greater commercial availability. The novel architectures of these highly branched polymers (both dendrimer and hyperbranched) have attracted much attention from a nanotechnology point of view, because of their diverse range of characteristic features such as, higher solubility, lower viscosity, and higher number of terminal groups, compared with those of the corresponding linear polymers (Table 15.1).[5]

However, the synthesis of dendrimers is a time-consuming process due to multi-step procedures of protection, deprotection, and purification. In contrast to dendrimers, hyperbranched polymers have an irregular structure, but they possess properties similar to the dendrimer and can be easily synthesized *via* one-pot polymerization. Anne-Marie Caminade[4] and co-workers collected up-to-date reviews on dendrimers and hyperbranched

Table 15.1 Comparison of linear, dendrimer and hyperbranched polymers.[a] [5]

Polymer	Linear	Dendrimer	Hyperbranched
Structure	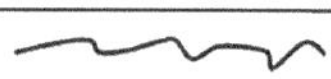	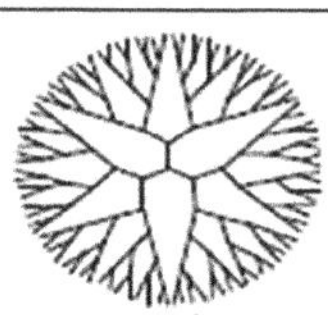	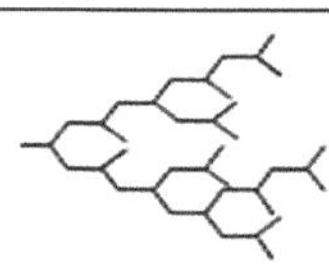
Topology	1D, linear	3D, regular	3D, irregular
Synthesis	One-step, facile	Multistep, laborious	One step, relatively facile
Purification	Precipitation	Chromatography	Precipitation or classification
Scaling-up	Already, easy	Difficult	Already, easy
Molecular weight	Discrepant	Identical	Discrepant
Dispersity *Đ*	>1.1	1.0 (<1.05)	>1.1
Degree of branching	0	1.0	0.4–0.6
Entanglement	Strong	Very weak or none	Weak
Viscosity	High	Very low	Low
Solubility	Low	High	High
Functional group	At two ends	On periphery (terminal units)	At linear and terminal ends
Reactivity	Low	High	High
Strength	High	Very low	Low

[a]Reproduced from ref. 5 with permission from The Royal Society of Chemistry.

polymers as a thematic issue that constitutes a comprehensive account of all the research carried out on dendrimers and hyperbranched polymers by the most prominent authors in the field. Since this chapter is concerned with the antimicrobial properties of dendrimers and hyperbranched polymers, we present only a brief discussion concerning their synthesis and various applications along with a detailed summary of anti-microbial properties.

15.2 Dendrimers

Dendrimers are the "polymers of the 21st century". Research on dendrimers has been explored and has attracted considerable attention around the world by synthesizing structurally perfect dendritic polymers compared with traditional linear polymers. The first synthesized dendrimers were poly-propyl amine[6] and polylysine dendrimer.[7] The first dendrimers to receive widespread attention were polyamidoamine dendrimers (PAMAM)[8] and "arborol" systems[9] (Figure 15.1).

15.2.1 Synthesis of Dendrimers

Dendrimers are synthesized in an iterative sequence of reaction steps that renders a unique, globular, three-dimensional macromolecular shape to the dendrimers. Due to their peculiar structural shape, dendrimers are highly soluble with low viscosity, adhesivity, and glass-transition temperatures corresponding to a linear analog. Dendrimers are synthesized either by divergent or by convergent methodology, as mentioned in the literature.[10–16] In a divergent synthesis the structure of dendrimer starts from the central core of the molecule and grows toward the periphery in a stepwise manner (Figure 15.2) by an iterative addition of monomer units. Tomalia and co-workers[8] used this protocol for the synthesis of PAMAM-NH_2 dendrimers by coupling of *N*-(2-aminoethyl) acrylamide monomers to an ammonia core. In a convergent synthesis, the growth of the dendrimer begins with the surface units that are coupled to additional building blocks from the periphery towards the focal point to construct a dendron (Figure 15.3), and then each dendron is coupled to a central core of the molecule for the formation of a complete dendrimer. The formed product in this protocol is simple to purify due to the difference from that of the reaction by-products. Hawker and Fréchet first published this method to synthesize polyether dendrimers by coupling of 3,5-dihydroxylbenzyl alcohol units to an activated benzyl bromide.[11]

Various types of dendrimers,[17] such as PAMAM, polypropylene imine (PPI), polyester, polylysine, poly(2,2-bis(hydroxymethyl) propionic acid, polypropyleneetherimine, peptide, carbohydrate, triazine, melamine, phosphorous, and tecto dendrimers are synthesized either by the convergent or divergent methods. Among these dendrimers, those most commonly used for biomedical applications are PAMAM and PPI (Figure 15.4).[17]

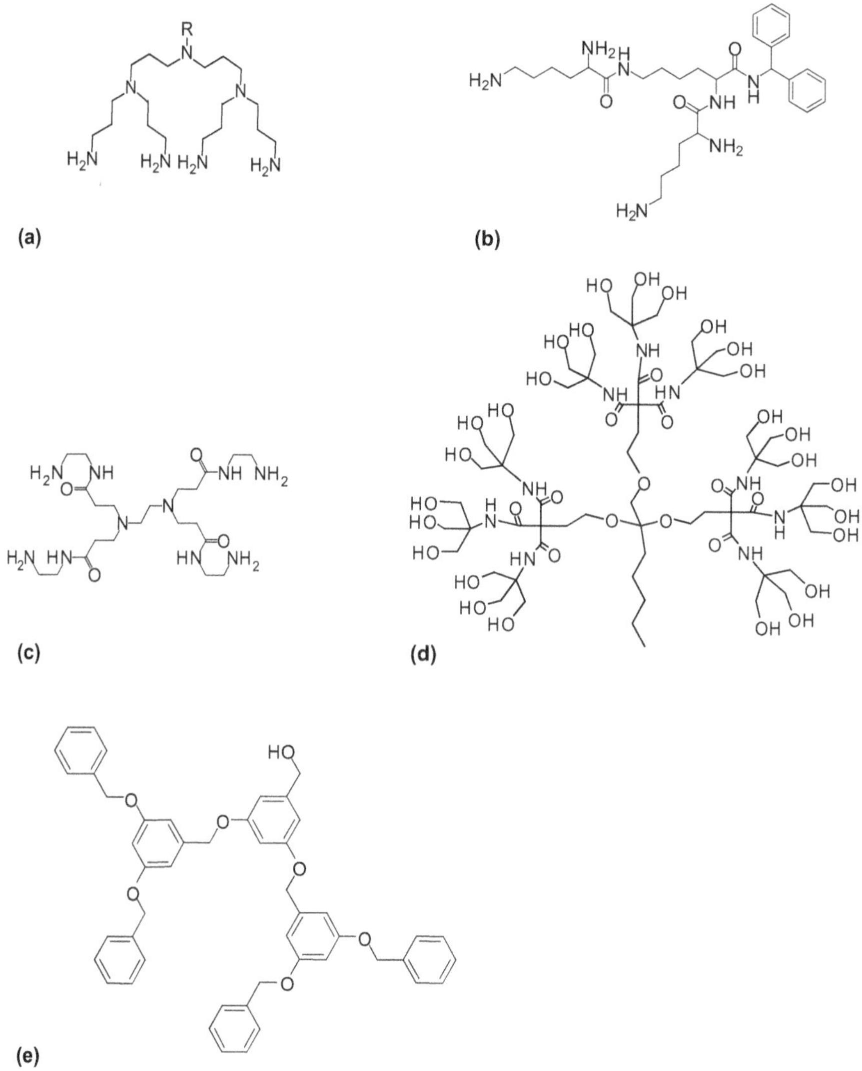

Figure 15.1 Various types of dendritic polymers: (a) Polypropyl amine dendrimer;[6] (b) polylysine dendrimer;[7] (c) PAMAM G-1 dendrimer;[8] (d) "arborol" system;[9] (e) Aromatic polyether dendrimers.[10]

15.2.2 Applications

Recently, researchers around the world have paid attention to the synthesis of dendrimers due to their structural perfection and unique properties. Commercial applications of these perfectly branched molecules, however, are scarce, since the tedious multi-step synthesis results in unacceptable costs for most applications. Dendrimers, therefore, are best used for high value-added applications[14,15] to justify high production costs. As a result,

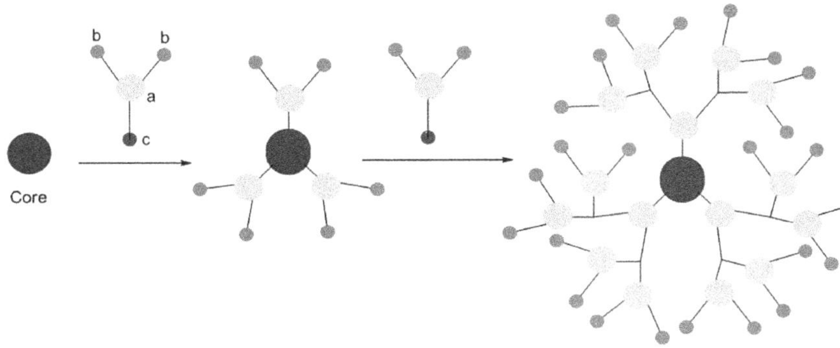

Figure 15.2 Divergent method for the synthesis of dendrimers.

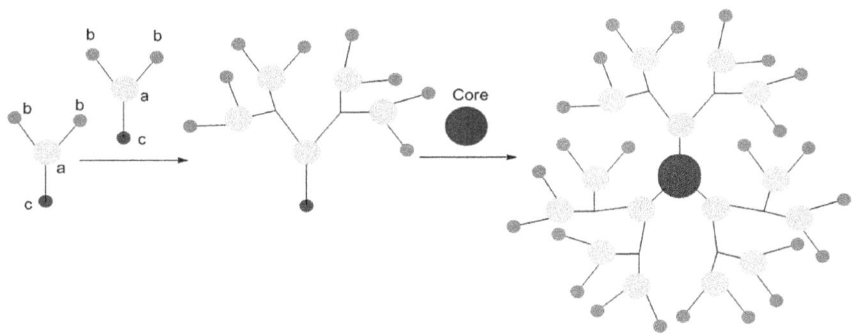

Figure 15.3 Convergent method for the synthesis of dendrimers.

researchers explored the synthesis of several accelerated dendrimers with a variety of applications[18–32] like encapsulation of guest molecules, enzyme mimics, processes of biological recognition, diagnostic agents, gene and drug delivery, chemical sensors, nanoscale catalysts, and unimolecular micelles, as well as in light harvesting and energy transfer. Despite the above-mentioned applications, in this chapter, the major emphasis is on antimicrobial activity studies of dendrimers.

15.2.3 Dendrimers as Antimicrobial Agents

Recently, several studies have been focused on the development of effective antimicrobial agents using nanotechnology to overcome resistant pathogens. This is an alternative approach to the currently available antibiotics-based approaches.[22] Bioactive nanomaterials with potential biological applications have been extensively reported in the literature.[23,24] In these studies, dendrimers have proved effective in drug and gene delivery and as contrast agents. With their well-defined, highly branched dendrimeric architecture, end-groups with various functionalities have found diverse biomedical applications.[25–27] Among the commonly available dendrimers (PAMAM and

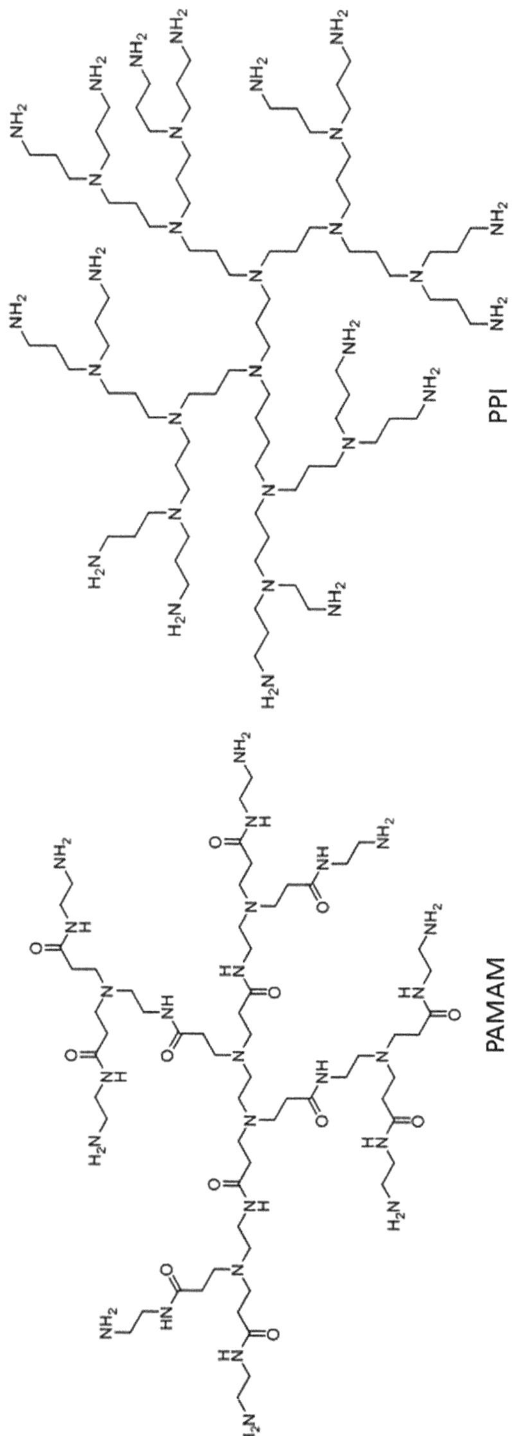

Figure 15.4 Structures of polyamidoamine (PAMAM) and polypropylene imine (PPI) dendrimers.

PPI), PAMAM dendrimers are mostly used for biomedical applications. However, knowledge, on the application of PPI dendrimers as antibacterial agents is scarce.[28–30] Certain types of dendrimers have not only shown potential antimicrobial properties, but they have also enhanced the antifungal or antibacterial activities of known drugs. Some important information on antimicrobial applications of dendrimers is presented in Table 15.2.

Natalia Wrońska and co-workers[30] studied the antibacterial activity of PPI dendrimers with amoxicillin, a β-lactam antibiotic commonly used in the treatment of bacterial infections.[31,32] These β-lactam antibiotics are detected mostly in aquatic environments of pharmaceutical contaminants.[33] Unfortunately, most Gram-negative bacteria are resistant to β-lactam antibiotics by producing β-lactamase, which is problematic for the medical application of these drugs.[34] Traces of these antibiotics in aquatic systems have an adverse effect and are essential to reduce the use of amoxicillin. This adverse effect can be minimized by simultaneous administration of amoxicillin- and dendrimer-type macromolecules. Recently, a series of studies has demonstrated that cationic dendrimers in high concentrations can be toxic to eukaryotic cells. Hence, to reduce the toxicity of PPI dendrimers Drzewińska *et al.*[35] chemically modified their surfaces to neutralize the cationic charge. They replaced some or all the cationic groups on fourth-generation PPI dendrimer surfaces by maltose and maltotriose to significantly reduce their toxicity. These modified glycodendrimers with a cationic core and a neutral surface charge are non-toxic to the cell lines studied.[36] Later, Wrońska *et al.*[30b] studied the antibacterial activity of amoxicillin in combination with PPI dendrimers on Gram-negative bacteria.[37] The authors used unmodified PPI dendrimers and their maltose derivatives in concentrations non-toxic to eukaryotic cells (Figure 15.5)[30b] against antimicrobial studies of two commonly used Gram-negative pathogens: *Escherichia coli* ATCC 25922 and *Pseudomonas aeruginosa* ATCC 15442. The infections caused by these bacteria are very difficult to cure. This is the first report of the simultaneous use of PPI dendrimers and β-lactam antibiotics against Gram-negative bacteria. The co-administration of amoxicillin and PPI dendrimer significantly reduces the growth of *P. aeruginosa*. The addition of unmodified PPI dendrimer and amoxicillin shows the inadequate growth of bacteria by 75%, even at the lowest dose of the dendrimers and the antibiotic, whereas for the dendrimer or the antibiotic applied alone, growth was inhibited by only 5% or 3%, respectively. In cultures containing maltose-modified dendrimers and the tested antibiotic, the reduction of bacterial growth was 30% higher compared to the control. The results clearly indicate that the antibiotic or the dendrimers (modified, unmodified) alone do not inhibit bacterial growth. Only the combination gives satisfactory results. *E. coli* strains also show promising results in the inhibition of bacterial growth. Results clearly indicate that modified PPI dendrimers, non-toxic to eukaryotic cells, show better activity against bacterial strains than unmodified PPI dendrimers when combined with amoxicillin.

Egambaram Murugan *et al.*[38] also studied the antimicrobial activity of PPI (G2)-silver nanoparticle hybrid dendrimers against *P. aeruginosa* and

Table 15.2 Information on antimicrobial applications of dendrimers.[a]

Dendrimer type/ components	Synthesis of dendrimer	Highlights	Antimicrobial action causing functionality	Antimicrobial action against	Application	Ref.
PPI-maltose dendrimers (PPI 100%malG3)	Using maltose and PPI	Non-toxic to eukaryotic cells Co-administration of PPI dendrimers and AMX enhanced the antimicrobial effect of AMX Possibility of reducing the doses of antibiotic applied, which reduces the environmental pollution caused by the widespread use of β-lactam antibiotics	*N*-substituted maltose groups	*E. coli* (ATCC 25922), *P. aeruginosa* (ATCC 15442)	Antimicrobial activity	30
PPI (G2)-silver nanoparticle hybrid dendrimers (MWCNTs-PPI(G2)-AgNPs nanohybdrids)	Using MWCNTs–COOH and PPI(G2)-Agþ complex	Reusability of catalyst, *viz* MWCNT–PPI (G2)–AgNPs-4 proved that the activity of the catalyst does not alter up to the fifth cycle.	–COOH groups and AgNPs	*P. aeruginosa* and *S. aureus*	Potent antimicrobial activity	38
Novel naphthalimide–poly(amidoamine) dendrimer fluorescent dyes	Using PAMAM dendrimers and naphthalimide dye	PAMAM dendrimers enhance the dyeing ability and improved the antimicrobial activity of the dyed samples Enhancing dyeing, fastness and	Quaternary ammonium salt structure	*S. aureus* and *E. coli*	Potent antimicrobial agents on wool fibers Promising tool in tailoring the different properties of	44

		antimicrobial properties of dyed wool fibers			naphthalimide dyes, being suitable for dyeing and antimicrobial finishing agents for wool fibers.	
Chitosan-PPI dendrimer hybrid	Using (PPI) (G = 2) and *N*-carboxyethyl chitosan	Dye uptake of nylon was considerably improved Antibacterial activity was an additional property imparted to nylon fabrics Eco-friendly finishing agent multifunctional modification of the nylon 6 fabrics	Free amine groups	*E. coli* and *S. aureus*	Dye removal from wastewater, antimicrobial finishing agent for textiles, and salt-free reactive dyeing of wool and cotton	41–43
G-1 TPGDA G1.0 (5) QOI dendrimer quaternary ammonium dendrimer of TPGDA	Using TPGDA, EDA, and OI	First report of integration of antimicrobial agent into the cement matrix as a co-monomer in the formulation Act as additives, including antibiotic drugs and silver materials, based formulations highly efficient at killing bacteria	Quaternary ammonium branches.	*E. coli* (ATCC 25922), *S. aureus* (ATCC 33807), *P. aeruginosa* (ATCC 27853)	Potential antimicrobial ingredient in acrylic bone cement	47
AMPDs, namely G3KL and G3RL	AMPDs were obtained by solid-phase peptide synthesis	In combination with biological bandages for improved treatment	Amino acid moiety (lysine/leucine molecules)	*P. aeruginosa* and *A. baumannii*	Use of polycationic AMPDs might inhibit	49, 51, 52

Table 15.2 (*Continued*)

Dendrimer type/components	Synthesis of dendrimer	Highlights	Antimicrobial action causing functionality	Antimicrobial action against	Application	Ref.
		with anti-microbial effects on burn wounds, anti-angiogenic effects, and tumor suppression			angiogenesis and thus risk skin regeneration progress Pharmacological formulations and potential clinical applications Antimicrobial agents within the biological bandage formulation	
PAMAM-NH_2 (G2 and G3)	Using methyl acrylate and ethylene diamine	PAMAM-NH_2 dendrimers G2 and G3 on physicochemical properties of hydrogels with erythromycin Greater effect on reducing the viscosity, hardness, cohesiveness, and adhesiveness of hydrogels Stable over 3 months of storage PAMAM-NH_2 improved the *in vitro* release of erythromycin	NH_2 groups	*S. erythreus* *S. aureus* (ATCC 29213) *E. faecalis* (ATCC 29212)	Used in topical applications for the treatment of acne rosacea, acne vulgaris, infections of skin and soft tissue, inflammation of the gums and eyelids	55

Organoiron melamine dendrimers capped with chloro, hydroxyl- and piperazine terminal end-groups	Arenecyclopentadienyliron cations were synthesized using the divergent method and melamine was used as a core, capped with piperazine, chloro- or hydroxyl-groups in the periphery	Biologically active against cancer and bacterial targets	Chloro, hydroxyl- and piperazine terminal end-groups and also melanine core	*S. aureus* *S. warneri* Vancomycin-resistant *E. faecium* *P. vulgaris* *P. aeruginosa* *C. albicans*	Against cancer and bacterial targets Screened against MCF-7 and HTB-26 breast cancer cell lines and dendrimer exhibited significant inhibitory activity	61
Peptide dendrimers (peptide derivatized-dendrimers), carbosilane dendrimers, polysulfated galactose functionalized glycodendrimers and PAMAM dendrimers		Dendrimeric treatments for mitigating the increasing prevalence of viral STIs associated with HS	Sulfonate groups in the periphery in carbosilane and glycodendrimers dendrimers Amino acid sequence in peptide dendrimers	HSV-2, HIV and HPV	Treatment of STIs: peptide dendrimers mainly used in biomedical Applications—design of new drug delivery systems, gene delivery, and their activity as therapeutic agents Anticancer drug delivery systems	62, 66
GATG dendrimers abbreviated as Gn-NH_2	Using gallic acid and triethylene glycol	Potentiality of GATG dendritic materials as a platform to develop new antimicrobials that can target microbial viability and/or virulence	–COOH and NH_2	*V. harveyi*	Antimicrobial applications	63

Table 15.2 (*Continued*)

Dendrimer type/components	Synthesis of dendrimer	Highlights	Antimicrobial action causing functionality	Antimicrobial action against	Application	Ref.
		Inhibit growth and increased membrane permeability in combination with cell clustering may be promising antibacterial features of these cationic dendrimers				
PAMAM dendrimer and PAMAM-EGDMA	For PAMAM-ethylene diamine and methyl acrylate For (PAMAM G1.0 EGDMA) -PAMAM G1.0 was added to a stirred solution of EGDMA	Disinfected 100% bacteria in water within a contact time of 5 minutes and maintained their activity even after 25 cycles	$-NH_2$, $-OH$	*S. aureus* *E. coli* (ATCC 25922), *S. aureus* (ATCC 33807), *P. aeruginosa* (ATCC 27853) and *B. subtilis* (ATCC 6633)	Antimicrobial agent Utilization in designing drinking water disinfection cartridge	65
Biocompatible PGSA dendrimer with anionic surfaces		New antimicrobial agent against human pathogens, in addition to drug delivery application and hemocompatibility	Anionic acid terminal groups	*S. aureus* *E. coli* *C. albicans*	Drug delivery applications	64

Novel amphiphilic peptide dendrimer (BALY) with potential application against multi-drug resistant bacteria			–	*S. aureus* (ATCC 25923), methicillin-resistant *S. aureus* (ATCC 43300) (MRSA), *E. coli* 25922 and *P. aeruginosa*	Dendrimers were able to destroy the membrane integrity through different mechanisms depending on the lipid phase and morphology. Drug delivery systems, especially in cancer, which continues to be a huge challenge for the therapeutics	72
Metallo phosphorus dendrimers (generation 3, 48 terminal groups	Complexation of phosphorus dendrimers bearing imino-pyridino end-groups with Au (III) or both with Au (III) and Cu (II)	Complexation of the dendrimer with Au (III) strongly increased (~30-fold) the anti-proliferative activities against both KB and HL-60	Metal atoms	*S. aureus* (ATCC 6538), *E. coli* (ATCC 8739), *P. aeruginosa* (ATCC 27853) *C. albicans* (ATCC 10231)	Anti-proliferative activities against both KB and HL-60 tumoral cell lines	73

[a] *A. baumannii*: *Acinetobacter baumannii*; *B. subtilis*: *Bacillus subtilis*; *C. albicans*: *Candida albicans*; *E. coli*: *Escherichia coli*; *E. faecalis*: *Enterococcus faecalis*; *E. faecium*: *Enterococcus faecium*; *P. aeruginosa*: *Pseudomonas aeruginosa*; *P. vulgaris*: *Proteus vulgaris*; *S. aureus*: *Staphylococcus aureus*; *S. erythreus*: *Streptomyces erythreus*; *S. warneri*: *Staphylococcus warneri*; *V. harveyi*: *Vibrio harveyi*. AgNPs: silver nanoparticles; AMPDs: antimicrobial peptide dendrimers; AMX: amoxicillin EDA: ethylene diamine; EGDMA: ethyleneglycol dimethacrylate; GATG: gallic acid-triethylene glycol; HIV: human immunodeficiency virus; HPV: human papillomavirus; HS: heparan sulfate; HSV2: herpes simplex virus type 2; MWCNT: multi-walled carbon nanotube; PAMAM: polyamidoamine; PGSA: phloroglucinol succinic acid; PPI: poly (propylene imine); QOI: quaternary octyl iodide; TPGDA: tripropylene glycol diacrylate; STI: sexually transmitted infection.

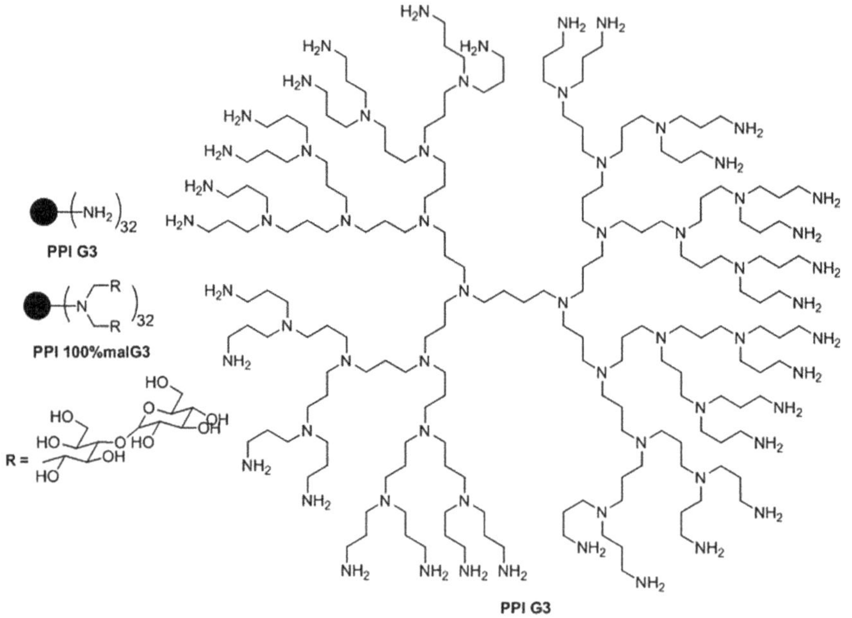

Figure 15.5 Structures of polypropylene imine (PPI) G3 dendrimers unmodified (PPI) and modified with maltose in 100% (PPI 100%malG3).[30b] Reproduced from ref. 30*b*, http://dx.doi.org/10.3390/molecules201019330, under the terms of the CC BY 4.0 licence, https://creativecommons.org/licenses/by/4.0/.

Staphylococcus aureus. They synthesized nanohybrid dendrimer catalysts with various silver concentrations. The *in vitro* antimicrobial studies of the newly prepared multi-walled carbon nanotube (MWCNT)-based silver nanoparticles (AgNPs) nanohybrid dendrimers (MWCNT-PPI (G2)-AgNPs nanohybrids) (Figure 15.6) prove that they are more highly active against *P. aeruginosa* and *S. aureus* bacterial pathogens than individual AgNPs.

Currently, colored textile wastewater creates serious environmental and ecological problems during the nylon dying process.[39] On the other hand, due to the restriction of terminal amine groups on nylon, the dye-absorbing sites obtained heavy shades.[40] To overcome this problem, researchers have recently modified nylon surfaces with chitosan and its derivatives, which may enhance some chemical and physical properties such as color fastness, dye-ability, wettability, antimicrobial properties, and shrink-proof characteristics. Moreover, modification of the surface appreciably decreases the amount of hazardous materials in the dye waste water. Chitosan-polypropylene imine dendrimer (CS-PPI) with numerous terminal amine groups has been used in dye removal from wastewater,[41] as an antimicrobial finishing agent for textiles,[42] and as a salt-free reactive dye of wool and cotton.[43] Antibacterial activity is imparted to nylon against *E. coli* and *S. aureus* microorganisms, when nylon is treated with CS-PPI.

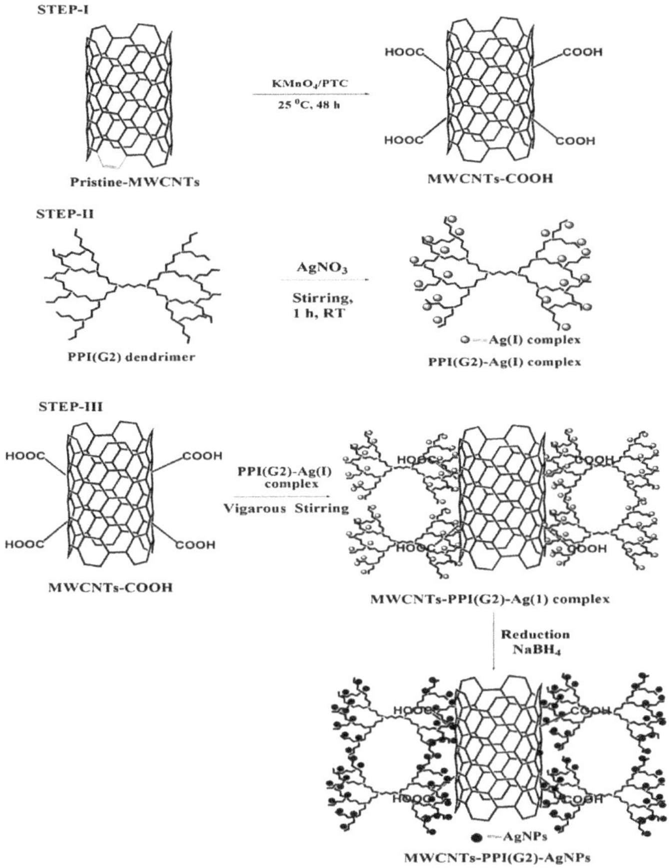

Figure 15.6 Multi-walled carbon nanotube (MWCNT)-PPI(G2) dendrimer stabilized silver nanoparticle hybrids (MWCNTs-PPI (G2)-AgNPs).
Reproduced from ref. 38 with permission from Taylor & Francis, Copyright 2016.

Mousa Sadeghi-Kiakhani *et al.*[44] synthesized naphthalimide–dendrimer hybrids as fluorescent dyes on wool fibers by the modification of naphthalimide derivatives with PAMAM dendrimers ($G = -0.5$ and $G = 0.5$). The antimicrobial activity of the synthesized dyes was targeted against *S. aureus* and *E. coli* bacteria. Wool fibers dyed with naphthalimide–dendrimer hybrids showed more effective antimicrobial activity than those dyed with naphthalimide alone.

Currently, acrylic bone cement is widely used in orthopaedic surgery for the anchoring of contiguous bones in their site. Preventing periprosthetic deep wound infections, commonly called prosthetic joint infection, by methicillin-resistant and -susceptible *S. aureus*, *Staphylococcus epidermis*, *Staphylococcus hominis*, and *P. aeruginosa*[45] is still a challenge during this process despite strict intraoperative infection control protocols. These infections persist as frequent and serious complications. To avoid this, antibiotics such as gentamicin sulfate are frequently used in bone cement to lower the risk of infection. However, 88% of infections in patients with primary arthroplasty fixed with gentamicin-loaded bone cement show that at least one of the infecting strains of staphylococcus is resistant to gentamicin. In addition, various bacterial infections are also associated with acrylic bone cement in revision surgery and are also gentamicin sulphate resistant.[46] Hence, it is essential to design new formulations of antimicrobial compounds with broad-spectrum killing properties. To counter these antimicrobial infections, Abid *et al.*[47] synthesized G-1 (generation 1) quaternary ammonium dendrimers of tripropylene glycol diacrylate (TPGDA) in controlled conditions *via* simple chemical procedures. This dendrimer acts as a potential antimicrobial ingredient in acrylic bone cement. The advantages over other approaches are hydrophobicity and compatibility with acrylic bone cement formulations. Significantly, these dendrimers work in a non-eluting mode without releasing any active materials. They have substantial potential as an antibacterial agent in cements. This dendrimer, even in 10% concentration in bone cement, is sufficient to destroy both Gram-positive and -negative bacteria. It retains its antimicrobial property even after 30 days.

Wound-healing biological bandages are used for their excellent wound-healing properties. *P. aeruginosa* is the most commonly found Gram-negative bacterium in burn wound infections, and it is resistant to many antibiotics.[48] In one study,[49] antimicrobial peptide dendrimers (AMPDs; namely G3KL and G3RL) were reported against multidrug-resistant (MDR) *P. aeruginosa*, which was resistant to at least two different classes of antibiotics. They have shown good results at minimum inhibitory concentrations. This novel AMPD has demonstrated *in vitro* activity against several Gram-negative strains, low toxicity to human red blood cells, stability in human serum, and easy preparation by standard solid-phase peptide synthesis.

These polycationic AMPDs are capable of having anti-angiogenic effects and exhibit tumor suppression applications.[50] The use of polycationic AMPDs might inhibit angiogenesis, and thus they can pose risks to the skin regeneration process. Abdel-Sayed *et al.*[51] used the two polycationic AMPDs, (G3KL and G3RL) in combination with biological bandages for an improved treatment with anti-microbial effects on burn wounds. At the lowest concentration, the AMPDs can inhibit *P. aeruginosa* growth in combination with the biological bandages, thus preventing its contamination with bacteria. The dendrimers of this study open an innovative opportunity as antimicrobial agents in the context of burn wound healing. João Pires *et al.*[52] also studied the antimicrobial activities of the novel antimicrobial peptide dendrimer G3KL against MDR strains such as *Acinetobacter baumannii* and *P. aeruginosa* strains, compared to standard antibiotics. The results clearly indicate that G3KL acts as a promising molecule with antibacterial activity against MDR and extensively drug-resistant strains.

Several studies have highlighted the importance of hydrogels as vehicles for dermal delivery of a wide range of drugs in the pharmaceutical industry.[53] Hydrogels are hydrophilic polymer networks with three-dimensional architecture capable of swelling in water or biological fluids, thus retaining a large amount of fluids in the swollen state.[54] The use of hydrogels in dermatology has several favourable features, *i.e.* adhesiveness, biocompatibility, spreadability, thixotropy, and simplicity of removal. Moreover, these are easily applicable and have better percutaneous absorption than other semisolid preparations. Magdalena Wróblewska and Katarzyna Winnicka[55] studied the effect of addition of dendrimers (PAMAM-NH_2 (G2 and G3)) on the physicochemical characteristics of hydrogels with erythromycin. They also examined the influence of PAMAM-NH_2 dendrimer on hydrogel stability and antimicrobial activity. *Streptomyces erythreus* (Gram-positive) strains produce erythromycin, which is a mixture of macrolide antibiotics soluble in alcohol and slightly soluble in water.[56] It is commonly used in topical applications for the treatment of acne rosacea, acne vulgaris, infections of skin and soft tissue, and inflammation of the gums and eyelids. In the commercial preparation of erythromycin, the ingredients present in these formulations have been reported as probable causes of adverse effects such as oiliness, local dryness, and desquamation.[57] To overcome these effects, hydrogels with erythromycin were designed and the effect of PAMAM-NH_2 (G2 or G3) on their physicochemical characteristics was evaluated. The researchers demonstrated that PAMAM dendrimers significantly enhance the solubility of erythromycin.[58] The particle size of erythromycin also decreases with the increase of PAMAM-NH_2 dendrimer concentration, due to the improved solubility of erythromycin in the presence of dendrimers. *In vitro* release studies show that erythromycin is released faster from hydrogels containing PAMAM-NH_2 than from hydrogels without dendrimers. Other researchers[59,60] worked on antibiotics such as nifedipine and clotrimazole, adding PAMAM-NH_2 dendrimer to hydrogels. These increase the *in vitro* release of the antibiotics.

Alaa S. Abd-El-Aziz *et al.*[61] synthesized novel organoiron melamine dendrimers capped with chloro, hydroxyl- and piperazine terminal end-groups and melamine as a core to determine the biological activities of these unique structures against cancer and bacterial targets. Dendrimers with a melamine core increase the solubility of some cancer drugs and also reduces the known hepatotoxicity associated with these drugs. These novel dendrimers inhibit the growth of both Gram-positive bacteria (methicillin-resistant *S. aureus* (MRSA), *Staphylococcus warneri*, vancomycin-resistant *Enterococcus faecium*), Gram-negative bacteria (*Proteus vulgaris* and *P. aeruginosa*), and yeast (*Candida albicans*).

One major recent challenge is sexually transmitted infections (STIs). Even though STIs are caused by more than 30 bacterial, viral, and parasitic pathogens, viruses still play a major role. The most commonly infected viral sexually transmitted diseases (STDs) are human immunodeficiency virus (HIV), herpes simplex virus (HSV) type 2, hepatitis C virus (HCV), and human papilloma viruses (HPV). However, sexual diseases associated with bacterial or parasitic pathogens with higher worldwide prevalence have a functional cure, whereas no cure has been found for the STIs associated with viral pathogens. There is, therefore, a need to deal with these viral infections to reduce economic as well as public health problems.

Rafael Ceña-Díez *et al.*[62] studied these STDs by highlighting the viral STIs associated with cell-surface heparan sulfate, which is a common link for STIs, with new dendrimeric treatments. They observed that the derivatives of peptide dendrimers with differences in the amino acid sequences of their surface groups (carbosilane dendrimers, polyamidoamine dendrimer derivatives with naphthalene disulfonic surface groups and polylysine dendrimers (polysulfated galactose functionalized glycodendrimer)) display potent *in vitro* antiviral activity against HSV-2 by inhibiting viral attachment to heparan sulfate. Polysulfated galactose-functionalized glycodendrimer (PS Gal 64mer) that contains two sulphate groups per galactose was synthesized and evaluated against HIV-1 infection. Based on the studies, the researchers proposed a mechanism of action for these dendrimers on heparan sulfate in several STIs. The studies reveal that heparan sulfate plays a major role in the case of HCV, HSV-2, and HPV viruses, but it is just an ancillary factor in HIV-1. The dendrimers show a common mechanism of action in all pathologies, acting at the level of viral entry into the target cell either binding to cellular co-receptors or directly blocking the viral particles that are meant to bind to the heparan sulfate.

Gallic acid-triethylene glycol (GATG) dendrimers functionalized with primary amines were synthesized.[63] Their antimicrobial properties on *Vibrio harveyi*, a marine pathogen, were investigated. GATG dendrimers are more active on pathogens compared to linear polymers of this category.

Murugesan Suresh Kumar[64] *et al.* synthesized a biocompatible phloroglucinol succinic acid (PGSA) dendrimer and studied its antibacterial activities against *S. aureus* and *E. coli* and fungal human pathogen *C. albicans*. The PGSA dendrimer with *S. aureus* shows better activities than two other pathogens, *E. coli* and *C. albicans*.

Abid *et al.*[65] prepared quaternary ammonium functionalized PAMAM G-1 dendrimer and PAMAM-ethyleneglycol dimethacrylate (EGDMA)G-1 dendrimer. These dendrimers show effective antimicrobial activities against *S. aureus* bacteria. The antibacterial activity of these dendrimers was also evaluated against *E. coli*, *S. aureus*, *P. aeruginosa*, and *B. subtilis*, which are the common Gram-negative and Gram-positive bacteria contaminants present in water. They proved highly efficient. The antimicrobial action of quaternary ammonium dendrimers was based on a contact killing mechanism. Even at low concentration these dendrimers are efficient to disinfect all types of tested bacteria in water. They maintain their activity even after 25 cycles. They are not toxic to mammalian cells, but are efficient to kill microorganisms.

Santos *et al.*[66] investigated the importance of peptide dendrimers in the design of new drug delivery systems, gene delivery, and as therapeutic agents. Urea function is interesting and may be employed as a core for the development of self-assembling dendrimers. Eggimann *et al.*[67] synthesized peptide dendrimers, which have efficient antimicrobial and antiviral properties. As antivirals, multiple antigen peptides dendrimers generally act in two ways, either binding to virus receptors on the cell surface or mimicking the cell receptor. Joshi *et al.*[68] used dendrimer heparin-sulfate biding peptide for the inhibition of HPV, human cytomegalovirus, and HSV types 1 and 2. Synthetic nucleoside analogs targeting viral DNA polymerase are a common use for the treatment of HSV infections. However, extended treatment may fail because of virus resistance. Tarallo and colleagues[69] investigated poly(amide)-based dendrimers as antiviral agents to overcome this problem. Some dendrimers received a gH625 peptide (membranotropic peptide sequence), which interacts with the membrane bilayer and possesses some antiviral activity. These dendrimers together with propidium iodide to distinguish apoptosis from necrosis were tested in African green monkey kidney cells (Vero). Results show that the inhibition of infectivity may be due to the formation of inactive aggregates. Moreover, gH625 peptide coupled to the dendrimer was effective to prevent viral entry and viral infectivity. HSV virus presents a consistent reduction in replication and more than 80% inhibition. In contrast, dendrimer without any peptide produces only 35% inhibition, suggesting that this structure has some antiviral activity. It is worth considering that at low toxicity approximately 90% of cells treated with both dendrimers survive.

Gupta *et al.*[70] prepared a dendrimer with glycerol and dimethyl ester of polyethylene glycol bis(carboxy methyl) ether (Figure 15.7) by employing immobilized *Candida antarctica* lipase (novozyme 435). Both the synthons used in the reaction are non-toxic, biocompatible, and readily available. The polymerization of the primary hydroxyl groups of glycerol occurs regioselectively by leaving the secondary hydroxyls for post-polymerization, chemical modifications, and also for attaching drugs/bioactive molecules. The secondary hydroxyl groups of these molecules were utilized for the synthesis of amphiphilic polymers by attaching alkyl chains using acylation. The resulting polymers have been evaluated for potential biomedical applications.

Stach and co-workers[71] synthesized AMPDs, varying with multiple short mono-, di-, or tripeptide branches that are resistant to proteolysis.

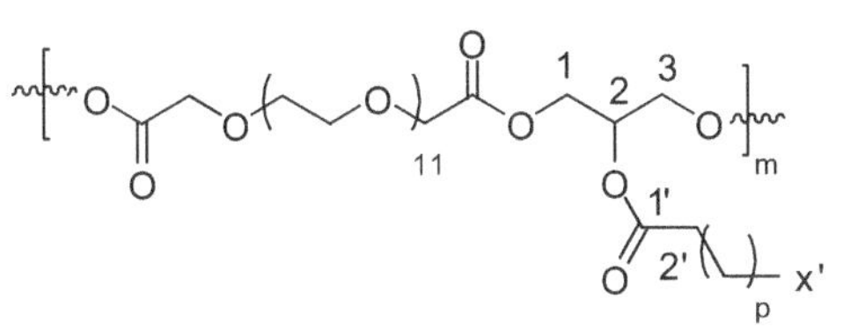

Figure 15.7 Glycerol and dimethyl ester of polyethylene glycol bis(carboxy methyl) ether dendrimer.[70]
Reproduced from ref. 70 with permission from John Wiley and Sons, Copyright © 2010 WILEY-VCH Verlag GmbH & Co. KGaA, Weinheim.

Antimicrobial activity was also studied against bacterium strains, namely *P. aeruginosa*, *E. coli*, and *B. subtilis*. Among the synthesized peptide dendrimers, G-3 dendrimer is active. In addition, activity of the peptide dendrimers depends up on the amino acid sequence. Change in the composition decreases activity and increases haemolytic resistance. It is worth mentioning that the AMPD is active in resistant strains. It is also stable, presenting strong activity in human serum.

In another study[72] a novel amphiphilic peptide dendrimer was synthesized with a potential application against MDR bacteria. The activity was evaluated in *S. aureus* ATCC 25923, methicillin-resistant *S. aureus* ATCC 43300 (MRSA), *E. coli* 25922, and *P. aeruginosa*. Results show that these dendrimers are able to destroy membrane integrity through different mechanisms depending on the lipid phase and morphology.

In another study, metallophosphorus dendrimers were synthesized[73] and studied for their antibacterial and antifungal activity against *S. aureus* ATCC 6538, *E. coli* ATCC 8739, *P. aeruginosa* ATCC 27853, and *C. albicans* ATCC 10231. They were also studied against drug-resistant clinical strains such as *S. aureus* ZMF KSK, *Enterococcus faecalis* ZMF BD 156, and yeasts *Candida glabrata* ZMF SZP 4. The results of this study indicate that all the dendrimers have a wide spectrum of bacteriostatic/fungistatic activity. In some cases, they also show bactericidal and fungicidal activity.

15.2.4 Antimicrobial Mechanism of Action of Dendrimers

Dendrimer structures contain a high density of positive charges that generate electrostatic interaction with negatively charged bacteria. When a dendrimer is in contact with bacteria, it first displaces surface divalent ions such as calcium and magnesium, and it immediately interacts with negatively charged phospholipids to cause membrane permeability. This interaction may neutralize the surface charge of bacteria. When the concentration of dendrimer is high, it may denature the membrane proteins and pass through the phospholipids bilayer, which in turn causes leakage of potassium ions. Even if there is a further increase in the concentration of dendrimer, it destabilizes the membrane structure and leads to complete disintegration of the bacterial cell.[74]

15.3 Hyperbranched Polymers

Hyperbranched polymers are a significant subclass of dendritic macromolecules having numerous branches connected by relatively short chains, resulting in the formation of highly branched tree-like structures with a three-dimensional dendritic architecture. Due to their non-crystalline behaviour and poor mechanical properties, they have not been considered as promising materials. However, since the discovery of dendrimers, they have gained importance due to their unique properties. Hyperbranched polymers are different from dendrimers, which are perfectly branched and

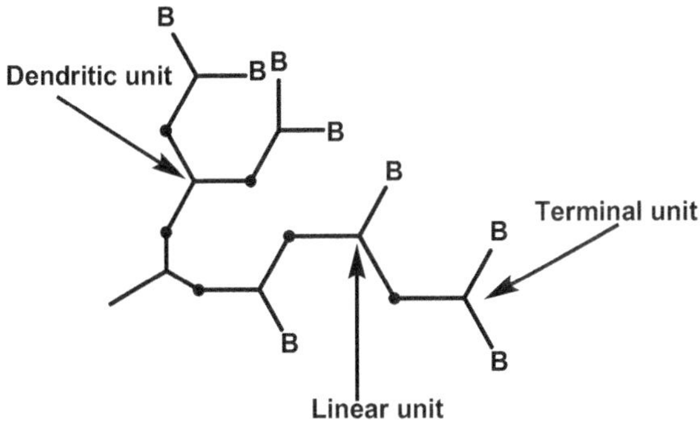

Figure 15.8 Schematic depiction of hyperbranched polymers.

monodisperse, consisting of dendritic units and terminal units. In contrast, hyperbranched polymers have a randomly branched structure with a lesser degree of branching composed of dendritic units, linear units, and terminal units (Figure 15.8). Dendrimers have been synthesized in a step-wise manner, whereas hyperbranched polymers can be synthesized in a one-step method.

Today, hyperbranched polymers have become an alternative to dendrimers due to their easy preparation and relatively lower cost. Hyperbranched polymers have a unique non-linear structure with more functional end-groups, which exhibits specific chemical and physical properties compared to linear analogs. These molecules, therefore, have attracted increasing attention leading to different applications. Even though these molecules possess some defects on the polymer backbone, they are believed to have properties similar to dendrimers. In particular, hyperbranched polymers exhibit unique properties and potential applications in diverse areas, mainly in the biomedical and biological fields. This is because of simple one-pot reactions in the preparation of hyperbranched polymers, which avoids complicated synthesis and purification procedures,[75–77] and thus reduces production costs. Another advantage is due to the presence of large numbers of functional end-groups in hyperbranched polymers, which provide a platform for the conjugation of functional groups and can also be employed to tailor-make the properties of hyperbranched polymers by enhancing their versatility in biological applications. In addition, hyperbranched polymers have excellent biocompatibility and biodegradability and the ability to incorporate a multiple array of guest molecules through covalent or non-covalent approaches, which yields potential for designing and producing biomaterials.

Synthesis of hyperbranched polymers depends on the applications, as discussed in the literature.[78–82] During the past decade there has been tremendous progress in hyperbranched polymers, but in terms of applications much remains to be accomplished. The structure of hyperbranched polymers has some limitations for potential applications. Hyperbranched polymers possess a less well-defined structure, and broad dispersity compared to dendrimers may be a drawback for certain applications. Hence, the controlled synthetic methodologies that yield hyperbranched polymers with reasonable dispersity are required to transfer this macromolecule from laboratory to clinic. Moreover, a continuous effort is required to modify and tailor hyperbranched polymers architectures to fit the future demands of biological and biomedical applications.

15.3.1 Synthesis of Hyperbranched Polymers

The synthesis of hyperbranched polymers involves two primary strategies: (a) single-monomer methodology, in which there is polymerization of an AB_n or a latent AB_n monomer; and (b) double-monomer methodology, in which there is direct polymerization of two types of monomers or a monomer pair. Detailed methodologies on the synthesis of hyperbranched polymers have been reported in the literature.[75,78,83,84]

15.3.2 Applications of Hyperbranched Polymers

Hyperbranched polymers contain highly reactive groups, large free volume and tailor-made properties, due to which they are highly soluble in solvents, have low viscosity, *etc.* They have been utilized in material science, nanotechnology, and in biomedicine.[85–103] This chapter, deals mainly with the antimicrobial applications of these hyperbranched polymers.

15.3.3 Antimicrobial Properties of Hyperbranched Polymers

Hyperbranched polymer-based antimicrobial agents are widespread due to their straight-forward preparation on a large scale, and they are also commercially available.[3] Some important information on the antimicrobial applications of hyperbranched polymers is discussed in Table 15.3.

Wang and co-workers prepared cationic hyperbranched poly(sulfone-amine)s with different branching for the preparation of poly(sulfone-amine)–Ag nanocomposites using an *in situ* approach.[104] The antimicrobial activity of poly(sulfone-amine)s and their polymer–silver nanocomposites has been studied. The degree of branching has great influence on antimicrobial activity, which is relatively different for poly(sulfone-amine)s and poly(sulfone-amine)–silver nanocomposites. The antimicrobial activity of poly(sulfone-amine)s decreases with an increase in the degree of branching. The reduced zeta-potential and low toxicity of these hyperbranched polymers is the main reason for this activity. In contrast, the nanocomposite shows an enhanced antimicrobial activity due to the high specific surface of small silver nanoparticles (NPs) with an increasing degree of branching.

Zhang and co-workers[105] prepared amine-functionalized silver NPs by a one-step reaction with amine-terminated hyperbranched polymers. Grafting was carried on the oxidized cotton fabric, showing excellent antibacterial

Table 15.3 Information on antimicrobial applications of hyperbranched polymers.[a]

Hyperbranched polymer type	Synthesis of hyperbranched polymer	Highlights	Antimicrobial action causing functionality	Antimicrobial action against	Application	Ref.
Hyperbranched PSA–Ag nanocomposites	PSAs with different branched architectures and $AgNO_3$ through an *in situ* approach	Showed an enhanced antimicrobial activity	PSA groups		Antimicrobial applications	104
(HP–NH_2)- AgNPs	One-step reaction with amino-terminated hyperbarnched polymer and AgNPs	Excellent antibacterial activity and laundering durability	NH_2		Antibacterial activity on cotton fabric.	105
HP–Ag (or Au) hybrid nanocomposites	Using stable AgNPs (or AuNPs) colloid system and amine-terminated HPAMAM (HPAMAM–NH_2) or dimethylamine terminated HPAMAMs–$N(CH_3)_2$	Exhibited excellent antibacterial activity	NH_2 and Ag/Au	Gram-positive and Gram-negative bacteria	Antimicrobial applications	110, 111
Hyperbranched glycoconjugated polymer	Using aminoglycoside through Michael-addition polymerization and incorporated gentamicin into the polymer backbone	High transfection and low cytotoxicity along with good antibacterial and antitumor activities.	—	—	Antimicrobial applications	112
Hyperbranched polyesters, polyethers,		Exhibit excellent *in vivo* stability			Used as drugs, as well as for dyes and other guest	113

Table 15.3 (*Continued*)

Hyperbranched polymer type	Synthesis of hyperbranched polymer	Highlights	Antimicrobial action causing functionality	Antimicrobial action against	Application	Ref.
polystyrenes, polyethylenes, poly(ureaurethanes), polyethylenimines, poly(amidoamine)s, polyethylenimines, polypeptides, polyphosphates, polyacrylates, and β-cyclodextrins					molecules by encapsulating the guest	
Poly ketal hyperbranched ethers	Incorporation of cyclic and acyclic ketal groups into the backbone of HPE		Ketal group		Great potentiality as multifunctional drug delivery vehicles	119
Hyperbranched polypeptides					Potential bioapplications	120
Hyperbranched polyester derivative (HBPE-St85)	Conversion of *Pongamia* oil to hyperbranched polyester and then styrene copolymerization further embedded with Ag-NP	Complex formation with proteins, then affect permeability and respiration leading to the death of the microbes		*B. pasteurii*, *E. coli*, *Streptococcus*, *A. niger*, *R. solani*, and *F. solani*	Good physico-mechanical properties and good antimicrobial properties	122
Bio-based hyperbranched poly(ester amide)/polyaniline		Potent efficacy against Gram-positive bacteria, as		*B. subtilis*, *S. aureus*, *P. aeruginosa*, *E. coli*, *A. niger*,	Good antimicrobial properties	123

nanofiber modified montmorillonite nanocomposites		compared to Gram-negative species Significant antifungal activity and antialgal activity		*C. capcii*, *F. oxysporum*, *Hormidium*, *Chlorella*, and *Cladophorella* species		
CuO-nanofibrillar cellulose/glycerol-based hyperbranched epoxy nanocomposite		Against various surgical site infections Efficient antimicrobial activity Acts as a potential sustainable scaffold for smooth muscle cell regeneration.		*S. aureus*, *E. coli* and *C. albicans*	Biodegradability with high performance and used as an antimicrobial scaffold material for reconstruction of muscles tissues	124
Betaine ester-wrapped hyperbranched polyethylenimine (BEHPEI)	—	Hand-washing lotion and mouth rinse	Betaine ester-shell		Antimicrobial and antibacterial applications	126
Hyperbranched PU/Ag nanocomposites	From hyperbranched polyurethane (HBPU-from *Mesua ferrea* L. seed oil) and AgNP	Ag ions enter into the bacterial cells by penetrating and consequently turn the DNA into a condensed form which reacts with the thiol group of the proteins; thus, it ultimately causes death of the cell	Due to the incorporation of the AgNPs	*S. aureus*, *E. coli* and the yeast *C. albicans*	Antimicrobial biomaterials	127
Hyperbranched epoxy nanocomposites	From *Homalomena aromatica* rhizome oil-modified bentonite as	Exhibits good toughness and high thermostability and			Antimicrobial applications	128

Table 15.3 (*Continued*)

Hyperbranched polymer type	Synthesis of hyperbranched polymer	Highlights	Antimicrobial action causing functionality	Antimicrobial action against	Application	Ref.
	well as organically modified montmorillonite clay	shows excellent biocompatibility Inhibits the growth of both Gram-positive and -negative bacteria as well as fungi				
Hyperbranched epoxy/Ag–RGO–Cur nanocomposite	From incorporation of Ag–RGO–Cur into hyperbranched epoxy resin	Effective antimicrobial properties These nanocomposites can be utilized to design a microbial fouling-resistant marine coating and in biomedical implants and devices	—	*S. aureus*, *C. albicans*, and *Chlorella*	Biomedical and marine applications	129
Cellulose-grafted hyperbranched polymers	Using bis(2-chloroethyl)amine and soluble cellulose tosylates	Exhibit antimicrobial activity on textiles	—	*E. coli*, *S. aureus*, *P. mirabilis*, *P. vulgaris*, *P. aeruginosa*, *E. aerogenes*, *B. thuringiensis*, *S. typhimurium*, and *S. mutans*	Great potential in biomedical applications, such as antimicrobial coating, medical devices or textiles	130

[a] *A. niger*: *Aspergillus niger*; *B. pasterurii*: *Bacillus pasteurii*; *B. subtilis*: *Bacillus subtilis*; *B. thuringiensis*: *Bacillus thuringiensis*; *C. albicans*: *Candida albicans*; *C. capsici*: *Coleotricum capcii*; *E. aerogenes*: *Enterobacter aerogenes*; *E. coli*: *Escherichia coli*; *F. oxysporum*: *Fusarium oxysporum*; *F. solani*: *Fusarium solani*; *P. aeruginosa*: *Pseudomonas aeruginosa*; *P. mirabilis*; *Proteus mirabilis*; *P. vulgaris*: *Proteus vulgaris*; *R. solani*: *Rhizoctonia solani*; *S. aureus*: *Staphylococcus aureus*; *S. mutans*: *Streptococcus mutans*; *S. typhimurium*: *Salmonella typhimurium*. Ag/AuNPs: silver/gold nanoparticles; HPAMAM: hyperbranched polyamidoamine; HPE: hyperbranched polyester; PSA: polysulphonic amine.

activity and laundering durability. Hyperbranched polymers are also applied to produce antimicrobial zinc oxide NPs and iron oxide (Fe_3O_4) NPs.[106,107] Also, hyperbranched polymer–metal hybrid nanocomposites, hyperbranched polymers containing cationic functionalities, and hyperbranched polymers with antibiotics are future antimicrobial agents. Quaternary ammonium functionalities are the most useful antiseptics and disinfectants. Therefore, these functionalities are used in the synthesis of antimicrobial polymers.[108,109] Stainless steel surfaces were incorporated with quaternary ammonium cations containing hyperbranched polymers and studied for antimicrobial and antibacterial properties. They show superior activity compared to stainless steel modified with linear functionalized polymers.[109] Excellent antibacterial activities have been observed against Gram-positive and Gram-negative bacteria when hyperbranched polymers–silver/gold hybrid nanocomposite system are used. This nanocomposite is prepared using stable silver/gold NPs colloid system with amine terminated polyamidoamine or dimethylamine terminated polyamidoamine.[110,111]

Multifunctional hyperbranched glycoconjugated polymers were synthesized from natural amino glycoside through Michael-addition, and gentamicin was loaded into the polymer backbone. This polymer matrix has displayed very good antibacterial and antitumor activity.[112]

Rainer Haag[113] and co-workers explored various hyperbranched polymers such as hyperbranched polyesters, polyethers, polystyrenes, polyethylenes, poly(ureaurethanes), polyethylenimines, poly(amidoamine)s, polyethylenimines, polypeptides, polyphosphates, polyacrylates, and β-cyclodextrins as water-soluble nanocarriers. These hyperbranched polymers exhibit excellent *in vivo* stability. Various hyperbranched polymers have been synthesized from polyglycerols, polypeptides, polyesters, polyethylene oxides, and polyphosphates. They are widely used in the biomedical applications.[114–118] Initially, hyperbranched polymers with high biocompatibility were developed.[118] Due to their non-biodegradable nature, the products were harmful to living cells. Therefore, it is important to prepare biocompatible polymer that can degrade under physiological conditions. Hyperbranched polymers containing ester functionalities were developed, since they undergo hydrolytic degradation. Boltron Hx ($x = 20$, 30, 40) is a commercially available hyperbranched polyester. Later hyperbranched polyesters were modified by incorporating cyclic and acyclic ketal groups to produce biodegradable and pH responsive poly ketal hyperbranched ethers. By adjusting the pH, degradation kinetics were altered from few minutes to days due to the presence of both acyclic and cyclic ketal groups.[119]

Chang and Dong proved that hyperbranched polypeptides are an important alternative to symmetrical analog peptide dendrimers due to the costly and difficult synthesis of the latter.[120] Biodegradable and biocompatible hyperbranched polyphosphates were synthesized. They reported significant biomedical applications.[121] Manawwer Alam and co-workers[122] prepared silver-nanocomposites (HBPE-St85) by the conversion of *Pongamia* oil into a hyperbranched polyester and then styrene copolymerization, which was further embedded with an Ag NP. These composites have good physicomechanical properties and good antimicrobial properties against *B. pasteurii*, *E. coli*, *Streptococcus*, *A. niger*, *Rhizoctonia solani*, and *Fusarium solani.* These nanocomposites undergo complex formation with proteins (cell wall) and then affect permeability and respiration, leading to the death of the microbes.

Sujata Pramanik[123] and co-workers fabricated bio-based hyperbranched poly(ester amide)/polyaniline nanofiber modified montmorillonite nanocomposites. The prepared nanocomposites show significant antimicrobial activity against Gram-positive bacteria (*B. subtilis* and *S. aureus*) compared to Gram-negative bacteria (*P. aeruginosa* and *E. coli*). They also exhibit significant antifungal activity against *A. niger*, *Coleotricum capcii* and *Fusarium oxysporum* and anti-algal activity against *Hormidium*, *Chlorella*, and *Cladophorella* species.

Shaswat Barua[124] and co-workers explored the potential of a "green" CuO–nanofibrillar cellulose/glycerol-based hyperbranched epoxy nanocomposite against the microorganisms *S. aureus*, *E. coli*, and *C. albicans* which are responsible for various surgical site infections. This nanocomposite is biodegradable with high performance and is used as an antimicrobial scaffold material for reconstruction of muscle tissues. The same researchers[125] isolated nanofibrillar cellulose (NFC) from the *Colocasia esculenta* plant, commonly found in India, by using a chemical method and decorated with Cu/CuONPs. This compound exhibits excellent biocompatibility and efficient antimicrobial activity with peripheral blood mononuclear cells and mammalian red blood cells. Later, this cellulose-based scaffolding became an interesting area for a number of biomedical applications. Shaswat Barua[124] *et al.* prepared a bio-based hyperbranched epoxy/CuO-NFC nanocomposite fabricated by decorating CuONPs on NFC surfaces to acquire efficient antimicrobial activity. This nanohybrid acts as a potential sustainable scaffold for smooth muscle cell regeneration.

Xin Zhou[126] and co-workers prepared a betaine ester wrapped hyperbranched polyethylenimine (BEHPEI) betaine ester-shell (Figure 15.9) which exhibits a dual role as antimicrobial and antibacterial, which has applications in hand-washing lotion, mouth rinse, *etc.*

Deka[127] *et al.* prepared hyperbranched polyurethane-silver nanocomposites as antimicrobial biomaterials and compared their antimicrobial activities with linear polyurethane/silver nanocomposite against bacteria such as *S. aureus*, *E. coli*, and the yeast *C. albicans*. The antimicrobial activities of the nanocomposites are due to silver NPs in the matrix, since the pristine hyperbranched polyurethane does not show any activity against the studied microorganisms. The Gram-positive bacteria *S. aureus* has a much stronger defense system compared to the Gram-negative bacteria *E. coli*, since the former possesses a thicker peptidoglycan cell wall. This thicker cell wall prevents the silver ions from penetrating into the cytoplasm of the organism. It has been suggested that silver ions enter into the bacterial cells by penetration and consequently turn the DNA into a condensed form that

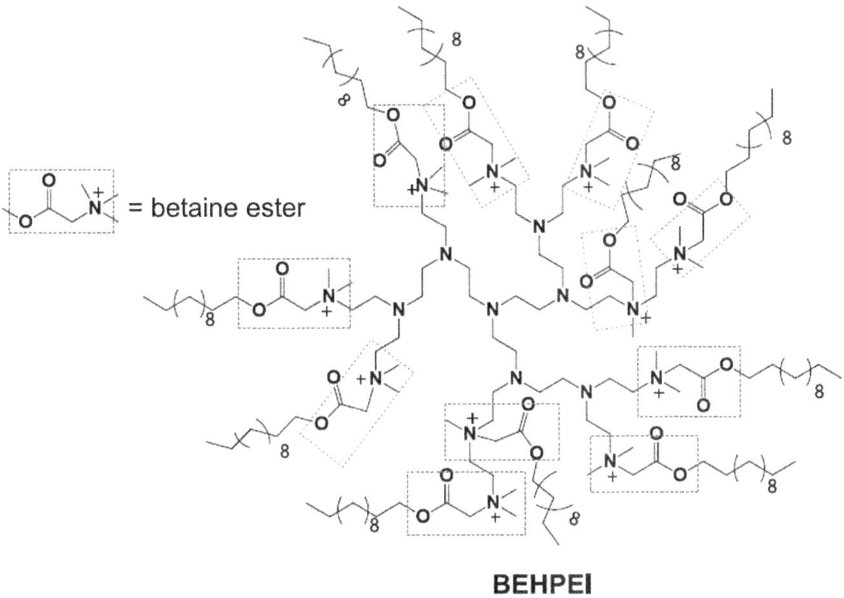

Figure 15.9 Betaine ester wrapped hyperbranched polyethylenimine (BEHPEI). Reproduced from ref. 126 with permission from Elsevier, Copyright 2014.

reacts with the thiol group of the proteins that ultimately causes death of the cell.

Barua[128] and co-workers prepared hyperbranched epoxy nanocomposites from *Homalomena aromatica* rhizome oil-modified bentonite as well as organically modified montmorillonite clay. These exhibit good toughness and high thermal stability and show excellent biocompatibility. Furthermore, they can inhibit the growth of both Gram-positive and Gram-negative bacteria as well as fungi. Barua[129] and co-workers prepared hyperbranched epoxy/silver–reduced grapheme oxide–curcumin nanocomposites by incorporating silver–reduced grapheme oxide–curcumin into hyperbranched epoxy resin. Due to the effective antimicrobial properties against *S. aureus* and *C. albicans* along with anti-algal properties against microalgal strain *Chlorella,* these nanocomposites can be used to design microbial fouling resistant marine coating.

Cellulose-grafted hyperbranched polymers were synthesized using bis-(2-chloroethyl)amine and soluble cellulose tosylates (Figure 15.10). Studies were conducted concerning their antimicrobial activities against various human pathogen Gram-negative and -positive bacteria: *E. coli*, *S. aureus*, *Proteus mirabilis*, *Proteus vulgaris*, *P. aeruginosa*, *Enterobacter aerogenes*, *Bacillus thuringiensis*, *Salmonella enterica* serotype *typhmurium*, and *Streptococcus mutans.*[130] The different behaviour of Gram-positive and Gram-negative bacteria is due to their different cell envelope structures. Gram-positive bacteria usually have loose cell wall structures, but gram Gram-negative bacteria have an outer membrane structure in the cell wall forming an additional

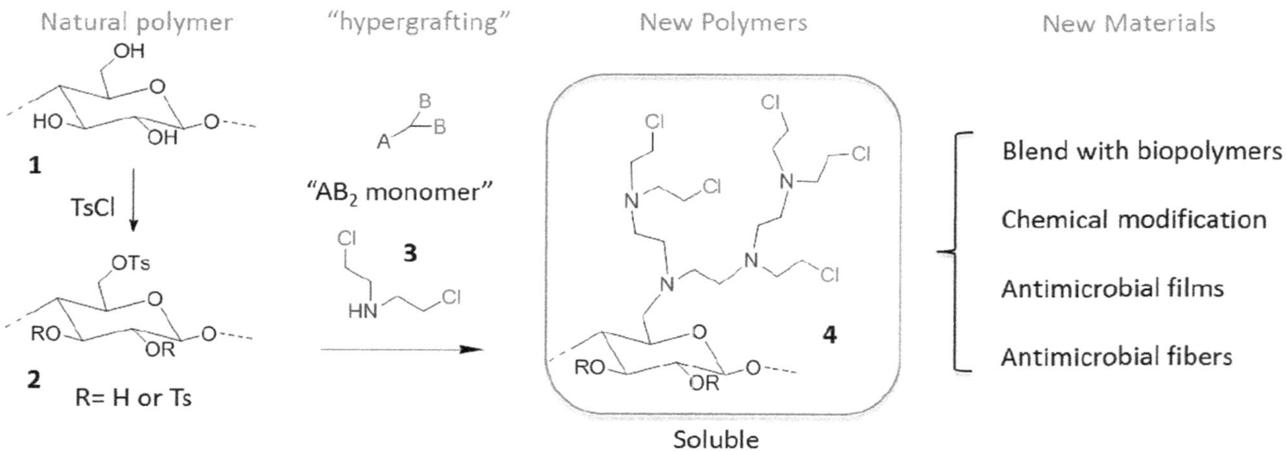

Figure 15.10 Synthesis of cellulose-grafted hyperbranched polyethyleneimine (PEI) with chlorine terminal groups. Reproduced from ref. 130 with permission from Elsevier, Copyright 2017.

barrier for antimicrobial agents. Therefore, interaction of the antimicrobial agents with the cell wall of Gram-negative bacteria is more complicated. In addition, these new materials exhibit moderate to strong antimicrobial activity against various bacteria and have great potential in biomedical applications, such as antimicrobial coating, medical devices, and textiles.

15.4 Conclusions

This chapter provides an overview of antimicrobial dendrimers and hyperbranched polymers. Dendritic polymers possess dendrimers and hyperbranched polymers. The main difference between these two classes of polymers is that the dendrimers contain a well-defined architecture with a branching point in each repeat unit, whereas hyperbranched polymers contain a less well-defined architecture. The advantage of hyperbranched polymers compared to dendrimers is that the synthesis is easy, and the materials can be prepared in bulk in a cost effective manner. Dendrimers and hyperbranched polymers both have their own advantages and have found their own applications as antimicrobial materials. During the past few years the application of these macromolecular structures in biological systems has seen rapid growth. However, there is still significant potential to develop novel antimicrobial dendrimers and hyperbranched polymers and to study their mechanism of action on microbes. Although these macromolecular structures have been studied for their antimicrobial action, they have not been researched under clinical conditions. Therefore, there is an urgent need to develop more robust methodologies to fulfil the needs of clinical trials.

References

1. Y. Zhou, W. Huang, J. Liu, X. Zhu and D. Yan, *Adv. Mater.*, 2010, **22**, 4567.
2. C. Gao and D. Yan, *Prog. Polym. Sci.*, 2004, **29**, 183.
3. D. Wang, T. Zhao, X. Zhu, D. Yan and W. Wang, *Chem. Soc. Rev.*, 2015, **44**(12), 4023.
4. A. M. Caminade, D. Yan and D. K. Smith, *Chem. Soc. Rev.*, 2015, **44**, 3870.
5. Y. Zheng, S. Li, Z. Weng and C. Gao, *Chem. Soc. Rev.*, 2015, **44**, 4091.
6. E. Buhleier, W. Wehner and F. Vogtle, *Synthesis*, 1978, 155.
7. R. G. Denkewalter, J. Kolc and W. J. Lukasavage, *U. S. Pat.*, 4289872, 1981.
8. D. A. Tomalia, H. Baker, J. Dewald, M. Hall, G. Kallos, S. Martin, J. Roeck, J. Ryder and P. Smith, *Polym. J.*, 1985, **17**(1), 117.
9. G. R. Newkome, Z. Q. Yao, G. R. Baker and V. K. Gupta, *J. Org. Chem.*, 1985, **50**(11), 2003.
10. R. Haag, A. Sunder and J. F. Stumbe, *J. Am. Chem. Soc.*, 2000, **122**, 2954.
11. C. J. Hawker and J. M. J. Fréchet, *J. Am. Chem. Soc.*, 1990, **112**, 7638.

12. C. J. Hawker and J. M. J. Fréchet, *J. Chem. Soc., Chem. Commun.*, 1990, 1010.
13. J. M. J. Fréchet, C. J. Hawker and A. E. Philippides, *U. S. Pat.*, 041 516, 1991.
14. S. H. Medina and M. E. H. El-Sayed, *Chem. Rev.*, 2009, **109**, 3141.
15. R. Touzani, *J. Mater. Environ. Sci.*, 2011, **2**, 201.
16. A. Carlmark, C. Hawker, A. Hulta and M. Malkoch, *Chem. Soc. Rev.*, 2009, **38**, 352.
17. U. Gupta and O. Perumal, *Natural and Synthetic Biomedical Polymers*, 2014, 243.
18. F. Aulenta, W. Hayes and S. Rannard, *Eur. Polym. J.*, 2003, **39**, 1741.
19. C. C. Lee, J. A. MacKay, J. M. J. Fréchet and F. C. Szoka, *Nat. Biotechnol.*, 2005, **23**, 1517.
20. X. D. Tang, Q. Z. Zhang, X. H. Fan and Q. F. Zhou, *Chin. J. Chem.*, 2004, **22**, 1366.
21. Y. Wang, W. Kong, Y. Song, Y. Duan, L. Wang, G. Steinhoff, D. Kong and Y. Yu, *Biomacromolecules*, 2009, **10**, 617.
22. N. Beyth, Y. Houri-Haddad, A. Domb, W. Khan and R. Hazan, *Evid.-Based Complementary Altern. Med.*, 2015, 246012.
23. W. Nakanishi, K. Minami, L. K. Shrestha, Q. Ji, J. P. Hill and K. Ariga, *Nano Today*, 2014, **9**, 378.
24. K. Minami, Y. Kasuya, T. Yamazaki, Q. Ji, W. Nakanishi, J. P. Hill, H. Sakai and K. Ariga, *Adv. Mater.*, 2015, **27**, 4020.
25. Y. E. Kurtoglu, M. K. Mishra, S. Kannan and R. M. Kannan, *Int. J. Pharm.*, 2010, **384**, 189.
26. S. Svenson and D. A. Tomalia, *Adv. Drug Delivery Rev.*, 2012, **64**, 102.
27. D. Chandrasekar, R. Sistla, F. J. Ahmad, R. K. Khar and P. V. Diwan, *Biomaterials*, 2007, **28**, 504.
28. A. Felczak, N. Wrońska, A. Janaszewska, B. Klajnert, M. Bryszewska, D. Appelhans, B. Voit, S. Różalska and K. Lisowska, *New J. Chem.*, 2012, **36**, 2215.
29. A. Felczak, K. Zawadzka, N. Wrońska, A. Janaszewska, B. Klajnert, M. Bryszewska, D. Appelhans, B. Voit and K. Lisowska, *New J. Chem.*, 2013, **37**, 4156.
30. (a) N. Wrońska, A. Felczak, K. Zawadzka, A. Janaszewska, B. Klajnert, M. Bryszewska and K. Lisowska, *New J. Chem.*, 2014, **38**, 2987; (b) N. Wrońska, A. Felczak, K. Zawadzka, M. Poszepczyńska, S. Różalska, M. Bryszewska, D. Appelhans and K. Lisowska, *Molecules*, 2015, **20**, 19330.
31. M. Homayoonfal and M. R. Mehrnia, *Sep. Purif. Technol.*, 2014, **130**, 74.
32. S. Xie, Y. Tao, Y. Pan, W. Qu, G. Cheng, L. Huang, D. Chen, X. Wang, Z. Liu and Z. Yuan, *J. Controlled Release*, 2014, **187**, 101.
33. M. K. Dail and S. P. Mezyk, *J. Phys. Chem.*, 2010, **114**, 8391.
34. A. J Huh and Y. J. Kwon, *J. Controlled Release*, 2011, **156**, 128.
35. J. Drzewińska, D. Appelhans, B. Voit, M. Bryszewska and B. Klajnert, *Biochem. Biophys. Res. Commun.*, 2012, **427**, 197.

36. M. Fischer, D. Appelhans, S. Schwarz, B. Klajnert, M. Bryszewska, B. Voit and M. Rogers, *Biomacromolecules*, 2010, **11**, 1314.
37. K. Poole, *Curr. Opin. Microbiol.*, 2001, **4**, 500.
38. E. Murugan, S. Arumugam and P. Panneerselvam, *Int. J. Polym. Mater. Polym. Biomater.*, 2016, **65**, 111.
39. M. Montazer, R. Malek and A. Rahimi, *Fiber Polym.*, 2007, **8**, 608.
40. K. Dodangeh, M. Gharanjig, S. Arami and S. Atashrouz, *Dyes Pigm.*, 2014, **111**, 30.
41. M. Sadeghi-Kiakhani, M. Arami and K. Gharanjig, *Iranian Polym. J.*, 2013, **22**, 931.
42. M. Sadeghi-Kiakhani, M. Arami and K. Gharanjig, *J. Appl. Polym. Sci.*, 2013, **127**, 2607.
43. A. Zargarkazemi, M. Sadeghi-Kiakhani, M. Arami and S. H. Bahrami, *J. Text. I.*, 2015, **106**, 80.
44. Mousa Sadeghi-Kiakhania and Siyamak Safapour, *Luminescence*, 2016, **31**, 1005.
45. D. Campoccia, L. Montanaroa and C. R. Arcioloa, *Biomaterials*, 2006, **27**, 2331.
46. B. Thomes, P. Murray and D. Bouchier-Hayes, *J. Bone Joint Surg. Br.*, 2002, **84**, 758.
47. C. K. V. Z. Abid, S. Jain, R. Jackeray, S. Chattopadhyay and H. Singh, *J. Biomed. Mater. Res., Part B*, 2017, **105**, 521.
48. P. Appelgren, V. Bjornhagen, K. Bragderyd, C. E. Jonsson and U. A. Ransjo, *Burns*, 2002, **28**, 39.
49. M. Stach, T. N. Siriwardena, T. Kçhler, C. Van Delden, T. Darbre and J. L. Reymond, *Angew. Chem., Int. Ed.*, 2014, **53**, 12827.
50. K. T. Al-Jamal, W. T. Al-Jamal, S. Akerman, J. E. Podesta, A. Yilmazer, J. A. Turton, A. Bianco, N. Vargesson, C. Kanthou, A. T. Florence, G. M. Tozer and K. Kostarelos, *Proc. Natl. Acad. Sci. U. S. A..*, 2010, **107**, 3966.
51. P. Abdel-Sayed, A. Kaeppeli, T. Siriwardena, T. Darbre, K. Perron, P. Jafari, J.-L. Reymond, D. P. Pioletti and L. A. Applegate, *Sci. Rep.*, 2016, **6**, 22020.
52. J. Pires, T. N. Siriwardena, M. Stach, R. Tinguely, S. Kasraian, F. Luzzaro, S. L. Leib, T. Darbre, J. L. Reymond and A. Endimiani, *Antimicrob. Agents Chemother.*, 2015, **59**, 7915.
53. K. Rehman and M. H. Zulfakar, *Drug Dev. Ind. Pharm.*, 2014, **40**, 433.
54. S. J. Buwalda, K. W. Boere, P. J. Dijkstra, J. Feijen, T. Vermonden and W. E. Hennink, *J. Controlled Release.*, 2014, **190**, 254.
55. M. Wróblewska and K. Winnicka, *Int. J. Mol. Sci.*, 2015, **16**, 20277.
56. Council of Europe, in *European Pharmacopoeia*, 6th ed., Council of Europe, Strasbourg, France, 2007, 2, 1801.
57. V. Jigar, G. Vishal, G. Tejas, Ch. Vishal and U. Umesh, *Int. J. Pharm Tech Res.*, 2011, **3**, 1714.
58. K. Winnicka, M. Wroblewska, P. Wieczorek, P. T. Sacha and E. A. Tryniszewska, *Molecules*, 2013, **18**, 8607.

59. B. Devarakonda, N. Li and M. M. de Villiers, *AAPS Pharm SciTech.*, 2005, **6**, 504.
60. K. Sosnowska and K. Winnicka, *Afr. J. Pharm. Pharmacol.*, 2013, **7**, 567.
61. A. S. Abd-El-Aziz, A. A. Abdelghani, S. K. El-Sadany, D. P. Overy and G. Russell, *Kerr Eur. Polym. J.*, 2016, **82**, 307.
62. R. Ceña-Díez, D. Sepúlveda-Crespo, M. Maly and M. A. Muñoz-Fernández, *RSC Adv.*, 2016, **6**, 46755.
63. E. Leire, S. P. Amaral, I. Louzao, K. Winzer, C. Alexander, E. Fernandez-Megia and F. Fernandez-Trillo, *Biomater. Sci.*, 2016, **4**, 998.
64. M. Suresh Kumar, M. Yuvaraj and P. Aruna, *Int. J. Polym. Mat and Polym. Biomat.*, 2015, **64**, 519.
65. C. K. V. Zainul Abid, R. Jackeray, S. Jain, S. Chattopadhyay, S. Asif and H. Singh, *J. Nanosci. Nanotechnol.*, 2016, **16**, 998.
66. S. S. Santos, R. V. Gonzaga, J. V. Silva, D. F. Savino, D. Prieto, J. M. Shikay, R. S. Silva, L. H. A Paulo, E. I. Ferreira and J. Giarolla, *Can. J. Chem.*, 2017, **95**(9), 907.
67. G. A. Eggimann, E. Blattes, S. Buschor, R. Biswas, S. M. Kammer, T. Darbre and J. L. Reymond, *Chem. Commun..*, 2014, **50**(55), 7254.
68. V. G. Joshi, V. D. Dighe, D. Thakuria, Y. S. Malik and S. Kumar, *Indian J. Virol.*, 2013, **24**, 312.
69. R. Tarallo, T. P. Carberry, A. Falanga, M. Vitiello, S. Galdiero, M. Galdiero and M. Weck, *Int. J. Nanomedicine*, 2013, **8**, 521.
70. S. Gupta, M. K. Pandey, K. Levon, R. Haag, A. C. Watterson, V. S. Parmar and S. K. Sharma, *Macromol. Chem. Phys.*, 2010, **211**, 239.
71. M. Stach, T. N. Siriwardena, T. Kohler, C. Van Delden, T. Darbre and J. L. Reymond, *Angew. Chem., Int. Ed.*, 2014, 53, 12827.
72. T. K. Lind, P. Zieli and L. E. T. Al, *Wiley Interdiscip. Rev.: Nanomed. Nanobiotechnol.*, 2013, **1**, 396.
73. S. M. Mignani, N. E. Brahmi, S. E. Kazzouli, R. Laurent, S. Ladeira, A. M. Caminade, E. P. Werbicka, E. M. Szewcyk, M. Bryszewska, M. M. Bousmina, T. Cresteil and J. P. Majoral, *Mol. Pharmaceutics*, 2017, **14**, 4087.
74. C. Z. Chen and S. L. Cooper, *Biometerials*, 2002, **23**, 3359.
75. C. R. Yates and W. Hayes, *Eur. Polym. J.*, 2004, **40**, 1257.
76. D. Konkolewicz, M. J. Monteiro and S. Perrier, *Macromolecules*, 2011, **44**, 7067.
77. M. Jikei and M. Kakimoto, *Prog. Polym. Sci.*, 2001, **26**, 1233.
78. C. Gao and D. Yan, *Prog. Polym. Sci.*, 2004, **29**, 183.
79. L. R. Hutchings, J. M. Dodds and S. J. Roberts-Bleming, *Macromolecules*, 2005, **38**, 5970.
80. M. Morell, D. Foix, A. Lederer, X. Ramis, B. Voit and A. Serra, *J. Polym. Sci., Part A: Polym. Chem.*, 2011, **49**, 4639.
81. D. Foix, X. Ramis, M. Sangermano and A. Serra, *J. Polym. Sci., Part A: Polym. Chem.*, 2012, **50**, 1133.

82. Y. Zheng, S. Li, Z. Weng and C. Gao, *Chem. Soc. Rev.*, 2015, **44**, 4091.
83. D. Wang, T. Zhao, X. Zhu, D. Yan and W. Wang, *Chem. Soc. Rev.*, 2015, **44**, 4023.
84. M. Jikei and Masa-aki Kakimoto, *Prog. Polym. Sci.*, 2001, **26**, 1233.
85. R. Mezzenga, L. Boogh and J. E. Manson, *Compos. Sci. Technol.*, 2001, **61**, 787.
86. T. Gurunathan, S. Mohanty and S. K. Nayak, *Polymer-Plastics Technology and Engineering*, 2016, **55**, 92.
87. J. Jang, J. H. Oh and S. I. Moon, *Macromolecules*, 2000, **33**, 1864.
88. T. J. Mulkern and N. C. B. Tan, *Polymer*, 2000, **41**, 3193.
89. K. Manczyk and P. Szewczyk, *Prog. Org. Coatings*, 2002, **44**, 99.
90. M. Johansson, E. Malmström, A. Jansson and A. Hult, *J. Coatings Technol.*, 2000, **72**, 49.
91. R. A. Van Benthem, *Prog. Org. Coat.*, 2000, **40**, 203.
92. S. W. Zhu and W. F. Shi, *Polym. Degrad. Stab.*, 2002, **75**, 543.
93. S. Wang, C. Wang and B. Zhang, *Mater. Lett.*, 2010, **64**, 9.
94. X. Zhao, Q. H. Zhang, D. J. Chen and P. Lu, *Macromolecules*, 2010, **43**, 2357.
95. C. J. G. Plummer, M. Rodlert, J. L. Bucaille, H. M. Grünbauer and J. E. Manson, *Polymer*, 2005, **46**, 6543.
96. M. Rodlert, C. J. G. Plummer, L. Garamszegi, Y. Leterrier, H. J. M. Grünbauer and J. A. E. Månson, *Polymer*, 2004, **45**, 949.
97. M. Seiler, C. Jork, A. Kavarnou, W. Arlt and R. Hirsch, *AIChE J.*, 2004, **50**, 2439.
98. J. Zou, Y. Zhao and W. Shi, *J. Membrane Sci.*, 2004, **245**, 35.
99. D. M. Sterescu, D. F. Stamatialis and M. Wessling, *J. Membr. Sci.*, 2008, **310**, 512.
100. A. Goswami and A. K. Singh, *React. Funct. Polym.*, 2004, **61**, 255.
101. S. Stiriba, H. Kautz and H. Frey, *J. Am. Chem. Soc.*, 2002, **124**, 9698.
102. C. J. G. Plummer, A. Luciani, T. Q. Nguyen and L. Garamszegi, *Polym. Bull.*, 2002, **49**, 77.
103. P. K. Maji, P. K. Guchhait and A. K. Bhowmick, *J. Mater. Sci.*, 2009, **44**, 5861.
104. R. B. Wang, L. Wang, L. Z. Zhou, Y. Su, F. Qiu, D. L. Wang, J. L. Wu, X. Y. Zhu and D. Y. Yan, *J. Mater. Chem.*, 2012, **22**, 15227.
105. D. S. Zhang, L. Chen, C. F. Zang, Y. Y. Chen and H. Lin, *Carbohydr. Polym.*, 2013, **92**, 2088.
106. D. S. Zhang, G. Y. Zhang, L. Chen, Y. F. Liao, Y. Y. Chen and H. Lin, *J. Appl. Polym. Sci.*, 2013, 3778.
107. B. Das, M. Mandal, A. Upadhyay, P. Chattopadhyay and N. Karak, *Biomed. Mater.*, 2013, **8**, 035003.
108. P. Ortega, B. M. Cobaleda, J. M. Hernandez-Ros, E. Fuentes-Paniagua, J. Sanchez-Nieves, M. P. Tarazona, J. L. Copa-patino, J. Soliveri, F. J. de la Mata and R. Gomez, *Org. Biomol. Chem.*, 2011, **9**, 5238.

109. W. J. Yang, K.-G. Neoh, E.-T. Kang, S. L.-M. Teo and D. Rittschof, *Polym. Chem.*, 2013, **4**, 3105.
110. Y. W. Zhang, H. S. Peng, W. Huang, Y. F. Zhou, X. H. Zhang and D. Y. Yan, *J. Phys. Chem. C*, 2008, **112**, 2330.
111. Y. W. Zhang, H. S. Peng, W. Huang, Y. F. Zhou and D. Y. Yan, *J. Colloid Interface Sci.*, 2008, **325**, 371.
112. M. S. Chen, M. Hu, D. L. Wang, G. J. Wang, X. Y. Zhu, D. Y. Yan and J. Sun, *Bioconjugate Chem.*, 2012, **23**, 1189.
113. S. Xu, Y. Luo and R. Haag, *Macromol. Biosci.*, 2007, **7**, 968.
114. A. M. Caminade, D. Yan and D. K. Smith, *Chem. Soc. Rev.*, 2015, **44**, 3870.
115. C. R. Yates and W. Hayes, *Eur. Polym. J.*, 2004, **40**, 1257.
116. Y. Zhou, W. Huang, J. Liu, X. Zhu and D. Yan, *Adv. Mater.*, 2010, **22**, 4567.
117. M. Hu, M. S. Chen, G. L. Li, Y. Pang, D. L. Wang, J. L. Wu, F. Qiu, X. Y. Zhu and J. Sun, *J. Biomacromol.*, 2012, **13**, 3552.
118. M. Calderon, M. A. Quadir, S. K. Sharma and R. Haag, *Adv. Mater.*, 2010, **22**, 190.
119. R. A. Shenoi, J. K. Narayanannair, J. L. Hamilton, B. F. Lai, S. Horte, R. K. Kainthan, J. P. Varghese, K. G. Rajeev, M. Manoharan and J. N. Kizhakkedathu, *J. Am. Chem. Soc.*, 2012, **134**, 14945.
120. X. Chang and C. M. Dong, *Biomacromolecules*, 2013, **14**, 3329.
121. J. Y. Liu, W. Huang, Y. F. Zhou and D. Y. Yan, *Macromolecules*, 2009, **42**, 4394.
122. M. Alam, M. R. Shaik and N. M. Alandis, *J. Chem.*, 2013, **962316**, 11.
123. Sujata Pramanik, Pranjal Bharali, B. K. Konwar and Niranjan Karak, *Mat. Sci. Eng., C*, 2014, **35**, 61.
124. S. Barua, B. Gogoi, L. Aidew, A. K. Buragohain, P. Chattopadhyay and N. Karak, *ACS Sustainable Chem. Eng.*, 2015, **3**, 1136.
125. S. Barua, G. Das, L. Aidew, A. K. Buragohain and N. Karak, *RSC Adv.*, 2013, **3**, 14997.
126. X. Zhou, Y. Chen, J. Han, X. Wu, G. Wang and D. Jiang, *Polymer*, 2014, **55**, 6261.
127. H. Deka, N. Karak, R. D. Kalita and A. K. Buragohain, *Polym. Degrad. Stab.*, 2010, **95**, 1509.
128. S. Barua, N. Dutta, S. Karmakar, P. Chattopadhyay, L. Aidew, A. K. Buragohain and N. Karak, *Biomed. Mater.*, 2014, **9**, 025006.
129. S. Barua, P. Chattopadhyay, M. M. Phukan, B. K. Konwar, J. Islam and N. Karak, *RSC Adv.*, 2014, **4**, 47797.
130. D. Demircan and B. Zhang, *Carbohydr. Polym.*, 2017, **157**, 1913.

CHAPTER 16

Antimicrobial Activities of Fatty Acids and their Derivatives

SANJIT KANJILAL* AND SHIVA SHANKER KAKI

Centre for Lipid Science & Technology, CSIR-Indian Institute of Chemical Technology, Uppal Road, Tarnaka, Hyderabad-500007, India
*Email: sanjit@iict.res.in

16.1 Introduction

A substance is called an antimicrobial agent due to its capability of killing (-cidal) or inhibiting the growth (-static) of microorganisms. Before the discovery of penicillin by Alexander Fleming, plant oils and fatty acids (FAs) were used traditionally as antimicrobial agents.[1] However, with the emergence of several conventional antimicrobial medicinal compounds such as penicillin over the past century and due to their excellent activity and selectivity, traditional therapeutic potential of plant oils and fatty acids slowly faded. Nevertheless, the wheel of research has taken a reverse trend due to the reduced susceptibility of organisms against drugs which were earlier found to be sensitive. Over-consumption of drugs and the ability of organisms to reorient and grow in an environment of inhibitors have been cited as the possible reason for the emergence of drug-resistant microbes. Such resistances were classified either as natural resistance or as acquired resistance of microbes.[2] Several hypotheses have been put forward by researchers for the occurrence of drug-resistant microbes.[3] In the backdrop of such occurrence, researchers rejuvenated their research area stressing more on

Biomaterials Science Series No. 5
Antimicrobial Materials for Biomedical Applications
Edited by Abraham J. Domb, Konda Reddy Kunduru and Shady Farah

Published by the Royal Society of Chemistry, www.rsc.org

renewable antimicrobial raw materials such as fatty acids. The present chapter emphasises the growing interest among researchers in understanding the structural and functional role of fatty acids and derivatives of fatty acids (FAD) in microorganisms, as evidenced by several reviews published since the late 1960s and 1970s, and also in recent times.[1,4–7]

16.2 Classification of Fatty Acids

Fatty acids are abundantly available in nature, typically found bound to other functional moieties such as glycerol (neutral lipid), carbohydrate (glycolipid) and phosphate head groups (phospholipid) to form three main classes of lipids. Thus, fatty acids are the main building blocks of different classes of lipids and play an important role in regulating many biological activities; they also serve as reserve sources of energy. Structurally, fatty acids are straight aliphatic carboxylic acids which include normally a chain length ranging from C4 to C22 and are responsible for hydrophobic nature of lipids. The presence of an ionisable hydrophilic carboxyl group attached to the long hydrophobic alkyl chain makes this molecule amphiphilic in nature. The general nomenclature, structure and composition of naturally occurring fatty acids are described in several reviews and books.[8,9] Fatty acids are classified based on carbon chain length, as short-chain (C4–C8), medium-chain (C10–C14) and long-chain (C16 and above) fatty acids. These are also classified based on unsaturation, into saturated fatty acid (SFA; no double bond), monounsaturated (MUFA; one double bond) and polyunsaturated fatty acids (PUFA; two or more double bonds). PUFAs are further classified as (ω-3) and (ω-6) based on the position of last unsaturation from the terminal methyl moiety (Figure 16.1). Edible oils and fats, milk and dairy products are

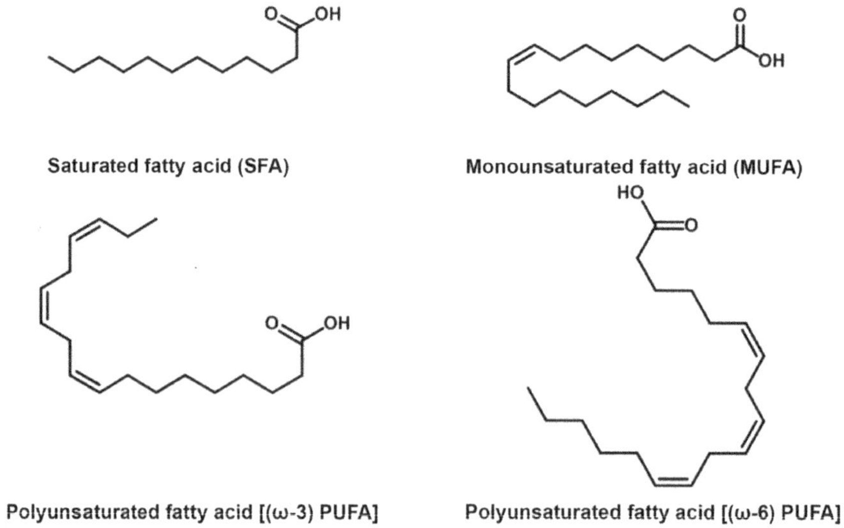

Figure 16.1 Structure of different classes of fatty acids.

major sources of fatty acids. Other than these commonly found fatty acids, some plant oils produce fatty acids with functional groups such as hydroxyl, cyano, cyclopropene, cyclopentene, cyclohexane, *etc.*, and are classified in this chapter as unusual fatty acids. Apart from being building blocks, fatty acids and FADs also help in host defences against potential pathogenic microorganisms. In this chapter, antimicrobial properties of both usual and unusual fatty acids of plant origin along with their derivatives are discussed. The defence mechanism of fatty acids against pathogens and microorganisms are found to be dependent on the structure of fatty acids, such as the carbon chain length, degree of unsaturation and also the presence of unusual functional moieties.[10]

16.3 Antimicrobial Activity of Fatty Acids

Shortest among all fatty acids is butyric acid (C4), obtained from cow and sheep milk-based products, especially from butter and ghee (clarified butter). A myriad bioactive features have been attributed to butyric acid and its sodium salts such, as anticancer, anti-inflammatory, hypocholesterolemic and hypolipidemic activities.[11] However, reports on antimicrobial activity of butyric acid are quite limited. With a view to obtain novel non-toxic and food-compatible disinfectants for application at food processing plants, Cabezas-Pizarro *et al.* studied the antimicrobial activity of sodium or potassium salts of short-chain fatty acids, namely butyric (C4), hexanoic (C6), octanoic (C8) and decanoic (C10) acids.[12] The authors reported mild inhibitory action of sodium or potassium salts of butyric and hexanoic acid against a host of food-borne pathogens. Another study reported mild antifungal activity of potassium salts of nine fatty acids (butyrate, caproate, caprylate, caprate, laurate, myristate, oleate, linoleate and linolenate) against fungal growth on orange rind fungus. Among the nine fatty acid salts tested, potassium caprate (C10) showed the highest antifungal activity against two *Penicillium* species.[13] In order to have a potential therapeutic approach to control oral bacteria, a dose-dependent study was carried out to determine inhibitory concentration (IC_{80}, 80% inhibition) of fatty acids (short-, medium- and long-chain) on the growth of Gram-positive and Gram-negative oral microorganisms.[14] The authors reported very limited anti-*Candida* activity of short- and long-chain fatty acids, but significant activity of medium-chain fatty acids, especially C12. Kabara *et al.*[7] studied 15 straight-chain usual fatty acids (caproic to arachidonic acid) against 10 susceptible organisms. They found no inhibitory action of short- and long-chain saturated fatty acids. The inhibitory action was observed from capric acid (C10) onwards, reaching maximum at lauric acid (C12), beyond which activity falls. Antibacterial activity of saturated (C8, C10, C12, C14) and unsaturated fatty acids (palmitioleic, C16 : 1 and oleic acid, C18 : 1) were also studied against common human pathogens and dependency of activity on the concentration of fatty acids tested was found. At 5 mM concentration, both C12 and C16 : 1 exhibited the highest /reduction of *Streptococcus* cell viability.[15] At lower concentrations, the activity of C12 falls drastically, but C16 : 1 retains intermediate activity until 0.31 mM. However, capric acid (C10) exhibited highest activity against *Staphylococcus aureus* at 10 mM concentration, also showing a falling pattern with decrease in concentration (at 5 mM). The antifungal activity of medium-chain saturated fatty acids (C8, C10, C12 and C14) against *Candida albicans* indicated the best activity for capric and lauric acids at 10 mM concentration.[16] However, at lower concentrations (5 mM), activity of lauric acid prevailed over capric acid. Another research group found both capric and lauric acids exhibited similar bactericidal activity against *Propionibacterium acnes*.[17] The therapeutic potential of lauric acid has also been tested by Nakatsuji *et al.* in treating acne as a natural antibiotic against *Propionibacterium acnes*.[18] Antibacterial activity of five saturated and five unsaturated long-chain fatty acids (C15–C22) against four marine pathogenic bacteria was also reported.[19] The authors reported that saturated acids with shorter chain lengths (C15) generally caused more cell death than those with longer chain lengths (C18), and unsaturated fatty acids are more active than the saturated fatty acids. Kabara and co-workers[7,10a] also showed that unsaturated fatty acids (myristoleic, C14 : 1 and palmitoleic acid, C16 : 1) exhibited better inhibitory action against studied bacterial strains compared to their saturated counterparts.

Based on literature reports, as presented in Table 16.1, it can be concluded that (a) C12 is the most active antimicrobial fatty acid among saturated fatty acids; (b) C16 : 1 is most bactericidal among monounsaturated fatty acids; (c) the bactericidal activity of fatty acids is more effective against Gram-positive bacteria compared to Gram-negative bacteria; (d) *trans* isomers are less active compared to their *cis* counterparts; (e) activity increases upon introduction of a second double bond, but decreases upon inclusion of a third double bond; and (f) fatty acids, mainly C10 and C12, are effective only against *Candida albicans* and ineffective against any other fungal strains. However, there are variations in the activity of fatty acids (saturated and monounsaturated) as reviewed by several authors.[10,20,21] This contrasting picture of the antimicrobial activity of fatty acids can be attributed to differences in the structure of fatty acid and its concentration, type of strain and study design.

An important parameter while studying the antibacterial activity of fatty acids is its pH dependence. At neutral pH, most fatty acids are not so active against many bacterial strains and start showing activity only when pH drops.[24,38] This is evident from the observed ability of fatty acid to create an unfavourable environment for bacterial growth on skin surface at acidic pH.[39] However, reports on optimum pH value for maximum bactericidal activity of fatty acids are contrasting in nature. In one report, C12 was reported to possess maximum bactericidal activity against *Mycobacterium smegmatis* at pH 5.0,[17] whereas the same fatty acid showed maximum activity against *Helicobacter pylori* at pH 7.4.[23] Thus, the role of pH on fatty acid-induced inhibition of cell growth is not clear mechanistically, largely dependent on targeted bacterium, structure of fatty acid and study design.[2,8,16,24,25,38–40]

Table 16.1 Antimicrobial activity of saturated and unsaturated fatty acids and monoglyceride of varying chain length (C4–C22).[a]

Microorganisms tested	Range of FA tested; activity	References
Antibacterial		
Helicobacter pylori (−)	*C8–C12 FA; C8–C12 MG* • C12 showed bactericidal activity against *H. pylori* • C10 MG and C12 MG possess nearly identical bactericidal activity	22 23
	C4–C16, C14:1, C16:1, C18:1, C18:2, C18:3 FA; C8–C18 MG • C12 most potent FA among saturates • C14:1 and C18:3 potent among unsaturates • C12 MG most potent among MGs	
Staphylococcus aureus (+), *S. epidermidis* (+), β-hemolytic *streptococci* (+), group D *Streptococcus* (+), *Bacillus subtilis* (+), *Sarcina lutea* (+), *Micrococcus* sp. (+), *Nocardia asteroides* (+), *Corynebacterium* sp. (+), and *Pneumococcus* (+)	*C6–C18; C14:1, C16:1, C18:1* (cis, trans), *C18:2* (cis, cis), *C18:2* (cis, trans), *C18:3, C20:4 FAs; C10 and C12 MGs* • C12 highest inhibitory activity among saturated FAs • Activity of C18:2 (*cis, cis*) more than C12 • Unsaturated FAs more active than their saturated counterparts • C10 and C12 MGs better agents than corresponding FA equivalents	7
Streptococcus group A and group B (+) and *Staphylococcus aureus* (+)	*C8–C14, C16:1, C18:1 FAs and MGs* • C10 exhibited highest activity at 10 mM • At 5 mM activity of C12 prevailed over C10 • C10 MGs showed significant activity against *S. aureus*	15
Clostridium perfringens (+)	*C2–C18, C18:1, C18:2 FAs* • C12 showed highest activity among FAs	24
Chlamydia trachomatis (−), *Neisseria gonorrhoeae* (−)	*C8–C14, C16:1, C18:1 FAs; C8–C12, C16:1, C18:1 MGs* • C10 caused fastest, effective killing • C12 most active at lower concentration • C12 and C16:1 most active against *N. gonorrhoeae* • C10 MGs most potent to kill *C. trachomatis* and *N. gonorrhoeae*	25

Table 16.1 (*Continued*)

Microorganisms tested	Range of FA tested; activity	References
Staphylococcus aureus (+), *Klebsiella pneumoniae* (−), *Haemophilus influenzae* (−), *Moraxella catarrhalis* (−), *Streptococcus faecalis* (+), *Corynebacterium equi* (+), *Listeria monocytogenes* (+), *Staphylococcus epidermidis* (+), *Bacillus megaterium* (+), *Lactobacillus acidophilus* (+), *Pseudomonas aeruginosa* (−), *Escherichia coli* (−), *Neisseria gonorrhoeae* (−)	*C20 : 4* • All Gram-positive bacteria susceptible to C20 : 4 • *L. acidophilus* is the most susceptible and *B. megaterium* is the least susceptible to C20 : 4	26
Listeria monocytogenes (+)	*C12–C18, C18 : 1, C18 : 2, C18 : 3 FAs; C12–C16 MGs* • C12, C18 : 2 and C18 : 3 exhibited strong activity • C12 MGs showed better activity than C12 FAs	27
Bacillus larvae (+)	*C6–C18, C14 : 1–C22 : 1, C18 : 2–C22 : 2, C18 : 3–C22 : 3, C20 : 4–C22 : 4, C22 : 6 FAs* • C12 and C13 most active among saturated FAs • C16 : 1 and C18 : 2 most active among unsaturated FAs • Hydroxy FAs such as ricinoleic acid exhibited moderate activity • Activity of C14 : 1 is equal to that of C12 • LC PUFA such as C20 : 4 and C22 : 6 also showed moderate activity	28
Bacillus megaterium (+), *Pseuodomonas phaseolicola* (−)	*C8–C18; C18 : 1* (cis, trans), *C18 : 2, C18 : 3 FAs* • C12 most active among saturates; order: C8 < C10 < C12 ≥ C14 > C16 ≥ C18 • Activity of C18 : 1 *cis* > C18 : 1 *trans*	29
Lactococcus garvieae (+), *Vibrio harveyi* (−), *Vibrio anguillarium* (−) and *Vibrio alginocolyticus* (−)	*C15–C22 (sat); C18 : 1, C18 : 4, C20 : 4, C20 : 5, C22 : 4, C22 : 5 FAs* • C15 caused more cell death than C18/22 • Unsaturated FAs are more active than saturated FAs	19
Staphylococcus aureus (+), *Streptococcus salivarius* (+), *Fusobacterium nucleatum* (+),	*C8–C12, C16 : 1 (Δ6 and Δ9) FAs* • C12 most active among saturates	20*a*

Table 16.1 (*Continued*)

Microorganisms tested	Range of FA tested; activity	References
Pseudomonas aeruginosa (−), *Escherichia coli* (−), *Propionibacterium acnes* (−), *Fusobacterium nucleatum* (−)	• C16 : 1 (Δ6 and Δ9) equally effective against *S aureus* and *S. salivarius* • C16 : 1 (Δ6 and Δ9) not effective against Gram-negative bacteria	
Propionibacterium acnes (−), *Staphylococcus aureus* (+), and *Staphylococcus epidermidis* (+)	*C12 FA* • Exhibited strong activity against these three strains	18
Staphylococcus aureus (+)	*C12 FA* • Exhibited activity against both MSSA and MRSA	20*b*
Staphylococcus aureus (+)	*C18 : 2 and C18 : 3 FAs* • C18 : 3 more active than C18 : 2	30
Streptococcus faecalis (+), *Streptococcus pyogenes* (+), *Staphylococcus aureus* (+), *Corynebacterium* sp. (+), *Nocardia asteroides* (+)	*C11, C12, C13 FAs and MGs* • C12 FA and MG most potent antibacterial compound • In general, MGs are more active than their corresponding FAs	10*c*
Mycobacterium smegmatis (+)	*C10–C20, C16 : 1, C18 : 1, C18 : 2, C20 : 4 FAs;* • C12 and C14 exhibited mycobacterial activity • C16 : 1, C18 : 2 and C20 : 4 showed strong bactericidal activity	31
Salmonella sp. (−)	*C2–C18, C18 : 1, C18 : 2 FAs* • Only C8 showed inhibitory activity	32
Escherichia coli (−)	*C2–C18, C18 : 1, C18 : 2 FAs* • C8 and C10 showed activity: C8 > C10	33
Bacillus cereus (+), *Staphylococcus aureus* (+), *Vibrio parahaemolyticus* (−), *Salmonella typhimurium* (−), *Salmonella enteritidis* (−), *Escherichia coli* (−)	*C18 : 3 FA; C12, C14 MGs* • C18 : 3 showed excellent inhibitory effect against *B. cereus* and *S. aureus* • C18 : 3 showed very mild activity against Gram-negative bacteria • Combination of C18 : 3 FA and C12/C14MG showed synergistic effect	34
Staphylococcus aureus (+), *Streptococcus pyogenes* (+), *Haemophilus influenzae* (−), *Pseudomonas aeruginosa* (−), Enterobacteriaceae (−)	*C12 FA; C12 MG* • Activity C12 MG is more than C12 FA • C12 MG is bactericidal against the Gram-negative bacteria and not C12 FA	35

Table 16.1 (*Continued*)

Microorganisms tested	Range of FA tested; activity	References
Cronobacter sakazakii (−), *Cronobacter malonaticus* (−)	*C8–C12 FAs and MGs* • C8 FA and C8 MG possess significant activity against *Cronobacter* sp.	36
Antifungal		
Candida albicans	*C6–C18; C14 : 1, C16 : 1, C18 : 1* (cis, trans), *C18 : 2* (cis, cis), *C18 : 2* (cis, trans), *C18 : 3, C20 : 4 FAs; C10 and C12 MG* • C12 most potent among saturated FAs • C16 : 1 and C18 : 2 showed best activity among unsaturated FAs • C12 MG showed good activity	7
	C8–C14, C16 : 1, C18 : 1 FA and MG • C10 caused fastest, effective killing • C12 most active at lower concentration • C12 MG showed good activity	16
	C1–C16 FAs • C6, C8 and C12 showed significant anti-*Candida* activity	37

[a](+): Gram-positive bacteria; (−): Gram-negative bacteria; FA: fatty acid; MG: monoglyceride; MRSA: methicillin-resistant *S. aureus*; MSSA: methicillin-susceptible *S. aureus*.

Both linoleic (18 : 2; ω-6; LA) and α-linolenic acid (18 : 3; ω-3; ALA) are classified under the category of essential fatty acids, since these fatty acids can't be synthesized by the human body and must be provided within the diet. Health benefits of LA and ALA and importance of their balance in the diet has been studied and well documented in the literature.[41] LA and ALA are precursor for the biosynthesis of very important (ω-6) and (ω-3) long-chain (LC)-PUFAs, respectively.[42] The intermediate lipids in the pathway of (ω-6) LC-PUFA are γ-linolenic acid (18 : 3; ω-6; GLA), dihomo-γ-linolenic acid (18 : 4; DGLA) and the arachidonic acids (20 : 4; AA). Similarly, in the pathway of (ω-3) LC-PUFA, the intermediate fatty acids are eicosapentaenoic (20 : 5; EPA), docosapentaenoic (22 : 5; DPA) and docosahexaenoic (22 : 6, DHA) acids. A plethora of bioactivity has been attributed to these two types of LC-PUFA[41,42] and very little attention has been paid to the use of LC-PUFA as an antimicrobial agent in clinical medicine. Literature reports on the bactericidal and fungicidal activity of the PUFA precursors LA and ALA are included in Table 16.1. Many literature reports indicated increased bactericidal activity upon the introduction of a second double bond (LA) and decreased activity upon inclusion of third double bond (ALA). However, the relationship between the degree of unsaturation and bactericidal activity is once again conflicting in nature. Bactericidal activity of C20 series fatty acid when studied against *Staphylococcus aureus* showed increased potency, with the increase in unsaturation reaching maxima for EPA.[26] Highest potency of EPA compared to other studied fatty acids was reported against *Mycobacterium* in a similar study.[43] Like other fatty acids, LC-PUFA showed more potency against Gram-positive than Gram- bacteria negative. However, EPA showed efficacy also against the peptic ulcer agent, Gram-negative *Helicobacter pylori*.[44] Thus, EPA has the potential to control the development of peptic ulcer. EPA was also reported to be as effective as DHA and ALA against the growth of ruminant Gram-positive bacteria, *Butyrivibro fibrisolvens*.[45] Antimicrobial activity of ALA, EPA and DHA was tested against various oral pathogens and showed excellent growth inhibitory potency against Gram-positive *Streptococcus mutans*, closely followed by Gram-negative *Porphyromonas gingivalis* and *Aggregatibacter actinomycetemcomitans*. In addition, in this study EPA showed supremacy over DHA and inhibitory potency was in the following order: EPA > ALA > DHA.[46] In fact, bactericidal potency of EPA is the highest among LC-PUFAs against many strains, but there is deviation in this general statement. This became evident when the bactericidal activity of four natural LC-PUFAs (GLA, DGLA, EPA and DHA) was studied against *Propionibacterium acnes* and *Staphylococcus aureus*.[47] DHA was found to be the most effective LC-PUFA in inhibiting the growth of *Propionibacterium acnes*, followed by GLA. Although LC-PUFAs in general are not so effective against *Staphylococcus aureus*, even then DHA showed supremacy in potency against this bacterium. Both these bacteria are responsible for skin infection and LC-PUFAs, especially DHA, have the potential to be used in topical application. Antibacterial activity of (ω-3) PUFAs, exemplified by ALA, EPA and DHA has also been assessed on pathogenic microorganisms, which are well known to cause infection through the creation of virulent factors, formation of biofilm and generation of resistance gene to survive in the host defence system. A review article summarizes various literature works on effectiveness of (ω-3) PUFAs, their mode of action, toxicity and therapeutic potential.[48] Like saturated and monounsaturated fatty acids, PUFAs do not exhibit antifungal activity. But there are reports of mild potency of EPA against *Candida* species[46,49] Thus, conflicting literature reports on antimicrobial activity of PUFAs stem from differences in the type of strains, structure of fatty acids and methodologies taken up for studying activity.

16.3.1 Antimicrobial Activity of Unusual Fatty Acids

Castor oil (*Ricinus communis*) has long been known as a medicinal oil, and was primarily used as purgative or laxative to counter constipation.[50] Castor oil is a rich source of a hydroxyl fatty acid (86–88%) called ricinoleic acid (12-hydroxy-9-*cis*-octadecenoic acid) and the laxative action is attributed to ricinoleic acid. Ricinoleic in free form showed mild antifungal activity against *Aspergillus niger* and in salt form showed antimicrobial activity against a host of microorganisms.[51] Ricinoleic acid was reported to have bactericidal activity against *Micrococcus luteus*, as otherwise it is found to be very mild against other studied strains.[52] However, on derivatization, ricinoleic acid exhibited a marked increase in its antimicrobial activity.[53] Ricinoleic acid on high temperature pyrolysis (550 °C) produces heptaldehyde and 10-undecenoic acid (Scheme 16.1).[54]

10-undecenoic acid (10-UDA) is a well known natural fungicide approved in many countries as therapeutic cure for different kind of skin disorders, infections, itching, burning and irritations. In 1949, 10-UDA was used to treat psoriasis, not knowing exactly whether its fungicidal role was responsible for its activity.[55] Later, its efficacy against three fungi, *Trichophyton mentagrophytes*, *Epidermophyton floccosum* and *Trichophyton rubrum*, responsible for the occurance of dermatomycosis, was proved.[56] A mixture of 10-UDA and its zinc salt was also used as an antimycotic active ingredient in the preparation of ointment for treating infection of glabrous skin, especially the hairy part of the skin, caused by *Microsporon audouim*.[57] Literature reports on the fungicidal activity of 10-UDA against some fungi, and its application or the use of its salts are given in Table 16.2.

The curative action of chaulmoogra oil in the treatment of leprosy dates back to 1916.[65] The oil and the crushed seeds have been used for a long time in southeast Asia to treat various skin diseases such as scabies, eczema, psoriasis, scrofula, ringworm and intestinal worms.[66] The therapeutic action of this oil has been attributed to two unusual cyclopentenyl fatty acids named chaulmoogric and hydnocarpic acids (Figure 16.2). Both these fatty acids are known to inhibit the multiplication of *Mycobacterium leprae*. Hadia Almahli reviewed thoroughly the history and biological activities of cyclopentenyl fatty acids.[67] An ayurvedic formulation containing hydnocarpic and chaulmoogric acid exhibited broad antibacterial activities, highest against *Staphylococcus aureus* (13.13 ± 0.41 mm) and lowest against *Pseudomonas aeruginosa* (11.5 ± 0.40 mm).[68] The presence of these two unusual fatty acids

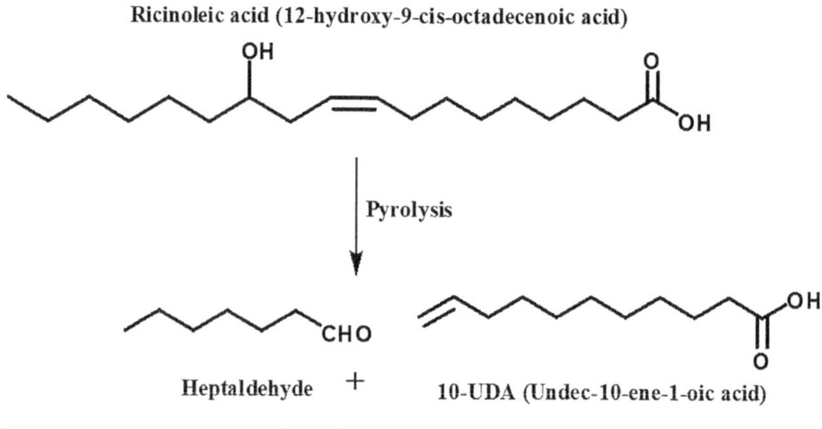

Scheme 16.1 Production of heptaldehyde and 10-UDA from ricinoleic acid.

Table 16.2 Antifungal activity and medicinal application of 10-UDA as such or its salts.[a]

Fatty acid/ derivatives	Disease	Fungal strain	References
10-UDA and zinc undecylenate	Dermatomycosis pedis (known as athlete's foot)	*Trichophyton mentagrophytes*, *Epidermophyton floccosum*, *Trichophyton rubrum*	56
10-UDA and zinc undecylenate	Infection of glabrous skin	*Microsporon audouim*	57
Zinc and calcium salts of 10-UDA and UDA	Dental plaque and infections of the teeth, gums and mouth Topical application	*Staphylococcus aureus*, *Staphylococcus .pyogenes*, *Candida albicans* and *Pseudomonas aeruginosa*	58, 59
10-UDA and zinc undecylenate with active ingredients	Treatment of mycotic (fungal) infections	*Candida albicans*, *Blastomyces dermatidis*, *Aspergillus fumigatus*, *Cryptococcus neoformans*, *Trichophyton mentagrophytes*, *Microsporum canis*	60
10-UDA	Denture stomatitis; chronic rhinosinusitis	*Candida albicans*	61, 62
10-UDA and its Calcium salt	Antifungal action on dermatophytes	*Trichophyton* sp., *Microsporum* sp. *Epidermophyton*, *Candida. albicans* and *Candida parakrusei*, *Staphylococcus aureus*, *Brevibacterium ammoniagenes*	63
8-hydroxy quinoline ester of 10-UDA	Used as medicaments against skin diseases caused by fungi	Active against organisms such as fungi, bacteria, protozoa, amoeba	64

[a]10-UDA: 10-undecenoic acid.

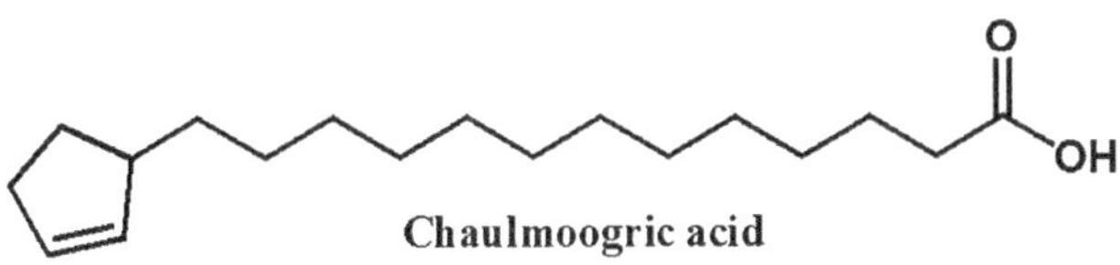

Chaulmoogric acid

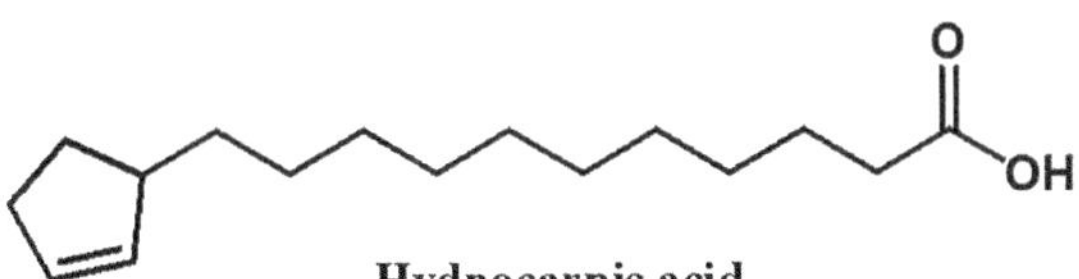

Hydnocarpic acid

Figure 16.2 Structure of chaulmoogric [13-(2-cyclopentene-1-yl) tridecanoic acid] and hydnocarpic [11-(2-cyclopentene-1-yl)undecanoic acid] acids.

makes chaulmoogra oil an active ingredient in several cosmetic and pharmaceutical formulations.[69]

Sterculia foetida seed oil belongs to the group of non-edible oils, as it contains two unusual cyclopropenoic acids, named sterculic and malvalic acids (Figure 16.3).[70] Several biological activities of cyclopropene fatty acids have been reviewed extensively in the literature.[71] Cyclopropene fatty acids are well known inhibitors of fatty acid biosynthesis in higher animal. There are no reports on antimicrobial activity of these two fatty acids except the one by Schmid and Patterson.[72] Similarity between animal and fungal desaturase enzyme systems prompted the authors to study the antifungal activity of cyclopropenoid fatty acids and they reported evidence of inhibition of desaturase enzyme in two fungi, *Fusarium oxysporum* and *Ustilago maydis*.

11-cyclohexylundecanoic acid, a minor fatty acid (0.1–0.2%) present in milk fat can be classified as an unusual fatty acid.[73] Structural similarity of this fatty acid with chaulmoogric/hydnocarpic acids led researchers to assume that this minor fatty acid may have antimicrobial activity. But due to its low content in milk fat and hence poor availability, organic chemists synthesized 11-cyclohexylundecanoic acid and evaluated its antimicrobial activity. It was found that 11-cyclohexylundecanoic acid possesses very good bactericidal activity against Gram-positive *Bacillus cereus* and Gram-negative *Escherichia coli*. Its fungicidal activity against *Fusarium culmorum* was reported to be good, but no activity was reported against *Saccharomyces cerevisiae*.[74]

Milk fat also contains a minor fatty acid, an unusual group of isomers of linoleic acid collectively called conjugated linoleic acid (CLA). The conjugation of the double bond is either in positions 9 and 11 or 10 and 12, and the configuration of double bond may be *cis* or *trans*. These are naturally occurring anticarcinogens synthesized from vaccenic acid (C18 : 1, *trans* 11) by endogenous desaturase enzyme in ruminant animals.[75] Among many isomers, 9 (*cis*), 11 (*trans*)-octadecenoic acid is the most active and predominant CLA, having many potent human health benefits.[76] Despite a host of studies reported in the literature on CLA, there are only a few reports on their efficacy as antimicrobial agent. Some reports suggest efficacy of CLA as

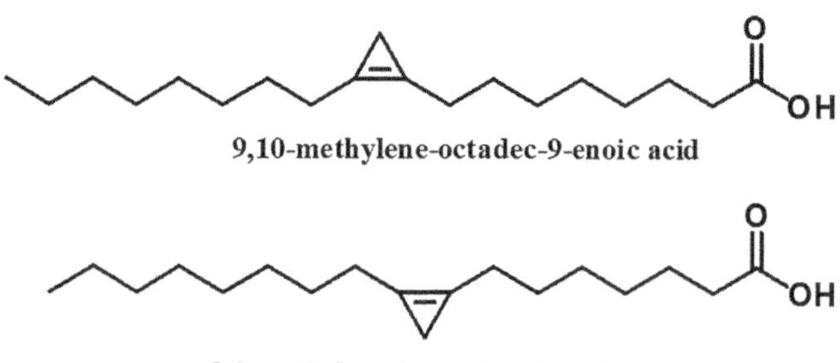

Figure 16.3 Structure of sterculic acid (9,10-methylene-octadec-9-enoic acid) and malvalic (8,9 methylene-heptadec-8-enoic acid) acids.

its potassium salt against some foodborne bacteria such as *Listeria monocytogenes*[27,77] and *Lactobacillus* species.[78] Bactericidal effects of CLA as its salt on some foodborne bacteria and pathogenic bacteria in a concentration-dependent manner are reported in the literature.[79] CLA salts are documented to possess growth inhibitory effect an both Gram-positive (*Bacillus cereus*, *Staphylococcus aureus* and *Streptococcus mutans*) and Gram-negative (*Salmonella typhimurium*, *Vibrio parahemolyticus*, *Pseudomonas aeruginosa*, *Klebsiella pneumoniae* and *Proteus mirabilis*) bacteria. However, the effect is more pronounced against Gram-positive than against Gram-negative species. Antifungal activity of CLA was reported only against *Candida albicans*, in modulating its hyphal growth.[80]

Eleostearic acid (9Z, 11E, 13E-octadecatrienoic acid), a conjugated linolenic acid (CLNA) found in the seed oil of *Momordica charantia* (commonly known as bitter melon) and also in Tung oil, is known to possess antitumor activity in cancer cells.[81] Several animal studies documented anti-inflammatory, anti-obesity, hypolipidemic activity of CLNA.[82] Although there is no literature report on antimicrobial activity of natural eleostearic acid, an isomer of eleosteraic acid produced microbially is documented to possess inhibitory activity against *Staphylococcus aureus* and *Helicobacter pylori*.[83]

16.3.2 Antimicrobial Activities of Fatty Acid Derivatives

Fatty acids are often conjugated to bioactive functional moieties in order to increase the activity of the latter. Such conjugation often results in increased antimicrobial activity. There are different derivatives comprising fatty acids with counterpart alcohols, amino acids, phenolics, sugars, *etc.* All such FADs are synthesized and evaluated as potential novel antimicrobial lipids. This section covers only those derivatives where there is a direct linkage between a fatty acid and/or a fatty alcohol with other compounds which can be of natural origin with potential antibiotic application (Figure 16.4).

16.3.2.1 1-MAG as a Fatty Acid Derivative

Esterification of fatty acid with monohydric alcohol such as methanol or ethanol decreases antimicrobial activity of the corresponding fatty acids.[7] However, such esterification when carried out with polyhydric alcohol such as glycerol results in increased efficiency and is the basis for 1-monoacylglycerol (1-MAG) to be the most studied FAD. A host of literature on antimicrobial activity of different 1-MAGs is given in Table 16.1. Among the different 1-MAGs studied, 1-monolaurin (C12-MAG) is reported to possess maximum antimicrobial efficacy against most of the studied strains.[7,10a,10c,16,22,23] The result is in line with that of fatty acids, where lauric acid is documented as having the highest antimicrobial efficacy. However, the activity of 1-monolaurin is found to be superior to lauric acid undoubtedly due to higher amphiphilic nature of 1-monolaurin compared to lauric acid. There are some studies where 1-monocaprin (C10-MAG) shows superiority over

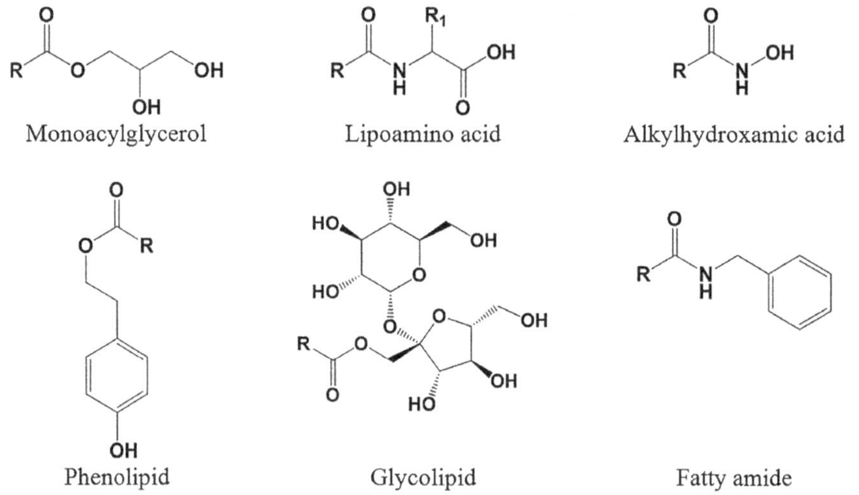

Figure 16.4 Structure of different derivatives of fatty acids. R: fatty acyl group.

1-monolaurin.[15,22,25] A human study of denture infection also showed efficacy of 1-monocaprin against *Candida* strains indicating its potential as topical agent.[84] However, there are deviations in results, as observed in a study by Marounek *et al.* wherein both caprylic acid (C8) and 1-monocaprylin (C8-MAG) exhibited better activity against *Cronobacter malonaticus* (Gram-negative) than 1-monolaurin and 1-monocaprin.[36] However, bactericidal efficacy of 1-MAG (C10, C11, C12 and C14) in free form is more active than in emulsion, as has been observed in a study by Petra *et al.*[85] The authors examined the antibacterial activity of 1-MAG as a micro-emulsion formulation against the growth of Gram-positive and Gram-negative bacteria and showed higher efficacy of neat 1-MAGs compared to its micro-emulsion form. A recent report from the author's laboratory on the synthesis and antimicrobial evaluation of 1-MAGs of unsaturated fatty acids suggested that MAG of γ-linolenic acid is the most potent antibacterial compared to other MAGs of unsaturated fatty acids.[86] Hence, 1-MAG is an efficient antimicrobial agent and promising alternative to synthetic antimicrobials. The US Food and Drug Administration declared GRAS (generally recognized as safe) status to both lauric acid and 1-monolaurin.[87]

16.3.2.2 *Phenolipid as a Fatty Acid Derivative*

Phenolics are well known natural antioxidants having many potent biological activities.[88] Lipoconjugated phenolics are known as phenolipids and are anticipated to have superior antioxidant activity in lipophilic media.[89] Ferulic acid on conjugation with C4 and C16 fatty acids exhibited good antibacterial activity against Gram-positive (*Bacillus subtilis* and *Staphylococcus aureus*) and Gram-negative bacteria (*Escherichia coli*, *Pseudomonas aeruginosa*, *Pseudomonas oleovorans*, *Klebsiella aerogenes*) and absolutely no activity against four studied fungal strains (*Candida albicans*, *Saccharomyces cerevisiae*, *Rhizopus oryzae* and *Aspergillus niger*).[90] However, bactericidal activity of shorter chain phenolipids was found to be better than longer chain homolog. Octyl gallate, another phenolipid, was found to possess antibacterial activity against *Salmonella choleraesuis* and antifungal activity against *Saccharomyces cerevisiae* and *Zygosaccharomyces bailii*.[91] In another study, lipophilic tyrosyl esters of fatty acids ranging from C2 to C18 : 1 were prepared and tested against eight Gram-positive and five Gram-negative strains, along with anti-leishmania activities. It was observed that only medium-chain derivatives (C8, C10 and C12) exhibited good antimicrobial and anti-leishmanial activities, with the C10 derivative showing maximum IC_{50} and minimum inhibitory concentration values.[92] Synthesis and antimicrobial activity of a range of phenolipids have been reported in a recent review article.[93]

16.3.2.3 Lipoamino Acid as a Fatty Acid Derivative

Combination of fatty acid and amino acids results in the formation of lipoamino acids, which are reported to possess excellent adsorption and aggregation properties, low toxicity and broad biological activity. Presence of amino acid residues as well as surface activity makes this molecule an obvious choice for biodegradable antimicrobial agent. *N*-stearoyl derivatives of amino acids, namely proline, phenyl alanine and tryptophan, were found to exhibit excellent activity against Gram-positive *Staphylococcus aureus*, *Micrococcus luteus* and *Bacillus cereus* and Gram-negative *Escherichia coli*, but no activity against *Pseudomonas aeruginosa* and mild antifungal activity against *Candida albicans*.[94] Synthesis of a series of lipoamino acids and their properties such as surface and biological activity has been reviewed in the literature.[95] An interesting observation is the enhancement in bactericidal properties of *N*-acyl leucine derivatives on inclusion of a functional moiety such as cyclopropane, hydroxyl groups or unsaturation in the acyl chain.[96]

16.3.2.4 Carbohydrate-lipid Conjugate as Fatty Acid Derivative

Another class of FAD is the carbohydrate fatty acid ester. Due to their biodegradability and eco-friendly nature, carbohydrate fatty esters are widely used in the food industry as emulsifiers. Although they are mostly used as a biodegradable mild surfactant, reports of their antimicrobial activity are documented in the literature.[97] Due to the higher potency of lauric and caproic acid, these two fatty acids are often conjugated to a carbohydrate moiety. Sucrose monolaurate is reported to possess strong bactericidal activity against *Staphylococcus aureus* and *Escherichia coli*.[98] In another study, antibacterial activity of six carbohydrate fatty acid esters was evaluated against these two bacterial strains.[99] All the studied compounds exhibited better bactericidal activity against *Staphylococcus aureus* than *Escherichia coli*. Among these esters, lauric ester derivatives of methyl-α-D-mannopyranoside and methyl-β-D-glucopyranoside showed higher activity compared to other ester derivatives against *Staphylococcus aureus*. This indicates that not only the chain length of the fatty acid, but also the nature of the carbohydrate is important in assessing the bactericidal efficacy of carbohydrate fatty acid esters, supporting the conclusion of another reported study.[100] Antimicrobial properties of four fatty acid fructose esters, namely caprinoylfructose, lauroylfructose, myristoylfructose and palmitoylfructose against *Bacillus cereus*, *Escherichia coli*, *Saccharomyces cerevisiae* and *Fusarium culmorum* showed that caprinoylfructose showed the highest antimicrobial activity. The inhibitory effect decreased rapidly as the chain length increased.[101] A series of lactose esters of fatty acid such as lactose monooctanoate, lactose monodecanoate, lactose monolaurate and lactose monomyristate were found to exhibit antibacterial activity against Gram-positive bacteria. The authors also observed that the chain length of the fatty acid ester influenced the antimicrobial activity of lactose esters towards Gram-positive bacteria. Lactose esters containing decanoate and laurate were more active than esters containing octanoate and myristate. However, no ester inhibited the growth of the Gram-negative bacteria, *Escherichia coli*.[102] Ricinoleic acid, when glycosidically linked to carbohydrate, showed a wide spectrum of antibacterial activity against Gram-positive bacteria such as *Staphylococcus aureus*, *Bacillus subtilis* and *Micrococcus luteus*.[103] Out of the 28 synthesized compounds, the authors found that the antibacterial activity of two compounds was better than norfloxacin.

16.3.2.5 Fatty Amides as Fatty Acid Derivatives

Fatty amides are another class of FAD having many potent biological activities.[104] Fatty amides are also reported to possess excellent antimycobacterial activity.[105] A series of fatty amide-conjugating different saturated and unsaturated fatty acids with cyclic and acyclic amines were tested as antituberculosis agents against *Mycobacterium tuberculosis*. Among the tested fatty acid amide derivates, ricinoleic acid-based fatty amides were the most potent and ricinoleylpyrrolidilamide was the best antimycobacterial agent. A recent article reviewed the antimicrobial activity of a series of ricinoleic-based fatty amides.[53a]

16.3.2.6 Hydroxamate Derivatives of Fatty Acid

Caprylohydroxamic acid, a hydroxamic acid derivative of octanoic acid is often used as chelating agent in skin care products.[106] This is primarily due to its potent activity against many Gram-positive and Gram-negative bacteria, yeasts and moulds. 10-Undecenehydroxamic acid, a structural analogue of caprylohydroxamic acid exhibited antimicrobial activity against a range of microorganisms. Ammendola *et al.* reported enhanced activity of 10-undecenehydroxamic acid compared to 10- UDA against Gram-negative (*Escherichia coli* and *Salmonella typhimurium*), Gram-positive bacteria (*Enterococcus faecalis*), yeast (*Candida albicans*) and fungi (*Aspergillus niger* and *Chaetomium globosum*).[107] The strong chelating and H-bond forming capabilities of hydroxamic functional moieties have been cited as the possible reason for their enhanced activity against microorganisms compared to 10-UDA.

16.4 Mechanism

Integrity of the cell membrane is crucial for the survival of microbes, as it offers the first barrier against any antimicrobial agent. Surface charge, viscosity and compactness of cell membranes are important for many vital cellular functions. Any perturbation in these biophysical properties of the cellular membrane leads to its malfunction. Bactericidal activity of fatty acids has been presumed to be due to such membrane disruption. However, diverse structural features of fatty acids such as difference in chain length, absence and presence of unsaturation, position/orientation and degree of unsaturation and also presence of functional group in the fatty chain makes it difficult to pinpoint the exact mechanism behind the bactericidal activity of fatty acids. The result is contrasting literature reports on possible mechanism behind antimicrobial activity of fatty acid. Some proposed mechanisms are based on diversity in microorganisms and some are based on diversity in fatty acid structure. Fatty acids, being amphiphilic in nature have the ability to be integrated into the cell membrane structure inducing alteration in its surface charge, viscosity, hydrophobicity and hence its overall integrity. Such alteration has been cited by several researchers as possible reason behind the antimicrobial activity of fatty acids.[39,41,48] By integrating with the bacterial cell membrane, fatty acids may interfere with many cellular functions such as electron transport chain, oxidative phosphorylation, nutrient uptake, cellular enzyme activity, cell lysis and may even generate bactericidal secondary oxidation products. Disruption in any one of these cellular processes will retard the growth of the organism.[108] Desbois and Smith explained nicely the implication of all these processes to bacterial cell growth.[10e]

Increase in chain length of saturated fatty acids increases their hydrophobicity *vis-a-vis* antimicrobial activity, reaching maxima at C12, beyond which the activity falls. This indicates that too much lipophilicity is not good for antimicrobial efficacy of fatty acids and a balanced lipophilic–hydrophilic nature is essential for cell disruption. Unsaturated fatty acids have better bactericidal efficacy than their saturated counterparts. This is because of two opposite roles played by saturated and unsaturated fatty acids while integrating with the cell membrane. Saturated fatty acids decrease the fluidity of the membrane, inducing autolysis,[109] whereas unsaturated fatty acids increase its fluidity, inducing cell leakage and eventually cell death.[7,10e,29] Natural monounsaturated fatty acids with *cis* orientation of double bonds exhibited higher efficacy than their *trans* isomers, presumably due to structural similarity of the *trans* isomer with the saturated counterpart.[7,28,29] Higher activity of 1-monolaurin compared to lauric has been attributed not only to its increased amphiphilic nature, but also to its ability to form intramolecular H-bonds.[110] Phenolipids are also proposed to exert their activity by integrating with phospholipids and changing the biophysical properties of membrane bilayer.[87]

Antimicrobial activity of polyunsaturated fatty acids is reported to follow an alternative pathway. It was earlier proposed that cellular integration of multiple *cis* double bonds of polyunsaturated fatty acids render the membrane more fluid and permeable resulting in cell leakage.[10e,46] Such cellular disruption leads to cell lysis and aggregation of cellular components. Such morphological change has been viewed by a recent scanning electron microscopic study, proving cellular distortions of two Gram-negative bacteria, *Porphyromonas gingivalis* and *Fusobacterium nucleatum* by EPA and DHA.[110] Later, most researchers proposed a mechanism in which they opined that the polyunsaturated fatty acids block the bacterial enoyl-acyl carrier protein reductase (FabI), an enzyme responsible for elongation process during the biosynthesis of fatty acids.[111–113] Such action resulted in a change in composition of fatty acids of membrane phospholipid triggering its structural distortion and subsequent malfunction.[42b] The proposed mechanism behind the antimycobacterial efficacy of cyclopropene fatty acid is unique and is based on structural similarity of this fatty acid with biotin. Jacobsen and Levy in their work with hydnocarpic acid proposed a mechanism in which the fatty acid inhibits the multiplication of *Mycobacterium* species either by interfering in a metabolic reaction of the organism requiring biotin as a coenzyme or inhibiting the pathway for biotin synthesis.[114] However, the authors could not correlate in conformity the relation between biotin deficiencies with antimicrobial activity. Another study proposed biosynthetic incorporation of these unusual fatty acids into membrane phospholipids as the possible mechanism behind alteration of membrane structure and its properties.[115]

Exhaustive work presented in this chapter has indicated higher potency of all fatty acids and various derivatives against Gram-positive bacteria and yeasts compared to that against Gram-negative bacteria. This could be due to their differences in cell wall structure. Lower activity of fatty acids and their derivatives against Gram-negative bacteria is due to the presence of an outer membrane, which is absent in Gram-positive bacteria. Most of the Gram-negative strains possess lipopolysaccharide in the outer membrane which acts as an effective barrier against hydrophobic substances, whereas those having lipo-oligosaccharides in the outer membrane are susceptible to fatty acids and derivatives.[116] Only exception is the monoglyceride (1-MAG), a non-ionic surface active derivative of fatty acids showing potency against Gram-negative bacteria. Monoglycerides can penetrate and get incorporated into outer membrane disrupting membrane permeability and nutrient transport. Fungi, a multicellular complex microorganism are found to be insensitive to fatty acids except 10-UDA, a derivative of ricinoleic acid and an excellent fungicide.

16.5 Conclusions

The ever increasing resistance of microbes towards antibiotics prompted researchers to investigate newer antimicrobial agents with less or no toxicity.

This in turn led researchers to look for biocompatible compounds where lipids occupy prominent positions. Few groups have been working on fatty acids and some derivatives with a view of understanding the behaviour of fatty acids in inhibition of microbes. Nevertheless, fatty acids and derivatives hold great potential as nontoxic ecofriendly antimicrobial agents and further research in this area could be of importance in view of the emerging resistance of microbes to antibiotics.

Acknowledgements

Dr Shiva Shanker Kaki gratefully acknowledges Science & Engineering Research Board (SERB), Department of Science & Technology, New Delhi, India for the financial assistance under 'Early Career Research Award' (ECR12017/000639).

References

1. H. Thormar, Antibacterial Effects of Lipids: Historical Review (1881 to 1960), in *Lipids and Essential Oils as Antimicrobial Agents*, ed. H. Thormar, John Wiley & Sons, 2011, ch. 2, pp. 25–45.
2. H. Nikaido, *J. Bacteriol.*, 1996, **178**, 5853.
3. G. A. Jacoby, History of Drug-resistant microbes, in *Antimicrobial Drug Resistance*, ed. D. L. Mayers, Humana Press, a part of Springer Science Business Media, 2009, ch. 1, pp. 3–7.
4. L. J. Mc Gaw, A. K. Jäger and J. van Staden, *S. Afr. J. Bot.*, 2002, **68**, 417.
5. A. P. Desbois, T. Lebl, L. Yan and V. J. Smith, *Appl. Microbiol. Biotechnol.*, 2008, **81**, 755.
6. H. N. Glassman, *Bacteriol. Rev.*, 1948, **12**, 105.
7. J. J. Kabara, D. M. Swieczkowski, A. J. Conley and J. P. Truant, Fatty Acids and Derivatives as Antimicrobial Agents, *Antimicrob. Agents Chemother.*, 1972, 23–28.
8. S. Charlie, Chemistry of fatty acids, in *Bailey's Industrial Oil and Fat Products*, 6th edn, ed. F. Shahidi, John Wiley & Sons, Inc., 2005, Six Volume Set, pp. 1–44.
9. *Food Lipids: Chemistry, Nutrition, and Biotechnology*, ed. C. C. Akoh and D. B. Min, 2nd ed., Marcel Dekker, Inc., New York, 2002.
10. (a) J. J. Kabara, Fatty acids and derivatives as antimicrobial agents: A review, *Symposium on the Pharmacological Effects of Lipids*, AOCS, Champaign, 1978, vol. I11, pp. 1–14; (b) J. J. Kabara, Fatty acid and esters as antimicrobial/insecticidal agents, *ACS Symp. Ser. (Ecol. Metab. Plant Lipids)*, 1987, 325, 220–238; (c) J. J. Kabara and R. Vrable, *Lipids*, 1977, **12**, 753; (d) L. Ababouch, A. Cahibi and F. F. Busta, *J. Food Prot.*, 1992, **55**, 980; (e) A. P. Desbois and V. J. Smith, *Appl. Microbiol. Biotechnol.*, 2010, **85**, 1629.
11. (a) K. N. Prasad, *Life Sci.*, 1980, **27**, 1351; (b) D. H. Teitelbaum, *J. Parenter. Enteral Nutr.*, 2011, **35**, 443; (c) V. Marcil, E. Delvin and

E. Seidman, *Am. J. Physiol.: Gastrointest. Liver Physiol.*, 2002, **283**, G340; (d) C. C. Roy, C. L. Kien, L. Bouthillier and E. Levy, *Nutr. Clin. Pract.*, 2006, **21**, 351; (e) G. Celasco, L. Moro, C. Aiello, K. Mangano, A. Milasi, C. Quattrocchi and R. D. Marco, *Biomed Rep.*, 2014, **2**, 559.

12. J. Cabezas-Pizarro, M. Redondo-Solano, C. Umana-Gamboa and M. L. Arias-Echandi, *Rev Argent Microbiol.*, 2018, **50**, 56.
13. M. Era, S. Sakai, A. Tanaka, T. Kawahara, T. Kanyama and H. Morita, *Jpn*. J. Food Engg., 2015, **16**, 99.
14. C. B. Huang, Y. Altimova, T. M. Myers and J. L. Ebersole, *Arch. Oral Biol.*, 2011, **56**, 650.
15. G. Bergsson, J. Arnfinnsson, O. Steingrímsson and H. Thormar, *APMIS*, 2001, **109**, 670.
16. G. Bergsson, J. Arnfinnsson, O. Steingrímsson and H. Thormar, *Antimicrob. Agents Chemother.*, 2001, **45**, 3209.
17. W.-C. Huang, T.-H. Tsai, L.-T. Chuang, Y.-Y. Li, C. C. Zouboulis and P. J. Tsai, *J. Dermatol. Sci.*, 2014, **73**, 232.
18. T. Nakatsuji, M. C. Kao, J.-Y. Fang, C. C. Zouboulis, L. Zhang, R. L. Gallo and C.-M. Huang, *J. Invest. Dermatol.*, 2009, **129**, 2480.
19. K. Benkendorf, A. R. Davis, C. N. Rogers and J. B. Bremner, *J. Exp. Mar. Biol. Ecol.*, 2005, **316**, 29.
20. (a) J. J. Wille and A. Kydonieus, *Skin Pharmacol. Appl. Skin Physiol.*, 2003, **16**, 176; (b) T. Kitahara, N. Koyama, J. Matsuda, Y. Aoyama, Y. Hirakata and S. Kamihira, *et al.*, *Biol. Pharm. Bull.*, 2004, **27**, 1321; (c) M. S. Rouse, M. Rotger, K. E. Piper, J. M. Steckelberg, M. Scholz and J. Andrews, *et al.*, *Antimicrob. Agents Chemother.*, 2005, **49**, 3187.
21. B. K. Yoon, J. A. Jackman, E. R. Valle-González and N.-J. Cho, *Intl. J Mol. Sci.*, 2018, **19**, 1114.
22. B. W. Petschow, R. P. Batema and L. L. Ford, *Antimicrob. Agents Chemother.*, 1996, **40**, 302.
23. C. Q. Sun, C. J. O'Connor and A. M. Roberton, *FEMS Immunol. Med. Microbiol.*, 2003, **36**, 9.
24. E. Skrivanova, M. Marounek, G. Dlouha and J. Kanka, *Lett. Appl. Microbiol.*, 2005, **41**, 77.
25. (a) G. Bergsson, J. Arnfinnsson, S. M. Karlsson, Ó. Steingrímsson and H. Thormar, *Antimicrob. Agents Chemother.*, 1998, **42**, 2290; (b) G. Bergsson, Ó. Steingrímsson and H. Thormar, *Antimicrob. Agents Chemo*ther., 1999, **43**, 2790.
26. H. R. Knapp and M. A. Melly, *J. Infect. Dis.*, 1986, **154**, 84.
27. L.-L. Wang and E. A. Johnson, *Appl. Environ. Microbiol.*, 1992, **58**, 624.
28. M. Feldlaufer, D. Knox, W. Lusby and H. Shimanuki, *Apidologie*, 1993, **24**, 95.
29. H. Galbraith, T. B. Miller, A. M. Paton and J. K. Thompson, *J. Appl. Bacteriol.*, 1971, **34**, 803.
30. R. W. Lacey and V. L. Lord, *J. Med. Microbiol.*, 1981, **14**, 41.
31. F. Kanetsuna, *Microbiol. Immunol.*, 1985, **29**, 127.

32. E. Skrivanová, O. Savka and M. Marounek, *Folia Microbiol.*, 2004, **49**, 199.
33. M. Marounek, E. Skrivanová and V. Rada, *Folia Microbiol.*, 2003, **48**, 731.
34. J.-Y. Lee, Y.-S. Kim and D-H. Shin, *J. Agric. Food Chem.*, 2002, **50**, 2193.
35. P. M. Schlievert and M. L. Peterson, *PLoS One*, 2012, 7, e40350.
36. M. Marounek, V. Putthana, O. Benada and D. Lukešová, *Czech J. Food Sci.*, 2012, **30**, 573.
37. C. B. Huang, Y. Altimova, T. M. Myers and J. L. Ebersole, *Arch. Oral Biol.*, 2011, **56**, 650.
38. H. Galbraith and T. B. Miller, *J. Appl. Bact.*, 1973, **36**, 635.
39. A. P. Desbois and V. J. Smith, *Appl. Microbiol. Biotechnol.*, 2010, **85**, 1629.
40. J. P. Fay and R. N. Farias, *J. Gen. Microbiol.*, 1975, **91**, 233.
41. G. D. Lawrence, *The fats of life: Essential fatty acids in health and disease*, Rutgers University Press, 2010, p. 11–43.
42. (a) I. Gill and V. R. Rao, *TIBTECH.*, 1997, **15**, 401; (b) N. M. Carballeira, *Prog. Lipid Res.*, 2008, **47**, 50; (c) U. N. Das, *Lipids in Health and Dis.*, 2008, 7, 9; (d) D. Richard, K. Kefi, U. Barbe, P. Bausero and F. Visioli, *Pharmacol Res.*, 2008, **57**, 451; (e) M. D. Barber, J. A. Ross and K. C. H. Fearon, n-3 Polyunsaturated Fatty Acids in the Treatment of Pancreatic Cancer Cachexia, in *Pancreatic Disease*, ed. A. C. D. Johnson and C. W. Imrie, 2nd edn, Springer, London, 1999, pp. 310–324; (f) S. M. Mir, S. Kanjilal, S. U. Ahmed, Omega-3 Fatty Acids in Inflammatory Diseases, in *Omega-3 Fatty Acids*, ed. M. Hegde, A. Zanwar and S. Adekar, Springer, Cham, 2016, pp. 141–155.
43. H. Saito, H. Tomioka and T. Yoneyama, *Antimicrob. Agents Chemother.*, 1984, **26**, 164.
44. L. Thompson, A. Cockayne and R. C. Spiller, *Gut*, 1994, **35**, 1557.
45. M. R. Maia, L. C. Chaudhary, C. S. Bestwick, A. J. Richardson, N. McKain, T. R. Larson, I. A. Graham and R. J. Wallace, *BMC Microbiol.*, 2010, **10**, 52.
46. C. B. Huang and J. L. Ebersole, *Mol. Oral Microbiol.*, 2010, **25**, 75.
47. A. P. Desbois and K. C. Lawlor, *Mar. Drugs*, 2013, **11**, 4544.
48. W. Chanda, T. P. Joseph, X.-F. Guo, W.-D. Wang, M. Liu, M. S. Vuai, A. Padhiar and M.-T. Zhong, *J Zhejiang Univ-Sci. B*, 2018, **19**, 253.
49. V. S. Thibane, J. L. F. Kock, R. Ells, P. W. J. van Wyk and C. H. Pohl, *Mar. Drugs*, 2010, **8**, 2597.
50. M. Hatice and M. A. R. Meier, *Eur. J. Lipid Sci. Technol.*, 2010, **112**, 10.
51. B. A. Black, E. Zannini, J. M. Curtis and M. G. Gänzle, *Appl. Environ. Microbiol.*, 2013, **79**, 1866.
52. B. Narasimhan, V. K. Mourya and A. S. Dhake, *Pharm. Chem. J.*, 2007, **41**, 16.
53. (a) S. Pabiś and J. Kula, *Curr. Med. Chem.*, 2016, **23**, 4037; (b) A. F. Novak, C. G. Clark and H. P. Dupuy, *J. Am. Oil Chem. Soc.*, 1961, **38**, 321; (c) Y. Mohini, R. B. N. Prasad, M. S. L. Karuna, Y. Poornachandra and

C. Ganesh Kumar, *Bioorg. Med. Chem. Lett.*, 2014, **24**, 5224; (d) B. Narasimhan, V. K. Mourya and A. S. Dhake, *Pharm. Chem. J.*, 2007, **41**, 133.
54. G. Das, R. K. Trivedi and A. K. Vasishta, *J. Am. Oil Chem. Soc.*, 1989, **66**, 938–941.
55. L. P. Ereaux and G. E. Craig, *Can. Med. Assoc. J.*, 1949, **61**, 361.
56. R. Q. Crotty and R. S. Weiss, *J. Invest. Dermatol.*, 1950, **14**, 313.
57. L. Reiner, and N. J. Bloomfield, Antimycotic composition, *U. S. Pat.* 2,680,088, 1954.
58. T. C. Miller Jr., Method for reducing dental plaque and pellicle precursor of plaque, *U. S. Pat.* 4,258,028, 1981.
59. R. Aly and H. I. Maibach, *J. Soc. Cosmet. Chem.*, 1981, **32**, 317–332.
60. B. Ayon-Covarrubias, Topical antifungal treatment, *U. S. Pat.* 6,080,744, 2000.
61. N. Mclain, R. Ascanio, C. Baker, R. A. Strohaver and J. W. Dolan, *Antimicrob. Agents Chemother.*, 2000, **44**, 2873.
62. S. Helms and A. L. Miller, *Altern. Med. Rev.*, 2006, **11**, 196.
63. H. N. Prince, *J. Bacteriol.*, 1959, **78**, 788.
64. R. Neher, Undecylenic acid esters of 8-hydroxyquinolines, *U. S. Pat.* 2,666,058, 1954.
65. E. L. Walker and M. A. Sweeney, *J. Infect. Dis.*, 1920, **26**, 238.
66. M. A. A. Amin Sikder, A. K. M. N. Hossian, A. B. Siddique, M. Ahmed, M. A. Kaisar and M. A. Rashid, *Pharmacognosy J.*, 2001, **3**, 72.
67. H. Almahli, *Curr. Topics Med. Chem.*, 2017, **17**, 2903.
68. S. P. Chaudhary, A. K. Sing and K. N. Dwivedi, *World J. Pharm. Res.*, 2015, **4**, 2492.
69. L. Wendy, Cosmetic or pharmaceutical formulation, *PCT Int. Appl*, WO 2012116391, 2012.
70. S. Herrera-Meza, A. J. Martínez, M. G. Sánchez-Otero, M. R. Mendoza-López, O. García-Barradas, G. R. Ortiz-Viveros and R. M. Oliart-Ros, *Grasas Y Aceites*, 2014, **65**, 1.
71. J. Salaun and M. S. Baird, *Curr. Med. Chem.*, 1995, **2**, 511.
72. K. M. Schmid and G. W. Patterson, *Lipids*, 1988, **23**, 248.
73. (a) J. C. M. Schogt and P. H. Begemann, *J. Lipid Res.*, 1965, **6**, 466; (b) H. Egge, U. Murawsky, P. Gyorgy and F. I. Zilliken, *FEBS Lett.*, 1969, **2**, 255.
74. J. Šmidrkal, T. Karlová, V. Filip, M. Zárubová and I. Hrádková, *Czech J. Food Sci.*, 2009, **2**, 463.
75. J. M. Griinari, B. A. Corl, S. H. Lacy, P. Y. Chouinard, K. V. V. Nurmela and D. E. Bauman, *Nutr. Metabol.*, 2000, **130**, 2285.
76. (a) S. Banni and J.-C. Martin, Conjugated linoleic acid and metabolites, in *Trans Fatty Acids in Human Nutrition*, ed. J. J. Sebedio and W. W. Christie, Oily Press, Dundee, Scotland, 1998, pp. 261–302; (b) M. W. Pariza, The biological activities of conjugated linoleic acid, in *Advances in Conjugated Linoleic Acid Research*, vol. 1, ed. M. P. Yurawecz, M. M. Mossoba, J. K. G. Kramer, M. W. Pariza, and G. J. Nelson, AOCS

Press, Champaign, IL, 1999, pp. 12–20; (c) T. E. Lehnen, M. R. da Silva, A. Camacho, A. Marcadenti and A. M. Lehnen, *J. Intl. Soc. Sports Nutr.*, 2015, **12**, 36; (d) J. M. Gaullier, J. Halse, K. Hoye, K. Kristiansen, H. Fafertun and H. Vik, *et al.*, *J. Nutr.*, 2005, **135**, 778.
77. E. Mbandi, M. Brywig and L. A. Shelef, *Food Microbiol.*, 2004, **21**, 815.
78. J. K. Jenkins and P. D. Courtney, *Can. J. Microbiol.*, 2003, **49**, 51.
79. J. I. Byeon, H. S. Song, T. W. Oh, Y. S. Kim, B. D. Choi, H. C. Kim, J. O. Kim, K. H. Shim and Y. L. Ha, *J. Agric. Food Chem.*, 2009, **57**, 3164.
80. J. Shareck, A. Nantel and P. Belhumeur, *Eukaryot Cell.*, 2011, **10**, 565.
81. R.-J. Zhou, F. Wang, X.-H. Zhang, J.-J. Zhang, J. Xu, W. Dong and Z.-Q. Zou, *Mol. Med. Rep.*, 2014, **9**, 993.
82. J. C. Andrade, T. A. P. Rocha-Santos, A. C. Duarte, A. M. Gomes, A. C. Freitas, Biotechnological production of conjugated fatty acids with biological properties, in *Handbook of food Bioengineering: Food Bioconversion*, vol. 2, ch. 4, ed. A. M. Grumezescu, A. M. Holban, Elsevier, United Kingdom, 2017, pp. 127–173.
83. A. A. Hennessyl, P. Ross, R. Devery and C. Stanton, *Eur. J. Lipid Sci. Technol.*, 2016, **118**, 1743.
84. T. O. Thorgeirsdóttir, T. Kristmundsdóttir, H. Thormar, I. Axelsdóttir and W. P. Holbrook, *Acta Odontol. Scand.*, 2006, **64**, 21.
85. S. Petra, K. Vera, H. Iva, H. Petr, K. Zdenka and B. Leona, *Eur. J. Lipid Sci. Technol.*, 2014, **116**, 448.
86. J. Johny, A. Jatla, V. K. Eruva, S. Misra and S. S. Kaki, *Grasas Aceites.*, 2019, **70**, e325.
87. J. J. Kabara, *J. Soc. Cosmet. Chem.*, 1980, **31**, 1.
88. M. Stasiuk and A. Kozubek, *Cell. Mol. Life Sci.*, 2010, **67**, 841.
89. F.-E. Maria-Cruz and P. Villeneuve, *J. Agric. Food Chem.*, 2005, **53**, 2779.
90. (a) S. Kanjilal, S. S. Kaki, K. K. Reddy, B. V. S. K. Rao, K. B. S. Kumar, M. L. Kantam and R. B. N. Prasad, *Eur. J. Lipid Sci. Technol.*, 2008, **110**, 1175; (b) S. S. Kaki, K. K. Reddy, S. Kanjilal and R. B. N. Prasad, *J. Oleo Sci.*, 2015, **64**, 845–852.
91. I. Kubo, P. Xiao and K. Fujita, *Bioorg. Med. Chem. Lett.*, 2001, **11**, 347–350.
92. I. Aissa, R. M. Sghair, M. Bouaziz, M. Laouini, S. Sayadi and Y. Gargouri, *Lipids Health Dis.*, 2012, **11**, 13.
93. E. Durand, J. Lecomte and P. Villeneuve, *Lipid Technol.*, 2017, **29**, 67–70.
94. A. Sivasamy, M. Krishnaveni and P. G. Rao, *J. Amer. Oil Chem. Soc.*, 2001, **78**, 897.
95. P. Clapes and M. R. Infante, *Biocatal. Biotrans.*, 2002, **20**, 215.
96. (a) J. Xia, Y. Xia and I. Nnanna, *J. Agric. Food Chem.*, 1995, **43**, 867; (b) S. Y. Mhaskar, G. Lakshminarayana and L. Saisree, *J. Am. Oil Chem. Soc.*, 1993, **70**, 23.
97. D. L. Marshell; L. B. Bullerman, in *Carbohydrate Polyesters as Fat Substitutes*, ed., C. C. Akoh, B. B. Swanson, Marcel Dekker, New York, 1994, pp. 149–167.

98. (a) A. Kato and K. Arima, *Biochem. Biophys. Res. Comm.*, 1971, **42**, 596; (b) S. Kato, H. Kobayashi and T. Watanabe, *J. Food Hyg. Soc. Jpn.*, 1987, **28**, 261.
99. A. Smith, P. Nobmann, G. Henehan, P. Bourke and J. Dunne, *Carbohydr. Res.*, 2008, **343**, 2557.
100. T. Watanabe, S. Katayama, M. Matsubara, Y. Honda and M. Kuwahara, *Curr. Microbiol.*, 2000, **41**, 210.
101. T. Karlová, L. Poláková, J. Šmidrkal and V. Filip, *Czech J. Food Sci.*, 2010, **28**, 146.
102. S.-M. Lee, G. Sandhu and M. K. Walsh, *Saudi J. Biol. Sci.*, 2017, **24**, 1483.
103. R. Kuppala, M. Govindarajan, R. Tambat, N. Patel, H. Nandanwar, K. K. Bhutani and K. P. R. Kartha, *RSC Adv.*, 2016, 3700–3713.
104. (a) V. Di Marzo, *Nat. Rev. Drug Discov.*, 2008, **7**, 438; (b) V. Micale, C. Mazzola and F. Drago, *Pharmacol. Res.*, 2007, **56**, 382; (c) S. Pillarisetti, C. W. Alexander and I. Khanna, *Drug Discovery Today*, 2009, **14**, 1098.
105. C. R. M. D'Oca, T. Coelho, T. G. Marinho, C. R. L. Hack, R. C. Duarte, P. A. Silva and M. G. M. D'Oca, *Bioorg. Med. Chem. Lett.*, 2010, **20**, 5255.
106. D. Winn, *Eur. Pat.* EP 2224973 B1, 2016; *U. S. Pat.* 8993641B2, 2015.
107. S. Ammendola, A. Lembo, A. Battistoni, P. Tagliatesta, C. Ghisalberti and A. Desideri, *FEMS Microbiol. Lett.*, 2009, **294**, 61.
108. (a) J. M. C. Gutteridge, P. Lamport and T. L. Dormandy, *J. Med. Microbiol.*, 1974, 7, 387; (b) S. Adolph, S. Bach, M. Blondel, A. Cueff, M. Moreau, G. Pohnert, S. A. Poulet, T. Wichard and A. Zuccaro, *J. Exp. Biol.*, 2004, **207**, 2935.
109. (a) T. Tsuchido, T. Hiraoka, M. Takano and I. Shibasaki, *J. Bacteriol.*, 1985, **162**, 42; (b) J. G. Kenny, D. Ward, E. Josefsson, I.-M. Jonsson, J. Hinds, H. H. Rees, J. A. Lindsay, A. Tarkowski and M. J. Horsburgh, *PLoS One.*, 2009, **4**, 4344; (c) L. E. Cybulski, D. Albanesi, M. C. Mansilla, S. Altabe, P. S. Aguilar and D. de Mendoza, *Mol. Microbiol.*, 2002, **45**, 1379.
110. H. Zhang, L. Zhang, L.-J. Peng, X.-W. Dong, D. Wu, V. C.-H. Wu and F.-Q. Feng, *J. Zhejiang Univ.-Sci. B.*, 2012, **13**, 83.
111. M. Sun, Z. Zhou, J. Dong, J. Zhang, Y. Xia and R. Shu, *Microb. Pathog.*, 2016, **99**, 196.
112. C. J. Zheng, J.-S. Yoo, T.-G. Lee, H.-Y. Cho, Y.-H. Kim and W.-G. Kim, *FEBS Lett.*, 2005, **579**, 5157.
113. Y.-M. Zhang and C. O. Rock, *Nat. Rev. Micro.*, 2008, **6**, 222.
114. P. L. Jacobsen and L. Levy, *Antimicrobial Agents Chemother.*, 1973, **3**, 373.
115. M. C. Cabot, *Lipids*, 1981, **16**, 146.
116. (a) H. Galbraith and T. B. Miller, *J. Appl. Bacteriol.*, 1973, **36**, 659; (b) C. W. Sheu and E. Freese, *J. Bacteriol.*, 1973, **115**, 869.

CHAPTER 17

Overview of Antimicrobial Resistance and Nanoparticulate Drug Delivery Approach to Combat Antimicrobial Resistance

MUTHUPANDIAN SARAVANAN,*[a] MELAKU ASHAGRIE,[a] OMAR ALI[a] AND BALAJEE RAMACHANDRAN[b]

[a] Department of Medical Microbiology and Immunology, Institute of Biomedical Sciences, College of Health Science, Mekelle University, Mekelle-1871, Ethiopia; [b] Department of Chemistry, University of Southern California, Los Angeles 90089, California, USA
*Email: bioinfosaran@gmail.com; Saravanan.muthupandian@mu.edu.et

17.1 Introduction

The rate of chronic carriers of bacterial pathogens is becoming a growing health concern worldwide.[1] The chemotherapy of diseases caused by intracellular bacteria shows numerous unusual challenges. Moreover, many bacteria have established a method to cause a 'silent' infection within the cells, which enables them to avoid contact with cellular antibacterial pathways. As a result, in such circumstances, cells are ineffective to destroy the

Biomaterials Science Series No. 5
Antimicrobial Materials for Biomedical Applications
Edited by Abraham J. Domb, Konda Reddy Kunduru and Shady Farah

Published by the Royal Society of Chemistry, www.rsc.org

intracellular bacteria and, moreover, will act as reservoirs, in turn contributing to the dissemination of the infection into other tissues and other parts of the body.

17.1.1 Overview of Antimicrobial Resistance and Its Mechanisms

Antimicrobial resistance among common bacterial pathogens is now threatening therapeutics and endangering critically ill patients. As a consequence, the World Health Organization (WHO) has declared antibiotic resistance to be the major public health threat of the 21st century.[2] A drastic increase in microorganisms that have developed resistance to currently available antimicrobial agents has become a major public health concern worldwide. Antibiotic resistance has emerged not only in the hospital environment, but is also often identified in community settings, suggesting that reservoirs of antibiotic-resistant bacteria are present outside the hospital. Adaptation and evolution influence the mutation, acquisition, and alteration of gene expression against the bacterial agents. These concepts are being studied to observe their root cause for clinical practice. Hence, cognizance on biochemical and genetic resistance patterns are of prime importance to design strategies to reduce the exposure of bacterial resistance and to innovative therapeutic approaches against multidrug-resistant organisms. Multi-drug resistance is considered to be an economic burden and costs about 20 billion dollars per year in the USA and leads to increased mortality compared to infections caused by susceptible bacteria.[3–5] The Centers for Disease Control and Prevention conservatively estimate that at least 23 000 people die annually in USA as a result of an infection with an antibiotic-resistant organism.[6] According to recent reports, antibiotic resistance is estimated to cause approximately 300 million premature deaths by 2050, with a loss of up to $100 trillion (£64 trillion) to the global economy.[7] This situation worsens due to a weakened robust antibiotic pipeline, resulting in the emergence of infections that are almost untreatable and leaving clinicians with no reliable alternatives to treat infected patients.

Apart from geographical data on infections and budget, the following points refer to some relevant concepts to deal with the problems of antimicrobial resistance. Firstly, antimicrobial resistance is a result of the interaction of many organisms with their environment. The naturally produced antimicrobial peptides evolve in the mechanism by extending their residence time to explore their action. Hence, these organisms are often considered to be 'intrinsically' resistant to one or more antimicrobials. Secondly, clinical practice mainly depends on the multiple layers of complexity. Most antimicrobial agents are derived from natural products naturally found in the soil. The antimicrobial molecule depends on size, sample source, physicochemical properties, and pharmacological parameters to investigate their barriers. *i.e.*, susceptible, intermediate and resistant, to detect their potential against targets. Furthermore, the susceptibility in the clinical framework depends on the availability of treatment procedures. In this review, the genetic basis mechanisms, evolutionary relationships, and biochemical pathways will be discussed in detail in order to understand the susceptibility of disease which was implemented in the clinical implications.

17.1.1.1 Evolutionary Relationships

The wide range of environmental threats and multidrug resistance may lead to genetic modification. This is because the bacteria are genetically organized and respond to the above threats. The genetic modification has matured through the evolutionary relationships to combat against harmful antibiotic molecules. Hence, bacteria have three major genetic strategies: (1) mutation of genes to undergo evolution; (2) horizontal gene transfer; and (3) acquisition of foreign DNA for resistance.

17.1.1.1.1 Mutational Occurrence. Resistant patterns have emerged through acquired mutational changes in various complex problems. One of the following mechanisms will alter the antibiotic action, resulting in mutations in antimicrobial resistance. (1) Modification of the antimicrobial target (to be estimated according to chemical kinetics and thermodynamics); (2) drug delivery or uptake; (3) efflux mechanisms; and (4) modulation of regulatory networks.

17.1.1.1.2 Acquisition of Foreign DNA. Acquisition of foreign DNA material is generally studied through the genetic exchange, *i.e.* horizontal gene transfer. The genetic exchange drives through bacterial evolution and is responsible for the development of antimicrobial resistance. This genetic exchange has been implicated in the dissemination of resistance to many frequently used antibiotics.

17.1.1.1.3 Biochemical Pathways. Antibiotics play a crucial role in bacterial resistance. They will resist bacterial dissemination instead of killing/destroying the bacteria. One class of bacterial cells will be targeted through multiple biochemical pathways. The three fundamental mechanisms of antimicrobial resistance are (1) enzymatic degradation of antibacterial drugs; (2) alteration of bacterial proteins that are antimicrobial targets; and (3) changes in membrane permeability to antibiotics. Antibiotic resistance can be either plasmid-mediated or maintained on the bacterial chromosome.[8] The most important mechanism of resistance to the penicillins and cephalosporins is antibiotic hydrolysis mediated by the bacterial enzyme β-lactamase. For example, the predominant mechanism of resistance to β-lactams in Gram-negative bacteria is the production of β-lactamases, whereas resistance to these compounds in Gram-positive organisms is mostly achieved by modifications of their target site, the penicillin-binding proteins (PBPs). It has been argued that this phenomenon is likely due to major differences in the cell envelope between Gram-negative and Gram-positive bacteria. In the former, the presence of an outer membrane permits the 'control' of the entry of molecules to the periplasmic space. Indeed, most β-lactams require specific porins to reach the PBPs, which are located in the inner membrane. Therefore, the bacterial cell controls the access of these molecules to the periplasmic space allowing the production of β-lactamases in sufficient concentrations to tip the kinetics in favour of the destruction of the antibiotic molecule.[9]

In addition, attributable to their being intracellular, bacteria are protected from antimicrobial activities and killing by the immune system and again from the bactericidal effect of the antibiotics present in the extracellular environment. Thus, even though there is a wide range of antibiotics that are active against isolated bacteria, unsatisfactory results are achieved against these bacterial pathogens in intracellular locations.[10] Typical examples of these intracellular bacteria include *Brucella abortus*, *Mycobacterium tuberculosis*, *Listeria monocytogenes* and *Salmonella enterica*. These organisms cause infectious diseases in the intracellular environment (brucellosis, tuberculosis, listeriosis and salmonellosis, respectively).[11] These intracellular pathogenic bacteria have the capability to replicate by creating relationships with the sensitive host.[12]

Today, regardless of our better understanding of the nature of microbial pathogenesis and the use of new and advanced therapeutics, infectious diseases still cause significant illnesses which contribute a high morbidity and mortality rate each year. Moreover, it is fact that many methods for diagnosing and treating intracellular bacterial infections exist presently; however, there is a pressing demand to discover new, improved and novel approaches for bacterial destruction. Despite recognized and predictable therapeutic efficacy, inefficient delivery of therapeutic drugs will be caused by an inadequate therapeutic index. It is nowadays obvious that a nanotechnology-driven approach using nanoparticles to specifically target and destroy intracellular pathogenic bacteria can be applied effectively. A nanotechnology drug delivery system is one goal to overcome challenges of other traditional drug delivery systems based upon the fabrication and development of minute nanostructures. The nanoparticles are highly effective in drug delivery methods where the surface area of the drug is more flexible to accommodate nanoparticles. Therefore, this interaction resulting in optimization of the interaction will boost the drug stability, solubility and pharmacokinetic parameters.[13] Due to these features, nanomedicines are easily administered as antimicrobial drugs.[14]

Recently, many types of nanoparticulate systems have been developed as potential drug delivery systems, including biodegradable polymeric nanoparticles, polymeric micelles, nano-capsules, fullerenes, solid lipid nanoparticles, liposomes, dendrimers, and metal nanoparticles. Nanoparticles have also been found to be very important in the establishment of oral, systemic, transdermal, pulmonary, and other routes of administration.[13]

This chapter focuses on nanoparticle-based drug delivery systems and their medical use for the treatment of a variety of intracellular bacterial infectious diseases, as well as their potential use in the field of medicine as a promising therapeutic option.

17.2 Types of Infections

Infectious disease is a clinical disorder caused by the existence of pathogenic agents such as viruses, parasites, bacteria or fungi. As the pathogens can be transferred from person to person (for instance tuberculosis, malaria) and sometimes from species to species (such as influenza), these diseases are described as infectious or communicable diseases. Infectious diseases can be usually categorized as (1) familiar diseases which are persistently there (for example tuberculosis, dengue, malaria); (2) new, previously unknown diseases (such as severe acute respiratory syndrome); and (3) emerging diseases that threaten to prevail in near future (for example, avian influenza). These diseases cause immense danger, accounting more than half of deaths worldwide, predominantly in developing countries.[15] A bacterial infection occurs when the pathogenic microorganism invades and mediates local and systematic damage of the host organism, resulting in acute or chronic infections.[16] As a result, infectious disease causes acute infection, a short period of severe illness, or chronic infection, which exists for a long time with low-grade clinical manifestations.[12]

17.3 Intracellular Bacterial Pathogens

Many bacteria can live and grow inside our cells either temporarily (facultative intracellular) or through their entire lives (obligate intracellular). These bacterial pathogens have the ability to adapt inside the hostile environment of the cells, including the first-line phagocytic cells, for survival and replication. Examples include *Mycobacterium tuberculosis* (tuberculosis), *Salmonella enterica* (typhoid fever and non-typhoidal salmonellosis), *Brucella abortus* (brucellosis) and *Listeria monocytogenes* (listeriosis).[12]

17.4 Antibiotic Treatment of Intracellular Bacterial Infections

Treatment using pathogen-immune serum was one of the important innovations and successes of the 19th century. Such a method of therapy encouraged medical scientists to produce chemicals capable of killing particular pathogens. It is obvious that attaining a raised level of a given antimicrobial drug inside infected cells is not sufficient to eradicate intracellular bacteria successfully. Intracellular pathogens, when inside cells, are protected from the extracellular environment such as antibodies and complement; moreover, they are prone to show resistance to antibiotics that are active *in vitro*.[17] Similarly, it must be kept in mind that the intracellular action of antibiotics may also alter, due to various reasons, such as pH change. As a result, the interaction between the antibiotic and the pathogen as forecast by *in vitro* studies may not be possible.[18,19]

To achieve success in treatment, the antibiotic employed must fulfil standard measures, such as the capability to enter and stay inside the cell, the ability to spread to specific intracellular targets, and show effective activity in the site at which the bacteria reside.

17.4.1 β-Lactams

Studies conducted regarding antibiotic treatment commonly report a lack of accumulation for all β-lactams both in phagocytic and non-phagocytic cells and tissues in general.[20,21] β-Lactam drugs show good diffusion capability through biological membranes. Moreover, all β-lactams exhibit a free carboxylic function, which is crucial for their antimicrobial activity.[22]

β-Lactam antibiotics inhibit the ability to synthesize the peptidoglycan of the bacterial cell wall. Studies demonstrate that the amount of this matter is generally lesser in acidic than in basic or neutral membrane-bound sections. β-Lactams show modest to low accumulation in cell lines (0.5 to 4.4 fold). This phenomenon attracts towards their respective activity in acidic pH. The corresponding outcomes provide the drugs with a therapeutic perspective regardless of their disparaging cellular pharmacokinetics.[21]

17.4.2 Aminoglycosides

Aminoglycosides have a wide range of broad antimicrobial activities, post-antibiotic effects and synergetic effects with β-lactam antibiotics. Nonetheless, aminoglycosides display restricted intracellular activity when compared to their solid bactericidal action extracellularly. They disseminate very slowly and poorly through cell membranes because of their high hydrophilicity; however, they can be integrated into macrophages by means of a fluid-phase pinocytosis method when applied at higher concentrations and with lengthened incubation times.[23] Additional studies have proved that intracellular aminoglycosides are nearly completely sequestered in the lysosomes, which they access for most cells *via* fluid-phase endocytosis.[24] This clarifies their slow rate of accumulation.

In contrast to the β-lactam antibiotics, aminoglycosides are weakly basic, hence once they have reached the interior of the cell, they are limited within the lysosomes, where the acidic pH might possibly suppress their activity. This diminished activity has been commonly related to the protonation of the molecule at acidic pH. Because an aminoglycoside must be actively transported to enter the bacteria, factors having a negative effect on this transport would diminish their antibacterial action.[25] An additional vital concern that limits the use of aminoglycosides is their illustrious toxicity. Nephrotoxicity is one of the main side effects of aminoglycosides, mainly as a result of renal cortical accumulation. Toxicity and neuromuscular blockage are also regarded as adverse effects.[25] Because of these adverse effects and narrow therapeutic range, aminoglycosides are not usually used in high concentrations.

17.4.3 Macrolides

Macrolides are weak organic bases that exhibit noticeable cellular penetration as well as tissue accumulation. Due to this feature, they are most commonly used to treat disease caused by intracellular bacterial pathogens. Macrolides have a marked intracellular buildup in nearly all cells when compared with β-lactams.[26,27]

The level of their accumulation, though, differs significantly among derivatives, with comparatively low values for erythromycin and one dose of macrolides, to high values for macrolides that comprise two essential functions. The comparisons over the antibiotics explore the similarity in their properties. The similarities of the antibiotics can be predicted by using one or more parameters; *e.g.*, molecular weight, log *P*, hydrogen bond donors and acceptors *etc.* This is to find their potential among the antibiotics against biological targets. Seral *et al.*, in 2003, demonstrated that the antibiotic macrolid, which is a weak base, has a major impact on drug concentration by responding well with membrane bound and acidic regions.[28]

17.4.4 Quinolones

Quinolones are another group of antibiotics that attain adequate levels inside both infected and non-infected cells.[23] They can diffuse easily to numerous cellular compartments with no association with cellular structure. The cellular concentration of fluoroquinolones is in general four- to 10-fold greater than the extracellular concentration. Most of the quinolones are commonly effective against intracellular bacteria such as *Legionella*[29] and *Listeria monocytogenes*.[23] However, some antibiotics show diminished activity at lower pH, which would explain the inconsistency between their intracellular and extracellular activities when combating pathogens confined within phagolysosomes (Figure 17.1).[30]

Furthermore, treatment becomes complicated as a result of the particular characteristics and features of the pathogenic bacteria; for instance, the existence of silent or metabolically inactive bacteria inside host that may not be susceptible to antimicrobial agents. Various features aid the bacteria to have the capability to develop such persistent infections. These include host and bacterial factors. Some pathogens are exceptionally able to escape the body's immune system and persist in the infected individuals for many years without showing sign and symptoms, for instance, *Salmonella typhi* and *Mycobacterium tuberculosis*.[31,32] Conversely, some pathogens like *Escherichia coli* or *Pseudomonas aeruginosa* result in both symptomatic chronic and acute infections, facilitating mechanisms to establish persistent infection.[33]

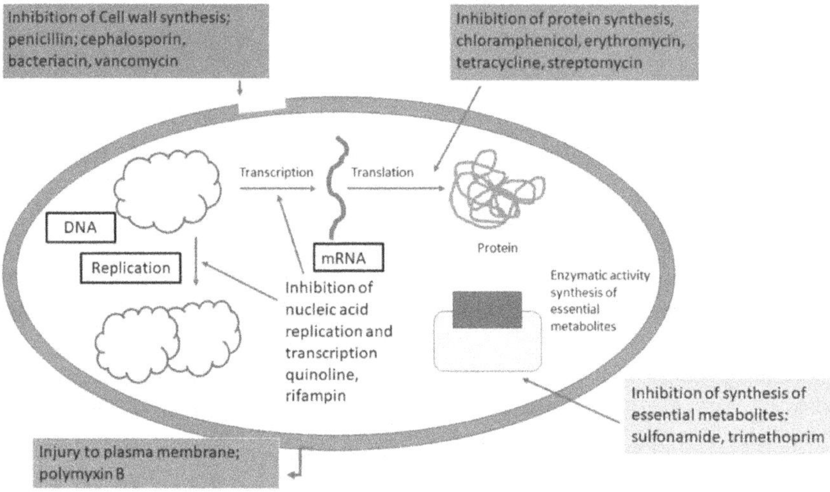

Figure 17.1 The main cellular targets and mechanisms of action of antibiotics within a bacterial cell.

Generally, it was indicated that the above antimicrobial drugs could affect the communication among phagocytic cells and microorganisms.[34] Antibiotic drug interaction with microorganisms inside endocytic organelles results in a multifaceted phenomenon, leading to the succeeding development of early and late endosomes, then true phagolysosomes, in which majority intracellular pathogens are digested due to the existence of local acidity of pH 5.0 and some acidic hydrolase enzymes.[35]

These antibiotics generally have the capability to change numerous roles of the antimicrobial action of host phagocytic cells. For instance, gentamicin boosts intracellular obliteration of pathogenic microorganisms by macrophages, whereas other aminoglycosides hinder such ability. Conversely, oxidative burst, chemotaxis, or cytokine production are not commonly influenced by aminoglycosides.[34]

However, various cellular, pathogenic, and environmental factors such as drug prescription and usage habits are contributing greatly to drug resistance in conventional antibiotic drugs intended for intracellular bacterial pathogen destruction.[24,36] Better drug delivery systems are needed to overcome the inhibition of cellular mechanisms in combating intracellular pathogens and subsequent drug resistance.

17.5 Challenges in Treating Infectious Diseases

Antibiotics have transformed the treatment of various medically significant bacterial illnesses, saving several millions of human lives from their discovery up to now. All antibiotic drugs, however, have a common weakness: the microorganisms develop resistance through a diverse group of mechanisms.[37]

The development of antimicrobial resistance is heightened mainly by the inappropriate use of antibiotics in humans. Many intracellular bacterial pathogens quickly develop mutations during the replication cycle. Different methods have been developed to prevail over the ubiquitous challenge of antimicrobial resistance. The emergence and the introduction of new and advanced antibiotic agents targeted at different microbial species, usually with high-tech screening of compounds,[38] is firming up and widening the therapeutic armamentarium. Moreover, combined therapies (*e.g.* multidrug treatment approaches for tuberculosis) have been shown to be effective in slowing down the emergence of resistance.

Another challenge in treating infectious disease is that the surprising flexibility of pathogens (mainly due to replicative and mutational capabilities) offers them an advantage against host immunity and antibiotic treatment pressures. These difficulties consist of ecological factors, antimicrobial drugs, and the human host immune response.[39] This struggle between microbial adaptation and human ingenuity remains a continuous challenge.[39,40]

Treatment failure is also a challenge in intracellular antibiotic therapy. Constant intracellular infection can diminish susceptibility to antimicrobials, resulting in higher rates of unsuccessful medication. For instance, persistent salmonellosis infection might decrease vulnerability to antimicrobials like ciprofloxacin and nalidixic acid.[41]

17.6 Targeted Therapy of Infections Using Nanoparticles

Some antibiotics need to be only in an aqueous solution to penetrate the cell, hence decreasing their bioavailability and bioactivity against intercellular bacteria. To overcome these difficulties, nanoparticle drug delivery systems (10–100 ng) can be used to deliver the drug to the specific target site.[42] This technology enables the drug to reach the site of infection with higher concentration. Two methods can be used for nanoparticle drug delivery; the first passive, where nanoparticles undergo extravasation at the site of infection, and the second is an active method, where nanoparticles contain ligands (for instance antibodies) that bind receptors (such as antigens) at sites of infection. Passive targeting nanoparticles encounter many obstacles, such as mucosal barriers and non-specific uptake and delivery of the particle.[42]

In case of active targeting methods, targeted nanoparticles facilitate receptor-mediated endocytosis, releasing therapeutic agents inside target cells, resulting in higher therapeutic efficacy and lower toxicity. Many active targeting strategies use the enhanced permeability and retention effect, so that active and passive targeting mechanisms act synergistically, leading to higher concentrations of nanostructures in the infected region.[42]

17.7 Antibiotic Nanocarriers in Drug Delivery Systems

In spite of the development of new antibiotics, intracellular infection management and treatment remains unresolved. The leading challenge for intracellular therapy is to plan and develop a transporter method and technique for antibiotics that are capable of being effectively endocytosed by phagocytic host cells and able to release the drug once inside the cells. In addition, conventional antimicrobial therapy mostly requires frequent administration and multiple dosages of drugs for an extended period. Nanoparticles successfully form new therapies to target intracellular regions of bacterial infections and are appropriate and suitable as vehicles for the delivery of antimicrobial agents since they typically provide a sustained drug release effect, minimize toxicity related to the loaded drugs and raise the total treatment effectiveness. Besides that, these drug delivery systems guard and keep the incorporated drug from premature immunological and enzymatic assaults and, in some cases, act synergistically with cellular bactericidal mechanisms.[19,43]

Nanoparticles are solid, colloidal particles comprising of macromolecular materials or substances that vary in size from 1 nm to 1000 nm (Figure 17.2).[44] Using nanotechnology, the drug of concern is principally melted, captured, adsorbed, attached, or encapsulated into the nanoparticle matrix. Based upon the preparation technique, nanoparticles or nanocapsules can be achieved with diverse properties and release characteristics for the encapsulated therapeutic agent.[44] In particular, use of nanoparticles for antibacterial therapy can sustain drug release over time, increase bioavailability and solubility, decrease aggregation, and improve efficacy.[45,46]

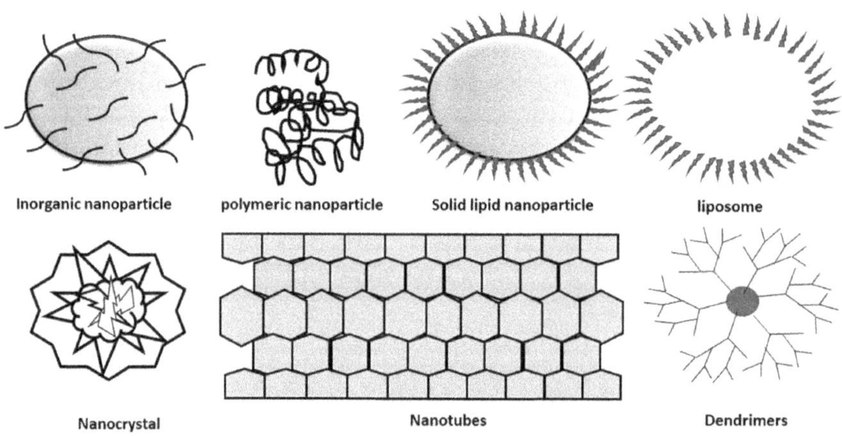

Figure 17.2 Different kinds of nanoparticles used in drug delivery and biomedical research.

In general, nanotechnology-based drug delivery systems are new approaches to advance (1) the therapeutic index, thus shortening the rate and dose of administration; (2) intracellular delivery to prevent the occurrence of drug-resistant bacteria;[47] and (3) targeted organ therapy to decrease systemic side effects.[48] The nanoscale size of these delivery systems is the basis for all these benefits.[49] As a result, it is assumed that drug delivery systems with heightened targeting features are greatly promising in increasing the efficacy and efficiency of therapy while simultaneously minimizing side effects.[50]

17.7.1 Polymeric Nanoparticles

Other antimicrobial agents are not able to locate the intracellular regions and the bacteria that are attached on the surface. However, polymeric nanoparticles have a better capacity to locate the specific site to attack the target.[51–53] The biologically active polymers play a vital role in the development of novel, biocompatible, active biomedical technologies that are capable of simplifying the burden on systemic antibiotics and preventing the emergence of antibiotic-resistant microbes.

Synthetic polymeric nanoparticles are of growing opportunities in drug delivery as a therapeutic agent. Polymers usually show improved pharmacokinetics in contrast to small-molecule drugs, having longer circulation time and the potential for tissue targeting. Polymers are being used as drug delivery systems as a polymeric drug alone or in combination with other small-molecule drugs or with biomacromolecules such as proteins and poly (nucleic acids).[54]

Nanoparticles were developed to resolve stability problems in time of storage and following administration in biological solutions. Such enhanced constancy and the likelihood of attaining a controllable release of the encapsulated drug are the main benefits of polymeric particles over liposomal carriers. These systems are being widely studied on the encapsulation of several active ingredients such as antibiotics, anticancer antigens and hormones attributable to advances in methods for encapsulation and development of novel and new polymers.[55]

The majority of polymeric nanoparticles are biocompatible, biodegradable, and are adopted as a chosen system for nanoparticle drug delivery. They also show a better capability for surface modification through chemical transformations, offer tremendous pharmacokinetic control, and are appropriate for the delivery of numerous therapeutic agents. Applicable nanoparticle forms include those prepared from chitosan, gelatin, poly(lactic-*co*-glycolic acid) copolymer, polyglycolic acid, polylactic acid, poly(alkyl cyanoacrylate), poly(methyl methacrylate), and poly(butyl)cyanoacrylate. Moreover, polymer-based coatings can be used onto other forms of nanoparticles to alter and advance their biodistribution characteristics. The biologically inert polymer polyethene glycol (PEG) has been covalently linked on the surface of different nanoparticles.[56] Such polymeric coating is believed to diminish immunogenicity, and prevent the phagocytosis of nanoparticles by the reticuloendothelial system, ensuring improved blood levels of the drug in various organs such as brain, intestine, and kidneys.[56]

Numerous approaches have been introduced to create polymeric micro- and nanoparticles.[57] Examples of the techniques that are widely used for polymeric particle preparation are emulsification-solvent removal (solvent evaporation), phase separation, interfacial polymerization, and spray-drying. The choice of a specific method is determined most commonly by the physicochemical properties of the drug of concern. Generally, the polymers used for drug encapsulation ideas are largely categorized into two clusters, based on their characters: synthetic and natural polymers. Natural polymers which are commonly human or bovine serum albumin, collagen, gelatin, chitosan, hyaluronan, starch *etc.* are polymers originated from natural sources. Their usage is reducing nowadays, due to their lack of homogeneity and purity and the risk of disease transmission. In contrast, synthetic biodegradable and biocompatible polymers (*i.e.* poly(a-hydroxy) acids), polyanhydrides like poly(sebacic acid) and poly(fatty acid dimer-sebacic acid), are widely used for the encapsulation of various drugs. Among these synthetic polymers, the aliphatic polyesters poly(lactic acid) (PLA), and copolymer poly(lactic-*co*-glycolic acid) (PLGA) are the most commonly explored class of polymers regarding toxicological and clinical data.[58] PLGA and PLA are biocompatible, non-toxic and biodegradable polymers approved by the US Food and Drug Administration for human consumption.[59,60]

17.7.2 Hydrogels

Hydrogels are polymeric substances that have a three-dimensional configuration with the ability to absorb a great quantity of water or other biological fluids. Their affinity to absorb water is mainly as a result of the presence of hydrophilic groups, for instance –OH, –CONH–, $-CONH_2-$, and $-SO_3H$ in polymers forming hydrogel structures.[61] As a result, hydrogels are high water content materials made from crosslinked polymers capable of providing persistent and local release of a range of therapeutic agents. Biocompatible, biodegradable hydrogels are prepared using natural polymers that are subject to enzymatic degradation or using synthetic polymers that possess hydrolysable moieties. Among these, the chitosan-based hydrogels have priority owing to their biocompatibility, degradability, and low toxicity.[62]

Hydrogels can keep the drug from thorny situations, such as enzymes and low pH in the stomach. Again, they are responsible for drug release by changing the gel structure in response to the different environmental stimuli. Such kinds of environment-sensitive hydrogels are also known as 'intelligent' or 'smart' hydrogels.[61,63] It has been established that among the numerous forms of drug delivery systems that have been developed to advance biocompatibility and effectiveness, hydrogels are enormously promising.[64]

In a general view, hydrogels can be categorized on the basis of different characteristics such as the nature of side groups (neutral or ionic), mechanical and structural features (affine or phantom), method of preparation (homo or copolymer), physical structure (amorphous, semicrystalline, hydrogen bonded, supermolecular, and hydrocollodial), and sensitivity to stimuli in the physiological environment (pH, ionic strength, temperature, electromagnetic radiation, *etc.*).[65] The polymers usually used for preparing hydrogels with pharmaceutical and biological applications are from synthetic or natural origins.[66]

Because the drug release mechanism from hydrogels mostly employed is inactive dispersal, molecules of different sizes and characteristics would generously disseminate into and out of the hydrogel matrix during the loading and storage periods. Generally, drug release techniques from hydrogels can be grouped as (1) diffusion-controlled; (2) swelling-controlled; and (3) chemically controlled.[61] Characteristically, hydrogels have been used to transport hydrophilic small-molecule drugs that have high solubility in both the hydrophilic hydrogel matrix and the aqueous solvent swelling the hydrogel.[67]

17.7.3 Liposomes

The liposomes are spherical vesicles having one or more lipid bilayers comprising hydrophilic heads, which encapsulate water-soluble drugs and hydrophobic tails, responsible for entrapping insoluble agents (Figure 17.3). Liposomes vary in size, composition, surface charge, and method of preparation. They can be single or in multiple bilayers, based on the lipid bilayer they possess; categorized into multilamellar, large unilamellar, and small unilamellar vesicles.[68] Nanoparticulate drug delivery systems using liposomes are most commonly used to improve the efficacy and targeting of drug and DNA delivery.[69]

The physicochemical characteristics of liposomes may be improved by altering the types of lipids, the contents and proportions of lipids in the liposomal formulation, the size of the liposome, the charge of the liposomal surface (positive, negative, or neutral), and the fluidity, as well as temperature- and pH-sensitivity of the liposomal membrane (fluid and rigid liposomes).[70]

Encapsulation of drugs in lipid vesicles is a better way of illustrating the necessary pharmacodynamic and pharmacokinetic characteristics. Liposomes have many advantages, such as being antibiotic carriers, having advanced pharmacokinetics and biodistribution, reduced toxicity, target selectivity, enhanced activity against intracellular as well as extracellular pathogens, and bacterial drug resistance.[70,71]

Numerous liposomal delivery systems present different formulation techniques or materials. This is mostly due to the variation in protein/peptide organization, charge, solubility, stability, and other characteristics.[72]

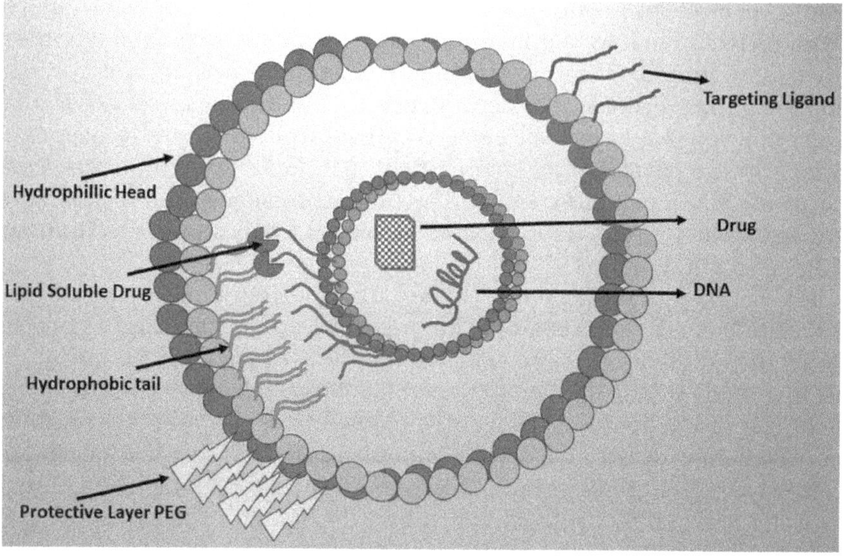

Figure 17.3 Diagram showing bilaminar liposome. The hydrophobic region traps drugs in the central core during the preparation of liposomes. The outer surface can be functionalized with ligands for active targeting.

17.7.4 Micelles

Micelles are submicroscopic masses of surfactant molecules assembly of amphiphilic block copolymers or polymer–lipid conjugates or other surface-active molecules that self-assemble in aqueous media to form structures with a hydrophobic core.[73] The capacity to functionalize the micelles and adapt the fragmentation behaviour by changing the copolymer structure is an advantageous parameter in making them drug carriers of choice. Due to their small size (1–50 nm), micelles are commonly ideal for intravenous delivery. Furthermore, they are more stable than liposomes, because of their chemically stable and biocompatible design.[74]

Micelles have a narrow size distribution in the nanometer range and are most commonly characterized by their unique core–shell architecture, in which hydrophobic segments are separated from the aqueous exterior. Micellar systems are most useful for the systemic delivery of water-insoluble drugs. Drugs can be partitioned in the hydrophobic core of micelles and the outer hydrophilic layer forms a stable dispersion in aqueous media, which can then be administered intravenously.[74] Polymeric micelles have been revealed to have a prolonged systemic circulation time following intravenous administration due to their smaller size and hydrophilic shell which is advantageous in minimizing their uptake by the reticuloendothelial system.[75]

Lately, substantial consideration has been given to several amphiphilic block copolymers, which can self-associate to form micelles in aqueous solution and have been widely studied as drug carriers.[76] Polymeric micelles have numerous benefits over conventional surfactant micelles, including their better thermodynamic stability in physiological solution, which makes polymeric micelles stable and prevents their rapid *in vivo* dissociation.[75] The preparation may lengthen the retention in blood circulation and attain a selective accumulation in the choroidal neovascularized lesions, but these features need further development.

17.7.5 Solid Lipid Nanoparticles

Solid lipid nanoparticles were discovered in the 1990s. They are lipid-based submicron colloidal carriers measuring between 50 nm and 1 μm. They are composed of a solid lipid matrix (Figure 17.4). Solid lipid nanoparticles helped to overcome the problems restricting the oral delivery of drugs that dissolve inadequately in water, where the drug is normally integrated in the submicron size range (below 1 μm) through (1) preservation of drug molecules in the molecularly dispersed (*i.e.* amorphous) form; (2) increased drug wettability within aqueous media due to the hydrophilic nature; and (3) action as precipitation inhibitors, thereby allowing supersaturated drug solubilization, which favors drug absorption.[77]

Lipid nanoparticles comprising solid matrix are generally two types. The first is solid lipid nanoparticles which are prepared from lipids that are solid at room temperature and at body temperature. The other is nanostructured lipid carriers, which are loaded in-between the fatty acid chains or in between lipid layers or in amorphous clusters in crystal imperfections within the solid lipid nanoparticle matrix.[78]

Solid lipid nanoparticles can be useful to carry drugs administered orally, topically, or *via* inhalation. They are commonly less toxic than liposomes and can be used to transport ciprofloxacin and ribafutin drugs. They can be helpful in the treatment of salmonella and tuberculosis infections.[56]

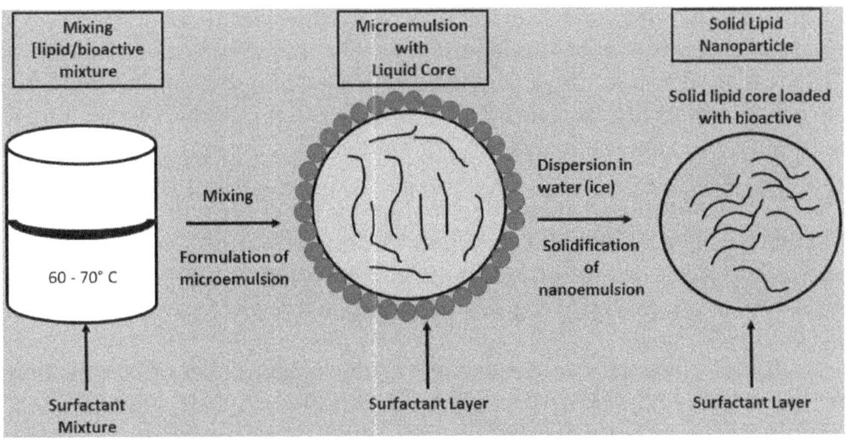

Figure 17.4 Formation of solid nanostructured silica–lipid hybrid microparticles using two different types of silica nanoparticles.

17.7.6 Fullerenes

The first fullerene, C60 fullerene, was discovered in 1985; since then, scientists have been searching for biomedical applications. Most are a form of carbon, although there are different forms such as diamond, graphite, and coal. For the first time, Friedman *et al.*[124] and Schinazi *et al.*[125] demonstrated that the hydrophobic cleft of the human immunodeficiency virus HIV genome I is able to successfully carry the C60 molecule. This discovery, and their small size, introduced the possibility that fullerenes could have a pharmaceutical function. C60 nanoparticles can be used to penetrate the cells walls of and kill many Gram positive bacterial strains, such as Staphylococcus species, Streptococcus species *etc.*, when compared to Gram negative bacterial strains such as *Klebsiella pneumoniae*, *Escherichia coli*, *Pseudomonas aeruginosa*, *Salmonella typhi*, *etc.* Moreover, the quinazoline–fullerene conjugate 18 was stated to have an inhibitory potential of 98.83%, with at least 1.562 $\mu g\ mL^{-1}$ that can prevent successful treatments for *Mycobacterium tuberculosis.*[12]

17.7.7 Dendrimers

Dendrimers are unique 'tree-like' structures (the word 'dendron' means tree in Greek) which contain polymers that have a branched architecture and nanoscale dimensions (Figure 17.5). These are synthetic macromolecules described by a huge number of branching points, three-dimensional globular shape, and nanometric size range. Poly (amidoamine) or PAMAM dendrimers are a class of hyperbranched polymers originally developed by Tomalia in 1979.[56] An essential distinctive feature of the dendrimers is that the dimension and surface charge of these constructs can be tailored exactly. The monodispersity of size, distinct structure, surface functionalization-ability, as well as stability are characteristics of dendrimers that enable them to be important future nanoparticle drug transporters.[56]

The first 'dendrimer family' that was synthesized, characterized, and commercialized was poly(amidoamine) (PAMAM) and have a measured size of 1.1 and 12.4 nm as their generations grow through 1–10.[42] These sizes are comparable with proteins (3–8 nm), linear polymer–drug conjugates (5–20 nm), and viruses (25–240 nm). Potential benefits for ocular use are (1) enhancement of the corneal residence time of drugs administered topically; (2) targeting of retinal neuroinflammation and provision of a targeted and constant level, sufficient for neuroprotection in retinal degeneration; (3) transport of medicine to the retina, leading to systemic administration; and (4) effectiveness as corneal glues to potentially replace sutures following corneal surgeries.[79]

In addition to ocular use, dendrimers can be used to treat other infections, such as SPL7013 (Gel HSV). It has been shown that it is safe when given vaginally in one dose and can protect monkeys against vaginal simian HIV infection. In addition, dendrimers also serve as carriers of immunostimulatory adjuvants and antigens that can be used for vaccine development.[80]

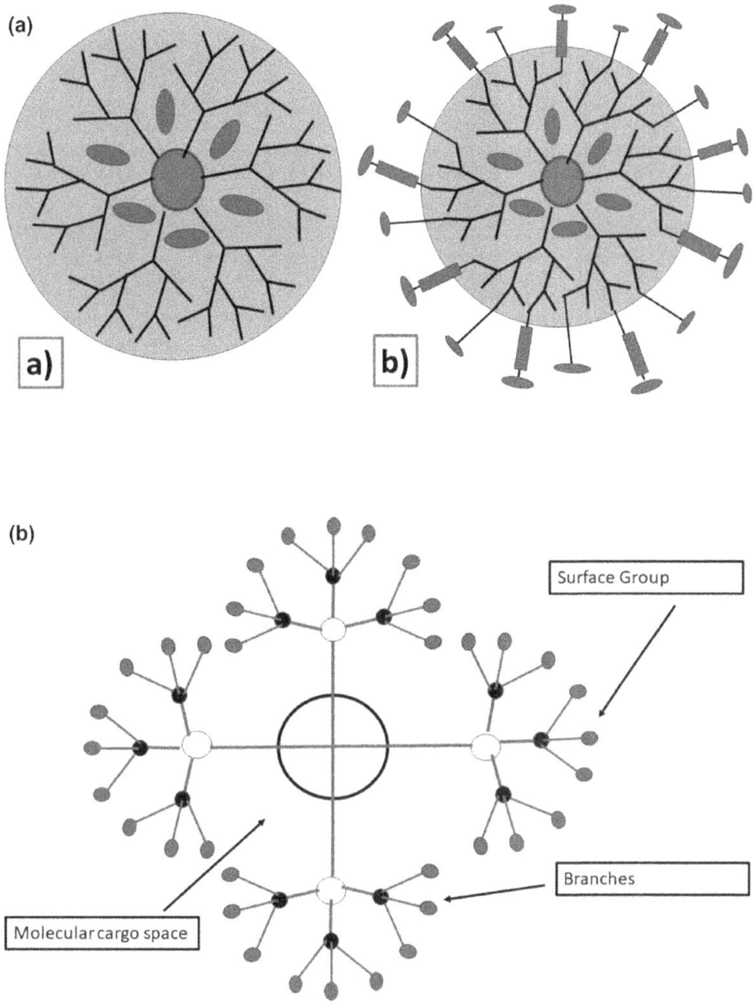

Figure 17.5 (a) Schematic representation of a) drug-encapsulated and b) drug-conjugated dendrimers; (b) schematic representation of dendrimer structure.

17.7.8 Metal Nanoparticles

A variety of metallic nanoparticles have captured much attention due to their effective application to various areas of science and technology. The non-essential metals, including silver, are toxic to bacteria and possess biocidal actions at extremely low concentrations. Furthermore, the essential metals, including copper, can also be very poisonous to the majority of bacteria and yeast at very low concentrations. Compounds that are made from metals, metal oxides or salts including copper and silver are among the most commonly used particles; for example, industries have been using vessels made of copper and silver for disinfecting water and preserving food since the time of the Persian kings. The biocidal actions are mostly triggered by (1) metal reduction and (2) metal donor atom selectivity and/or speciation (Figure 17.6).[80,81]

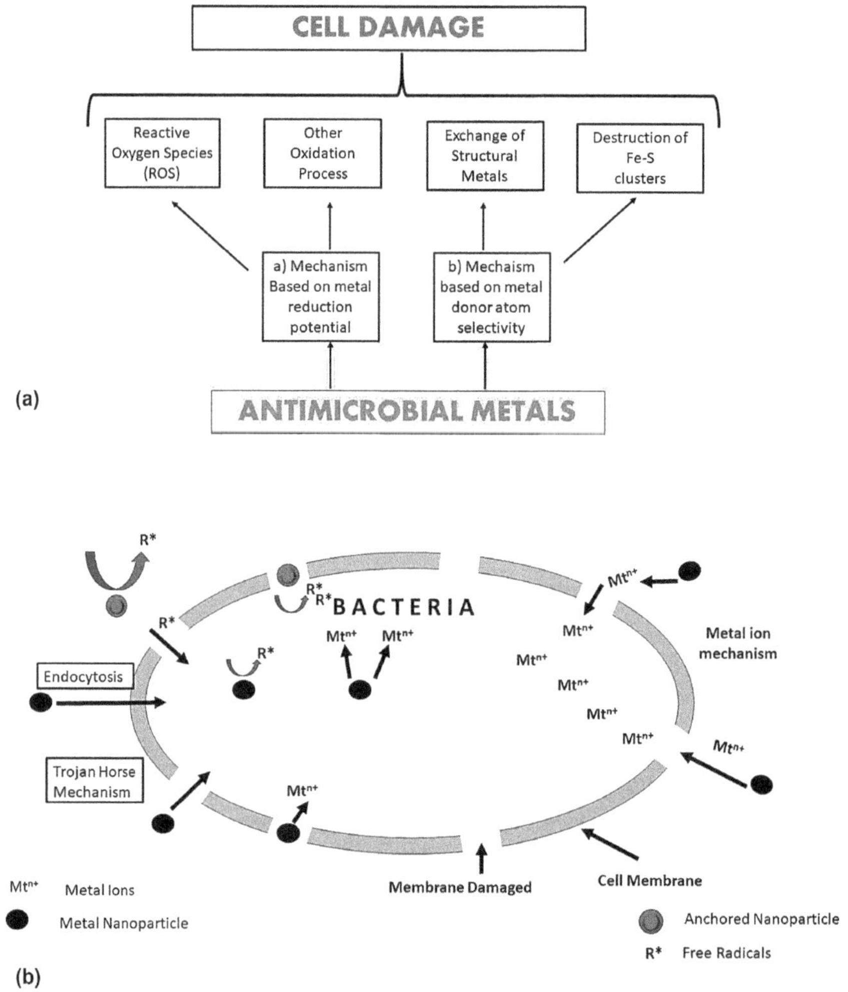

Figure 17.6 (a) An outline of the major antimicrobial mechanisms of metals according to the particular metal characteristics responsible for this action: a) reduction potential; b) donor atom selectivity and/or speciation; (b) diagram showing mechanisms involved in the antimicrobial action of metal nanoparticles: "Trojan-horse" effect due to endocytosis processes; attachment to the membrane surface; catalyzed radical formation; and release of metal ions.

17.8 Mechanism of Nanoparticulate Drug Delivery for Intracellular Infection

Based on the physicochemical properties of the nanocarrier and the nature of the targeted cells, two internalization pathways occur. These are phagocytosis (Figure 17.7a) or endocytic pathways (*i.e.* clathrin and caveolae-mediated endocytosis). Remarkably, based on the drug physicochemical properties, the internalization pathway, besides the intracellular fortune of the nanocarrier, is a crucial concern for the drug to be effective. The exposure of the drug into the enzymatic atmosphere of the lysosomes or the cell cytoplasm will have a significant influence on the pharmacological action.

Moreover, the release properties of polymeric nanoparticles are vital features of the preparation of the drug/polymer due to the subsequent application in continual drug delivery. The release rate of the entrapped drug can be affected by numerous influential factors. Size is one of these vital features. Smaller particles have a higher initial burst release and shorter continued release than larger particles. Moreover, greater burst and faster release rate are a result of greater drug loading.[82]

An effective nanoparticle system has a high loading capacity to yield the required amount of carrier necessary for administration.[83] Drug loading into nanoparticles can be performed by inserting into a solution that comprises formerly ready nanoparticles or through adding them into the reaction mixture during the polymerization process. These two techniques offer a solid solution intended for drug encapsulation and dispersion in the polymer, adsorption of the drug onto the nanoparticle surface, as well as chemical binding of the drug to the polymer. Type of interaction of drug and nanoparticles, as well as the quantity of the bound drug, depends on the chemical structure of the drug plus the polymer and the conditions of drug loading.[84]

17.9 Treatment of Experimental Infections Mediated by Drug Delivery Systems

17.9.1 Tuberculosis

Tuberculosis (TB) is one of the most important causes of death around the globe and is responsible for over three million deaths per year. The responsible agent is *Mycobacterium tuberculosis* and it is believed that this bacterium has infected around 30% of the world's population, with Asian and African regions comprising the highest number of new cases each year. Furthermore, tuberculosis is the major opportunistic infection in immunodeficiency syndrome (AIDS) patients, because previously acquired dormant bacilli reactivate, and the susceptibility to infection increases when immunity is weakened. Treatment failure is now becoming common in

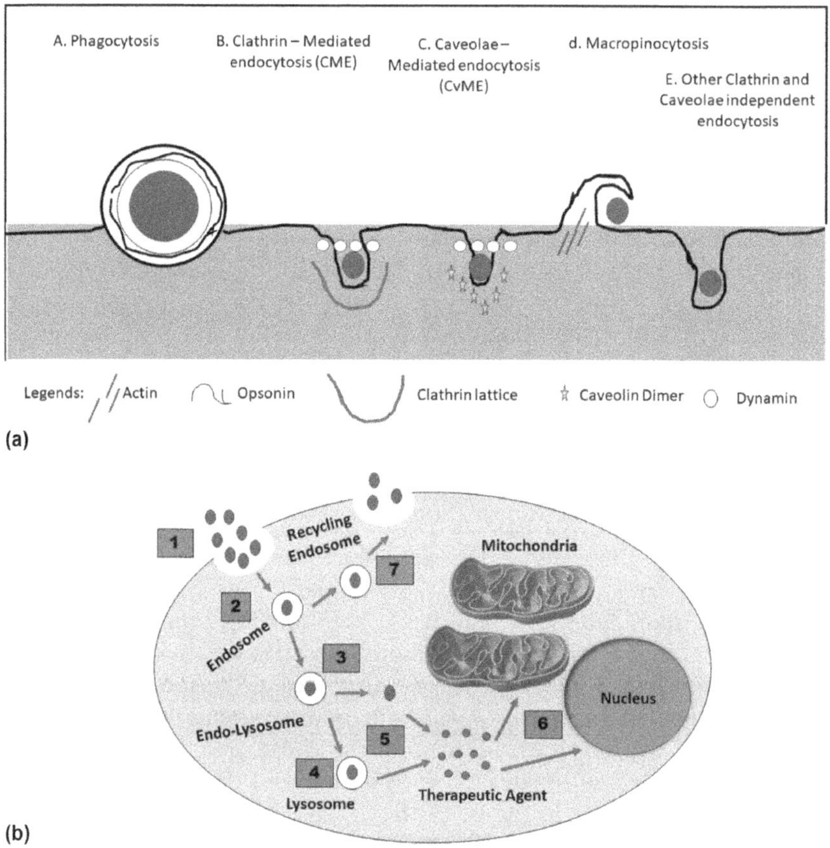

Figure 17.7 (a) Major nanocarrier internalization mechanisms in mammalian cells. a) Phagocytosis is an actin-based pathway taking place principally in phagocytes, including macrophages; b) clathrin-mediated endocytosis is a common pathway, based on the establishment of a clathrin lattice based on the GTPase dynamin; c) caveolae-mediated endocytosis takes place in distinctive membrane invaginations covered with caveolin dimers; d) macropinocytosis is a form of endocytosis which is an actin-based mechanism in which a large fluid-filled vesicle is pinched off from the cell membrane to engulf nanoparticles and extracellular milieu having poor selectivity. (b) Steps showing the cytosolic delivery of therapeutic antimicrobial agents through nanoparticle carriers; (1) cellular association of nanoparticles; (2) nanoparticle internalization through endocytosis; (3) escape of nanoparticles from the endosome; (4) nanoparticle degraded by lysozyme; (5) antimicrobial therapeutic agent generously circulates into cytoplasm; (6) transport to target organelle; (7) exocytosis of nanoparticles.

developing countries, and in addition to first-line multidrug-resistant tuberculosis, extensively drug-resistant TB, which is resistant to second-line TB drugs, is a growing concern.

Up to now, the treatment of TB requires taking several anti-TB drugs per day for a minimum of 6 months, and as a result of that many patients find it difficult to adhere to the treatment schedule and break the regular daily dose. Thus, a treatment failure occurs, which eventually leads to the development of antimicrobial drug-resistant *Mycobacterium tuberculosis.*

Therefore, a nanoparticle loaded with anti-TB drugs administered once, which might possibly keep the drug active in a sufficient concentration for a longer period could be a solution to achieve successful treatment. One of the most important benefits of the encapsulation of anti-TB drugs include reduced drug toxicity levels and a reduction of the treatment dose and dose duration needed to attain a good therapeutic result.

The primary goal of this drug delivery system-mediated therapy is to get constant active drug levels at the target location where the infection exists. With this method, a better result has been found. With the current practice of using formulated forms, rifampicin and isoniazid have a low concentration in serum and are retained in the circulation for <24 hours. On the other hand, doses of 12 $mg\,kg^{-1}$ and 10 $mg\,kg^{-1}$ liposome-encapsulated isoniazid and rifampicin, respectively, given intravenously have maintained the therapeutic drug levels of these two drugs for up to 5 and 7 days in serum and organs, respectively.[85] This alteration of the pharmacokinetics of these drugs might also allow a treatment dosage on a weekly basis, instead of the present daily regimen. In addition, the chemotherapeutic efficiency of liposome-encapsulated isoniazid and rifampicin given once weekly has been shown in a murine TB model. From this experiment, it is noted that simply a third of the dosage of isoniazid and rifampicin was needed to attain therapeutic effects when used in the nanoparticulate form.

Nanocarriers are able to transport particulates that have the ability for higher drug encapsulation and increase the biostability of drugs in oral delivered drugs. The reason behind this is because it is able to pass and withstand the biological barriers and can easily target bacterial cellular components of *Mycobacterium tuberculosis.* Examples of natural polymers that are commonly used for carrying anti-TB drugs are alginate and chitosan, and these could act as a possible substitute for drug delivery and cost effectiveness.[14]

The presence of a rich layer of mycolic acid in the cell wall of *Mycobacterium tuberculosis* makes the conventional treatment of TB challenging. It makes it difficult for potential anti-TB drugs to enter into the infected cells. Nanoparticulate drug delivery through biomaterials such as dendrimers, liposomes, *etc.* have made many significant biochemical alterations, which favor their entry into the target cells.[86]

Having seen that the majority of the TB infections occur in the respiratory system, delivery to the lungs by inhalation routes has been given much attention, to target the cellular reservoir of TB infection (alveolar macrophages), giving a further advantage of decreasing the systemic adverse effects of the drugs.[87] The current nanoparticle-encapsulated drug delivery methods for TB include oral, intravenous and inhalation administration forms.[88]

17.9.2 Brucellosis

Brucellosis is a transmittable illness resulting from four bacterial organisms: *Brucella abortus*, *Brucella melitensis*, *Brucella suis*, and *Brucella canis*. The four *Brucella* cause infection in humans and each one is related with a different natural host animal. These are small coccobacilli which are capable of living intracellularly in the phagocytic cells, hence the difficulty of treatment in most cases; although the antibiotics show high efficiency *in vitro*, they do not easily pass through cellular membranes of the bacteria.[89]

It is for this reason that an extended duration of therapy and the combination of many drugs are needed. The use of oral doxycycline in combination with rifampicin by mouth for 45 days has been the best treatment option, since it does not cause more side effects and has good patient compliance. Nevertheless, this treatment is not effective and a 15% treatment failure rate is reported, mostly in complicated cases.[19] Relapse of *Brucella* infections are common, since drugs are not effective and treatment is disagreeable to patients. Therefore, other options must be used to ease these treatment difficulties, for example, nanoparticulate drug delivery systems to accomplish higher intracellular bactericidal action of the drugs.

Gentamicin, loaded in different types of liposomes, has been assessed against murine monocytes infected with *Bacillus abortus* and showed good results. Rifampicin-encapsulated mannosylated dendrimers have also achieved specific pH-dependent delivery of this antibiotic to rat alveolar macrophages.[90] In recent times, gentamicin-loaded poly(D,L-lactide-*co*-glycolide)(PLGA) has been found by a number of emulsion solvent evaporation methods for the treatment of brucellosis.[91]

Hence, this new option for drug delivery allows an improvement in intracellular bactericidal activity and is promising. The likely use of drug delivery systems including aminoglycosides is also one of the most suitable therapeutic enhancements in the treatment of human brucellosis at the present time. However, despite the hopeful results mentioned above, there are significant disadvantages that are credited to these vesicles, such as their instability in the presence of blood lipoproteins and their osmotic fragility that can undermine them, causing leakage of the loaded drug. Moreover, liposomes cannot survive long-term in storage and this instability during storage is unresolved.

17.9.3 Salmonellosis

Infections caused by invasive non-typhoidal and typhoidal *Salmonella* are increasing. Furthermore, the drug resistance of *Salmonella* spp. is a growing concern. Reasons for the drug resistance include mutation of target proteins such as gyrA having a plasmid that carries resistance gene. Nevertheless, *Salmonella* spp. are facultative intercellular species, and this characteristic is one way of developing drug resistance. For example, when ingested, *Salmonella typhi* crosses the stomach and attaches to the epithelium of

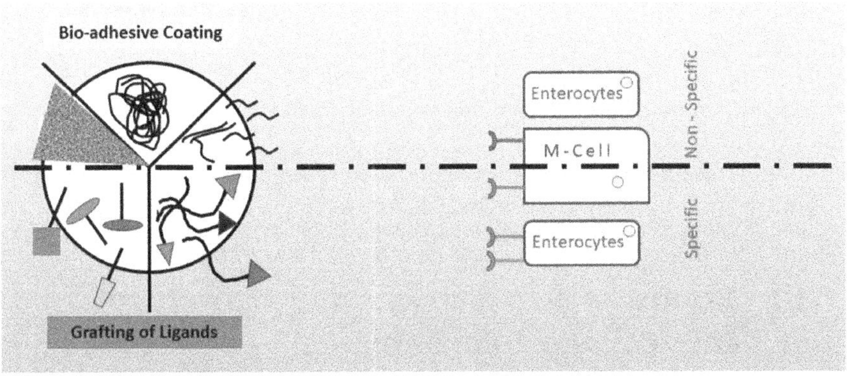

Figure 17.8 Surface modifications of the nanoparticle to increase their uptake by M cells and enterocytes. Non-specific strategies: mucoadhesive polymers coating or forming nanoparticles; PEG.

small intestine to attack the M (microfold) cells located in Peyer patches and the enterocytes. It stays there by replicating in the endocytic vacuoles, which helps to avoid the immune system and antimicrobial activity (Figure 17.8).[92,93]

Due to minor pitfalls in formulations, many antibiotics which are used to treat bacterial infections cause severe adverse effects. A good solution can be using carrier systems, for example, ceftriaxone coupled with chitosan nanoparticles shows significant improvement in an animal model. In addition, a better result was obtained with other drugs, including ciprofloxacin and ampicillin when encapsulated with a nanoparticle. Thus, this alternative method provides higher protecting and bactericidal activity and must be kept in mind as future appropriate treatment for *Salmonella*-caused illnesses.[94,95]

17.9.4 Listeriosis

A *Listeria monocytogenes* bacterium is a cause of listeriosis and results in opportunistic infections in high-risk individuals such as immune-compromised patients, new-borns, the elderly, and pregnant women. It causes increased mortality despite timely initiation of antibiotic treatment. *Listeria* is a facultative intracellular parasite that causes meningitis as well as septicemia. It has the ability to target the central nervous system and fetus. Due to the fact that it is a common foodborne pathogen, even persons that do not show risk factors can be affected by *Listeria* infection and is a common faecal carrier in 1–10% of the community. The encapsulation of ampicillin in liposomes reduces the survival of *Listeria monocytogenes* in mouse peritoneal macrophages in diverse magnitudes, based upon the confirmation of the liposomes. A week after treatment, ampicillin-loaded liposomes had shown a reduction in the infection by 3.2 logs in the liver and 2.8 logs in the spleen, whereas free ampicillin was found to be ineffective.[12,96]

The last foodborne outbreak of listeriosis occurred in Northern Spain in 2014. This outbreak was caused by a different strain that has altered its intracellular ability in dendritic cells. Youssef *et al.* found that ampicillin loaded onto nanoparticles of polyisohexylcyanoacrylate (PIHCA) was more effective than free ampicillin in a mouse model.[97] Ghasemian *et al.* also found that copper-loaded nanoparticles were significantly active against biofilm formation by *Listeria* in healthcare settings.[98]

17.10 Routes of Nanocarrier Drug Delivery for Intracellular Infections

Nanoparticle suspensions are applied *via* various routes due to their impressive features, such as increased saturation solubility, increased dissolution speed, improved bioadhesivity, flexibility in surface modification, and modest post-production processing. The main problem with the intravenous administration of nanoparticles is their interaction with the reticular–endothelial system.[99] The application of nanosuspensions *via* oral and parenteral routes have been widely explored. Moreover, pulmonary and ocular delivery systems have been discovered, but their applications in buccal, topical, and nasal delivery still need exploration.[100]

17.10.1 Oral Delivery

Oral routes of nanoparticle delivery are a relatively better choice, due to their oral use and no invasive procedure is used. Furthermore, the procedures are patient-friendly, suppress the risk of disease transmission, reduce costs, and increase patient compliance, hence they can be used for chronic therapy.[101] However, anti-cancer oral nanoparticle drugs are not bioavailable, and exemplify little chance to reach the tumour site through the blood.[102]

When exposed to the gastrointestinal tract, alteration of pH levels and the presence of enzymes or bile salts results in degradation of polymeric nanocarriers. Therefore, *in vitro* tests in intestinal and gastric-simulated fluids are needed to examine where and how these active molecules will be released. The solubility and permeability properties are two important parameters that influence the oral absorption of nanoparticles *via* passive diffusion. In summary, the oral route is the most recommended for drug delivery, but most of the drugs are not appropriate for oral administration mainly because of poor stability, solubility, and bioavailability. Encapsulation of such drugs in nanoparticles can avoid such limitations and permits the possibility of targeted delivery in the gastrointestinal tract.[103]

17.10.2 Pulmonary Delivery

Drug delivery to the lung seems to be easier compared to other routes, due to the less risky procedure *via* inhalation of aerosols. In an experimental study, anticancer drugs such as doxorubicin and curcumin loaded with Janus nanoparticles were found to be effective against lung cancer.[104,105] Many diseases, such as chronic obstructive pulmonary disease, asthma, tuberculosis, and genetic disorders affecting the airways (cystic fibrosis, α-1-antitrypsin deficiency) are treated using drugs loaded onto inhaled PLGA nanoparticles. One study found an increased bioavailability of rifampicin, isoniazid, and pyrazinamide (13-fold, 33-fold, and 15-fold, respectively) for treating tuberculosis compared to the corresponding oral administration. Various nanoparticles can be used to deliver drugs into the lungs. The most studied ones are lysosomes to treat lung cancers, obstructive lung diseases (asthma), and vaccination. Areas of concern in the oral delivery of nanoparticles are how to optimize nanoparticle deposition in lung tissues, minimizing immunological activation *in vivo*, maintaining their integrity and bioactivity of the core compound, as well as how to avoid toxicity of inhaled nanoparticles (Figure 17.9).[106–108]

17.10.3 Ocular Delivery

Local and systemic drug administration are commonly employed to reach ocular tissues. The tissue barriers limit the entry of drugs to their targets (Figure 17.10).[109]

During the administration of topical eye drops, the amount that absorbed into the eye is less than 5% of the dose.[110] The dose is commonly absorbed through conjunctival and nasal blood vessels into the blood circulation. For instance, at least 70% of the timolol dose is absorbed systemically within 5 mins.[111] Hampered by corneal limitations, ocular absorption is only increased by prolonging the duration of ocular contact. Due to the wide-ranging conjunctival systemic absorption, the maximal possible ocular absorption is only about 10% of the administered dose.

Ocular delivery is advantageous for nanosuspension drugs that have poor solubility in lachrymal fluids. Various strategies including suspension and ointment approaches have been proposed for the delivery of such drugs. Most commonly, the drug's inherent decomposition rate in lachrymal fluid directs its release and ocular bioavailability.[112]

Poly-cationic polymers are valuable penetration enhancers for ocular drug delivery. It was shown that the cyclosporin-A loaded nanoparticles are vital for the treatment of extraocular illnesses such as keratoconjunctivitis sicca or dry eye disease. It is reported that the benefits of such systems in ocular drug delivery is mainly attributed to their ability to contact closely the conjunctival as well as the corneal surface, thus raising delivery to external ocular tissues without conceding systemic drug exposure and inner ocular structures.[113] For instance, chitosan nanoparticles have been observed to be delivered to the ocular surface of rabbits, penetrating conjunctival epithelial cells.[114]

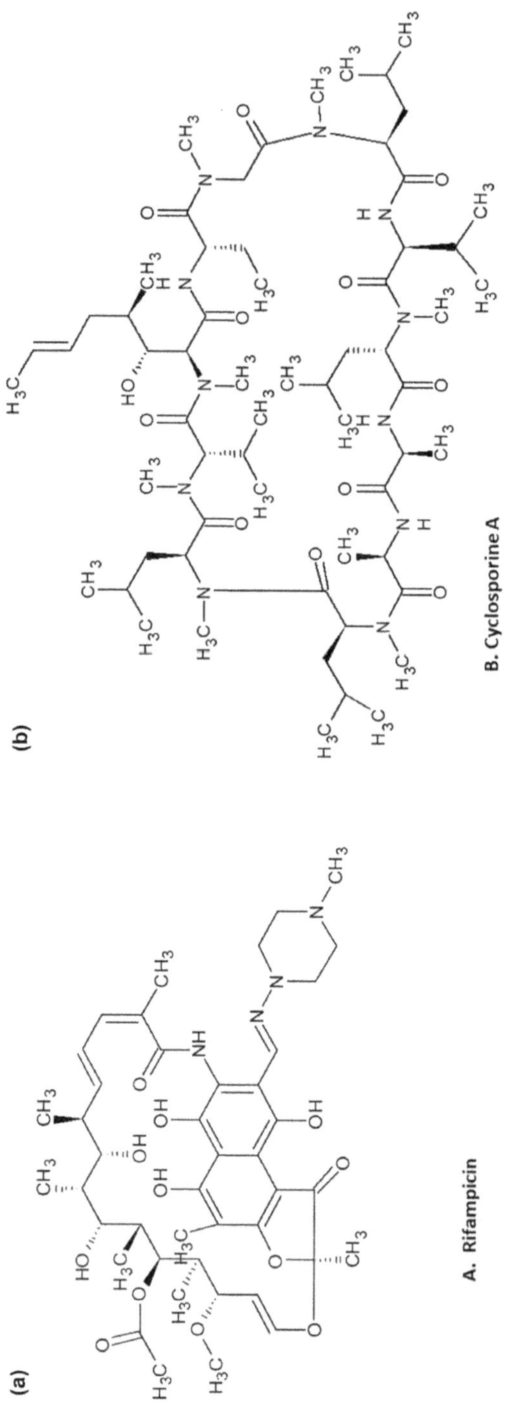
(a)
A. Rifampicin
(b)
B. Cyclosporine A

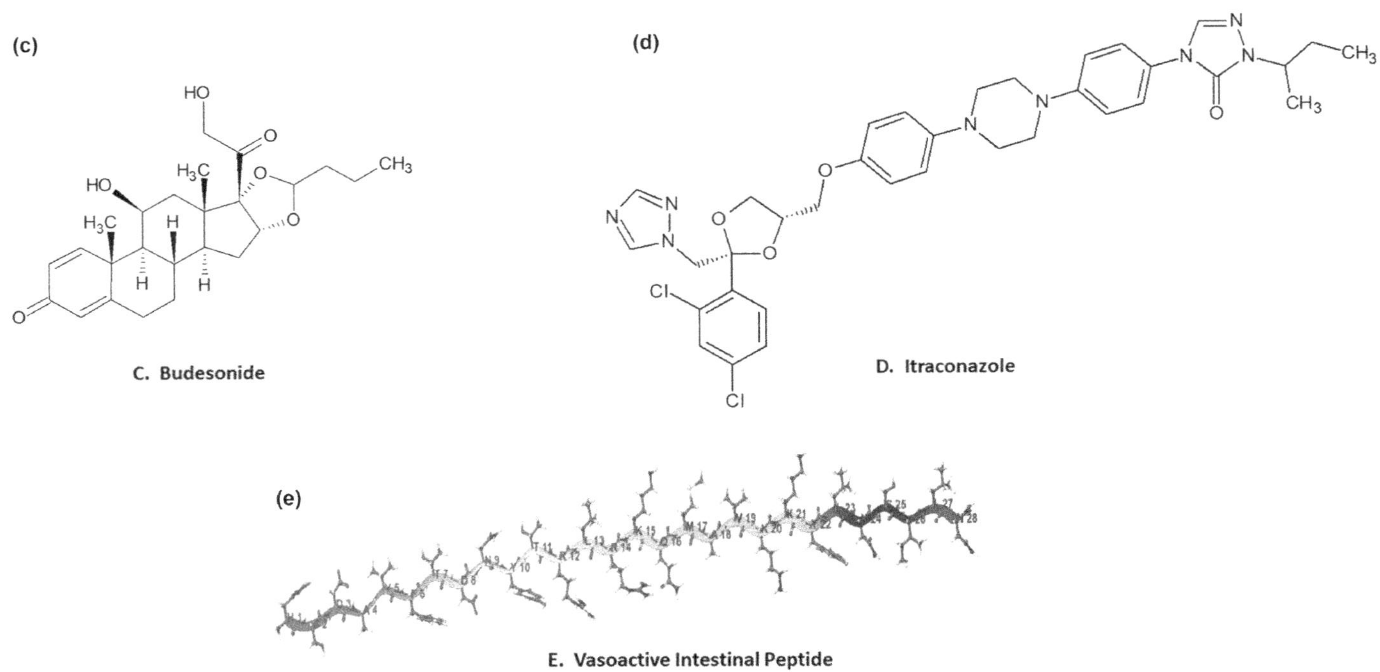

Figure 17.9 Example of inhaled nanoparticles. A. Rifampicin; B. cyclosporine A; C. budesonide; D. itraconazole; E. vasoactive intestinal peptide.

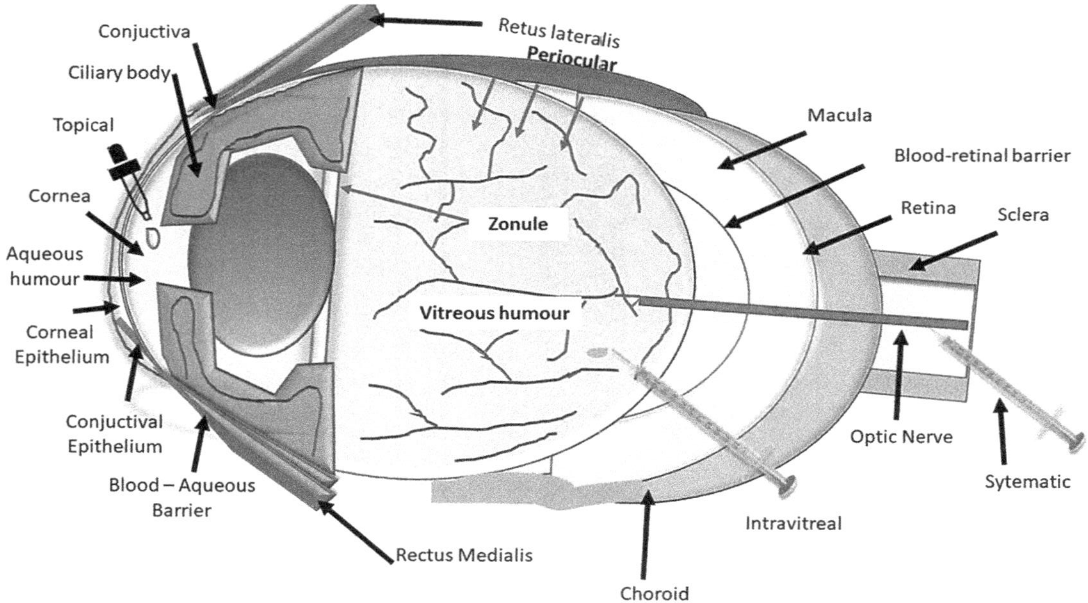

Figure 17.10 Diagrammatic representation of the ocular structure with the routes of drug kinetics. (1) Drug passes through cornea from lacrimal fluid into the anterior chamber; (2) drug passes into anterior uvea through conjunctiva and sclera; (3) drug delivery from blood circulation into anterior chamber through blood–aqueous barrier; (4) removal of drug from the anterior chamber to Schlemm's canal and trabecular meshwork; (5) drug elimination into the systemic uveoscleral circulation from aqueous humor; (6) drug dissemination to posterior eye from blood circulation crossing blood–retina barrier; (7) intravitreal administration of drug; (8, 9) drug elimination mechanisms from the vitreous through posterior route across the blood–retina barrier, and anterior route to the posterior chamber, respectively.

17.10.4 Brain-targeted Delivery

Due to the reduced ability of therapeutic antimicrobial agents to reach the central nervous system (CNS), successful management and treatment of neurological diseases are commonly limited. Most drugs and biotechnological agents cannot easily enter into the brain parenchyma, mainly due to the existence of the blood–brain barrier (BBB) and blood–cerebrospinal fluid barrier. As a result, the main significant challenge for CNS drug development and advance is the unavailability of effective brain-targeting technology. However, recent developments in nanotechnology, including polymeric nanoparticles, solid lipid nanoparticles, liposomes, micelles, and dendrimers have provided encouraging solutions to such challenges and have been well studied for CNS therapeutic delivery.[115]

Brain drug delivery is challenging, mainly due to the presence of the BBB, which is made up of endothelial cells that efficiently cover brain entry pathways by acting together with perivascular elements, perivascular neurons and pericytes (Figure 17.11).[116] Nanoparticles can cross the BBB and have the potential for targeting molecules or can bind blood molecules that are recognized through receptors present on brain endothelium such as receptors for iron transferrin, insulin, glutathione, low-density lipoprotein, mediating endocytosis, or transcytosis. Antimicrobial drugs conjugated with such receptors or with other peptides are widely studied[117] and in recent times started to be used for nanocarrier targeted delivery.

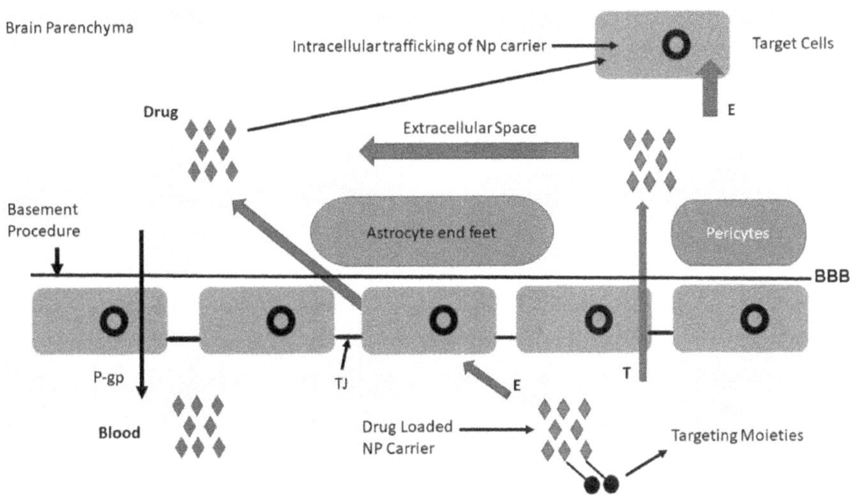

Figure 17.11 Diagrammatic illustration of nanoparticle (NP)-mediated drug delivery to the brain through intact blood–brain barrier (BBB) in the healthy condition. P-gp: P-glycoprotein-mediated drug efflux system; diamonds: free drug; ovals: NP-containing the drug; E: endocytosis; T: transcytosis; TJ: tight junctions.

Generally, nanocarrier drug delivery for brain-infecting intracellular pathogens includes intracerebral (intraparenchymal) delivery, intraventricular delivery (transcranial drug delivery), intrathecal delivery (intra-cerebrospinal fluid drug delivery), and the most commonly used BBB disruption.[118]

17.11 Nanoparticle Toxicity

Nanoparticles have been studied for cell toxicity, immunotoxicity, and genotoxicity. Animal and human studies have shown that after inhalation and through oral exposure, nanoparticles are distributed to the liver, heart, spleen, and brain, in addition to lungs and gastrointestinal tract.[2,119]

In order to clear the nanoparticles from the body the components of the immune system are activated. The estimated half-life of nanoparticles in human lungs is about 700 days, posing a consistent threat to the respiratory system. During metabolism, some of the nanoparticles congregate in the liver tissues.[120] Nanoparticles are more toxic to human health in comparison to the large-sized particles of the same chemical substance, and it is usually suggested that toxicities are inversely proportional to the size of the nanoparticles (Figure 17.12).[121]

Different kinds of nanoparticles pose various toxicities to the human body. Nanoparticles of metallic substances and fullerenes are among the common toxicants; nanoparticles of metallic substances disturb cell viability, damage and alter mitochondrial function, increase oxidative stress,

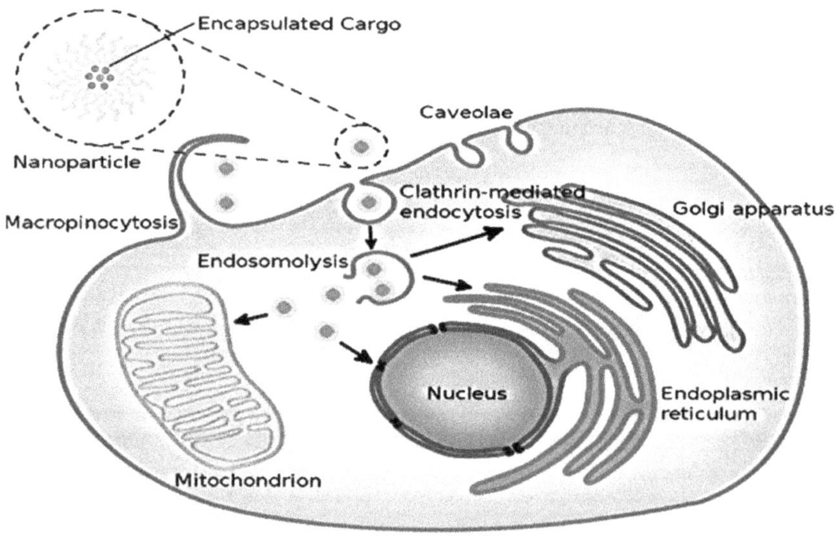

Figure 17.12 Nanotoxicological mechanisms due to nanoparticle interaction with cells and intracellular targets.
Reproduced from ref. 122 with permission from the Royal Society of Chemistry.

mediate inflammatory biomarkers, mediate apoptosis, and alter tight-junction protein expression of the BBB. Moreover, they result in dose-dependent genotoxicity and DNA damage. Similarly, fullerenes cause DNA strand breakage and chromosomal damage.[120,122,123]

17.12 Concluding Remarks and Perspectives

Most of the antibiotic therapy we use has not been able to fully eradicate most of the intracellular bacterial pathogens such as *Brucella*, *Mycobacterium*, *etc.* In addition, prolonged exposure and longtime usage of combined antibiotics are needed for prevention of disease relapses. In this scenario, drug delivery systems have an overwhelming role in the treatment and management of intracellular infections. Nanoparticulate drug delivery systems appear to be a vital and promising approach for the biopharmaceutical industry. They have numerous benefits over commonly used conventional drug delivery systems. This benefit is attributed to their ability to improve and boost the solubility, permeability, and bioavailability of various effective drugs; they can reduce side effects of drugs as well as raise patient compliance; hence making healthcare delivery cost effective. In such a scenario, the potential use of antibiotic-loaded drug delivery systems has become one of the most appropriate therapeutic improvements in intracellular bacterial disease treatment. The ability of nanoparticulate drug delivery systems to reduce the frequency of drug dosage and to raise patient compliance and other promising features make them the preferred method for developing several drugs that are predominantly characterized by weak permeability, aqueous solubility, and diminished bioavailability. Therefore, it is likely that intracellular disease treatment and management will become easier and fully successful once these antibiotic-loaded drug delivery systems are widely studied and used in humans worldwide.

Abbreviations

BBB	blood–brain barrier
CNS	central nervous system
DNA	deoxyribose nucleic acid
PAMAM	polyamidoamine dendrimers
PEG	polyethylene glycol
PLA	polylactic acid
PLGA	poly D,L-lactide-*co*-glycolide

Authors' Contributions

MS and MA contributed to the drafting and final version of the of the book chapter. All authors read and approved the final version of the book chapter.

Conflict of Interests

The authors declare that they have no conflict of interests.

References

1. J. S. Gunn, G. Gonzalez-Escobedo and J. M. Marshall, *Nat. Rev. Microbiol.*, 2011, **9**, 9.
2. *Antimicrobial Resistance: Global Report on Surveillance 2014*, World Health Organization, 2014, Downloaded from http://www.who.int/drugresistance/documents/surveillancereport/en/, last accessed on March 4, 2015.
3. S. E. Cosgrove, *Clin. Infect. Dis.*, 2006, **42**(Suppl 2), S82.
4. C. A. Diaz Granados, S. M. Zimmer, M. Klein and J. A. Jernigan, *Clin. Infect. Dis.*, 2005, **41**(3), 327.
5. E. R. Sydnor and T. M. Perl, *Clin. Microbiol. Rev.*, 2011, **24**(1), 141.
6. *Antibiotic Resistance Threats in the United States*, Centers for Disease Control and Prevention, 2013, Downloaded from http://www.cdc.gov/drugresistance/threat-report-2013/index.html, last accessed on March 9, 2015.
7. *Antimicrobial Resistance: Tackling a Crisis for the Future Health and Wealth of Nations*, 2014, Downloaded from http://amr-review.org/, last accessed on March 11, 2015.
8. L. A. Dever and T. S. Dermody, *Arch. Intern. Med.*, 1991, **151**(5), 886.
9. J. M. Munita and C. A. Arias, *Microbiol. Spectr.*, 2016, **4**(2), 1.
10. H. W. Boucher, G. H. Talbot and J. S. Bradley, *Clin. Infect. Dis.*, 2009, **48**, 1.
11. E. Pamer, Immune responses to intracellular bacteria, in *Fundamental Immunology*, ed. W. E. Paul, Lippincott Williams & Wilkins, Philadelphia, 2008, p. 1165.
12. M. Salouti and A. Ahangari, *Nanoparticle based Drug Delivery Systems for Treatment of Infectious Diseases. Application of Nanotechnology in Drug Delivery*, IntechOpen, United Kingdom, 2014, p. 155.
13. P. Ghosh, G. Han, M. De, C. K. Kim and V. M. Rotello, *Adv. Drug Delivery Rev.*, 2008, **60**, 1307.
14. M. Saravanan, K. Duche, T. Asmelash, A. Gebreyesus, L. Negash and A. Tesfay, *Pharm. Nanotechnol.*, 2015, **3**, 5.
15. M. Alkilany, L. B. Thompson and S. P. Boulos, *Adv. Drug Delivery Rev.*, 2012, **64**, 190–199.
16. P. Tallury, A. Malhotra and L. Byrne, *Adv. Drug Delivery Rev.*, 2010, **62**, 424–437.
17. Alonso and F. Garcia-del Portillo, *Int. Microbiol.*, 2004, 7(3), 181–191.
18. J. Alanis, *Arch. Med. Res.*, 2005, **36**, 697.
19. E. Imbuluzqueta, C. Gamazo, J. Ariza and M. J. Blanco-Prieto, *Front. Biosci.*, 2010, **15**, 397.
20. F. Van Bambeke, J. M. Michot and P. M. Tulkens, *J. Antimicrob. Chemother.*, 2003, **51**, 1067.

21. M. Barcia-Macay, C. Seral and M. P. Mingeot-Leclercq, *Antimicrob. Agents Chemother.*, 2006, **50**, 841.
22. T. Carryn, H. Chanteux, C. Seral, M. P. Mingeot-Leclercq, F. V. Bambeke and P. M. Tulkens, *Infect. Dis. Clin. North Am.*, 2003, **17**, 615.
23. S. Carryn, F. V. Bambeke, M. P. Mingeot-Leclercq and P. M. Tulkens, *Antimicrob. Agents Chemother.*, 2002, **46**, 2095.
24. S. S. Grant and T. H. Deborah, *Virulence*, 2013, **4**, 273.
25. C. Schmitz, J. Hilpert and C. Jacobsen, *J. Biol. Chem.*, 2002, **277**, 618.
26. D. Tyteca, P. Van Der Smissen and M. Mettlen, *Exp. Cell Res.*, 2002, **281**, 86.
27. D. Tyteca, K. Leys and P. M. Tulkens, *Eur. J. Cell Biol.*, 2001, **80**, 466–478.
28. C. Seral, F. Van Bambeke and P. M. Tulkens, *Antimicrob. Agents Chemother.*, 2003, **47**, 2283.
29. L. Baltch, L. H. Bopp, R. P. Smith, P. B. Michelsen and W. J. Ritz, *J. Antimicrob. Chemother.*, 2005, **56**, 104.
30. H. A. Nguyen, J. Grellet, D. Paillard, V. Dubois, C. Quentin and M. C. Saux, *J. Antimicrob. Chemother.*, 2006, **57**(5), 883.
31. J. E. Gomez and J. D. McKinney, *Tuberculosis*, 2004, **84**, 29.
32. D. M. Monack, A. Mueller and S. Falkow, *Nat. Rev. Microbiol.*, 2004, **2**, 747.
33. M. C. Chifiriuc, A. M. Haloban, C. Curitiu, L. M. Ditu, G. Mihaescu and A. E. Oprea, *IntechOpen*, 2016, DOI: 10.5772/61327.
34. M. Cuffini, V. Tullio, N. Mandras, J. Roana, G. Banche and N. A. Carlone, *Curr. Med. Chem.: Anti-Infect. Agents*, 2004, **3**, 1.
35. M. Maurin and D. Raoult, *Antimicrob. Agents Chemother.*, 2001, **45**, 2977.
36. B. Spellberg, R. Guidos, D. Gilbert, J. Bradley, H. W. Boucher and W. M. Scheld, *Clin. Infect. Dis.*, 2008, **46**, 155.
37. J. Davies and D. Davies, Origins and evolution of antibiotic resistance, *Microbiol. Mol. Biol. Rev.*, 2010, **74**, 417.
38. K. Agarwal and C. W. Fishwick, Structure-based design of anti-infectives, *Ann. N. Y. Acad. Sci.*, 2010, **1213**, 20.
39. M. Morens, G. K. Folkers and A. S. Fauci, *Lancet Infect. Dis.*, 2008, **8**, 710.
40. J. Lederberg, *Science*, 2000, **288**, 287.
41. M. Parry, H. Vinh and N. T. Chinh, *PLoS Neglected Trop. Dis.*, 2011, **5**, e1163.
42. M. Salouti and A. Ahangari, *IntechOpen*, 2014, DOI: 10.5772/58423.
43. A. Ranjan, N. Pothayee and M. N. Seleem, *FEMS Microbiol. Lett.*, 2012, **332**, 1.
44. Y. Kim, J. T. Rutka and W. C. Chan, *N. Engl. J. Med.*, 2013, **63**, 2434.
45. S. Gelperina, K. Kisich, M. D. Iseman and L. Heifets, *Am. J. Respir. Crit. Care Med.*, 2005, **172**, 1487.
46. K. Dillen, J. Vandervoort, G. Van den Mooter and A. Ludwig, *Int. J. Pharm.*, 2006, **314**, 72.
47. B. D. Brooks and A. E. Brooks, *Adv. Drug Delivery Rev.*, 2014, **78**, 14.

48. H. Brüssow, *Microbiology*, 2005, **151**, 2133.
49. A. Mahapatro and D. Singh, *J. Nanobiotechnol.*, 2011, **9**, 55.
50. M. O. Emeje, I. C. Obidike, E. I. Akpabio and S. I. Ofoefule, *IntechOpen*, 2012, 70.
51. A. Cavallaro, S. Taheri and K. Vasilev, *Biointerphases*, 2014, **9**, 5.
52. K. Vasilev, J. Cook and H. J. Griesser, *Expert Rev. Med. Devices*, 2009, **6**, 553.
53. K. Vasilev, V. Sah and K. Anselme, *Nano Lett.*, 2010, **10**, 202.
54. D. Schmaljohann, *Adv. Drug Delivery Rev.*, 2006, **58**, 1655.
55. R. C. Mundargi, V. R. Babu, V. Rangaswamy, P. Patel and T. M. Aminabhavi, *J. Controlled Release*, 2008, **125**(3), 193.
56. M. Faraji and P. Wipf, *Bioorgan. Med. Chem.*, 2009, **17**, 2950.
57. F. Siedenbiedel and J. C. Tiller, *Polymers*, 2012, **4**, 46.
58. L. Timofeeva and N. Kleshcheva, *Appl. Microbiol. Biotechnol.*, 2011, **89**, 475.
59. F. Tewers, F. Boury and J. P. Benoit, Biodegradable Microspheres: Advances in Production Technology, in *Microencapsulation: Methods and Industrial Applications*, ed. S. Benita, Marcel Dekker, New York, 2006.
60. K. Kuroda and G. A. Caputo, *Wiley Interdiscip. Rev.: Nanomed. Nanobiotechnol.*, 2013, **5**, 49.
61. M. Hamidi, A. Azadi and P. Rafiei, *Adv. Drug Delivery Rev.*, 2008, **60**, 1638.
62. N. Bhattarai, J. Gunn and M. Zhang, *Adv. Drug Delivery Rev.*, 2010, **62**, 83.
63. H. Omidian, K. Park and J. G. Rocca, *J. Pharm. Pharmacol.*, 2007, **59**(3), 317.
64. C. T. Schwall and I. A. Banerjee, *Materials*, 2009, **2**, 577.
65. T. Coviello, P. Matricardi, C. Marianecci and F. Alhaique, *J. Controlled Release*, 2007, **119**, 5.
66. C. C. Lin and A. T. Metters, *Adv. Drug Delivery Rev.*, 2006, **58**, 1379.
67. T. R. Hoare and D. S. Kohane, *Polymer*, 2008, **49**, 1993.
68. M. R. Mozafari and N. O. Sahin, Manufacturing methods and mechanism of formation of lipid vesicles, in *Nanoliposomes: From Fundamentals to Recent Developments*, Trafford Publishing Ltd., Oxford, UK, 2005, p. 39–48.
69. V. P. Torchilin, *Nat. Rev. Drug Discovery*, 2005, **4**, 145.
70. Z. Drulis-Kawa and A. Dorotkiewicz-Jach, *Int. J. Pharm.*, 2010, **387**, 187.
71. T. M. Allen and F. J. Martin, *Semin. Oncol.*, 2004, **31**, 5.
72. M. L. Tan, P. M. Choong and C. R. Dass, *Peptides*, 2010, **31**, 184.
73. G. Gaucher, M. H. Dufresne, V. P. Sant, N. Kang, D. Maysinger and J. C. Leroux, *J. Controlled Release*, 2005, **109**, 169.
74. N. Nishiyama, Y. Bae, K. Miyata, S. Fukushima and K. Kataoka, *Drug Discovery Today: Technol.*, 2005, **2**, 21.
75. R. Savic, *Science*, 2003, **300**, 615.
76. H. Kagaya, M. Oba and Y. Miura, *Gene Ther.*, 2012, **19**, 61.
77. T. J. Dening, S. Rao, N. Thomas and C. A. Prestidge, *AAPS J.*, 2016, **18**(1), 23.

78. S. Das and A. Chaudhury, *AAPS PharmSciTech*, 2011, **12**(1), 62.
79. G. Thiagarajan, S. Sadekar, K. Greish, A. Ray and H. Ghandehari, *Mol. Pharmaceutics*, 2013, **10**(3), 988.
80. M. Kalomiraki, K. Thermos and N. A. Chaniotakis, *Int. J. Nanomed.*, 2016, **11**, 1.
81. H. Palza, *Int. J. Mol. Sci.*, 2015, **16**, 2099.
82. M. L. Hans and A. M. Lowman, *Curr. Opin. Solid State Mater. Sci.*, 2002, **6**, 319.
83. A. Kumari, S. K. Yadav and S. C. Yadav, *Colloids Surf., B*, 2010, **75**, 1.
84. D. B. Shenoy and M. M. Amiji, *Int. J. Pharm.*, 2005, **293**(1–2), 261.
85. S. Labana, R. Pandey, S. Sharma and G. K. Khuller, *Int. J. Antimicrob. Agents*, 2002, **20**(4), 301.
86. Y. Cheng, Z. Xu and M. Ma, *J. Pharm. Sci.*, 2008, **97**(1), 123.
87. A. Sosnik, A. M. Carcaboso, J. Romina, R. J. Glisoni, M. A. Moretton and D. A. Chiappetta, *Adv. Drug Delivery Rev.*, 2010, **62**, 547.
88. S. Gelperina, K. Kisich, M. D. Iseman and L. Heifets, *Am. J. Respir. Crit. Care Med.*, 2005, **172**(12), 1487.
89. S. Prior, B. Gander, C. Lecaroz, J. Irache and C. Gamazo, *J. Antimicrob. Chemother.*, 2004, **53**, 981.
90. P. V. Kumar, A. Asthana, T. Dutta and N. K. Jain, *J. Drug Targeting*, 2006, **14**(8), 546.
91. C. Lecaroz, C. Gamazo, M. J. Renedo and M. J. Blanco-Prieto, *J. Microencapsulation*, 2006, **23**, 782.
92. P. R. Murray, K. S. Rosenthal and M. A. Pfaller, *Medical Microbiology*, 7th edn, Philadelphia, Elsevier Saunders, 2013, ISBN: 978-0-323-08692-9.
93. A. Crump, M. Sjölund-Karlsson, M. A. Gordon, C. M. Parry, *Clin. Microbiol. Rev.* 2015, **28**(4), 901.
94. N. A. Zaki and M. M. Hafez, *AAPS PharmSciTech*, 2012, **13**(2), 411.
95. N. Hussain, *Int. J. Pharm.*, 2016, **514**(1), 15.
96. R. Calderon-Gonzalez, M. Marradi, I. Garcia, N. Petrovsky and C. Alvarez-Dominguez, *Hum. Vaccines Immunother.*, 2015, **11**(10), 2501.
97. M. Youssef, E. Fattal and M. J. Alonso, *Antimicrob. Agents Chemother.*, 1988, **32**(8), 1204.
98. E. Ghasemian, A. Naghoni, H. Rahvar, M. Kialha and B. Tabaraie, *Jundishapur J. Microbiol.*, 2015, **8**(5), e17430.
99. P. Aggarwal, J. B. Hall, C. B. McLeland, M. A. Dobrovolskaia and S. E. McNeil, *Adv. Drug Delivery Rev.*, 2009, **61**, 428.
100. V. B. Patravale, A. Date and R. M. Kulkarni, *J. Pharm. Pharmacol.*, 2003, **56**, 827.
101. L. Plapied, N. Duhem, A. Rieux and V. Préat, *Curr. Opin. Colloid Interface Sci.*, 2011, **16**, 228.
102. Y. Win and S. S. Feng, *Biomaterials*, 2005, **26**, 2713.
103. M. Ensign, R. Cone and J. Hanes, *Adv. Drug Delivery Rev.*, 2012, **64**, 557.
104. M. Beck-Broichsitter, O. M. Merkel and T. Kissel, Controlled pulmonary drug and gene delivery using polymeric nano-carrieers, *J. Controlled Release*, 2012, **161**, 214.

105. B. Garbuzenko, J. Winkler, M. S. Tomassone and T. Minko, *Langmuir*, 2014, **30**, 12941.
106. J. C. Sung, B. L. Pulliam and D. A. Edwards, *Trends Biotechnol.*, 2007, **25**(12), 563.
107. H. M. Mansour, Y. S. Rhee and X. Wu, *Int. J. Nanomed.*, 2009, **4**, 299.
108. T. Florence, *Drug Discovery Today: Technol.*, 2005, **2**(1), 236.
109. A. Urtti, *Adv. Drug Delivery Rev.*, 2006, **58**(11), 1131.
110. L. Pitkanen, *Invest. Ophthalmol. Visual Sci.*, 2005, **46**, 641.
111. V. P. Ranta and A. Urtti, *Adv. Drug Delivery Rev.*, 2006, **58**(11), 1164.
112. R. C. Nagarwal, S. Kant and T. Singh, *J. Controlled Release*, 2009, **136**, 2.
113. M. DeCampos, Y. Diebold, E. S. Carvalho, A. Sanchez and M. J. Alonso, *Pharm. Res.*, 2004, **21**(5), 803.
114. E. De Salamanca, Y. Diebold, M. Calonge, G. Garcia-Vazquez, S. Callejo and M. J. Alonso, *Invest. Ophtlamol. Visual Sci.*, 2006, **47**, 1416.
115. H. L. Wong, X. Y. Wu and R. Bendayan, *Adv. Drug Delivery Rev.*, 2012, **64**(7), 686.
116. D. J. Begley, *Pharmacol. Ther.*, 2004, **104**, 29.
117. R. Gabathuler, *Neurobiol. Dis.*, 2010, **37**, 48.
118. M. I. Alam, S. Beg and A. Samad, *Eur. J. Pharm. Sci.*, 2010, **40**, 385.
119. A. Elsaesser and C. V. Howard, *Adv. Drug Delivery Rev.*, 2012, **64**, 129.
120. H. Bahadar, F. Maqbool, K. Niaz and M. Abdollahi, *Iran. Biomed. J.*, 2016, **20**(1), 1.
121. S. Mostafalou, H. Mohammadi, A. Ramazani and M. Abdollahi, *Daru, J. Pharm. Sci.*, 2013, **21**(1), 14.
122. Y. Wang, A. Santos, A. Evdokiou and D. Losic, *J. Mater. Chem. B*, 2015, **3**, 7153.
123. J. S. Macwan, A. Hirani and Y. Pathak, *Challenges in Ocular Pharmacokinetics and Drug Delivery. Nano-Biomaterials for Ophthalmic Drug Delivery and Implications for Cancer Therapy*, Springer, 2016, pp. 593–611.
124. S. H. Friedman, D. L. DeCamp and R. P. Sijbesma, *et al.*, *J. Am. Chem. Soc.*, 1993, **115**, 6506–6509.
125. R. F. Schinazi *et al.*, *Antimicrob. Agents Chemother.*, 1993, **37**(8), 1707–1710.

Subject Index